The Regional Geography of Canada

The Regional Geography of Canada

Fifth Edition

Robert M. Bone

OXFORD
UNIVERSITY PRESS

OXFORD
UNIVERSITY PRESS

8 Sampson Mews, Suite 204, Don Mills, Ontario M3C 0H5
www.oupcanada.com

Oxford University Press is a department of the University of Oxford.
It furthers the University's objective of excellence in research, scholarship,
and education by publishing worldwide in

Oxford   New York
Auckland   Cape Town   Dar es Salaam   Hong Kong   Karachi
Kuala Lumpur   Madrid   Melbourne   Mexico City   Nairobi
New Delhi   Shanghai   Taipei   Toronto

With offices in
Argentina   Austria   Brazil   Chile   Czech Republic   France   Greece
Guatemala   Hungary   Italy   Japan   Poland   Portugal   Singapore
South Korea   Switzerland   Thailand   Turkey   Ukraine   Vietnam

Oxford is a trade mark of Oxford University Press
in the UK and in certain other countries

Published in Canada
by Oxford University Press

Library and Archives Canada Cataloguing in Publication

Bone, Robert M.
The regional geography of Canada / Robert M. Bone.—5th ed.

Includes bibliographical references and index.
ISBN 978-0-19-543373-9

1. Canada—Geography—Textbooks. I. Title.
FC76.B66 2010   917.1   C2010-903396-5

Cover images: Nicolas Loran/iStockphoto, Zubin Li/iStockphoto

This book is printed on permanent (acid-free) paper ∞
which contains a minimum of 10% post-consumer waste.

Printed and bound in the United States of America.

1 2 3 4 – 14 13 12 11

BRIEF CONTENTS

CONTENTS

Regions of Canada

Canada's Physical Base

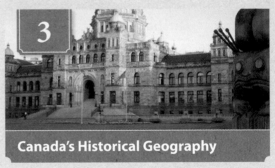

Canada's Historical Geography

4 Canada's Human Face

5 Ontario

6 Québec

7 British Columbia

Western Canada

Atlantic Canada

The Territorial North

Canada: A Country of Regions

FIGURES

TABLES

PREFACE

The purpose of this book is to introduce university students to Canada's regional geography. In studying the regional geography of Canada, the student not only gains an appreciation of the country's amazing diversity but also learns how its regions interact with one another. By developing the central theme that Canada is a country of regions, this text presents a number of images of Canada, revealing its physical, cultural, and economic diversity as well as its regional complexity. A number of features such as photos, maps, vignettes, tables, graphs, further readings, chapter bibliographies, and glossary are designed to facilitate and enrich student learning.

The Regional Geography of Canada divides Canada into six geographic regions: Ontario, Québec, British Columbia, Western Canada, Atlantic Canada, and the Territorial North. Each region has a particular regional geography, history, population, and a unique location. These factors have determined each region's character, set the direction for its development, and created a sense of place. In examining these themes, this book underscores the dynamic nature of Canada's regional geography. Part of the dynamism of Canada's regional geography is the changing relationship between Canada's six regions. Trade liberalization provides one element of change while regional tensions provide another element of change.

Following World War II, the liberalization of world trade and the creation of the North American trading bloc exposed Canada's economy to new challenges and dangers. The consequences of trade liberalization were sixfold:

- Canada's old spatial economic structure was transformed from one that was national to one that is continental.
- Canada's economy was integrated into the North American market.
- Canada's manufacturing sector was restructured.
- Rates of unemployment first increased and then declined.
- Canada's regional economies were reoriented to adjust to the larger North American market.
- Canada's economic growth and regional development have become tied more and more to trade with the United States.

This book explores both the national and regional implications of these economic changes and proposes scenarios for greater stability in the new global economic order.

Regional tensions have always existed in Canada. These tensions often place pressure on the existing social and political systems. This pressure for change reflects the 'potential' for a redress of the division of power within Canada's federation. From this perspective, regional tensions can lead to a shift of power within a region and may even alter relations between regions and/or the federal government. In this book, four tensions or faultlines are examined. These tensions occur between Aboriginal and non-Aboriginal Canadians, French and English Canadians, centralist and decentralist forces, and recent immigrants (newcomers) and those born in Canada (old-timers). Each of these tensions exists in all parts of Canada, but because of spatial variations in Canada's human and/or physical geography, each appears more prominently in a particular region or regions. For instance, the Aboriginal/non-Aboriginal faultline is most deeply felt in the Territorial North, where a larger proportion of the population is Aboriginal. The French/English faultline manifests itself most commonly in Québec. The centralist/decentralist faultline is frequently played out between the federal government and Western Canada, often over resources. The impact of immigration is heavily concentrated in the three major cities of Toronto, Vancouver, and Montréal. This book explores the nature of these faultlines, the need to reach compromises, and the fact that these faultlines provide the country with its greatest strength—diversity. Ultimately, these faultlines are shown to be not divisive forces but forces of change that ensure Canada's existence as a country of regions.

Organization of the Text

This book consists of 11 chapters. Chapters 1 through 4 deal with general topics related to Canada's national and regional geographies—Canada's physical, historical, and human geography—thereby setting the stage for a discussion of the six main geographic regions of Canada. Chapters 5 through 10 focus on these six geographic regions. The core/periphery model provides a guide for the ordering of these six regions. The regional discussion begins with Ontario and Québec, which represent the traditional demographic, economic, and political core of Canada. The core regions are followed by British Columbia, Western Canada, Atlantic Canada, and the Territorial North. Chapter 11 provides a conclusion.

Chapter 1 discusses the nature of regions and regional geography, including the core/periphery model and its applications. Chapter 2 introduces the major physiographic regions of Canada and other elements of physical geography that affect Canada and its regions. Chapter 3 is devoted to Canada's historical geography, such as its territorial evolution and the emergence of regional tensions and regionalism. This discussion is followed, in Chapter 4, by an examination of the basic demographic, economic, and social factors that influence both Canada and its regions. This chapter explores the national and global forces that have shaped Canada's regions as well as the features of its population (size, urbanization, etc.). To sharpen our awareness of how economic forces affect local and regional developments, four major themes running throughout this text are introduced in these first four chapters. The primary theme is that Canada is a country of regions. Two secondary themes—the integration of the North American economy and the changing world economy—reflect the recent shift in economic circumstances and its effects on regional geography. These two economic forces, described as continentalism and globalization, exert both positive and negative impacts on Canada and its regions, and are explored through the core/periphery model.

In Chapters 5 to 10, the text moves from a broad, national overview to a more regional focus. Each of these six chapters profiles one of Canada's large geographic regions: Ontario, Québec, British Columbia, Western Canada, Atlantic Canada, and the

Territorial North. These regional chapters explore the physical and human characteristics that distinguish each region from the others and that give each region its special sense of place. To emphasize the economic specialty of each region, a predominant economic activity is identified and explored through a 'Key Topic'. They are: the automobile industry in Ontario; Hydro-Québec in Québec; forestry in British Columbia; agricultural transition in Western Canada; the fishing industry in Atlantic Canada; and megaprojects in the Territorial North. From this presentation, the unique character of each region emerges. The book concludes with Chapter 11, which discusses the future of Canada as a country of regions.

Fifth Edition

For Canada's regions, the consequences of recent global economic developments—notably the increased demand for natural resources and the remarkable industrialization of Asian countries, especially China—are twofold. First, Western Canada (led by Alberta) and British Columbia are enjoying high prices for their resources. The resulting boom has meant that the economies of Alberta and British Columbia are attracting record numbers of newcomers: migrants from other parts of Canada and immigrants from abroad. With the spillover of Alberta's economic boom into Saskatchewan, this traditionally 'have-not' province is undergoing its own mini-boom. High oil prices have also benefited Newfoundland and Labrador, though its population continues to decline.

Second, while Ontario and Québec remain the economic and population pillars of Canada, a shift of regional power seems to be in the wind. In recent years, Ontario and Québec have suffered a decline in their manufacturing sectors. This decline began with the loss of most of their textile firms at the beginning of trade liberalization. Often these firms relocated offshore where labour costs were significantly lower. Now many other manufacturing firms are following the same path—moving offshore but selling their goods in the Canadian market. Canadian manufactured goods also are troubled by the so-called 'Dutch disease'—a combination of high energy prices and a rising Canadian dollar—which has made their production and export more difficult, thus magnifying the problems facing the industrial heartland of Canada.

Capitalism has its ups and downs. By 2009, the global economic crisis had hit home, causing havoc in Canada and its regions. Ontario's automobile industry was badly wounded and its recovery is crucial to Ontario and Canada. At the same time, a federal program to deal with global warming remained a work in progress, partly because of the focus on emissions from Alberta's oil sands and partly because of the importance of harmonizing Canada's program with that of the United States. These two issues have had an enormous impact on Canada's regions and its spatial framework based on the core/periphery model. For that reason, many of the changes in this fifth edition stem from those two issues.

Reviews play an important role. This time, the six reviewers of the fifth edition have provided me with new insights and fresh challenges. How to recast Canada's regions within the global economic crisis was one challenge. The call for a more sophisticated text was made, but I am conscious of the undergraduate audience, many of whom come from the general student population. Perhaps there is a place for a more advanced regional text with a greater overburden of theory, historiography, and jargon for advanced students, yet I fear regionalism in geography is not well. As I read the literature, geographers have become more and more specialized while others—political scientists, sociologists, even journalists—have staked out a claim on this very geographic area that purports to interpret geographic space in a thoughtful, understandable, and considered manner. My aim, in this new edition, is to focus on who we are, where we have been, and where we are headed—individually, collectively, and as a country of regions—all from a regional perspective.

Besides many new photos, tables, figures, and examples throughout the text, the fifth edition has several other new features: a complete glossary is now included at the end of the book for ease of reference and study; likewise, the chapter bibliographies have been streamlined and are placed together at the end; new 'Think About It' questions are placed throughout the text as sidebars, to challenge readers to pause and reflect on the ramifications and relationships of events and geographical facts; handy 'Cross-Chapter References' also are included as sidebars, to direct the reader to other places in the text where specific issues are discussed.

Acknowledgements Past and Present

I have continued to rely on the help of colleagues across Canada. With each edition, I have benefited from the constructive comments of anonymous reviewers selected by Oxford University Press. I especially owe a debt of thanks to one of those anonymous reviewers who spiced his critical comments with words of encouragement that kept me going.

In the first edition, professors at the University of Saskatchewan were particularly helpful—several provided me with their comments on specific matters that fall under their areas of specialization while some were pressed into reviewing draft versions of regional chapters. Most contributed photographs. These colleagues and friends include Alec Aitken, Bill Archibold, William Barr, Dirk de Boer, K.I. Fung, the late Walter Kupsch, Lawrence Martz, the late John McConnell, John Newell, Jim Randall, Jack Stabler, and Mike Wilson. Professors from other institutions who played a role in shaping this book include Jim Miller, Robert MacKinnon, and Gilles Viaud from Cariboo College; Hugh Gayler and Dan McCarthy from Brock University; Ben Moffat from Medicine Hat College; and Keith Storey from Memorial University.

Thanks to a sabbatical leave granted to me by the University of Saskatchewan, I was able to spend the fall term of 1996 at the Département de géographie at Université Laval. My generous hosts not only put up with my limited command of French but, more importantly, contributed significantly to the design and content of my chapter on Québec. Among those contributing to my education were Louis-Edmond Hamelin, who must be considered the Dean of Canadian geographers; Jean-Jacques Simard, a well-known sociologist who has written extensively on Québec's north; and Benoit Robitaille, the Chair of the Département de géographie at Université Laval. Discussions of critical issues with Jacques Bernier, Marc St-Hilaire, and Eric Waddell proved extremely fruitful. For me, one of the highlights at Laval was the opportunity to present my ideas on Québec at one of the Friday afternoon seminars organized by Dean Louder. The constructive and stimulating responses from the cultural geographers attending that seminar were especially helpful to me.

Maps are a critical element in geography. I must thank Keith Bigelow, the cartographer in the Department of Geography at the University of Saskatchewan, for the help he provided on the maps. The library staff in Government Documents at the University of Saskatchewan was particularly helpful in my search for detailed information on Canada's regions.

The National Atlas of Canada and Statistics Canada have created important websites for geography students. These provide access to a wide range of geographic data and maps. The National Atlas of Canada has placed many of its printed maps on its website <www.atlas.gc.ca>. Statistics Canada publishes elements of its census data by provinces and territories on its website <www.statcan.ca>. Statistics Canada also publishes key statistics for all urban places on another website <www2.statcan.ca>. Students can also access annually published statistical data and population estimates from the Statistics Canada site under the heading CANSIM (Canadian Socio-economic Information Management System). Joel Yan of Statistics Canada was particularly helpful in identifying suitable data from CANSIM used in this book.

The staff at Oxford University Press, but particularly Phyllis Wilson, made the preparation of the fifth edition a pleasant and rewarding task. Allison McDonald, Developmental Editor, who worked with me in the process of revising the text and selecting new photographs, deserves special thanks. Richard Tallman, who diligently and skilfully has edited my manuscripts into polished finished products for each of the last four editions, deserves special mention. He has become an old friend who often pushes me to expand my ideas.

Finally, a special note of appreciation to my wife, Karen, who put up with me disappearing into my study.

IMPORTANT FEATURES OF THIS EDITION

The fifth edition of *The Regional Geography of Canada* incorporates a wide range of resources for students that complement and enhance the text. Building on the strengths of previous editions while updating discussions to reflect current issues, the text takes into account key factors in human geography such as recent economic trends and crises, severe weather challenges, and relevant environmental issues.

Highlights include:

- **New** and updated vignettes throughout the text focus on issues specific to each chapter
- **New** and updated colour photographs engage the reader and provide strong visual references tied to the material
- **New** and updated maps provide key visual aids that highlight the characteristics of various regions across Canada
- **New** 'Think About It' questions placed throughout the text prompt students to analyze the material both in and out of the classroom
- **New** cross-chapter references highlight the interconnectedness of content across chapters to ensure a comprehensive study of the material

The result is a new edition that retains the strengths that have made *The Regional Geography of Canada* a best-selling text while introducing new concepts and exploring topics of interest to today's student.

REGIONS OF CANADA

INTRODUCTION

Geography helps us understand our world. Since Canada is such a huge and diverse country, its geography is best understood from a regional perspective. In fact, the mental image of Canada as 'a country of regions' runs deep in the Canadian psyche.

Canada can best be understood as consisting of six regions with each having a distinct location, physical geography, and historical development. A strong sense of regional identity exists and these identities were shaped over time as people came face to face with challenges presented by their economic, physical, and social environments. Then, too, our geographic proximity and close working relationships with the United States have impacted each region of Canada differently.

Chapter 1 presents Canada's geographic reality, which consists of three fundamental parts. First and foremost, geography and history have forged Canada into a complex and diverse set of regions. Second, powerful and sometimes negative tensions exist between the regions and Ottawa. Lastly, each region has a unique economic position within Canada, North America, and the world. The challenge of interpreting Canada's regional diversity, the changing demographic and economic strengths of the regions, and their trade relationships within Canada and with foreign nations is not an easy task. A spatial conceptual framework based on the core/periphery model helps us to understand both the nature of and the socio-economic processes behind Canada's regionalization. At the same time, the concept of faultlines identifies and addresses deep-rooted tensions in Canadian society that often mould each region's special identity.

CHAPTER OVERVIEW

Geographic analysis and synthesis of our world require an intuitive grasp of its regional nature and spatial relationships, especially as Canada's regions relate to contiguous US regions. The following topics are examined in Chapter 1:

- Geography as a discipline.
- Regional geography.
- Regionalism.
- Canada's geographic regions.
- Dynamic nature of regions.
- Four faultlines within Canada.
- Power of place.

- Sense of place.
- Core/periphery model.
- Canada–US trade.
- Thickening of the border.
- Implications of Canada–US relations.
- Understanding Canada's regions.

Trans-Canada Highway through Rogers Pass, Mount Tupper, Glacier National Park, British Columbia. Photo: John E Marriott/All Canada Photos.

Geography as a Discipline

Geography provides a description and explanation of lands, places, and peoples beyond our personal experience. Geography also determines life's opportunities. De Blij and Murphy (2006: 3) believe that 'Geography is destiny', meaning that for most people, place is the most powerful determinant of their life chances, experiences, and opportunities. In this instance, place refers to the community where one was born and raised. Not surprisingly, then, geographers have always been curious about distant places and foreign lands (Vignette 1.1). Living, working, and sharing together in a common space inevitably leads to the formation of a **regional identity** and **consciousness**. Both are products of a region's physical geography, historical events, and economic situation. As such, they form one of the cornerstones of regional geography. **Regional self-interest**, a logical outcome of regional identity and consciousness, shows its face in the centralist/decentralist faultline.

Regional Geography

The geographic study of a particular part of the world is called **regional geography**. In such studies, people, interacting with their economic, physical, and social environments, place their imprint on the landscape. In layperson's terms, the goal of regional geography is to find out what makes a region 'tick'. By achieving such an understanding, we gain a fuller appreciation of the complexity, diversity, and interconnectivity of our world.

Regional geography has evolved over time.[1] Originally, geographers focused their attention on the physical aspects of a **region** that affected and shaped the people and their institutions. Today, geographers place more emphasis on the human side because the physical environment is largely mediated through culture, economy, and technology (Agnew, 2002; Paasi, 2003). Over time, a multitude of profound and often repeated extreme experiences, whether they are economic, natural, or political events, mark the people, forcing them to respond. In turn, these responses help create a common sense of regional belonging and regional consciousness. Other expressions of regional belonging and consciousness are **sense of place** and **power of place**. While power of place is closely linked to globalization, sense of place emphasizes local control over regional and community affairs.

Regional self-interest, in Canada, often results in conflicts between the provincial and federal governments. A striking example of how a bitter dispute unfolds is illustrated by the recent public brawl between Ottawa and St John's (and to a lesser degree Halifax) over the clawing back of equalization payments because of larger provincial energy royalties. Compromise prevailed with the 2004 agreement reached between Ottawa and St John's/Halifax. The consequences for the well-being of the country caused by this dispute over equalization payments pales in comparison to the 1995 referendum dealing with Québec's place in Confederation. As a threat to national unity and regional harmony, tensions between Québec and the rest of the country, and especially between Québec and Ottawa, rank as the top concern. Indeed, the closeness of this referendum caused so much bad feeling and consternation that the mood of the country and its regions reached a low point. The political cartoonist Brian Gable presented a spatial image in his 1995 version

Vignette 1.1   Curiosity: The Starting Point for Geography

Curiosity about distant places is not a new phenomenon. The ancient Greeks were curious about the world around them. From reports of travellers, they recognized that the earth varied from place to place and that different peoples inhabited each place. Stimulated by the travels, writings, and map-making of scholars such as Herodotus (484–c. 425 BC), Aristotle (384–332 BC), Thales (c. 625–c. 547 BC), Ptolemy (AD 90–168), and Eratosthenes (c. 276–c. 192 BC), the ancient Greeks coined the word 'geography' and mapped their known world. By considering both human and physical aspects of a region, geographers have developed an integrative approach to the study of our world. This approach, which is the essence of geography, separates geography from other disciplines. The richness and excitement of geography are revealed in Canada's six regions—each region is the product of its physical setting, past events, and contemporary issues that combine to produce a set of unique regional identities.

Figure 1.1   **Gable's regions of Canada.**
Following the 1995 referendum, the heated political scene cooled somewhat. Yet relationships within Canada remained strained. Political cartoonist Brian Gable has aptly depicted this low point in relationships between provinces and territories with his map of Canada. The root of the conflict is that Québec perceives its place within Canada as more of a partnership while the other nine provinces see all provinces as equal.

of each region's pessimistic attitude stemming in large part from the troubled times associated with the referendum and the often misunderstood aspirations and hopes of Québec (Figure 1.1).

 See Chapter 3, 'Equalization and Transfer Payments', page 96, and Vignette 9.5, 'The Atlantic Energy Accord', page 376, for more on these subjects.

 For the 1995 Québec referendum, see Vignette 3.10, 'The Results of the 30 October 1995 Referendum', page 125.

Geographers often refer to such regional identities as a sense of place. Within Canada, there is no stronger sense of place than that exhibited by the Québécois, who translate their social identity into political aspirations. While Québec is one of 10 provinces, most

Québécois perceive their place within Canada as more of a partnership with the rest of Canada, while fervent nationalists continue to dream of and plan for a day when their nation achieves independence from Canada. Consequently, this divide on the nature of Confederation is a recurring political issue and represents the most critical social/political faultline, with its origins reaching back to 1759. The competing concepts of Canada as two nations or as 10 provinces are discussed more fully in Chapter 4 ('The French/English Faultline') and in Chapter 6 ('The French/English Faultline in Québec').

Regionalism

Regionalism divides countries into different parts. Canada is particularly prone to regionalization, making it a country of regions. In the minds of Canadians, Canada consists of several distinct regions. Figure 1.1 illustrates a humorous version of Canada's regions. This book uses a slightly different set of regions, as shown in Figure 1.2. Regardless of the particular set of regions, the question remains: Why has regionalism exerted—and continued to exert—such a powerful influence over Canadian affairs? Put differently, why is Canada not a more 'homogeneous' state like the United States? Several factors quickly come to mind.

- Canada's vast geographic size and physical geography create natural regional divisions.
- The country's physiographic regions have a north/south orientation that encourages internal divisions within Canada and, on a North American scale, **continentalism**.
- Each region experienced a different pattern of historic settlement and relationship with Aboriginal peoples, which together provided a distinct cultural base. In turn, the more recent immigration of people from around the world has created a **pluralistic society** that contains a significant **visible minority**.
- Québec, which is a product of early French settlement, provides a cornerstone in Canada's regional identity.
- The British North America Act gave considerable powers to provinces, such as education, health, and natural resources, which lend a political dimension to Canadian regionalism.
- Canada's uneven population distribution and economic activities concentrate 'power' in Central Canada.
- The federal government formulates policies and programs designed to promote the national interest, but the impact of Ottawa's efforts flow mainly to the heavily populated areas, namely Central Canada.

Sociologists and political scientists see the last factor functioning under the guise of the **corporate and political elite**, who are concentrated in Toronto, Montreal, and Ottawa and who co-ordinate their efforts to maintain the economic status quo by formulating policies, such as the National Energy Policy of the early 1980s, that promote economic development in Central Canada at the expense of the rest of the country (Hiller, 2000: 131–3).

Canada's Geographic Regions

The geographer's challenge is to divide a large spatial unit like Canada into a series of 'like places'. To do so, a regional geographer selects the critical physical and human characteristics that logically divide a large spatial unit into a series of regions and that distinguish each region from adjacent ones. Towards the margins of a region, its core characteristics become less distinct and merge with those characteristics of a neighbouring region. In that sense, boundaries separating regions are best considered transition zones rather than finite limits.

In this book, we examine Canada as composed of six geographic regions (Figure 1.2):

- Atlantic Canada
- Québec
- Ontario
- Western Canada
- British Columbia
- Territorial North

The six regions were selected for several reasons. First, a huge Canada needs to be divided into a set of manageable segments. Too many regions would distract the reader from the goal of easily grasping the basic nature of Canada's regional geography. Six regions allow us to readily comprehend Canada's regional geography and to place these regions within a conceptual framework based on the core/periphery model, discussed later in this chapter. This is not to say that there are not internal regions or sub-regions. In Chapter 5, Ontario provides such an example. Ontario is subdivided into southern Ontario (the industrial **core** of Canada) and northern Ontario (a resource **hinterland**). Southern Ontario is Canada's most densely populated area and contains the bulk of the nation's manufacturing industries. Northern Ontario, on the

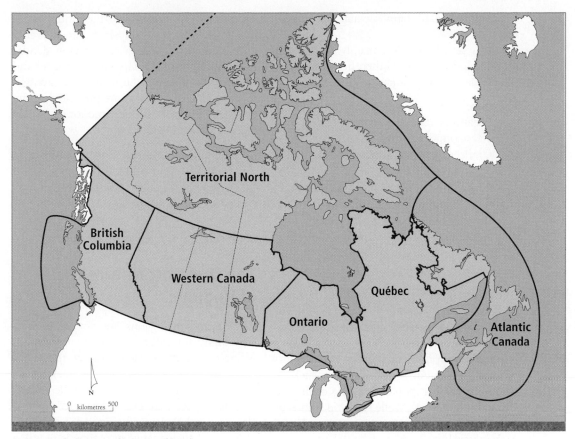

Figure 1.2   The six geographic regions of Canada.
The coastal boundaries of Canada are recognized by other nations except for the 'sector' boundary in the Arctic Ocean.

other hand, is sparsely populated and is losing population because of the decline of its mining and forestry activities.

Second, an effort has been made to balance these regions by their geographic size, economic importance, and population size, thus allowing for comparisons (see Table 1.1). For this reason, Alberta is combined with Saskatchewan and Manitoba to form Western Canada, while Newfoundland and Labrador along with Prince Edward Island, New Brunswick, and Nova Scotia comprise Atlantic Canada. The Territorial North, consisting of three territories, makes up a single region. Three provinces, Ontario, Québec, and British Columbia, have the geographic size, economic importance, and

Table 1.1 General Characteristics of the Six Canadian Regions, 2006					
Geographic Region	**Area* (000 km²)**	**Area (%)**	**Population (000s)**	**Population (%)**	**GDP (%)**
Ontario	1,076.4	10.8	12,160.3	38.5	38.7
Québec	1,542.1	15.4	7,546.1	23.9	19.7
British Columbia	944.7	9.5	4,113.5	13.0	12.5
Western Canada	1,960.7	19.6	5,406.9	17.1	22.6
Atlantic Canada	539.1	5.4	2,278.3	7.2	6.0
Territorial North	3,909.8	39.3	101.3	0.3	0.5
Canada	9,972.8	100.0	31,612.9	100.0	100.0

*Includes freshwater bodies such as the Canadian portion of the Great Lakes.
Sources: Statistics Canada (2006a, 2007a, 2007b).

population size to form separate geographic regions.

This set of regions is readily understood by Canadians because of their frequent use in the media. These regions:

- are associated with distinctive physical features, natural resources, and economic activities;
- reflect the political structure of Canada;
- facilitate the use of statistical data;
- are linked to regional identity;
- are associated with reoccurring regional complaints and disputes;
- reveal regional economic strengths and cultural presence.

The six geographic regions form the heart of this book (Chapters 5–10). The critical question is: What distinguishes each region? Certainly geographic location and historical development play a key role. Equally important are contemporary elements such as variations in area, population, and economic strength (Table 1.1), while the proportions of French-speaking and Aboriginal peoples in each region form another essential part of the puzzle (Table 1.2). These basic geographic elements provide a start to understanding the

nature of the six regions. Further understanding is provided by analysis of the key economic activity found in each region. Under the subheading of *Key Topic*, fuller analysis and detailed insights into the nature and strength of each regional economy are presented, as are the challenges faced by these economies. These key economic activities are:

- Ontario: automobile manufacturing
- Québec: hydroelectric power
- British Columbia: forest industry
- Western Canada: agriculture
- Atlantic Canada: fisheries
- The Territorial North: megaprojects

The Dynamic Nature of Regions

Canada and its regions are not static entities. Many forces are behind the dynamics of regions. Culture, demography, the economy, immigration, politics, and technology are leading forces of change. In the future, climate change may play a role, too. Change also alters relationships between regions. One measure of such change over time is revealed in the growth of population (which also underscores economic growth) in Western

Think About It

In which geographic region do you live? Is this an important part of your self-identity?

Table 1.2	Social Characteristics of the Six Canadian Regions, 2006			
Geographic Region	**French (000s)**	**French (% by region)**	**Aboriginal Peoples (000s)***	**Aboriginal Peoples (% by region)**
Ontario	510,240	4.2	242,495	2.0
Québec	5,916,840	79.6	108,430	1.4
British Columbia	58,890	1.4	196,075	4.8
Western Canada	127,060	2.4	505,650	9.4
Atlantic Canada	276,640	12.1	67,010	2.9
Territorial North	2,555	1.1	53,135	52.5
Canada	6,892,230	22.1	1,172,790	3.7

*These statistics represent **Aboriginal identity** rather than **Aboriginal ancestry**. In 2006 the census asked two questions to determine the number of Aboriginal peoples. The 2006 Aboriginal population, as determined by a question about identity, was 1,172,790, comprised of North American Indian (698,025), Métis (389,780), and Inuit (50,480); the 2006 Aboriginal population, as determined by a question based on ethnic origin, was 1,678,235. The difference in number is due to the more restrictive nature of the definition of identity compared to the ancestry definition, which is concerned with ethnicity, i.e., the cultural origins of a person's ancestors. Statistics Canada reports the number of Aboriginal peoples using the results from the identity definition in the 2006 Census of Canada. However, Statistics Canada uses the results from the ancestry figures for North American Indian, Métis, and Inuit when reporting on the ethnicity of the Canadian population.
Sources: Statistics Canada (2007c, 2008).

Canada and British Columbia (Figure 1.3). Population figures for regional populations were first recorded in the 1871 Census of Canada. As a percentage of Canada's population, the most remarkable gains took place in Western Canada and British Columbia. While populations in all regions increased over this time period, the growth rates for Atlantic Canada, Québec, and Ontario were smaller than those for Western Canada and British Columbia. In 1871, for instance, Atlantic Canada contained 21 per cent of Canada's population while Western Canada had less than 3 per cent. By 2006, the population figures were almost reversed with Western Canada accounting for 17 per cent and Atlantic Canada for 7 per cent.

Change is continuous, economically, socially, and physiographically. For example, the Canada of the 1960s (see Hare, 1968) is gone. Events that changed Canada and its regions include:

- The Auto Pact, signed in 1965, had a profound impact on Ontario's economy and signalled Ottawa's intention to look more favourably on continental trade relations with the United States.
- The emergence of the Parti Québécois gave political expression to the desire for political autonomy/separation in Québec and, at the same time, threatened the unity of Canada.
- The end of the baby boom has meant greater reliance on immigration to keep Canada's population increasing; with rapid growth in urban Canada fed by immigration, our cities have become much more pluralistic, reflected by federal support for multiculturalism beginning in the 1970s.
- The 1969 White Paper calling for 'equality' for Aboriginal Canadians, repeal of the Indian Act, and an end to treaty benefits led to a strong reaction and the emergence of Aboriginal political power.

Change over the past several decades raises a central question: What events might

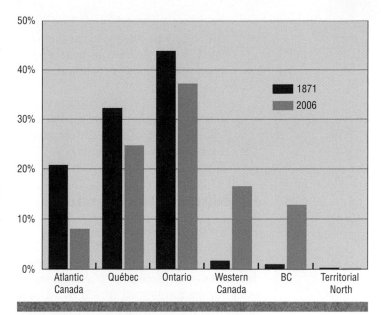

Figure 1.3   Regional populations by percentage, 1871 and 2006.
Sources: Statistics Canada (1997, 2007a).

change the nature of Canada and its regions in the future? Among various future scenarios, five possible events, all related to the economy, might emerge over the next 10–20 years:

- A shift in the manufacturing economies of Ontario and Québec to higher **value-added production** requiring highly skilled labour and advanced technology, coupled with a downsizing of the automobile industry in Ontario.
- Increased energy production from the oil sands in Alberta, plus a solution (possibly sequencing) to its release of greenhouse gases; more mega-hydroelectric developments in BC, Manitoba, Labrador, and Québec to satisfy the growing demand for 'clean' energy in the neighbouring regions in the United States and Ontario.
- More diversified and profitable agriculture in Western Canada due to growing world demand for foodstuffs and the accompanying increase in world prices for these products.

- Ocean-going ships passing through the Northwest Passage in the summer, and the St Lawrence Seaway staying open year-round as a consequence of global warming.
- The recovery of the US housing market revitalizing the stagnant forest industry, which is especially important for BC, Ontario, Québec, and New Brunswick.

Faultlines within Canada

Tensions exist within Canada, especially between its regions. In 1993, *Globe and Mail* columnist Jeffrey Simpson applied the term '**faultlines**'—the geological phenomenon of cracks in the earth's crust caused by tectonic forces—to the economic, social, and political cracks that divide regions and people in Canada and threaten to destabilize Canada's integrity as a **nation**. In this text, 'faultlines' refers to four divisions within Canada. For long periods of time, these faultlines can remain dormant, but they can shift at any time, dividing the country into wrangling factions.

While many tensions within Canadian society emanate from the plight of the disabled, the homeless, and the poor, our discussion is confined to four major faultlines that have had profound regional consequences and that have challenged our national unity. These four faultlines represent struggles between: centralist and decentralist visions of Canada; English and French; old and new Canadians; and Aboriginal and non-Aboriginal Canadians. Each faultline has played a fundamental role in Canada's historic evolution, and they remain critical to Canada in the twenty-first century. These cracks within Canadian society can threaten and have threatened the cohesiveness of Canada and, by doing so, have shaken the very pillars of federalism. Forced to seek a compromise between these differing positions, Canada, over time, has become what John Ralston Saul (1997: 8–9) describes as a 'soft' country, meaning a society where conflicts are, more often than not, resolved through discussion and negotiations.[2] Disagreement over the nature of Canada—is it a partnership between the two so-called founding societies of the

French and the English or is it composed of 10 equal provinces?—has troubled the country since Confederation in 1867 and in recent years has come to a head twice with the sovereignty-association and independence referendums in Quebec in 1980 and 1995, the latter of which was won by the federalist side by the narrowest margin—a mere percentage point. At the height of the referendum campaigns, uneasy relationships tore at the very fabric of Canada. But the ensuing dialogue and goodwill between Québec and the rest of the country have led to an unspoken compromise that cements the country and its regions together.

How can we conceptualize such complex events? As Canadian society moves through time, history provides a view of the past, but the future is much less certain. While all societies change, Canada is evolving at a more rapid rate. Four factors contribute to this rapid change. First, population and economic power are shifting from Atlantic Canada, Québec, and even Ontario to Western Canada and British Columbia. The shift is related to a slowdown in manufacturing centred in Ontario and Québec, to a depletion of resources in Atlantic Canada, and to the world demand for resources found in Western Canada and British Columbia. In spite of the global economic crisis that began in late 2008, the economic situation in these two western regions remains more positive than in Central Canada. Second, the pluralistic nature of Canadian society represents another significant factor. Pluralism is a relatively new phenomenon in Canada, superseding the biculturalism represented by the French and English. Pluralism is an expanding phenomenon because of the flood of non-Western immigrants into Canada. Adjustments and accommodation of the newcomers and their cultural/religious practices are necessary for continued social harmony, but so is their need to fit into Canadian society. In Québec, fears that the traditional Québécois way of life was slipping away came to a head with Hérouxville's 'code of standards' (Vignette 1.2). Aboriginal peoples form the third factor. Canada's First Peoples have struggled to find a place within a Canadian society in which they have been marginalized, 'othered', and subjugated under the Indian Act of 1876 and its numerous revisions. The federal government's assimilation policy of the nineteenth century and much of the twentieth century resulted in the misguided and hurtful Indian residential school program. Fast forward to the twenty-first century: outstanding land claims, the rapidly growing Aboriginal population, the ever-increasing numbers of Native people in Canadian cities, and the education/income gap between Aboriginal and non-Aboriginal Canadians are four issues that demand attention. The last factor is the place of Québec within Canada. With its distinct culture and French language, Québec, as a people and as a government, views its place within Canada and North America through a different lens from that used in other regions of Canada. For the first time, the House of Commons in 2006 declared that the Québécois are a nation within Canada. Does this declaration support the concept of two founding nations or is it an empty political statement? For more on this subject, see the subsection 'One Country, Two Visions' in Chapter 3

Continuity with the past is essential for successfully navigating into the future. For Canada, this desire for continuity must be tempered by a need to accommodate the many new cultures that have become a presence in Canada in recent years. In simple terms, Canada must grip tightly its core values while necessarily reinventing itself to meet new circumstances. This situation is of special significance in the discussion of these four faultlines.

Centralist/Decentralist Faultline

Canada's heterogeneous nature often forms the basis of regional quarrels. These disputes often flare up between particular regions and Ottawa, but at other times disputes are inter-regional. Why is this so? Demography favours Central Canada because the majority of Canadians reside in Ontario and Québec. From a political perspective, these two regions often benefit from federal policies that support the 'national interest'. Adding to their demographic superiority and political clout, Ontario and Québec have long been the manufacturing centre of Canada. These demographic, economic, and political advantages illustrate the dominant position of these provinces in the nation's affairs and support the notion of a core region. Not surprisingly, other regions have perceived Ottawa as favouring Ontario and Québec. While Ottawa may maintain that its considerable subsidies to manufacturing industries in these provinces are in the national interest, the rest of Canada perceives such subsidies as 'special treatment' designed to gain the favour of voters in Ontario and Québec. Because no political party can form a majority government without strong support from these two provinces, the regions in the rest of Canada feel powerless to get a fair hearing in Ottawa. Over the years, this has provided a deep foundation for regional alienation.

Similarly, tensions arise between the regions and Ottawa when regions believe the federal government is seeking to exploit them for its own advantage. Thus, an example of tension between Ottawa and Alberta is fuelled by Alberta's fear that Ottawa will once again intrude into provincial power over natural resources to take royalty revenues from the province with a new version of the 1980 National Energy Program. Royalties from oil and gas have made Alberta extremely rich, allowing the province to eliminate its debt and to maintain lower tax rates. Now other provinces—Saskatchewan, British Columbia, Newfoundland and Labrador, and Nova Scotia—hope to profit from their petroleum deposits. As the economies of petroleum-producing provinces gain new strength, will Ottawa again seek a share of their natural wealth?

Besides tensions between the regions and Ottawa, inter-regional conflicts fester in the Canadian polity. One of the most bitter examples of regional tension is between Newfoundland and Labrador and Québec over the controversial Churchill Falls hydroelectric agreement (1969). The original contractual agreement, signed by a private company, the Churchill Falls Labrador Company, and the Crown corporation, Hydro-Québec, gave Québec a very advantageous position. Within a few years, the unfairness of this agreement—at least to Newfoundlanders—caused extreme political animosity between the two provinces.

Think About It

Do any faultlines exist in your region?

Newfoundlanders see the agreement as a form of exploitation of their resources.

English-speaking/French-speaking Canadians

Québec, with its culture and French language, forms a distinct region within Canada. Tensions between English-speaking Canada and Québec have often erupted over language issues. Within Québec, an internal faultline exists between separatists and federalists. Separatists argue that Québec cannot fulfill its aspirations unless it becomes a nation. Federalists take the position that Québec can best flourish within the Canadian federation. This argument was played out in the 1980 and 1995 referendums. The struggle to maintain its culture and language has been long and hard, going back to the defeat of the French army on the Plains of Abraham in 1759. Until 1969, English was Canada's only official language. This fact underscores the difficult struggle francophones have had in finding a place within Canada.

A key to that struggle is the French language. The political and cultural desire to maintain French as a viable language in a principally English-speaking continent resulted in the controversial Bill 101 in 1977, which proclaimed French as the official language in Québec for just about every facet of life: government, the judicial system, education, advertising, and business. In 1982, the Supreme Court of Canada struck down Bill 101. However, the Liberal government of Robert Bourassa overturned this ruling by employing the 'notwithstanding clause' (section 33 of the 1982 Charter of Rights and Freedoms). While language is much less of an issue in Québec today, language disputes remain a hot button between English and French Canadians residing in Québec.

Aboriginal Peoples and the Non-Aboriginal Majority

As the first people to occupy the territory now called Canada, Aboriginal peoples have had to deal with the settlement of their lands, with subjugation as colonized peoples, and with Ottawa's restrictive Indian Act. The Métis mounted two rebellions to defend their land (see the subsection 'Manitoba: The Métis against the Newcomers' in Chapter 3 for a more complete discussion). While Aboriginal Canadians reside in all provinces, they form a majority in the Territorial North and in the northern extremes of the other geographic regions. While treaty Indians were assigned to reserves, many have relocated to cities where more opportunities are available.

Modern land-claim agreements have led to the possibility of self-government on reserves.[3] But what does self-government mean? Would it take the form of ethnic government or public government? Could it contain provincial-like powers or just municipal powers? Some First Nations have obtained self-government in the form of an ethnic government with powers at least at the municipal level but not at the provincial level. Public government within an area dominated by Aboriginal peoples provides another approach. Nunavut is one example while Nunavik in Arctic Québec may provide another example. In Nunavik, the Inuit constitute over 85 per cent of the total population and almost 100 per cent of residents born and raised there. The case for self-government is strong—as it was in Nunavut—but the political ramifications of creating a new level of public government within a province are huge. Negotiations among the federal government, the Québec government, and Makivik Corporation (which represents the Inuit of Nunavik) are well underway and may result in an innovative political development—a semi-autonomous region within a province. The concept of 'nested federalism' would break new ground in Canada's political realm and could have ramifications for other provinces.

Newcomers and Old-timers

Canada is a land of immigrants. Early French and English settlers had a very difficult time adjusting to the New World. In the early twentieth century, immigrants came to settle the prairie lands of Western Canada. Finding a place—a comfort level that would allow them to be themselves and gain a sense of belonging—was not as easy for the many non-English speaking immigrants to Canada from Central Europe and czarist Russia as it was for those from Great Britain who were English-speaking and who found strong similarities in Canadian society and its institutions to those in their home country. The Doukhobors, for example, who practised collective farming, failed to see their lifestyle gain acceptance.

See Chapter 3, page 112, for an elaboration of the Doukhobor struggle to find a place in Canada.

The rubbing and bumping between those whose cultural roots are in overseas homelands and those whose roots developed in North America necessarily require adjustments and compromises. Old-timers are not always prepared to give ground and sometimes, as in the case of the code of standards created by the municipal council in Hérouxville, Québec, they react strongly to protect their status quo (Vignette 1.2). Yet, these sometimes unpleasant interactions between newcomers and old-timers are a necessary fact of Canadian life and signal that a dialogue is taking place. Generally speaking, the second and subsequent generations of immigrant groups, by being born and raised in Canada, have had a much easier time feeling connected to Canada. Yet, as with those

nicknamed CBCs (Chinese born in Canada), a cultural gap emerges within the diasporic community. Now the same challenge faces immigrants from Asia, Africa, and the Middle East.

Unlike before the 1960s, when the vast majority of immigrants to Canada came from the British Isles and from Northern Europe, most immigrants in more recent years have come from non-European countries. This has made integration into Canadian society more difficult, especially for visible minorities who may face both open and subtle forms of racial discrimination. Now that Canada is entering the second decade of the twenty-first century, racism, while still present, has hopefully been addressed and Canadians have become more open to newcomers, regardless of their origins and beliefs. Nonetheless, friction continues because some newcomers wish for more control over their 'people'. For example, when a former Ontario attorney-general floated the idea of the use of Islamic or **shariah law** to settle family disputes, the idea was strongly supported by conservative Muslim groups. (Since the early 1990s Ontario had allowed Catholic and Jewish faith-based tribunals to settle family law matters on a voluntary basis.) Many protested, including more moderate Muslims, and the much deeper issues of the boundary between church and state and the place of Muslim women in Canadian society

---

Vignette 1.2 The Challenge of Cultural Change: Québec and Muslim Canadians

All cultures change with time. What is different in Canada is the constant addition, with each wave of immigrants, of new cultures to the pluralistic mix. Cultures brought to Canada by immigrants from distant lands impact on local, regional, and Canadian cultures and values. This interaction forms a key aspect of the old-timers/newcomers faultline. Those born and raised in Canada watch such changes with both pleasure and trepidation. Pleasure comes from the enrichment of Canadian society while trepidation comes from fear of losing control of their world, whether resulting from so-called 'monster houses' built by wealthy immigrant Chinese in Vancouver or from the demand for space for Muslim prayers in Montréal universities. In 2007,

matters came to a boil in Québec, when the Hérouxville municipal council announced its 'code of standards'. The purpose of the code is 'to inform the new arrivals that the lifestyle that they left behind in their birth country cannot be brought here with them and they would have to adapt to their new social identity' (Hamilton, 2007: A7). The resulting firestorm within Québec society caused the provincial government to launch the Bouchard-Taylor Commission[4] to address the issue of 'accommodation practices related to cultural differences in response to public discontent concerning reasonable accommodation'. The Commission's report received mixed reviews and some felt that its recommendations were unreasonably accommodating.

CP Photo/Adrian Wyld

Photo 1.1

The rights of women in a secular society are challenged by those advocating the application of *shariah* law to Muslims. During a protest in Toronto against *shariah* law on 8 September 2005 a woman points out that the book Mubin Shaikh is referring to about women was written by a man.

Think About It

Do you favour a strict separation of state and church? If so, where do you stand on public support for Roman Catholic schools?

surfaced (photo 1.1). In September 2005, Ontario Premier Dalton McGuinty rejected this request and, at the same time, rescinded approval of Catholic and Jewish tribunals. Even so, another form of difficulty faces Muslim Canadians today. In the current socio-political climate, Muslim Canadians still must deal with the fallout from the 2001 terrorist attacks on New York and Washington, the broader struggles in the Muslim world about the place of women in Canadian society, and Canada's role in Afghanistan. American 'no fly' policy affects airline passengers originating in Canada and this US policy poses travel restrictions for some Canadian citizens who were born in the Middle East and who hold dual passports.

The concentration of new immigrants in Canada's major cities has both advantages and disadvantages. Newcomers have more cultural anchors to support them, such as family and friends who speak their native tongue, restaurants that serve traditional foods, and religious institutions that provide both spiritual and community support. A disadvantage may be a sense of isolation from other Canadians. But adjustment to Canadian life and the elusive Canadian identity comes with the next

generation, whose attachment to the 'new country' is stronger and whose roots to the 'old country' are much shallower than those of their parents. This process of adaptation may be more difficult for visible minorities. For example, recent research in the Toronto area suggests that some visible minorities born in Canada do not feel connected to Canadian society (Lewington, 2007).

Power of Place

Power of place is a concept derived from economic geography and has deep Canadian roots going back to the 1930s and the classic **staples thesis** of economic historian Harold Innis. Contemporary scholars, including Barnes, Britton, Hayter, Norcliff, and MacLachlan, to name but a few, have enriched the field of economic geography and our understanding of the relationship between place and power. Power of place is closely associated with the global economy and, in that sense, is the opposite of a sense of place. While power of place is applicable to all geographic levels, it represents a mix of natural and economic components, emphasizing at the regional level natural wealth, geographic situation, and access to global markets. Traditionally, hinterlands with resource-oriented economies were at a disadvantage compared to highly industrialized core areas. The explanation is complex, but is linked to the historic fact that prices of manufactured goods increased more rapidly over time than did prices of commodities (with the exception of oil). Price differentiation between manufactured goods and commodities is a key element in the **core/periphery model**. In addition, hinterlands with narrow resource-based economies are subject to rapid increases in economic activity in good times (booms) and rapid decreases in bad times (busts). **Boom-and-bust cycles** are most evident in the Territorial North, where the economy depends most heavily on natural resources.

Let us examine the relationship between power of place and changes in global prices over time more closely. Over the long term, world prices tend to rise and fall in accordance with demand. Also, since the beginning of the Industrial Revolution, world prices have favoured manufactured goods over resources, that is, prices for manufactured goods have risen more than prices for resources. This price relationship underlies the nature of the core/periphery theory, namely that manufacturing areas gain more wealth over time than do resource-based regions. But the world economy may have reached a tipping point and entered what is called the super cycle, thus reversing this price relationship. The argument supporting the **super cycle theory** is that, in the current economic downturn, commodity prices declined but bottomed at higher levels than ever before—in fact, resource prices bottomed above the peaks of the previous cycle because of sustained demand from rapidly industrializing countries such as China and India (Parkinson, 2009). If this trend continues, could Canada's economic power shift from manufacturing-based regions like Ontario and Québec to resourced-based regions, especially Western Canada?

Sense of Place

In our globalized world, place still matters. The term 'sense of place' embodies this perspective. Sense of place has deep roots in cultural and regional geography. Leading scholars in this area include Agnew, Cresswell, Paasi, Relph, and Tuan. While 'sense of place' has been defined and used in different ways by geographers, other social scientists, and architects, in this text it reflects a deeply felt attachment to a region or area by local residents who have, over time, bonded to their environment and resulting institutions. As such, sense of place provides some protection from the landscapes produced by economic and cultural **globalization**. These landscapes are associated with a sense of **placelessness** (Relph). Sense of place, on the other hand, involves a powerful psychological bond between people and locale; globalized or generic landscapes do not have local roots. Local roots stem from the physical nature, human activities, and institutional bodies found in a region. Yellowknife, for example, located on the rocky shores of Great Slave Lake, exhibits a uniquely northern character (see photo 1.2).

With its vast spaces and magnificent diversity of physical settings, Canada lends itself to strong regional bonds. Atlantic Canada and Western Canada provide two examples. The sea exerts a profound influence on all who live in Atlantic Canada. The fishery sparked early settlement in the region and, until recently, provided the economic basis for many coastal communities. The tragic loss of the northern cod fishery in the 1990s brought great economic, social, and personal dislocation to the small fishing communities in Atlantic Canada, but especially to the outports in Newfoundland (for more on this subject, see Chapter 9). The vast open expanses of the Prairie region also exert a powerful influence on its inhabitants. Its dry continental climate, however, has made farming difficult at times, and climatic events can make the memories of farmers bitter. For example, the dust bowl of the 1930s deeply marked Prairie farmers, forcing many to abandon their homesteads. While the drought conditions of that earlier era have faded into past collective memory, this semi-arid environment continues to influence life on the Prairies, with drought a real threat to Prairie agriculture. Indeed, two years of arid growing conditions (the summers of

Think About It

Would Alberta have become a 'have' province back in the 1970s if the Organization of Petroleum Exporting Countries (OPEC) had not restricted world oil supply, thus increasing prices dramatically?

Photo 1.2

Yellowknife is no longer a rough-and-tumble mining town. This city has evolved into a sophisticated centre with expensive housing along its waterfront. In the warm summer months, pleasure craft, sailboats, and float planes are moored along its sheltered coves on the north shore of Great Slave Lake.

Terry Parker/NWTT

2001 and 2002) followed by three years of untimely frosts and wet falls (2003, 2004, and 2005) left many western grain farmers deep in debt or forced them to declare bankruptcy (for more on this topic, see Chapter 8). In both regions, poor economic circumstances have forced families to relocate to major cities, especially those in Alberta and British Columbia, or have resulted in heads of households effectively commuting across the country, as many Newfoundlanders do in working in the Alberta oil patch.

The concept of a sense of place, then, recognizes that people living in a region have undergone a collective experience that leads to shared aspirations, concerns, goals, and values. Over time, such experiences develop into a social cohesiveness among those people living within a spatial unit. Often these experiences are coloured by extreme weather events. The rainstorm and subsequent flooding in the Saguenay Valley of Québec in July 1996 was such an experience. Ice storms often have devastating effects by knocking out regional electrical systems. In early February 2003, New Brunswick was hit with a severe ice storm that dropped 60 millimetres of frozen rain, leaving some residents from Saint John to Moncton without power for several days while temperatures were hovering around −10°C.

A sense of place can also evolve from a region's history and geography. Early in its history, British Columbia was isolated from the rest of Canada by the Rocky Mountains, which form the largest mountain system in Canada, extending 1,200 km from the US–Canada border north towards 60° N. While this formidable physical barrier, considered by many to be the backbone of Canada, has been surmounted by modern transportation systems, it remains an important physical and cultural feature in British Columbia, and the promise of a trans-Canada rail line, the Canadian Pacific Railway, was instrumental in BC joining Confederation in 1871. In *The Unknown Country*, Bruce Hutchison (1942: 266) described the effect of the Rocky Mountains on people as he observed changes in the landscape from a passenger car on the Canadian Pacific train: 'Unlike the life of the

Prairies or the East—so unlike it that, crossing the Rockies, you are in a new country, as if you had crossed a national frontier. Everyone feels it, even the stranger feels the change of outlook, tempo and attitude. What makes it so, I do not know.' Feelings of alienation run deep in British Columbia. This theme reverberates in Philip Resnick's *The Politics of Resentment: British Columbia, Regionalism and Canadian Unity* (2000). Resnick supports Hutchison's contention that BC is a distinct region, but he also stresses that a sense of alienation from Ottawa is widespread within the province.

An equally powerful expression of place and alienation exists in Québec, where history and geography have had four centuries to nurture a strong sense of place and to give birth to a nationalist movement that has sought to separate Québec from the rest of Canada. New France was born and grew within the confines of the St Lawrence Lowland. Since Québec's early days, the St Lawrence River has played a key role in its settlement and economic development. Prior to the British Conquest of New France, formalized in 1763 by the Treaty of Paris, this mighty river promoted the interests of the French colony by providing a supply route to France, allowing the fur traders (coureurs du bois) to penetrate far inland to trade with distant Indian tribes, and giving French explorers access into the heart of North America.

A sense of place, however, is much more than bad memories, resentment, and alienation. Indeed, in its truest sense, it is about home and fond memories, about security and the strength to endure the vagaries of the natural world in a particular locale. Peggy's Cove in Nova Scotia provides an example of a community with close ties to the sea. Originally a tiny fishing village, Peggy's Cove, with its sense of history and its location on the sea, has become an important tourist destination (see photo 1.3).

A region, then, is a synthesis of physical and human characteristics that, combined with its distinctiveness from surrounding regions, produces a unique character, including a sense of place and power. People living and working in a region are conscious of belonging to that place and frequently demonstrate an attachment and commitment to their 'home' region. A sense of place, falling somewhere between regional tribalism and commitment to Canada, unites people on common issues and challenges, and compels them to seek solutions to these issues and challenges. For the Inuit, a sense of place is embodied in their natural homeland of the Arctic and in their search for a place within Canadian society; for new Canadians, a sense of place may seem elusive at first, but, with time, they too will develop such feelings as they (or more likely their children) sink their roots into Canadian society. Indeed, the theme of this book is that Canada is a country of regions, each of which has a strong sense of regional pride, but also a commitment to Canadian federalism.

Photo 1.3

Peggy's Cove, Nova Scotia. Along its deep but sheltered waters, small fishing boats are moored while lobster traps line the dock. Brightly painted houses and buildings provide for a picturesque background.

Scott Goldsmith/GetStock.com

The Core/Periphery Model

Development theory provides a conceptual framework for understanding the world. As Peet and Hartwick (2009) point out, many theories exist, each reflecting different ideologies. What is the advantage of a theoretical framework for Canada's regional geography? Simply put, a modified version of the core/periphery model provides a conceptual basis for understanding the origins and economic structures of Canadian regions as well as interactions between regions and Ottawa. The core/periphery model has its intellectual roots in Innis's staples thesis and in Wallerstein's **capitalist world-systems theory**. By their very nature, however, development theories must simplify the real world to fit the past and present economic circumstances. In doing so, when applied to actual regions, they often lose touch with the complexity and variety of economic, political, and social forces at play. The model may even suggest that regions are not dynamic places, when in fact change is constantly taking place. On the world stage, for example, European nations are no longer the only 'industrial core regions' in the world. Japan and China are considered leading industrial countries.

An important question in this book is: *Can such spatial changes occur in Canadian regions?*

Applying the Core/Periphery Model to Canada

John Friedmann (1966) devised a national version of the core/periphery model as it related to Venezuela (Table 1.4). In the 1960s, Friedmann, an American scholar, was working as a regional planner in Venezuela and applied the core/periphery concept to regions within a single country. He viewed core regions as territorially organized subsystems of society that have high capacity for generating and absorbing innovative change, and peripheral regions as subsystems whose development is determined chiefly by core region institutions on which they depend. From a modernization perspective, Friedmann conceived an industrial core and three types of peripheral regions, all of which, in different ways, are dominated by the core.

Think About It

Do you have a stronger sense of belonging to your province or to Canada?

Table 1.3   Key Steps Leading to a More Open Canadian Economy

Date	Event
1947	The General Agreement on Tariffs and Trade (GATT) was signed by Ottawa and 22 other governments. These countries agreed to reduce tariffs and eliminate import quotas among members through multilateral negotiations and agreements.
1965	The Auto Pact between Canada and the United States created a continental market for automobiles. An important provision for Canada was that its share of automobile production would be maintained. The three large American car manufacturers (General Motors, Ford, and Chrysler) could now expand their production to a continental scale. In this process, Canadian branch plants ceased to produce a wide variety of automobiles for the Canadian market and began to produce only a few models (but more of them) for the continental market.
1989	The Canada–US **Free Trade Agreement** (FTA) was approved. The aim of the FTA was to integrate the two economies. Unlike the Auto Pact, however, there were no assurances that Canada's share of economic activities would be maintained. Instead, the North American marketplace would determine the type and amount of economic activity in Canada.
1994	Superseding the FTA, the **North American Free Trade Agreement** (NAFTA) broadened the geographic area of the FTA to include Mexico. The North American economy now had a member with much lower wage rates, causing many labour-intensive industries to close their operations in Canada and reopen in Mexico.
1995	The final GATT treaty (the so-called Uruguay Round) was ratified by 124 governments on 1 January 1995. This agreement, which established the World Trade Organization (WTO), a permanent institution with powers to enforce trade rules and assess penalties against members, is the most far-reaching trade pact in world history and moves closer to the idea of global free trade.
2000	The WTO rules that the Auto Pact discriminates against foreign companies. In 2001, the Auto Pact is dissolved.
2001	Following the suicide mission of Al-Qaeda terrorists who commandeered and then crashed two passenger planes into the twin towers of the World Trade Center in New York City and another one into the Pentagon in Washington, DC, the US government created a Department of Homeland Security, promoted a **North American Security Perimeter**, and announced the Enhanced Border Security and Visa Entry Reform Act that, together, had enormous implications for Canadian access to the United States.
2002	Canada signed the Kyoto Protocol that committed 38 industrialized countries to cut their emissions of greenhouse gases between 2008 and 2012 to levels 5.2 per cent below 1990 levels.
2009	The United States continued with its North American Security Perimeter, so that, after some delays and modifications, all cross-border travellers—by air, land, and water—must carry passports.

In applying Friedmann's model to Canada, a core region and three **periphery** regions can be identified. They are:

- core region (Ontario and Québec);
- periphery region: upward transitional region (British Columbia and Western Canada);
- periphery region: downward transitional region (Atlantic Canada);
- periphery region: resource frontier (the Territorial North).

The broad brush of history and geography provides one explanation for the original classification of Ontario and Québec as the core—their favourable agricultural base coupled with their head start in historic settlement, rapid population growth, and development of a manufacturing base. For those reasons, plus high tariffs on imported manufactured goods imposed by the National Policy (1879), Ontario and Québec formed the industrial and population core in the first years following Confederation.

But such a model is not set in stone. For instance, what are the implications of economic and population growth in British Columbia and Western Canada outpacing that found Québec? Another key question is: Can Ontario recover from its current economic malaise? In other words, Canada's six regions can change their position within this theoretical classification scheme.

One measure of Friedmann's spatial model is found in Canada's **economic structure**, which consists of primary, secondary, and tertiary economic activity (Table 1.4). Each region has a different mix of primary, secondary, and tertiary sectors, which provides empirical support for the classification of the six regions into core, upward transition, downward transition, and resource frontier.

Table 1.4   National Version of the Core/Periphery Model

Regions	Characteristics
Core Region	The core region is the focus of economic, political, and social activity. Most people live in the core, which is highly urbanized and industrialized. The core has a high capacity for innovation and economic change. Innovations and economic advances are disseminated downward through the national urban hierarchical system to the periphery. Ontario and Québec form Canada's core region.
Upward Transitional Region	The upward transitional region's economy and population are growing as both capital and labour flow into this rapidly developing area. While initial development occurred in the primary sector, there is now a greater emphasis on manufacturing and service activities. British Columbia and Western Canada are examples of upward transitional regions.
Downward Transitional Region	In a downward transitional region, the economy is declining, unemployment is rising, and out-migration is occurring. Often this is an 'old' region dependent on resource development for its economic growth. Now that these resources have passed their prime or have been exhausted, the regional economy has stalled. Atlantic Canada is an example of a downward transitional region.
Resource Frontier	Located far from the core region, few people live in this frontier and little development has taken place. Resource companies are just beginning to penetrate into this remote area. As energy and mineral deposits are discovered, the prospects for economic growth are enhanced. The Territorial North represents Canada's last resource frontier.

Sources: Friedmann (1966); Wallerstein (1979).

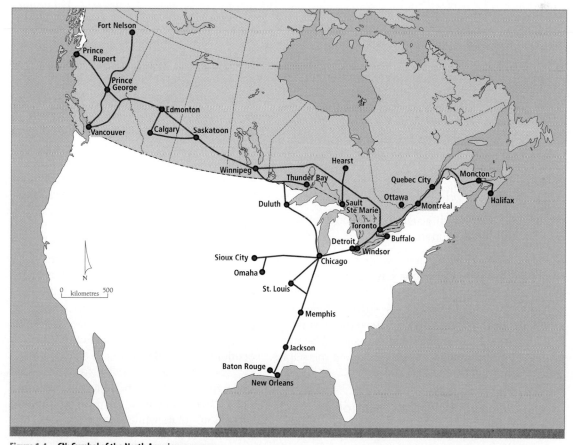

Figure 1.4   CN: Symbol of the North American economy.
With the Free Trade Agreement, CN switched its emphasis from a predominantly Canadian east–west railway system to a North American continental system. CN now operates in eight Canadian provinces and 16 US states.

Table 1.5 Economic Structure of Canada's Six Geographic Regions, 2006						
Economic Structure	Ontario	Québec	British Columbia	Western Canada	Atlantic Canada	Territorial North
Primary	2.2	2.8	3.6	10.1	5.4	13.0
Secondary	22.5	21.2	17.7	16.7	16.0	2.0
Tertiary	75.3	76.0	78.7	73.2	78.6	85.0
Total	100.0	100.0	100.0	100.0	100.0	100.0

Sources: Statistics Canada (2007d); for Territorial North, author's estimates based on Yukon (2006) and Northwest Territories (2006).

But how does Statistics Canada measure economic structure? The answer lies in the labour force because it is accurately recorded and is easily assigned to a particular activity. For example, Ontario has the highest proportion of its labour force engaged in secondary (or manufacturing) activities (22.5 per cent in 2006). At the other end of the scale, the Territorial North has the lowest proportion of its labour force involved in the secondary sector (an estimated 2 per cent in 2006).

An examination of Table 1.4 leads us to pose a fundamental question:

Is the classification of Canada into core and periphery regions supported by the data on their economic structures?

This question and two others underlie much of the discussion in the six regional chapters, and a hint of the answers is found in Vignette 1.4. The other two questions are:

Does each geographic region have a different economic structure?

Is there a trend whereby workers are shifting from one category to another?

While Friedmann's spatial model provides a framework for classifying Canada's six geographic regions, students should be aware of other spatial models, including the staples thesis (Vignette 1.3).

Canada–US Trade Relations

Since the Auto Pact agreement in 1965, Canada's trade relations with its superpower neighbour have dominated Canada's political agenda, and this domination is reflected in its

efforts to gain access to US markets through the Free Trade Agreement (1989) and the North American Free Trade Agreement (1994). Partly because of close, somewhat integrated economies, the value of Canadian–US trade is the largest among any two countries in the world. As President Kennedy remarked in his speech to Canada's Parliament in 1961:

Geography has made us neighbors. History has made us friends. Economics has made us partners. And necessity has made us allies.

Since Canada's economic well-being is strongly affected by trade, Canada's foreign policy has emphasized trade agreements designed to obtain greater access to markets in the United States and the rest of the world (see Table 1.3 for a list of major trade agreements). Yet, in spite of trade agreements between Canada and the United States, Washington has imposed trade barriers against Canada to protect its domestic producers. Trade disputes, especially in agricultural and forestry products, are relatively frequent and have sometimes turned nasty.[5]

Why does Canada seek trade agreements? First, with a relatively small domestic market, Canada's production, whether in commodities or manufactured goods, depends heavily on exports. Trade permits Canadian industries to achieve low costs of production through **economies of scale**.

Second, Canada seeks assurances that trade agreements will be followed. With the FTA, Canada obtained **dispute settlement provisions** that, while not perfect, reduced the threat of arbitrary use of protectionist US measures against Canadian exports (Hart, 1991; Burney, 2007). NAFTA includes a similar dispute settlement provision.

Vignette 1.3   The Staples Thesis

Early in the twentieth century, Harold Innis (1930) presented his theory of Canada's economic development to explain why Canada developed distinct regional economies and political institutions. Known as the staples thesis, Innis based his theory on resource development in the hinterland (Canada) and trade with the **heartland** (Great Britain). Over time, the exploitation of these staples led to regional economic diversification and the formation of institutions that defined the political culture of each region. The staples thesis accomplished two important objectives. First, Innis articulated an explanation for Canada's economic development, which was initially based on the exploitation of resources (staples). Over time, the development of primary extraction industries led to a more broadly based economy.

Second, Innis argued that each resource had a different impact on its region, its institutions, and its political culture. In Western Canada, for instance, early economic development was driven by the wheat staple. With a strong demand for grain in Great Britain, the Prairie wheat economy expanded rapidly, creating a dense network of railways, hundreds of villages and towns, and most important of all, a need for co-operation among the many thousands of farmers.

Farmers felt hard done by institutions based in Ontario and Québec, including the Canadian Pacific Railway, Canadian National Railways, the grain companies, banks, and even the federal government. Fighting back, they formed co-operatives to pool both their product and, ultimately, their political power. In 1913, farmers banded together to form the United Grain Growers, and in the 1921 federal election the Progressive Party, with roots in the Prairie West, came in second at the polls, well ahead of the Conservatives. In 1932, the first politically viable socialist party in Canada took root in Saskatchewan. The Co-operative Commonwealth Federation or CCF represented a coalition of worker and farmers.

The staples thesis, then, provided Innis with a Canadian version of regional development within the context of a core/periphery framework (Watkins, 1963, 1977; Barnes, 1993; Hayter and Barnes, 2001). In his 1963 article, Watkins introduced three concepts to the staples thesis: **backward**, **forward**, and **final demand linkages**; in his 1977 article, Watkins took a more pessimistic interpretation of regional development in a frontier area where he felt a **staple trap**, that is, a collapse of an economy based on non-renewable resources, was inevitable.

By 2008, Canada's largest trading partners were the United States and China. Senator Pamela Wallin, Canada's former consul general in New York, provided a yardstick for Canada's two main trading partners: 'We do more business with the US in one week than we do with China in a year', she said. 'And we do more trade with the head office of Home Depot in Atlanta, Georgia, than we do with the country of France' (Simco, 2009). Trade between the two North American countries has boomed since the signing of the FTA in 1989. More than $620 billion worth of merchandise was traded between Canada and the United States in 2006—an average of $1.7 billion crossing the Canada–US border every day (Statistics Canada, 2006b).

With a continental economy in place, a North American core/periphery spatial framework has been superimposed on the national Canadian version. The increasing integration of the North American economy has had both positive and negative impacts. Improved access to the American market has boosted Canadian production and trade and, consequently, Canadian living standards. One telling measure of the degree of that integration is revealed in Canada's international merchandise trade. In 2006, 79 per cent of Canadian exports went to the United States while 66 per cent of its imports came from the US (Statistics Canada, 2007e). From 2005 to 2006, exports to the United States declined by 2.3 per cent (Table 1.6). The main reasons were the slowdown in the US economy and the high Canadian dollar. The three export sectors affected were the automotive industry, forestry, and natural gas (Statistics Canada, 2007f). Table 1.6 outlines value of Canadian exports for 2006.

Think About It

Of Canada's six regions, which one would be most prone to Watkins's staple trap?

Table 1.6 Canadian Exports, 2006			
Country	$ millions	%	% Change since 2005
United States	361,310	78.9	–2.0
European Union	33,556	7.3	16.2
Other OECD countries	18,379	4.0	20.6
Japan	10,760	2.3	2.8
All other countries	34,161	7.5	14.3
Total Trade	458,166	100.0	

Source: Statistics Canada (2007e).

Think About It

Statistics Canada reported that from 2003 to 2008, Canadian exports to the United States fell from 82 per cent to 76 per cent. Does this signal a shift in Ottawa's trade policy, or is it simply a reflection of the woeful state of the American economy?

The decline in trade with the United States has bought to the surface an old idea—the desire to diversify Canada's trade with other countries. In Prime Minister Pierre Elliott Trudeau's time, the appeal was to the European Community. Now it is to the rapidly growing Asian countries, but especially China. According to Statistics Canada (Khondaker, 2007), by 2003 China had emerged as Canada's second-largest trading partner, surpassing the UK and Japan. With nearly 1.4 billion people, an ever-expanding industrial base, and an increasing standard of living, China has become a colossus that Canadian companies and governments seek to court, and Canada is clearly benefiting from China's demand for natural resources and foodstuffs. The downside of trade with China has two parts: first, we chiefly sell our natural resources to the Chinese rather than finished goods; and second, a large trade imbalance exists between the two countries—Canada imports much more from China than it sells to China.

While Canadian relations with the rest of the world, including China, are important, the long-term trend is clear—Canada's geopolitical interests, cultural development, and economic priorities are, for better or worse, linked to its superpower neighbour, the United States. One advantage of sharing the North American continent with our powerful American neighbour is proximity to the American market, while many would argue that another is the shared defence of the continent. Economic and military agreements such as NAFTA and the North American Aerospace Defence Command (**NORAD**) have, for the most part, worked well for Canada. Both of these agreements

foster a closer co-operation between the two countries. While common interests have led Canada and the US to sign numerous agreements and to undertake joint ventures, Canada remains a distinct society (Adams, 2003). Evidence of such independence was revealed in Ottawa's decision not to join the US-led invasion of Iraq but to participate in the UN-sanctioned and NATO-led military intervention in Afghanistan. Yet, living next door to the most powerful country in the world caused Prime Minister Trudeau to remark in 1969 that 'living next to you is in some ways like sleeping with an elephant: No matter how friendly and even-tempered the beast, one is affected by every twitch and grunt' (Colombo, 1987: 54).

Thickening of the Border

Access to the US market is essential for Canadian businesses, especially for the automobile assembly and parts companies that operate on a **just-in-time principle**, whereby manufacturers do not maintain large inventories but expect delivery of parts when they are needed. Following the Al-Qaeda attacks of 11 September 2001, a new era of security imperatives was initiated by the United States resulting in the 'thickening of the border'. As described by Konrad and Nicol (2008), Washington took unprecedented measures to protect itself against future attacks that, unfortunately for Canada, created delays in border crossing. Under the Obama government, Canada–US border regulations continue to 'thicken'. In an April 2009 interview with CBC reporter Neil Macdonald, Homeland Security Secretary

Janet Napolitano clearly stated that the US is determined to create a safe but 'workable' border:

> We have borders both north and south, and we have to have some standards and implement them at the border, so we're not going to, we're no longer going to have this fiction that there's no longer a border between Canada and the United States. That is very different, however, from not having a workable border, and that's where my goal is. We're going to have a working border where families can go back and forth, a hockey team that has a game that has to be on one side of the border or the other, that we facilitate that. That we use technology to make those lines move more quickly, and fast lanes and sentry lanes and all the other kinds of things that you can set up. To me that's really the future that we are building. (Macdonald, 2009)

In sum, the US border restrictions reshaped the architecture of Canada–US relations, transformed the 'unguarded' Canada–US border into a quasi-military checkpoint, and raised the issue of a harmonized North American security perimeter. Canada's manufacturing and tourist industries, and especially Ontario's automobile industry, which is based on the just-in-time delivery of auto parts, are affected by the thickening of the border. A slowing of the flow of goods and people across the border at Windsor and Detroit (photo 1.4) would have a negative impact on Ontario's automobile assembly and parts plants. In addition, an unknown number of Canadians, particularly those born in Middle Eastern countries, are listed on the US Homeland Security's 'no-fly' list, which

Photo 1.4   Ambassador Bridge

The Ambassador Bridge over the Detroit River connecting Windsor, Ontario, and Detroit, Michigan. Remarkably, this economic lifeline connecting Canada to the United States is privately owned by a Detroit-based billionaire and his family; with one exception (a rail bridge at Fort Frances, Ontario), all of the other major bridges connecting the two countries are owned by Ottawa (Potter, 2010).

Vignette 1.4   Is Continentalism Inevitable?

Living in the northern half of North America, Canadians have long recognized two facts: first, that the 'grain of the land'—the topography, from the Appalachians to the Rockies—has a north–south orientation; and second, that exports to the US are essential for Canadian industries. For those reasons alone, continentalism has always lurked just below the surface of serious political thought. Since the 1960s, Canada has sought a form of economic continentalism. The debate is whether or not continentalism will erode Canadian sovereignty and overpower its fragile (except for Québec) regional cultures. Americans naturally view North America from their perspective. Joel Garreau's *The Nine Nations of North America* (1981) represents one such view. As the editor of the *Washington Post*, Garreau recognized that North America did not respect political boundaries but took on an easily recognizable set of cultural/economic regions. In his widely read book, he argues that conventional national and provincial/state borders are largely artificial and irrelevant, and that his 'nine nations' (Figure 1.5) provide a more accurate way of understanding the true nature of North American society. While Garreau's book is clearly a product of popular regionalization, could his nine nations just be a latent version of US-style continentalism?

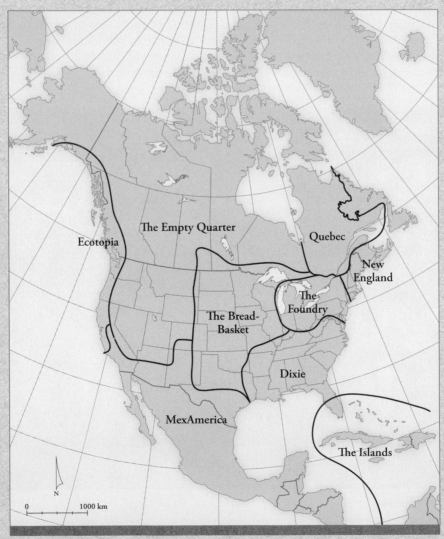

Figure 1.5   The nine nations of North America.
Source: Adapted from Garreau (1981: following 204).

not only prevents them from travelling by air to the United States but also from crossing American air space, as do many flights originating in Canada or coming to Canada from other countries.

What is at stake? Will these border restrictions cripple trade between Canada and the US? Are foreign and Canadian manufacturing companies less likely to locate in Canada? And, for that matter, have these strict policies of recent years cooled the generally warm relations between Canada and the United States? The US government remains focused on preventing terrorist attacks. By 2010, the virtual fence stretching nearly 9,000 km along the Canada–US border was partly in place. Already, high-tech monitoring equipment, including unmanned aircraft designed to spot illegal crossings, is in place. Beyond the issue of a thick border, Washington has called for a North American security perimeter that includes a common position on immigration, military, and trade policies. All of these measures are designed to reduce the risk of terrorist attacks. Clearly, **harmonization** has gained favour in Washington and it nicely complements US security concerns. For Ottawa, harmonization may lead to an erosion of Canadian sovereignty. Does Canada face a **Hobson's choice**—either embrace this North American security perimeter or run the risk of economic damage because of even greater delays at the US border? Hopefully, a middle ground is available and Ottawa is able to subscribe to those elements of the North American security perimeter that lead to economic rationalization between the two countries but not to cultural or political continentalism (Vignette 1.4). This challenge is not new and has taken different forms in the past. As many prime ministers may have mused, dancing with the American elephant generates economic advantages for Canada and its regions, but it also brings with it the threat of diminished Canadian sovereignty and of having one's toes stepped on. The following two quotations from leading Canadian political figures stress two points:

The extent to which Canada should be drawn into a fading American empire certainly is open to debate, and this debate lies at the root of historic Canada–US relations. John Manley, former Deputy Prime Minister in the Jean Chrétien Liberal government and a continentalist, stated that:

> Canada has no choice but to co-operate with American anti-terrorism efforts to keep the border open and the economy running smoothly. I think we have to be . . . a little bit less religious about the issue of sovereignty and be a little bit more practical about it. (Blackwell, 2006: A7)

Yet, compare this 2006 view with the position of former Prime Minister and Liberal leader John Turner, who, in the 1988 leaders' debate during a federal election campaign, declared:

> We built a country east and west and north. We built it on an infrastructure that deliberately resisted the continental pressure of the United States. For 120 years, we've done it. With one signature of a pen [to the FTA], you've [Prime Minister Brian Mulroney] reversed that, thrown us into the north–south influence of the United States and will reduce us, I am sure, to a colony of the United States, because when the economic levers go the political independence is sure to follow. (Azzi, 2006: A18)

While the Auto Pact (1965), the Free Trade Agreement (1989), and the North American Free Trade Agreement (1994) did boost Canada's economic performance, did these trade agreements erode Canada's sovereignty? Derek Burney, Chief of Staff to Prime Minister Brian Mulroney from 1987 to 1989 and ambassador to the United States from 1989 to 1993, argues that Canada's political independence has not been affected by FTA: 'None of the dire predictions about vanishing social programs ever materialized, either. Canada's approach to healthcare certainly has problems, but they are homegrown, and not due to free trade' (Burney, 2007).

Think About It

Given the 'thickened' border today and economic and demographic changes that have occurred in the past 30 years, how would you redraw the Canadian section of Garreau's map?

Regional Implications of US Trade

Twitches and grunts from the American elephant often resonate more strongly in one or more regions of Canada. The explanation is simple: each region of Canada has a different set of natural resource and manufacturing activities. Consequently, the type of trade with the United States varies from one region to the next. Some Canadian exports to the US are welcomed—hydroelectricity from Québec to energy-deficient New England; potash from Saskatchewan to US farm states; and oil and gas from Alberta to the US Midwest. Some are not—lumber from BC; hogs from Manitoba; pulp and paper from Ontario. Exports to the United States reveal that primary and semi-processed products dominate trade from British Columbia, Western Canada, Atlantic Canada, and the Territorial North, while manufactured goods are key exports from Ontario and Québec.

Trade disputes affect regions differently. Washington may wish to punish certain Canadian industries, but the impacts of American-erected trade barriers take on a highly regional character. Forest industries across Canada suffered from the US duties on softwood lumber, an issue that has impacted trade in wood products for the past generation, but British Columbia, Québec, and Ontario were most severely affected, and the 2006 agreement remains less than satisfactory at least for some producers. Western Canada is just recovering from a US ban on cattle. The blow to the cattle industry was triggered by a case of bovine spongiform encephalopathy (BSE) in an Alberta steer. Exports were halted, causing the industry to lose over 60 per cent of its markets and to create a glut of cattle in Canada. Prices fell dramatically for cattle producers, causing some to declare bankruptcy. In July 2005, the US border reopened to Canadian cattle exports, causing prices for Canadian cattle to increase.

How can Ottawa interpret US trade policy? It all depends on US interests and needs. Take oil as an example. On the one hand, environmentalists in the US call Alberta's oil sands 'dirty oil' and are pressuring Washington to ban the import of this oil. On the other hand, Washington wants to reduce US dependency on oil imports from the Middle East. The facts of the matter are that the US is importing more and more Canadian oil and much of it comes from Alberta's oil sands. Perhaps T. Boone Pickens, an Oklahoma oilman, captured Washington's position on energy security designed to lessen dependency on foreign oil when, speaking in Calgary in 2009, he said: 'I can only see that the only loser in this deal is foreign oil. And I don't call you foreign. You don't look foreign' (VanderKlippe, 2009).

Understanding Canada's Regions

As a prelude to understanding the many interrelated physical and human dimensions of Canada's geographic regions, the next three chapters focus on Canada's physical geography, history, and human geography. Chapter 2 ('Canada's Physical Base') explores the wide range of natural and geomorphic processes that account for wide variations in landforms across Canada. Seven physiographic regions are described with brief accounts of their geology, soils, natural vegetation, and climate. A major theme is the physical environment's influence on human occupancy of the land and the effects of human occupancy on the physical environment. Chapter 2, therefore, is more than a mere backdrop for the remainder of this text; rather, it illuminates how human activities and human decision-making both shape and are shaped by the physical environment in which they occur, and it introduces the extent of environmental problems resulting from industrial development, as described in each of the regional chapters.

Chapter 3 ('Canada's Historical Geography') focuses on the arrival of Canada's First Peoples, early French and British settlements, and the territorial evolution of Canada. The four faultlines discussed in this chapter have affected Canada and its six geographic regions differently and at different times throughout this country's history. The tensions surrounding these faultlines often lie dormant but then flare up unexpectedly, and these tensions provide a background for the subject of Chapter 4 ('Canada's Human Face'), which is directed to contemporary issues dealing with demography (baby boom and bust), Canada's pluralistic society (the impact of a more open immigration policy), and a continental economy (the effect of the Canada–US Free Trade Agreement). This chapter also looks at the population trends associated with the tensions that arise from centralist and decentralist visions for Canada, French and English views of 'nation', immigration and Canada's pluralistic society, and claims to the land and their place on it from Aboriginal and non-Aboriginal populations.

SUMMARY

Canada's six regions provide a vehicle to explore the geographic essence of Canada. To simplify the complexities of space, regional geographers divide the world and countries into regions, which vary by scale but often are interrelated in a hierarchical order. A regional geographer selects critical physical, historic, and human characteristics that logically divide a large spatial unit into a series of regions. Towards the margins of a region, its main characteristics become less distinct and merge with those of a neighbouring region. For that reason, boundaries are best considered as transition zones. Canada is a country of regions. Shaped by its history and physical geography, Canada is distinguished by six geographic regions: Ontario, Québec, British Columbia, Western Canada, Atlantic Canada, and the Territorial North. Within Canada and its regions, four key tensions exist that, through their interaction, demonstrate the very essence of Canada as a 'soft' nation, where conflicts are usually resolved or ameliorated through compromise rather than by political or military power.

The essential foundation for studying regional geography is to conceptualize places and regions as components of a constantly changing global system. The core/periphery model provides an abstract spatial framework for understanding the general workings of the modern capitalist system. It consists of an interlocking set of industrial cores and resource peripheries. This model can function at different geographic scales and serves as an economic framework for interpreting Canada's regional nature. In addition to the core region, three types of regions devised by Friedmann extend our appreciation of the diversity of the Canadian periphery. They are: (1) an upward transitional region, (2) a downward transitional region, and (3) a resource frontier.

Time affects geographic regions. Perhaps the most dramatic events to impact Canada and its regions have been (1) changes to immigration regulations and (2) the shift over time from a closed economy to an open one. Liberalization of immigration policy began in 1967 when Canada no longer favoured European applicants by deciding to accept applicants on merit from around the world. Although Canada was one of the original signatories to the General Agreement on Tariffs and Trade in 1947, which worked towards freer trade over the course of eight rounds of negotiations spanning nearly a half-century, Canada truly opened its economy to greater international influence with the 1989 Canada–US Free Trade Agreement, the signing of NAFTA in 1994, and the establishment of the World Trade Organization in 1995. Canada's place in that world has changed. Trade is one measure of those changes. International trade clearly dominates the regional economies in Canada, while interprovincial trade remains an important element, especially for Ontario and Québec. Canadian regions reflect these changes, with resource-rich regions now benefiting from high prices due to growing world demand for **primary products**. On the other hand, Québec and Ontario manufacturers are facing stiff competition from other countries where costs of labour are low.

CHALLENGE QUESTIONS

1. What does Saul mean by referring to Canada as a 'soft' country?
2. Given that culture is an essential element of human societies, does the concept of 'place-lessness' reflect favourably or unfavourably on the future well-being of the diversity of global cultures?
3. Does the core/periphery model as applied to Canada measure up to Nobel laureate Paul Samuelson's observation that 'Every theory, whether in the physical or biological or social sciences, distorts reality in that it oversimplifies. But if it is a good theory, what is omitted is outweighed by the beam of illumination and understanding thrown over the diverse empirical data.'
4. De Blij and Murphy state that 'Geography is destiny.' Are they implying that if you were born and raised on a First Nation reserve in a remote part of northern Ontario, you would have a different set of life opportunities compared to those who were raised in a large Canadian city like Toronto?
5. One notable version of the core/periphery theory was an attempt by the Marxist scholar Immanuel Wallerstein to explain the capitalist world-system. By using a spatial version of the core/periphery for Canada, do you think that the author is interpreting Canada's regional geography from a Marxist perspective?

FURTHER READING

Bourne, Larry S., and Damaris Rose. 2001. 'The Changing Face of Canada: The Uneven Geographies of Population and Social Change', *Canadian Geographer* 45, 1: 105–19.

Over the last 50 years, Canada and its regional geography have seen dramatic changes. Larry Bourne and Damaris Rose, two prominent Canadian geographers, have identified the demographic and social forces that caused these changes and argue that these forces are not yet spent. In fact, they predict that these demographic and social forces will continue to shape the country's future more so than economic and political forces. However, their impact is not evenly spread across Canada, but, in a synergetic manner, they come together in particular places at particular times to reshape the regional and local rural/urban landscape. In their analysis, Bourne and Rose identify processes of change that, in combination, have recast Canadian society into a different format. To prove their case, the authors focus their attention on four critical forces:

- population and spatial demography, including the changing components of population growth;
- lifestyles, families, and living arrangements, including changes in family structure, domestic relations, and household composition;
- social diversity in the urban system caused by immigration and migration;
- the labour market nexus, including shifts in the linkages between the domestic or household sphere and the sphere of work and production, and the changing nature of the state and civil society.

CANADA'S PHYSICAL BASE

INTRODUCTION

The earth provides a wide variety of natural settings for human beings. For that reason, physical geography helps us understand the regional nature of our world. The basic question posed in this chapter is: Why is Canada's physical geography so essential to an understanding of its regional geography? The answer lies in the regional character of Canada's physical geography and in the interrelationships between physical geography and human settlement, activity, and culture.[1] Physical geography, but especially our northern climate and physiographic regions, is an underlying factor in shaping Canada's national and regional character, and it provides a fundamental explanation for the distribution of population within Canada (Figure 4.1). In fact, population differences between Canada and the United States can be attributed, in part, to physical geography (Vignette 2.1). In this text, physical geography also provides the raison d'être for the basis of the core/periphery model. The argument is a simple one: regions with a more favourable physical base are more likely to develop into core regions, while regions with less favourable physical conditions have less opportunity to encourage settlement and economic development. Canada's ecumene—which stretches along the US border—supports this argument.

CHAPTER OVERVIEW

This chapter provides a basic introduction to Canada's physical base, emphasizing the interrelationships between the influence of physical geography on human occupation and humans' impact on the land. Here we will examine the following topics:

- How physical geography has shaped the regional nature of Canada.
- The geological structure, origins, and characteristics of Canada's physical base and seven physiographic regions.
- The main climatic controls that influence climate.
- The concept of extreme weather events and how these shape regional consciousness.

- How physical conditions influence human occupation of the land, with reference to the core/periphery model.
- The relationship between human activities and the physical environment.
- The impact of human activities on Canada's natural environment and the potential threat of global warming.

Surf and sandstone cliffs, Cavendish Beach, PEI National Park, Prince Edward Island. Photo: Barrett & MacKay/All Canada Photos.

Physical Variations within Canada

The physical geography varies across Canada. Climate provides one example. While Canada has a cold, northern climate, the types of climate vary from place to place. The Maritimes, for instance, has a mild, wet climate, while the Arctic has a cold, dry climate. Climate also affects the shape of landforms (mountains, plateaus, and lowlands) through a variety of weathering and erosional processes. Millions of years of weathering and erosion have produced the Appalachian Upland, which in distant geological times was a young, rugged mountain range similar in many ways to the Rocky Mountains. Major landforms form seven physiographic regions in Canada and they illustrate the regional distinctiveness of Canada's physical geography. For instance, the flat to gently rolling topography of the Prairies is totally different from that found in the Canadian Shield, which consists of rugged, rocky, hilly terrain.

Geographers perceive an interaction between people and the physical world. This interactive two-way relationship is a fundamental component of regional geography. Favourable physical conditions can make a region more attractive for human settlement. The combination of a mild climate and fertile soils in the Great Lakes–St Lawrence Lowlands encourages agricultural settlement, while the St Lawrence River and the Great Lakes provide low-cost water transportation to local, American, and world markets. The favourable physical features of this region have allowed it to become Canada's industrial heartland.

As scientists who study the spatial aspects of nature and the processes that shape nature, physical geographers are concerned with all aspects of the physical world: **physiography** (landforms), bodies of water, climate, soils, and natural vegetation. Regional geographers, however, are more interested in how physical geography varies and subsequently influences human settlement of the land. The Rocky Mountains, for instance, offer few opportunities for agricultural settlement, but the spectacular scenery has led to the emergence of an economy based on tourism. Nature often works slowly and change may take centuries, but nature can work quickly. Floods and storms have had sudden and dramatic impacts on human occupation of the land while climate warming is an example of relatively rapid change to our environment that could impact the human landscape.

Regional geographers also are concerned about the effect of human activities on the natural environment. In most cases, humans have a negative impact on the environment. For example, within the Bow Valley of the Rocky Mountains, extensive land developments have reduced the size of the natural habitat of wild animals such as bears and elk. Ironically, if

Vignette 2.1   Two Different Geographies

Canada and the United States occupy the northern and central parts of North America, yet the two countries have strikingly different geographies. Canada, while larger in geographic area, has a much smaller area suitable for agriculture and settlement. A significant portion of Canada lies in high latitudes where polar climates and permafrost place these lands far beyond the limits of commercial agriculture and settlement. Consequently, most Canadians live in a narrow zone close to the border with the United States (see Figure 4.1). Here, more temperate climates prevail. Geography, therefore, has been kinder to the United States, if this is to be judged by carrying capacity—how many people the land can sustain—and economic potential. The US simply has more suitable physical space for settlement, which has allowed its population to reach 300 million in 2006, compared to 33 million for Canada. Thus, Canada has a population density of 3 people per square kilometre compared to 29 in the United States. This physical reality, best described as Canada's northern handicap, limits the areas suitable for settlement. Recent immigrants to Canada have recognized this geographic fact and most take up residence in one of Canada's three largest cities (Toronto, Montréal, and Vancouver).

more land is converted into golf courses, resort facilities, and housing developments, the animals that make this wilderness region so unique and attractive to tourists may no longer be able to survive. Another example is urban sprawl, which has gobbled up some of Canada's best farmland in the Niagara Peninsula, the Fraser Valley, and the Okanagan Valley. In our contemporary world, therefore, humans are the most active and, some would say, the most dangerous agents of environmental change.

The discussion of physical geography in this chapter and in the six regional chapters is designed to provide basic information about the natural environment and its essential role in the regional geography of Canada. To that end, the following points are emphasized:

- Physical geography has distinct and unique regional patterns across Canada.
- Physiographic regions are one aspect of this physical diversity.
- Climate, soils, and natural vegetation are another aspect and provide the basis for biodiversity.
- The impact of human activity is changing the natural environment and, in the case of air, soil, and water pollution, there are long-term negative implications for all life forms.
- Physical geography has a powerful impact on Canadians by making certain areas more attractive for settlement and urban/industrial development. This relationship between the natural environment and the human world forms the basis of the core/periphery model.

We begin our discussion of physical geography by examining the nature and origin of landforms.

The Nature of Landforms

The earth's surface features a variety of landforms: mountains, plateaus, and lowlands. These landforms are subject to change by various physical processes. Some processes create new landforms while others reduce them. The earth, then, is a dynamic planet, and its surface is actively shaped and reshaped over very long periods of time. For instance, the process known as **denudation** gradually wears down mountains by erosion and weathering. The Appalachian Uplands are a prime example of the denudation of an ancient mountain chain. How did this happen? First, over millions of years, **weathering** broke down the solid rock of these ancient mountains into smaller particles. Second, **erosion** transported these smaller particles by means of air, ice, and water to lower locations where they were deposited. The result was a much subdued mountain chain from what once resembled the Rocky Mountains. Denudation and **deposition**, then, are constantly at work and, over long periods of time, dramatically reshape the earth's surface.

The earth's crust, which forms less than 0.01 per cent of the earth and is its thin solidified shell, consists of three types of rocks: igneous, sedimentary, and metamorphic. When the earth's crust cooled about 3.5 billion years ago, **igneous rocks** were formed from molten rock known as magma. Nearly 3 billion years later, **sedimentary rocks** were formed from particles derived from previously existing rock. Through denudation (weathering and erosion), rocks are broken down and transported by water, wind, or ice and then deposited in a lake or sea. At the bottom of a water body, these sediments form a soft substance or mud. In geological time, they harden into rocks. Hardening occurs because of the pressure exerted by the weight of additional layers of sediments and because of chemical action that cements the particles together. Since only sedimentary rocks are formed in layers (called **strata**), this feature is unique to this type of rock. **Metamorphic rocks** are distinguished from the other two types of rock by their origin: they are igneous or sedimentary rocks that have been transformed into metamorphic rocks by the tremendous pressures and high temperatures beneath the earth's surface. Metamorphic rocks are often produced when the earth's crust is subjected to folding and faulting. Lava from volcanoes constitutes a metamorphic rock and basalt is a product of the cooling of a thick lava flow. **Faulting** is a process that fractures the earth's crust, while **folding** bends and deforms the earth's crust. A **fault line** in geological terms refers to a crack in the earth's crust.

The earth's crust is broken into at least 14 huge slabs or plates, each moving in response

Think About It

In Chapter 1, the term 'faultline' is used. How is this term similar to yet different from the geologic term 'fault line'?

to the currents of molten material just below the crust. This motion, originally described as **continental drift**, is now known as plate tectonics. It can result in the folding and faulting of the earth's crust.[2] For example, as these huge plates drift, they may collide with one another, thus causing earthquakes. Over sufficient geological time, these plates have compressed parts of the earth's crust into mountain chains.

Physiographic Regions

The earth's surface can be classified into a series of physiographic regions. A **physiographic region** is a large area of the earth's crust that has three key characteristics:

- It extends over a large, contiguous area with similar relief features.
- Its landform has been shaped by a common set of geomorphic processes.
- It possesses a common geological structure and history.

Canada has seven physiographic regions (Figure 2.1). The Canadian Shield is by far the largest region, while the Great Lakes–St Lawrence Lowlands is the smallest. Perhaps the most spectacular and varied topography (i.e., the landforms of the earth's surface) occurs in the Cordillera, while the Hudson Bay Lowland has the most uniform relief. The remaining three regions are the Interior Plains,

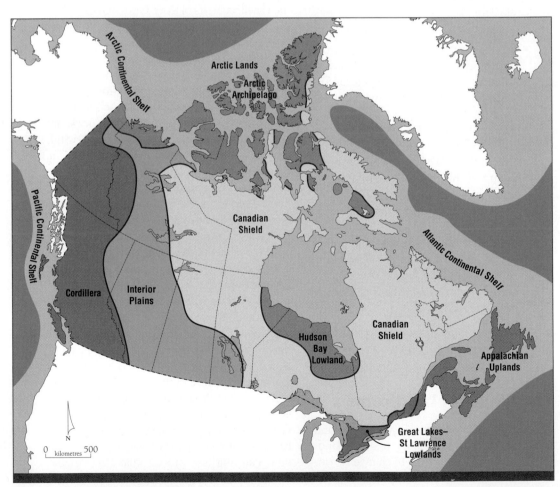

Figure 2.1   Physiographic regions and continental shelves in Canada.
The Arctic Lands consist of a dozen large islands and numerous small islands that together are known as the Arctic Archipelago. The Canadian Shield is the largest physiographic region and extends beneath the Interior Plains, the Hudson Bay Lowland, and the Great Lakes–St Lawrence Lowlands. (Further resources: *Atlas of Canada*, 1967, 'Physiographic Regions', at: <atlas.nrcan.gc.ca/site/english/maps/environment/land/arm_physio_reg>.)

Table 2.1   Geological Time Chart

Geological Era	Geological Time (millions of years ago)	Physiographic Region(s) Formed
Precambrian	600 to 3,500	Canadian Shield
Paleozoic	250 to 600	Appalachian Uplands, Arctic Lands
Mesozoic	100 to 250	Interior Plains
Cenozoic	0 to 100	Cordillera
Quaternary		The Great Lakes–St Lawrence Lowlands

Arctic Lands, and Appalachian Uplands. Most significantly, three of these physiographic regions—the Cordillera, Interior Plains, and Appalachian Uplands—display a strong north–south orientation to the topography of North America.

Each physiographic region has a different geological structure. These structural differences have produced a particular set of mineral resources in each physiographic region. For example, formed from the solidification of the earth's crust about 3.5 billion years ago, the Precambrian crystalline rock that makes up the Canadian Shield contains deposits of copper, diamonds, gold, nickel, iron, and uranium. Other physiographic regions were formed much later, as shown in the geological time chart (Table 2.1). The formation of the Interior Plains began about 500 million years ago when rivers deposited sediment in a shallow sea that existed in this area. Over a period of about 300 million years, more and more material was deposited into this inland sea, including massive amounts of vegetation and the remains of dinosaurs and other creatures. Eventually, these deposits were solidified into layers of sedimentary rock 1 to 3 km thick. As a result, the Interior Plains have a sedimentary structure that contains oil and gas deposits. Such variation in the geological structure of each physiographic region has produced unique mineral resources. Furthermore, as these regions developed their various resources, differences in regional economies began to take shape.

Each physiographic region has its own topography, and each region varies according to its surface material, with some material more resistant to physical and chemical

breakdown and other material more readily broken down into smaller particles. At the same time, erosional and depositional agents, such as wind, water, and ice, are more active in some regions than others.

One common natural force affecting the topography of all of the Canadian physiographic regions was the last ice advance of the Wisconsin glaciation, which began some 30,000 years ago at the end of the **Pleistocene epoch** (Table 2.1). The late Wisconsin ice advance consisted of two major ice sheets, the Laurentide and the Cordillera. The Laurentide Ice Sheet was centred in the Hudson Bay area. As its mass increased, the sheer weight of the ice sheet caused it to move, eventually covering much of Canada east of the Rocky Mountains. In the Cordillera, a series of alpine glaciers coalesced into the Cordillera Ice Sheet, which spread westward into the continental shelf off the Pacific coast and eastward, eventually merging with the Laurentide Ice Sheet, which reached its maximum southern extent about 18,000 years ago (Figure 2.2). Gradually, the global climate began to warm and the grip of colossal ice sheets began to weaken. Ice sheets retreated first in the Interior Plains and much later in Ontario and Québec.

Glaciation from these two huge ice sheets radically altered the geography of Canada. Glacial scouring and deposition took place everywhere (see photo 2.1). The Laurentide Ice Sheet slowly pushed southward across the Canadian Shield, stripping away its surface material and depositing it much further south. When the ice sheet began to melt, it deposited material in situ, and meltwaters formed glacial lakes, including Lake Agassiz. As the world's greatest glacial lake, it created

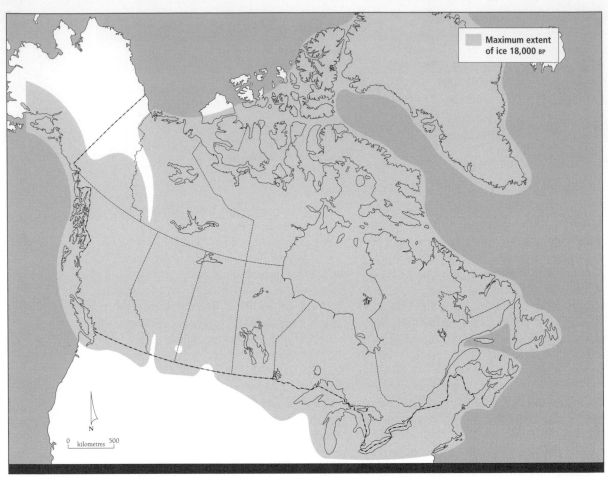

Figure 2.2   Maximum extent of ice, 18,000 BP.
The last advance of the late Wisconsinan ice age (the combined Laurentide and Cordillera Ice Sheets) covered almost all of Canada and extended into the northern part of the United States around 18,000 BP. Geologists believe that the present 'warm' climate is an interlude before the next ice advance. (Further resources: *Atlas of Canada*, 2003, 'Retreat of the last ice sheet', at: <atlas.nrcan.gc.ca/site/english/maps/archives/4thedition/environment/land/031_32>.)

the nearly flat topography of the Manitoba Lowland in Western Canada. The Great Lakes provide another example of the impact of glaciation. These lakes were formed by glacial scouring and then filled with meltwater from the receding ice sheet. Only a few remnants remain of these huge ice sheets. Today, the largest glaciers in Canada are in the mountains of Ellesmere Island.

The Canadian Shield

As noted previously, the Canadian Shield is the largest physiographic region in Canada. It extends over nearly half of the country's land mass. The Canadian Shield forms the ancient

geological core of North America. More than 3 billion years ago, molten rock solidified into the Canadian Shield (Table 2.1). Today, these ancient Precambrian rocks are not only exposed at the surface of the Shield but also underlie many of Canada's other physiographic regions.

During the last ice advance, the surfaces of the Canadian Shield and those of other physiographic regions were subjected to glacial erosion and deposition (Vignette 2.2). **Glacial erosion** and deposition are caused by giant ice sheets slowly grinding over the earth's surface (see photo 2.1). As the ice sheet moved over the surface of the Canadian Shield, the ice scraped, scoured, and scratched its massive rock surface. During the movement of an ice sheet, a variety of loose materials such as sand, gravel, and boulders were trapped within the ice sheet. As the ice sheet reached its maximum extent, the edge of the ice sheet melted, depositing rocks, soil, and other debris. This debris is called **till**. Towards the end of the **ice age**, these ice sheets melted in situ, depositing whatever debris they contained. Sometimes the water from the melting ice was blocked from reaching the sea by the retreating ice sheet. These waters then formed temporary lakes. Once this ice was removed, these waters surged towards the sea.

Evidence of the impact of these processes on the surface of the Canadian Shield is widespread. Drumlins and eskers, both depositional landforms, are common to this region. **Drumlins** are long, low hills composed of till (material deposited and shaped by the movement of an ice sheet), while **eskers** are long, narrow mounds of sand and gravel deposited by meltwater streams found under a glacier. There are also **glacial striations**, which are scratches in the rock surface caused by large rocks embedded in the slowly moving ice sheet.

The Canadian Shield consists mainly of a rugged, rolling upland. Shaped like an inverted saucer, the region's lowest elevations are along the shoreline of Hudson Bay, while its highest elevations occur in Labrador and on Baffin Island, where the most rugged and scenic landforms of the Canadian Shield are found. The Torngat Mountains in northern Labrador, for instance, provide spectacular scenery with a coastline of fjords (photo 2.2). These mountains reach elevations of 1,600 m, making them the highest land east of the Rocky Mountains. They also form the boundary between northern Québec and Labrador.

Another area of the Canadian Shield, known as the Laurentides, is located just north of Montréal. It has many lakes and hills that are the summer and winter playground for local residents as well as for tourists from Ontario and New England.

Other areas of the Canadian Shield are dotted by single-industry mining towns, such as the iron-mining town of Labrador City in

Photo 2.1

Exposed glaciated bedrock of the Canadian Shield, Melville Peninsula, Nunavut. The Laurentide Ice Sheet dramatically altered the landscape of the Canadian Shield by scouring, scratching, and polishing its surface. Rocks and boulders left behind when the ice sheet melted are visible.
Reproduced with the permission of Natural Resources Canada 2010, courtesy of the Geological Survey of Canada (Photo 2002-456 by Lynda Dredge).

Newfoundland and Labrador, the nickel centre of Sudbury in Ontario (although Sudbury has become a regional service centre in recent years), and the copper mining and smelter town of Flin Flon in Manitoba. Often located in remote places, resource towns such as these are vulnerable to major out-migration or even closure if the mining operation ceases.

Photo 2.2

Arctic landscape with tundra vegetation in the foreground of Saglek Fjord in Torngat National Park Reserve, Labrador.

Wolfgang Kaehler/Alamy

As shown in Chapter 9 (photo 9.2, page 364), the Torngat Mountains form an impressive mountain chain extending in a north–south direction in Labrador and the adjacent area of Québec.

The Cordillera

The Cordillera, a complex region of mountains, plateaus, and valleys, occupies over 16 per cent of Canada's territory. With its north–south alignment, the Cordillera extends from southern British Columbia to Yukon; its western border is the Pacific Ocean (photo 2.3). The Cordillera, classified as a young geological structure, was formed about 40 to 80 million years ago (Table 2.1) when the North American tectonic plate slowly moved westward, eventually colliding with the Pacific plate. The collision compressed sedimentary rocks into a series of mountains and plateaus now known as the Cordillera. These faulted and folded sedimentary rock strata can be seen on exposed mountain sides in the Rockies. Along the Pacific coast, tectonic plate movement continues, making the coast of British

Think About It

If you were Canada's first Prime Minister, John A. Macdonald, would you have seen the Canadian Shield as a bridge or a barrier to the Red River Settlement (now part of Manitoba)?

Vignette 2.2   Alpine Glaciation

While glaciers still exist in the Rocky Mountains, they are slowly melting and retreating. During the late Wisconsin ice advance about 18,000 years ago, these glaciers grew in size and eventually covered the entire Cordillera. At that time, alpine glaciers advanced down slopes, carving out hollows called **cirques**. As the glaciers increased in size, they spread downward into the main valleys, creating **arêtes**, steep-sided ridges formed between two cirques. As these glaciers advanced, they eroded the sides of the river valleys, creating distinctive U-shaped glacial valleys known as **glacial troughs**. The Bow Valley is one of Canada's most famous glacial troughs. Cutting through the Rocky Mountains, the Bow Valley now serves as a major transportation corridor. It has also developed into an international tourist area. The centre of this tourist trade is the world-famous ski resort of Banff.

Barrett & MacKay Photography

Photo 2.3

Located along the Continental Divide between British Columbia and Alberta, the Athabasca Glacier forms part of the massive Columbia Icefield. Known as the 'mother of rivers', the meltwaters from the Columbia Icefield nourish the Saskatchewan, Columbia, Athabasca, and Fraser river systems, the waters of which empty into three oceans—the Atlantic, Arctic, and Pacific oceans.

Columbia vulnerable to both earthquakes and volcanic activity. With the vast majority of the population and human-built environment of this region clustered along the coast in the cities of Vancouver, Victoria, New Westminster, and Nanaimo, the damage and loss of life from a major earthquake (measuring 7.0 or greater on the Richter scale) would be the worst natural disaster to strike Canada. The strongest earthquake ever recorded in Canada shook the sparsely populated Queen Charlotte Islands (Haida Gwaii) in August 1949. This earthquake measured 8.1 on the Richter scale.

In more recent geological times, the Cordillera Ice Sheet altered the landforms of the region. Over the last 20,000 years, alpine glaciation has sharpened the features of the mountain ranges in the Cordillera and broadened its many river valleys (Vignette 2.2). The Rocky Mountains are the best known of these mountain ranges. Most have elevations between 3,000 and 4,000 metres. Their sharp, jagged peaks create some of the most striking landscape in North America. The highest mountain in Canada—at nearly 6,000 metres—is Mount Logan, part of the St Elias Mountain Range in southwest Yukon (photo 2.4).

The Interior Plains

The Interior Plains region is a vast sedimentary plain that covers nearly 20 per cent of Canada's land mass. The Interior Plains are wedged between the Canadian Shield and the Cordillera, extending from the Canada–US border to the Arctic Ocean. Within the Interior Plains, most of the population lives in the southern area where a longer growing season permits grain farming and cattle ranching. To those who are not native to this region, its topography often seems featureless.

Millions of years ago, a huge shallow inland sea occupied the Interior Plains. Over the course of time, sediments were deposited into this sea. Eventually, the sheer weight of these deposits produced sufficient heat and pressure to transform these sediments into sedimentary rocks. The oldest sedimentary rocks were formed during the Paleozoic era, about 500 million years ago (Table 2.1). Since

Photo 2.4

In Kluane National Park Reserve in southwest Yukon, the St Elias Mountains form Canada's highest mountains.

then, other sedimentary deposits have settled on top of them, including those associated with the Mesozoic and Jurassic eras when dinosaurs roamed the earth.

Tectonic forces have had little effect on the geology of this region. For that reason, the Interior Plains is described as a stable geological region. For example, sedimentary rocks formed millions of years ago remain as a series of flat rock layers within the earth's crust. Geologists have used such sedimentary structures as geological time charts.

AllCanadaPhotos/Wayne Lynch

Photo 2.5

The Alberta Badlands represent a geological wonder, as land that dates back to the late Cretaceous Period has been exposed by erosion at the end of the last ice age when vast amounts of meltwater flowed through the Red Deer River and its tributaries. These quick-moving waters easily cut through soft sedimentary rocks of the Interior Plains to reach rock layers that date back to the days of dinosaurs, some 70 million years ago. The Dinosaur Trail that explores these badlands and the Royal Tyrrell Museum of Paleontology are located near Drumheller, Alberta.

In Alberta and Saskatchewan, rivers have cut deeply into these soft rocks, exposing Cretaceous rock strata. The Alberta Badlands provide an example of this rough and arid terrain (photo 2.5). Archaeologists have discovered many dinosaur fossils within these Mesozoic rocks in southern Alberta and Saskatchewan.

Beneath the surface of the Interior Plains, valuable deposits of oil and gas are in sedimentary structures called **basins**. Known as fossil fuels, oil and gas deposits are the result of the capture of the sun's energy by plants and animals in earlier geologic time. The storage of this energy in the form of hydrocarbon compounds takes place in sedimentary basins. The Western Sedimentary Basin is the largest such basin. Most oil and gas production in Alberta comes from this basin. Fossil fuels are non-renewable resources, meaning that they cannot regenerate themselves. Renewable resources, such as trees, can reproduce themselves.

As the Laurentide Ice Sheet melted and began to retreat from the Interior Plains about 12,000 years ago, the surface of the region was covered with as much as 300 m of debris deposited by the ice sheet. Huge glacial lakes were formed in a few places. Later, the meltwater from these lakes drained to the sea, leaving behind an exposed lakebed. **Lake Agassiz**, for example, was once the largest glacial lake in North America and covered much of Manitoba, northwestern Ontario, and eastern Saskatchewan—its lakebed is now flat and fertile land that provides some of the best farmland in Manitoba. When glacial waters escaped into the existing drainage system, they cut deeply into the glacial till and sedimentary rocks, creating huge river valleys known as **glacial spillways**. The geological history of the Interior Plains accounts for the great variety of landforms found within the area.

Just north of Edmonton, the Interior Plains slope towards the Arctic Ocean. The Athabasca River marks this northward course as it eventually enters the Mackenzie River. Edmonton lies on the banks of the North Saskatchewan River, which flows eastward into Lake Winnipeg. These waters then enter the Nelson River, which empties into Hudson Bay. Across this section of the Interior Plains, the land slopes towards Hudson Bay. Elevations decline from 1,200 m just east of the Rocky Mountains to about 200 m near Lake Winnipeg. These changes in elevation create three sub-regions within the Canadian Prairies: the Manitoba Lowland, the Saskatchewan Plain, and the Alberta Plateau. Typical elevations are 250 m in the Manitoba Lowland, 550 m on the Saskatchewan Plain, and 900 m on the Alberta Plateau. The Cypress Hills, however, provide a sharp contrast to the flat or rolling terrain of the Canadian Prairies (Vignette 2.3), as do the deeply incised river valleys of the Peace River country.

Vignette 2.3   Cypress Hills

The Cypress Hills, a sub-region of the Interior Plains, consist of a rolling plateau-like upland that is deeply incised by fast-flowing streams. Situated in southern Alberta and Saskatchewan, this area is the highest point in Canada between the Rocky Mountains and Labrador. The hills are an erosion-produced remnant of an ancient higher-level plain formed in the Cenozoic era (see Table 2.1). With an elevation of over 1,400 m, these hills rise 600 m above the surrounding plain. During the maximum extent of the Laurentide Ice Sheet about 18,000 years ago, the Cypress Hills were enclosed by the ice sheet but the higher parts remained above the ice sheet. Known as **nunataks**, these areas served as refuge for animals and plants. As the alpine glacier melted, streams flowing from the Rocky Mountains deposited a layer of gravel up to 100 m thick on these hills.

Today, the Cypress Hills area is a humid 'island' surrounded by a semi-arid environment and has entirely different natural vegetation compared to the area surrounding it. Unlike the grasslands, the Cypress Hills have a mixed forest of lodgepole pine, white spruce, balsam poplar, and aspen. The Cypress Hills also contain many varieties of plants and animals found in the Rocky Mountains. This area was and is considered a sacred place by Plains Aboriginal peoples.

The Hudson Bay Lowland

The Hudson Bay Lowland comprises about 3.5 per cent of the area of Canada. It lies mainly in northern Ontario, though small portions stretch into Manitoba and Québec. This region extends from James Bay along the west coast of Hudson Bay to just north of the Churchill River. The northern section lies beyond the treeline.

Water is everywhere in the short summer. Much of the ground's surface consists of wet peatland known as **muskeg** (photo 2.6). Low ridges of sand and gravel are interspersed between these extensive areas of muskeg. These ridges are the remnants of former beaches of the **Tyrrell Sea**. Because of its almost level surface and the presence of permafrost, much of the land is poorly drained. Underneath the peatland are recently deposited marine sediments mixed with glacial till. With few resources to support human activities, the region has only a handful of tiny settlements. From this perspective, the Hudson Bay Lowland is one of the least favourably endowed physiographic regions of Canada. Moosonee (at the mouth of the Moose River in northern Ontario) and Churchill (at the mouth of the Churchill River in northern Manitoba) are the largest settlements in the region. Each has a population of just over 1,000 people. These two settlements, formerly fur-trading posts, are now the termini of two northern railways (the Ontario Northland Railway and the Hudson Bay Railway, respectively).

The Hudson Bay Lowland was formed by two events. First, a warmer climate appeared some 15,000 years ago, causing the Laurentide Ice Sheet to retreat. By 12,000 years ago, the ice-free coastal plain now known as the Hudson Bay Lowland was submerged by waters from the Atlantic Ocean. Second, with the huge weight

Photo 2.6

The Hudson Bay Lowland is a vast wetland where the lack of slope and the presence of permafrost restrict the development of a drainage system. Consequently, this lowland is dotted with myriad ponds and lakes. Muskeg prevails while black spruce occupies the higher, better-drained land made up of **terraces** (old sea beaches) and drumlins. The northern half of the Hudson Bay Lowland lies beyond the treeline.
Reproduced with the permission of Natural Resources Canada 2010, courtesy of the Geological Survey of Canada (Photo 2001-124 by Lynda Dredge).

Vignette 2.4   Isostatic Rebound

At its maximum extent about 18,000 years ago, the weight of the huge Laurentide Ice Sheet caused a depression in the earth's crust. When the ice sheet covering northern Canada melted, this enormous weight was removed, and the elastic nature of the earth's crust allowed it to return to its original shape. This process, known as isostatic rebound or uplift, follows a specific cycle. As the ice mass slowly diminishes, the isostatic recovery begins. This phase is called a **restrained rebound**. Once the ice mass is gone, the rate of uplift reaches a maximum. This phase is called **postglacial uplift**. It is followed by a period of final adjustment called the **residual uplift.** Eventually the earth's crust reaches an equilibrium point and the isostatic process ceases. In the Canadian North, this process began about 12,000 years ago and has not yet completed its cycle.

of the ice removed, the earth's crust began to rise, thus causing the coastal plain to emerge from beneath the waters. This inland extension of the Atlantic Ocean began some 12,000 years ago and has been named the Tyrrell Sea.

This inland saltwater sea reached its maximum extent about 7,000 years ago, extending over much of the lowlands surrounding

Photo 2.7

Basalt on Axel Heiberg Island, Nunavut. Basalt is a hard, black volcanic rock that, when cooled, can form various shapes, including tabular columns. Because they are resistant to erosion, basalt columns often form prominent cliffs. These weathered basalt columns date to the Cretaceous Period of the Paleozoic Era (Table 2.1). At that geological time, North America, Greenland, and Eurasia broke into separate landmasses.

Hudson and James bays. With the huge weight of the ice sheet removed, the earth's crust began to rise, forcing the Tyrrell Sea to retreat. This process is called **isostatic rebound** (Vignette 2.4). Slowly the isostatic rebound caused the seabed of the Tyrrell Sea to rise above sea level, thus exposing a low, poorly drained coastal plain (most of which is called the Hudson Bay Lowland). This process of isostatic rebound has slowed over time from 600 cm per century to 100 cm per century. This relatively recent geomorphic process makes the Hudson Bay Lowland the youngest of the physiographic regions in Canada (Table 2.1).

Arctic Lands

The Arctic Lands region stretches over nearly 10 per cent of the area of Canada. Centred in the Canadian Arctic Archipelago, this region lies north of the **Arctic Circle**. It is a complex composite of coastal plains, plateaus, and mountains. The Arctic Platform, the Arctic Coastal Plain, and the Innuitian Mountain Complex are the three principal physiographic sub-regions. The Arctic Platform consists of a series of plateaus composed of sedimentary rocks. This sub-region is in the western half of the Arctic Archipelago around Victoria Island. The Arctic Coastal Plain extends from the Yukon coast and the adjacent area of the Northwest Territories into the islands located in the western part of the Beaufort Sea. The third sub-region, the Innuitian Mountain Complex, is located in the eastern half of the Arctic Archipelago. It is composed of ancient sedimentary rocks. Like the Rocky Mountains, its sedimentary rocks were folded and faulted. However, unlike the Rocky Mountains, the plateaus and mountains in the Innuitian sub-region were formed in the early Paleozoic era (Table 2.1). During this geological time, volcanic activity took place as the world island broke into North America, Eurasia, and Africa, leaving behind vast areas of basaltic rocks exposed at the surface (photo 2.7). At 2,616 m, Mount Barbeau on Ellesmere Island is the highest point in the Arctic Lands region.

Across these lands, the ground is permanently frozen to great depths, never thawing, except at the surface, even in the short summer. This cold thermal condition is called **permafrost**. Physical weathering, consisting mainly of differential heating and frost action, shatters bedrock and produces various forms of patterned ground. **Patterned ground** consists of rocks arranged in polygonal forms by minute movements of the ground caused by repeated freezing and thawing. Patterned ground and **pingos** (ice-cored mounds or hills) give the Arctic Lands a unique landscape.

The climate in this region is cold and dry. In the mountainous zone of Ellesmere Island, glaciers are still active. That is, as these alpine glaciers advance from the land into the sea, the ice is 'calved' or broken from the glacier, forming icebergs. On the plains and plateaus, it is a polar desert environment. The term 'polar desert' describes barren areas of bare rock, shattered bedrock, and sterile gravel. Except for primitive plants known as lichens, no vegetation grows. Aside from frost action, there are no other geomorphic processes, such as water erosion, to disturb the patterned ground.

Most people live in the coastal plain in the western part of this physiographic region. The three largest settlements are situated at the mouth of the Mackenzie River. Inuvik has a population of almost 3,000, while Aklavik and Tuktoyaktuk are smaller communities.

The Appalachian Uplands

The Appalachian Uplands region represents only about 2 per cent of Canada's land mass. Sometimes known as Appalachia, this physiographic region consists of the northern section of the Appalachian Mountains (stretching south to the eastern United States), though few mountains are found in the Canadian section. With the exception of Prince Edward Island (Vignette 2.5), its terrain is a mosaic of rounded uplands and narrow river valleys. Typical Appalachian Uplands terrain is found in Cape Breton, Nova Scotia (photo 2.8). These weathered uplands are either rounded or flat-topped. They are the remnants of ancient mountains that underwent a variety of weathering and

Ivy Images

Photo 2.8

The Appalachian Uplands have sustained much erosion and the resulting landscape represents 'worn-down' mountains. In rugged Cape Breton Highlands National Park, a table-like surface or peneplain lies between steep valleys carved by streams.

Vignette 2.5　Prince Edward Island

Unlike other areas of Appalachian Uplands, Prince Edward Island has a flat to rolling landscape. While the island is underlain by sedimentary strata, these rocks consist of relatively soft, red-coloured sandstone that is quickly broken down by weathering and erosional processes. Occasionally, outcrops of this sandstone are exposed, but for the most part the surface is covered by reddish soil that contains a large amount of sand and clay. The heavy concentrations of iron oxides in the rock and soil give the island its distinctive reddish-brown hue. Prince Edward Island, unlike the other provinces in this region, has an abundance of arable land.

erosional processes over a period of almost 500 million years. Together, weathering and erosion (including transportation of loose material by water, wind, and ice to lower elevations) has worn down these mountains, creating a much subdued mountain landscape with **peneplain** features. The highest elevations are on the Gaspé Peninsula in Québec, where Mount Jacques Cartier rises to an elevation of 1,268 m. The coastal area has been slightly submerged; consequently, ocean waters have invaded the lower valleys, creating bays or estuaries. The result is a number of excellent small harbours and a few large ones, such as Halifax harbour. The island of Newfoundland consists of a rocky upland with only pockets of soil found in valleys. Like the Maritimes, it has an indented coastline where small harbours

Vignette 2.6　Champlain Sea

About 12,000 years ago, vast quantities of glacial water from the melting ice sheets around the world drained into the world's oceans. Sea levels rose, causing the Atlantic Ocean to surge into the St Lawrence and Ottawa valleys, perhaps as far west as the edge of Lake Ontario. Known as the Champlain Sea, this body of water occupied the depressed land between Québec City and Cornwall and extended up the Ottawa River Valley to Pembroke. These lands had been depressed earlier by the weight of the Laurentide Ice Sheet. About 10,000 years ago, as the earth's crust rebounded, the Champlain Sea retreated. However, the sea left behind marine deposits, which today form the basis of the fertile soils in the St Lawrence Lowland.

abound. The nature of this physiographic region favoured early European settlement along the heavily indented coastline where there was easy access to the vast cod stocks. With the demise of the cod stocks, these tiny settlements are declining or, like Great Harbour Deep, have been abandoned (see Vignette 9.6).

The Great Lakes–St Lawrence Lowlands

The Great Lakes–St Lawrence Lowlands physiographic region is small but important. Extending from the St Lawrence River near Québec City to Windsor, this narrow strip of land is wedged between the Appalachians, the Canadian Shield, and the Great Lakes.[3] Near the eastern end of Lake Ontario, the Canadian Shield extends across this region into the United States, where it forms the Adirondack Mountains in New York state. Known as the Frontenac Axis, this part of the Canadian Shield divides the Great Lakes–St Lawrence Lowlands into two distinct sub-regions.

As the smallest physiographic region in Canada, the Great Lakes–St Lawrence Lowlands comprises less than 2 per cent of the area of Canada. As its name suggests, the landscape is flat to rolling. This topography reflects the underlying sedimentary strata and its thin cover of glacial deposits. In the Great Lakes sub-region, flat sedimentary rocks are found just below the surface. This slightly tilted sedimentary rock, which consists of limestone, is exposed at the surface in southern Ontario, forming the Niagara Escarpment. A thin layer of glacial and lacustrine (i.e., lake) material, deposited after the melting of the Laurentide Ice Sheet in this area about 12,000 years ago, forms the surface, covering the sedimentary rocks.

In the St Lawrence sub-region, the landscape was shaped by the Champlain Sea, which occupied this area for about 2,000 years. It retreated about 10,000 years ago and left broad terraces that slope gently towards the St Lawrence River (Vignette 2.6). The sandy to clay surface materials are a mixture of recently deposited sea, river, or glacial materials. For the most part, this sub-region's soils are fertile, which, when combined with

Joseph Gareri/iStock

Photo 2.9

In the lower Great Lakes section of the Great Lakes–St Lawrence Lowlands, the last glaciation formed a hummocky moraine landscape with widespread till deposits. Stranded ice blocks from the last glaciation left behind depressions that became filled with water. The gently rolling landscape is underlain by limestone.

a long growing season, allows agricultural activities to flourish.

The physiographic region lies well south of 49° parallel, which forms the US–Canada border west of Ontario. The Great Lakes sub-region extends from 42° N to 45° N, while the St Lawrence sub-region lies somewhat further north, reaching towards 47° N. As a result of its southerly location, its proximity to the industrial heartland of the United States, and its favourable physical setting, the Great Lakes–St Lawrence region is home to Canada's main ecumene and manufacturing core.

The Impact of Physiography on Human Activity

Not only do physiographic regions provide a basic understanding of the physical shape and geological structure of Canada, they have also exerted a powerful influence over the geographic pattern of early settlers' land selection. In some instances, settlers were attracted to certain types of land while they avoided other types. Two examples illustrate this point. In the seventeenth century, the St Lawrence Lowland was an attractive area for the establishment of a French colony because of its agricultural lands and its accessibility by water to France. Few settlers ventured beyond this favoured area. To the north was the rocky Canadian Shield, while to the south were the Appalachian Uplands. Neither of these surrounding physiographic regions offered attractive land for farming. Instead, Aboriginal groups, forced from their more productive land, occupied these rather marginal lands where they hunted game for food and trapped furs to barter with French traders for European goods.

Physical features can also create barriers to settlement. The Rocky Mountains were

Think About It

Which physiographic region do you live in?

such a barrier in the nineteenth century. In 1867, when the Dominion of Canada was formed, the Rocky Mountains isolated the small British colony on the southern tip of Vancouver Island from the settled area of Canada. At that time, communications and trade with adjacent American settlements along the Pacific coast proved much easier and quicker than the overland route used by fur traders to reach Montréal. Until the completion of the CPR in 1885, the Rocky Mountains were such an imposing physical barrier that many residing on Vancouver Island felt closer to Americans living along the Pacific coast than to Ontarians, Québecers, and Maritimers, and a vocal minority openly favoured joining the United States rather than the Dominion of Canada.

In our contemporary world, such physical barriers are no longer the obstacles they once were. Technological advances in transportation and communications have greatly reduced the **friction of distance**, the term used to describe how interactions between two points decrease as the distance between them increases. The obstacles presented by physical barriers and distance have been greatly diminished.

Geographic Location

Canada is bounded by the Arctic, Atlantic, and Pacific oceans, making it a maritime nation (Vignette 2.7). Shipping routes are crucial for a trading country like Canada. At present, ocean shipping is not possible in the Arctic Ocean because, even in the short Arctic summer, much of the Arctic Ocean is covered by a slow-moving permanent ice pack, ice floes, and ice fields. In the long winter, fast ice (**sea ice** that has frozen along coasts and extends out from land to the ice pack) completes the ice cover of the Arctic Ocean. Only small areas of open water, called **polynyas**, occur in the winter. In the late summer, fast ice disappears, leaving a narrow stretch of open water that makes coastal shipping possible. All this could change if predictions of global warming, resulting in more open water and perhaps even an ice-free Arctic Ocean, are accurate. Under these conditions, the **Northwest Passage** would become the world's most important ocean route between Asia and Europe.

A measure of geographic location on the earth's surface is provided by latitude and longitude. Because the earth is a spherical body, this measure is given in degrees. By **latitude**, we mean the measure of distance north and south of the equator. For example, Ottawa is 45 degrees 24 minutes north of the equator. Degrees and minutes are expressed as ° and ', respectively. Since the distance between each degree of latitude is about 110 km, Ottawa is about 5,000 km north of the equator. By **longitude**, we mean the distance east or west of the prime meridian. As the equator represents zero latitude, the prime meridian

---

Vignette 2.7   Facts about Canada as a Maritime Nation

- At 243,792 km, Canada has the longest coastline in the world, forming 25 per cent of the world's coastline.
- With an offshore economic zone stretching seaward some 200 nautical miles and comprising 3.7 million km², Canada has the largest offshore zone in the world.
- Canada, with two million lakes and rivers covering 755,000 km² or 7.6 per cent of the country's landmass, has the largest freshwater system in the world.
- From the Gulf of St Lawrence to Lake Superior, Canada's inland waterway extends over 3,700 km, making it the longest in the world.
- The Arctic Archipelago covers 1.4 million km², making it the largest archipelago in the world.
- Approximately 7 million Canadians live along its coastal littoral.

Source: Adapted from Fisheries and Oceans Canada (2003) <www.dfo-mpo.gc.ca/communic/facts-info/facts-info_e.html>.

represents zero longitude. It is an imaginary line that runs from the North Pole to the South Pole and passes through the Royal Observatory at Greenwich, England. Canada lies entirely in the area of west longitude. Ottawa, for example, is 75°28′ west of the prime meridian. Within Canada, latitude and longitude vary enormously. The variation in latitude has considerable implications for the amount of solar energy received at the surface of the earth, and, to a lesser degree, for climate, which in turn affects the types of soils, natural vegetation, and wildlife found in each climatic zone. Examples of the range of latitudes and longitudes found in Canada are shown in Table 2.2.

Climate

Our physical world encompasses more than just landforms, physiographic regions, and geographic location. Climate, for instance, is a central aspect of the physical world. **Climate** describes average weather conditions for a specific place or region over a very long period of time, while weather refers to the current state of the atmosphere with a focus on weather conditions that affect people living in a particular place. In short, climate is what we can expect while weather is what we get. Canadians are winter-bound for much of the year. NASA has captured a satellite image of Canada firmly in the grip of winter. Photo 2.10 shows the maximum extent of snow cover and ice conditions for 28 February 2009. At that time, only BC's narrow coastal littoral had escaped winter's grip. Most of the Great Lakes were ice-bound, as was the St Lawrence River.

Without a doubt, Canada has a cold environment. As seen in Figure 2.6, most of Canada is associated with two northern climatic types—the Arctic and Subarctic. Both have extremely long, cold winters. Although the two maritime climatic types—along the coast of BC and in Atlantic Canada—have relatively short periods of cold weather, winter is a fact of life for Canadians. While many Canadians take winter holidays to tropical places, artists, musicians, and novelists have found this northern theme appealing. In

Table 2.2	Latitude and Longitude of Selected Centres	
Centre	**Latitude**	**Longitude**
Windsor, Ontario	42°18′ N	83° W
Alert, Northwest Territories	82°30′ N	62°20′ W
St John's, Newfoundland and Labrador	47°34′ N	52°43′ W
Victoria, British Columbia	48°26′ N	123°20′ W
Whitehorse, Yukon	60°41′ N	135°08′ W

'Mon pays', Gilles Vigneault, one of Québec's best-known chansonniers, refers to Québec, his country, in the opening line: *Mon pays ce n'est pas un pays, c'est l'hiver* (My country is not a country, it is winter). Such a sentiment resonates equally well in all regions of Canada. Another appealing theme expresses the physical challenge of a cold and harsh land faced by early explorers and settlers. Stan Rogers's heroic song, 'Northwest Passage', captures that spirit. In fact, Rogers's moving lyrics have become one of Canada's unofficial anthems.

Since climate is relatively stable over a long period of time, it plays a key role in the formation of soils and natural vegetation. One outcome is the emergence of global patterns of soils and natural vegetation. Climatic conditions vary around the world and within Canada. In the Köppen climatic classification scheme for the world, for example, 25 climate

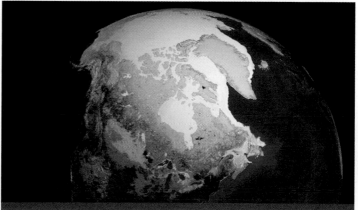

Photo 2.10

This NASA image shows Canada in the grip of winter, 28 February 2009. Canada is a northern nation with winter its dominant season. Yet, most Canadians live close to the Canada–US border, where a more temperate climate prevails.

NASA Goddard Scientific Studio

Table 2.3 Climatic Types		
Köppen Classification	**Canadian Climatic Zone**	**General Characteristics**
Marine West Coast	Pacific	Warm to cool summers, mild winters Precipitation throughout the year with a maximum in winter
Highland	Cordillera	Cooler temperature at similar latitudes because of higher elevations
Steppe	Prairies	Hot, dry summers and long cold winters Low annual precipitation
Humid continental	Great Lakes–St Lawrence Lowlands	Hot, humid summers and short, cold winters Moderate annual precipitation with little seasonal variation
Humid continental, cool	Atlantic Canada	Cool to warm, humid summers and short, cool winters
Subarctic	Subarctic	Short, cool summers and long, cold winters Low annual precipitation
Tundra	Arctic	Extremely cool and very short summers; long, cold winters Very low annual precipitation

Sources: Adapted from Christopherson (1998); Hare and Thomas (1974).

Think About It

Is the climate in your area changing?

types reflect different temperature patterns and seasonal precipitation patterns. Seven of Köppen's climatic types are found in Canada and the equivalent Canadian climatic type is shown in Table 2.3.

Climatic Controls

Three dominant climatic controls affect Canada's weather and climate, and these are related to the global atmospheric and oceanic circulation system:

- Variations in the amount of solar energy reaching different parts of the earth's surface correspond with latitude and temperature. That is, lower latitudes receive more solar energy and therefore have higher temperatures than higher latitudes.
- The global circulation of air masses causes a westerly flow of air across Canada, although invasions of air masses from the Arctic and the Gulf of Mexico can temporarily disrupt this general pattern of air circulation.
- Continental effect refers to the effect of distance from oceans on temperature and precipitation. That is, as distance from oceans increases, the annual temperature range increases and the annual amount of precipitation decreases.

Global Circulation System

Regional climates are controlled by the amount of solar energy absorbed by the earth and its atmosphere and then converted into heat. The amount of energy received at the earth's surface varies by latitude. Low latitudes around the equator have a net surplus of energy (and therefore high temperatures), but in high latitudes around the North and South poles, more energy is lost through re-radiation than is received, and therefore annual average temperatures are extremely low. Canada, where settlements extend from 42° N (Windsor) to 83° N (Alert), experiences great variation in the amount of solar energy received (and therefore great variation in temperatures).

The **global circulation system** redistributes this energy (i.e., energy transfers) from low latitudes to high latitudes through circulation in the atmosphere (system of winds and air masses) and the oceans (system of ocean currents). For example, the Japan Current warms the Pacific Ocean, bringing milder weather to British Columbia. On Canada's east coast, the opposite process occurs as the Labrador Current brings Arctic waters to Atlantic Canada. While Halifax, at 44°40′ N, lies about 500 km closer to the equator than Victoria, at 48°26′ N, Halifax's winter temperatures, on average, are much lower than those experienced in Victoria.

The atmospheric circulation system travels in a west-to-east direction in the higher latitudes of the northern hemisphere, causing air masses that develop over large water bodies to bring mild and moist weather to adjacent land masses. Such air masses are known as **marine air masses**. In this way, energy transfers ultimately determine regional patterns of global weather and climate (see Tables 2.3 and 2.4). Air masses originating over large land masses are known as **continental air masses**. These air masses are normally very dry and vary in temperature depending on the season. In the winter continental air masses are cold, while in the summer they are associated with hot weather.

Canada experiences warmer and moister weather in its lower latitudes and colder and drier conditions in its higher latitudes. However, Canada's coastal areas (particularly its Pacific coast) experience smaller ranges of seasonal temperatures and more annual precipitation than do inland or continental areas at the same latitude (Figures 2.3, 2.4, and 2.5). Winnipeg, for example, experiences a much greater daily and annual range in temperature than does Vancouver, even though both lie near 49° N. The principal reason is Vancouver's greater proximity to the ameliorating effects of the Pacific Ocean.

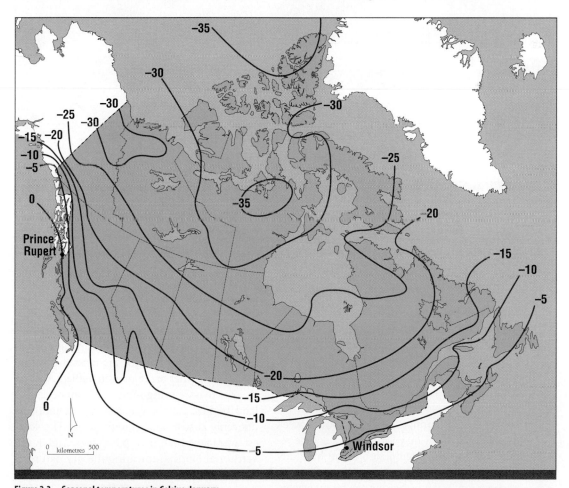

Figure 2.3   Seasonal temperatures in Celsius, January.
The moderating influence of the Pacific Ocean and its warm air masses is readily apparent in the 0 to −5°C January isotherm. For example, Prince Rupert, located near 55° N, has a warmer January average temperature (0°C) than Windsor (−2°C), which is located near 42° N. (Further resources: *Atlas of Canada*, 'January Mean Daily Minimum and Maximum Temperatures', at: <atlas.nrcan.gc.ca/site/english/maps/environment/climate/temperature/temp_winter>.).

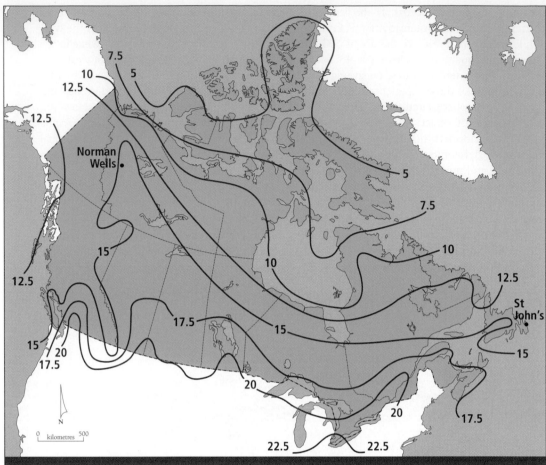

Figure 2.4   Seasonal temperatures in Celsius, July.
The continental effect results in very warm summer temperatures that extend into high latitudes, as illustrated by the 15°C July isotherm. For example, Norman Wells, located near the Arctic Circle, has warmer July temperatures than St John's. (Further resources: *Atlas of Canada*, 2007, 'July Mean Daily Minimum and Maximum Temperatures', at: <atlas.nrcan.gc.ca/site/english/maps/environment/climate/temperature/temp_summer>.)

Air Masses

Air masses are large bodies of air with similar temperature and humidity characteristics. They form over large areas with uniform surface features and relatively consistent temperatures. Such areas are known as source regions. The Pacific Ocean is a marine source region, while the interior of North America is a continental source region. During a period of about a week or so, an air mass may form over a source region, taking on the temperature and humidity characteristics of that source region. Canada's weather is affected by five air masses associated with the northern hemisphere. For example, Pacific air masses bring mild, wet weather to British Columbia's coast for most of the year.

These air masses are much stronger in the winter, so British Columbia normally experiences greater precipitation in winter than in summer. In some years, British Columbia can have a relatively dry summer. The general characteristics of the five major air masses affecting Canada's weather are shown in Table 2.4.

Air masses bring moisture from oceans to land bodies. Across Canada, precipitation is unevenly distributed (Figure 2.5). The lowest average annual precipitation occurs in the Territorial North, indicating the dry nature of the Arctic air masses that originate over the ice-covered Arctic Ocean. The highest average annual precipitation takes place along the coast of British Columbia. Here, much of the precipitation falls as frontal and orographic

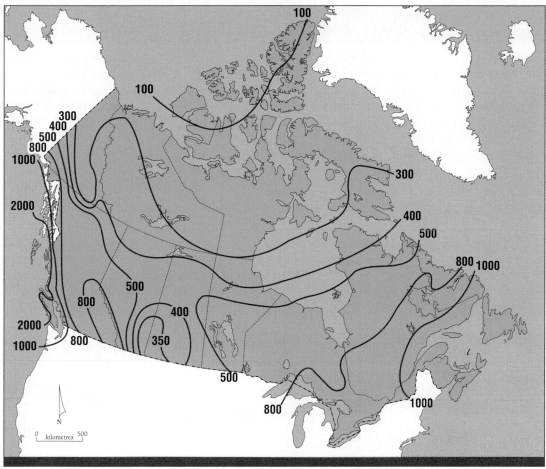

Figure 2.5   Annual precipitation in millimetres.
The lowest average annual precipitation occurs in the Territorial North, indicating the dry nature of the Arctic air masses that originate over the ice-covered Arctic Ocean. The highest average annual precipitation occurs along the coast of British Columbia due to the moist marine air masses and the coastal mountains. (Further resources: *Atlas of Canada*, 2007, 'Mean Total Precipitation', at: <atlas.nrcan.gc.ca/site/english/maps/environment/climate/precipitation/precip>.)

rainfall due to two factors: (1) the warm Pacific Ocean serves as a source for eastward-moving Pacific air masses that often contain large quantities of water vapour, and (2) along the British Columbia coast, precipitation occurs either as the warm Pacific air mass rises over a colder one or as this same air mass must rise over the coastal mountain ranges. In both cases, the water vapour condenses and falls as rain or, at higher elevations, as snow. The three principal types of precipitation are discussed in Vignette 2.8.

Think About It

Which air mass dominates the summer climate where you live?

Table 2.4	**Air Masses Affecting Canada**		
Air Mass	**Type**	**Characteristics**	**Season**
Pacific	Marine	Mild and wet	All
Atlantic	Marine	Cool and wet	All
Gulf of Mexico	Marine	Hot and wet	Summer
Southwest US	Continental	Hot and dry	Summer
Arctic	Continental	Cold and dry	Winter

Vignette 2.8   Types of Precipitation

As an air mass rises, its temperature drops. This cooling process triggers condensation of water vapour contained in the air mass. With sufficient cooling, water droplets are formed. When these droplets reach a sufficient size, precipitation begins. Precipitation refers to rainfall, snow, and hail. There are three types of precipitation. **Convectional precipitation** results when moist air is forced to rise because the ground has become particularly warm. Often this form of precipitation is associated with thunderstorms. **Frontal precipitation** occurs when a warm air mass is forced to rise over a colder (and denser) air mass. **Orographic precipitation** results when an air mass is forced to rise over high mountains. However, as the same air mass descends along the leeward slopes of those mountains (that is, the slopes that lie on the east side of the mountains), the temperature rises and precipitation is less likely to occur. This phenomenon is known as the **rain shadow effect**.

Climate, Soils, and Natural Vegetation

As noted earlier, climate affects the development of soils and the growth of natural vegetation. In fact, the interdependency of climate, soils, and natural vegetation is so strong that physical geographers have identified an orderly and interrelated global pattern of climatic, soil, and natural vegetation zones. This relationship is revealed in Figures 2.6, 2.7,

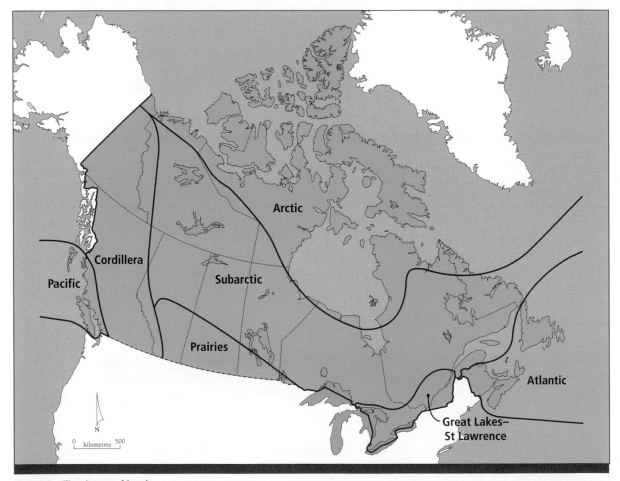

Figure 2.6   Climatic zones of Canada.
Each climatic zone represents average climatic conditions in that area. Canada's most extensive climatic zone, the Subarctic, is associated with the boreal forest and **podzolic** soils.

Table 2.5   Canadian Climatic Zones

Canadian Climatic Zone	Natural Vegetation Type	Soil Order
Pacific	Coastal rain forest	Podzolic
Cordillera	Montane and boreal forests	Mountain complex
Prairies	Grassland and parkland	Chernozemic
Great Lakes–St Lawrence	Broadleaf and mixed forests	Luvisolic
Atlantic	Mixed and boreal forests	Podzolic
Subarctic	Boreal forest	Podzolic
Arctic	Tundra and polar desert	Cryosolic

Note: See Figures 2.6, 2.7, and 2.8. Also see Glossary for definitions of soil orders.

and 2.8. As shown in Table 2.5, climate determines to a large extent the **soil order** and native vegetation in a given region and hence influences land use, such as crop cultivation, forestry, or grazing. Together with topography, climate determines the land's suitability for human settlement.

Climate has a direct impact on many economic activities. Long, cold winters cause people to use more energy to heat their

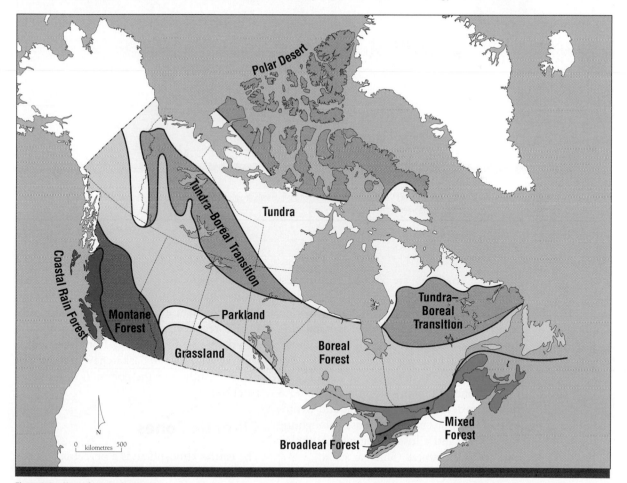

Figure 2.7   Natural vegetation zones.
These natural vegetation zones have 'core' characteristics, which diminish towards their edges. Transitions exist between natural vegetation zones. Two major transition zones shown here are the Tundra–Boreal Transition and the Parkland. (Further resources: *Atlas of Canada*, 1957, 'Natural Vegetation and Flora', at: <atlas.nrcan.gc.ca/site/english/maps/archives/3rdedition/environment/ecology/038>.)

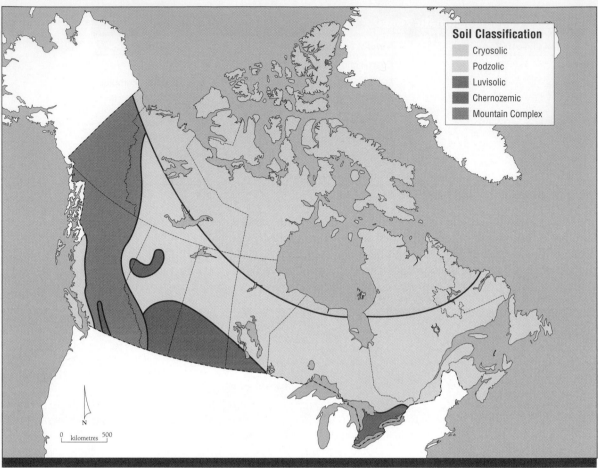

Figure 2.8   Soil zones.
Most agricultural land is in **luvisolic** and **chernozemic** soil zones that together comprise about 5 per cent of Canada's land base. (Further resources: *Atlas of Canada*, 1972, 'Soils 1972', at: <atlas.nrcan.gc.ca/site/english/maps/archives/4thedition/environment/land/041_42>.)

homes; Canadians are among the highest consumers of energy in the world. Variations in precipitation affect economic activities. Several years of below-normal precipitation often have a negative effect on agriculture, forestry, and hydroelectric production. For example, the 1988 drought in Canada cost the national economy approximately $1.8 billion in decreased agricultural and hydro-electric output, increased costs of fighting forest fires, and a loss of commercial timber and wildlife habitat (Shabbar et al., 1997: 3016). Certainly, the drought of 2001 and 2002 wreaked havoc on farmers and ranch-ers in Alberta and Saskatchewan. The lack of rainfall made pastures next to useless and

grain farmers faced crop failure. The bounti-ful hay harvest in Ontario, Québec, and the Maritimes demonstrated the regional nature of this recent Prairie drought and, with the voluntary shipment of hay from these prov-inces to Alberta and Saskatchewan in the fall of 2002, indicated the cohesion of Canada's rural society.

Climatic Zones

The earth's atmosphere is a perpetually mov-ing global system of air circulation that works to adjust the differences in pressure and tem-perature over different parts of the globe. This global circulation system, together with ocean bodies and major topographic features, affects Canada's weather. There is a climatic order within our complex and dynamic atmosphere. This order is expressed in several ways, including climatic zones. A **climatic zone** is an area of the earth's surface where similar weather conditions occur. Long-term data describing annual, seasonal, and daily temperatures and precipitation are used to define the extent of a climatic zone. Similar weather conditions occur in a particular area for complex reasons. For example, land near large water bodies usually receives more precipitation than land far from large water bodies. As well, coastal settlements have a small range of seasonal temperatures due to the sea's cooling effect in the summer and its warming effect in the winter. Land-locked places, however, do not benefit from the influence of the sea and have a much wider range of annual, seasonal, and daily temperatures.

Canada lies in the northern half of North America. This geographical location has several consequences for Canadians:

- Canada, located in the middle and high latitudes, receives much less solar energy than the continental United States and Mexico. It therefore has shorter summers and longer winters.
- Canada is noted for its long, cold winters, which affect Canadians in many ways. 'Coldness', wrote French and Slaymaker (1993: i), 'is a pervasive Canadian characteristic, part of the nation's culture and history.' They go on to state that winter's effects include not only low absolute temperatures but also exposure to wind chill, snow, ice, and permafrost.
- Continental climates are widespread in Canada's interior. These climates, but notably the Subarctic climate, are formed over large areas of the interior of Canada and are characterized by cold, dry winters and warm, dry summers. Except for the Pacific and Atlantic climates, Canadians live in areas with continental-type climates.
- Marine climates are limited in their geographic extent to the Pacific coast of British Columbia and to Atlantic Canada.

Canada has seven climatic zones (Figure 2.6): the Pacific, Cordillera, Prairies, Great Lakes–St Lawrence, Atlantic, Subarctic, and Arctic. The Subarctic zone is the largest climatic zone. It extends over much of the interior of Canada and is found in each geographic region. Though the Arctic climate exists along the Labrador coast and in the extreme northern reaches of Québec, the Subarctic climate prevails in the northern areas of Atlantic Canada, Québec, Ontario, and Western Canada, and it is present in northeast British Columbia. As well, the Subarctic climate is found in the Territorial North and is the principal climate in the Northwest Territories. The Subarctic climatic zone extends into much higher latitudes in northwest Canada than in northeast Canada because of warmer temperatures in the northwest. In northwest Canada, the average July temperature often reaches or exceeds 10°C, thus permitting the growth of trees. In similar latitudes of northeast Canada, summer temperatures are much lower. In the extreme north of Québec, for example, the average July temperature is below 10°C, thus resulting in tundra rather than a tree vegetation cover. The Subarctic climatic zone therefore has a southeast to northwest alignment (Figure 2.6). This alignment, somewhat modified in Québec, is caused by three factors:

- A continental effect means that land heats up more rapidly than water in the summer, resulting in higher summer temperatures in continental areas, such as Yukon and the Mackenzie Valley, compared to coastal areas, such as the coast of Hudson Bay and Baffin Island, at the same latitude. Also, oceans warm up more slowly than land because oceans reflect more solar energy and because solar energy is distributed throughout the water body.
- The snow cover in the western section of the Subarctic is much thinner than in the eastern half. Pacific air masses dominate the weather pattern in this

area in the spring and bring warmer weather. A thinner snow cover and warmer spring temperatures cause snow to disappear more quickly in the western Subarctic. Once the snow is gone, temperatures rise sharply.

- The Atlantic Ocean (including Hudson Bay) cools northern Québec and Labrador (thus breaking the pronounced southeast to northwest alignment of the Subarctic climatic zone). Part of that cooling effect is due to the Labrador Current, which brings Arctic waters to the middle latitudes of Atlantic Canada, and to the marine air masses that originate over the Atlantic Ocean. Combined with the deeper snow pack, this keeps spring temperatures low in the eastern sub-region.

Each climatic zone has a particular natural vegetation type and soil (Table 2.5; Figures 2.7 and 2.8). The Subarctic climate, for instance, is associated with the boreal forest and podzolic soils (Table 2.5 and Vignette 2.9). The

Think About It

Which climatic zone do you live in?

core characteristics of each of these climatic zones are presented in the appropriate regional chapter, e.g., the Pacific and Cordillera climatic zones are presented in the chapter on British Columbia.

Extreme Weather Events

Extreme weather events—such as blizzards, droughts, and ice storms—are also part of climate and often have very powerful impacts on humans. In fact, extreme weather events constitute the most serious of natural hazards, whether they take the form of droughts, floods, ice storms, or tornados. Often, extreme weather events occur without warning and result in heavy losses of property and sometime lives. Hurricanes are such an extreme weather event. Atlantic Canada has been the site of many destructive tropical cyclones. The most recent one took place on 29 September 2003, when Halifax felt the full fury of Hurricane Juan. As Conrad (2009:163–5) notes, the impact was devastating, with power outages in Nova Scotia and Prince Edward Island lasting for up to two weeks and damage estimates

Vignette 2.9   Subarctic Climate Type

The Subarctic climate extends over most of Canada. As a continental climate, the Subarctic has the greatest seasonal variation in temperatures of all the climatic types in Canada, with long, cold winters and short, warm summers. As is typical of continental climates, extremely cold winter temperatures occur. January minimum daily temperatures often drop to –40°C and sometimes even to –50°C. Winters are influenced by Arctic air masses and are therefore extremely dry. In the short summer period, daily temperatures often exceed 20°C and occasionally reach 30°C. During the summer, Pacific air masses usually dominate this weather pattern, providing most of the precipitation in the Subarctic zone. Under these air mass conditions, the annual temperature range is quite broad, perhaps as much as 80°C difference between the coldest and warmest days of the year.

There are also important variations in annual precipitation within the Subarctic zone. In the western

Subarctic, annual precipitation is low—about 40 cm—due to the rain shadow effect of the Cordillera. In the eastern subzone, annual precipitation is much higher, sometimes exceeding 80 cm, most of which is provided by the Atlantic and Gulf of Mexico air masses.

The warm but short summers provide adequate growing conditions for coniferous trees. For example, average monthly summer temperatures exceed 10°C, thereby promoting tree growth. Black and white spruce are the most common species in the Canadian boreal forest. Birch and poplar also occur, especially along the southern edge of the boreal forest. Stands of Jack pine indicate an area recovering from a forest fire. Beneath this coniferous forest are podzolic soils. Wetlands are widespread: much of the land is poorly drained due to the disrupted drainage pattern caused by glaciation and permafrost. Wetlands contain numerous lakes, peat bogs, and marshes. Canadians often refer to this type of poorly drained land as muskeg.

ranging from $100 million to $150 million; eight deaths occurred. In addition, the well-treed Point Pleasant Park in Halifax was particularly hard hit and lost most of its trees. Conrad (2009: 1) places such weather in a broader context:

> Canada's climate is typified by extremes, and thus Canadians are interested in the weather out of necessity and concern. With the inevitable changes in our global climate, scientists as well as the general public are concerned with the impact such change will have on extreme weather events in Canada.

Not surprisingly, extreme weather events often have a cultural impact by providing a common threat and, as people struggle against this threat, creating a common bond. As indicated in Chapter 1, natural disasters have contributed to people's sense of belonging to a region, and often they are recurring phenomena, such as floods in a flood-prone area. As de Loë (2000: 357) explains, 'Floods are considered hazards only in cases where human beings occupy floodplains and shoreland.' Heavy rainfall combined with rapid snowmelt often triggers catastrophic floods. An excellent example is found in the flat Manitoba Lowland where the normally benign Red River winds its way from North Dakota in the United States northward to Lake Winnipeg. Since 1950, residents of Winnipeg and other communities along the Red River have suffered through seven major spring floods, in 1950, 1979, 1996, 1997, 2001, 2006, and 2009. In 1950, the Red River flood drove over 100,000 people from their homes. Following that disaster, the Red River Floodway, a wide channel nearly 50 km long, was constructed. Its purpose was to divert the flood waters around the city of Winnipeg. However, small communities in the Red River Basin remained vulnerable to flooding. In 1997, the largest flood in the twentieth century occurred (Rasid et al., 2000). While the Red River Floodway saved Winnipeg, the towns of Emerson, Morris, Ste Agathe, and St Adolphe and the surrounding farm buildings and lands were less fortunate. In April 2006

Photo 2.11

Flood waters from the Red River surround a farmstead near St Jean Baptiste, Manitoba, 10 April 2009.

CP PHOTO/Winnipeg Free Press-Ken Gigliotti

and again in April 2009, the Red River flooded (photo 2.11).

Permafrost

Permafrost is a relic from a very cold Pleistocene climate (Table 2.1). This distinctive feature of Canada's physical geography is permanently frozen ground with temperatures at or below zero for at least two years. The vast extent of permafrost in Canada and, in places, its great depth provide a measure of the size of the country's cold environment (Figure 2.9). Permafrost exists in the Arctic and Subarctic climatic zones and occurs at higher elevations in the Cordillera zone. Overall, permafrost is found in just over two-thirds of Canada's land mass.

In more northerly regions, permafrost extends far into the ground. North of the Arctic Circle, the permafrost may be several hundred metres deep. Further south, permafrost is less frequent and where it occurs, it rarely penetrates more than 10 m into the ground. Permafrost is found in all six of Canada's geographic regions and reaches its most southerly position along 50° N in Ontario and Québec. Along the southern edge of permafrost, there is a transition zone where small pockets of frozen ground have a depth of less than 1 m. Further south, these pockets of permafrost disappear.

Think About It

Even though the Red River floods regularly, Winnipeg is able to avoid serious flooding. Why?

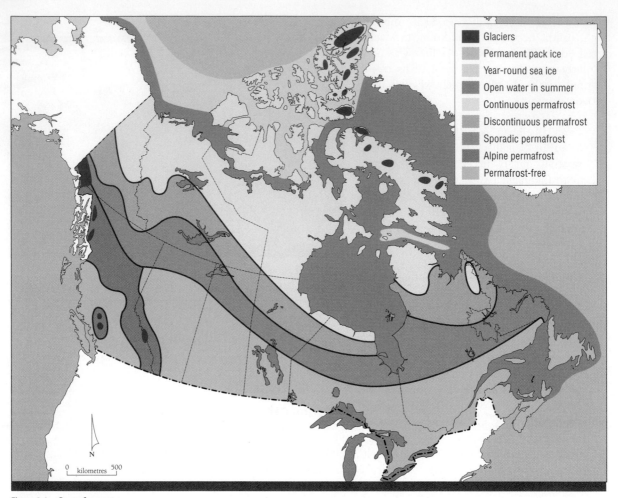

Figure 2.9   Permafrost zones.

Canada's cold environment is demonstrated by the permanently frozen ground that extends over two-thirds of the country. Sea ice varies in thickness and duration. The most durable and thickest ice is found in the permanent ice pack. In the area of open water in summer, sea ice disappears first in the Great Lakes and in the offshore waters of Atlantic Canada, and last in the Arctic Ocean. In September 2007, satellite imagery indicated that the extent of open water in the Arctic Ocean was greater than in previous decades. (Further resources: *Atlas of Canada*, 2003, 'Permafrost', at: <atlas.nrcan.gc.ca/site/english/maps/environment/land/permafrost>.)

Permafrost is divided into four types. **Alpine permafrost** is found in mountainous areas and takes on a vertical pattern as elevations of a mountain increase. Over most of Canada, however, permafrost follows a zonal pattern, which does not correspond to latitude but rather to the annual mean temperatures that fall below zero.[4] The zonal pattern has a northwest to southeast alignment, that is, from Yukon to central Québec (see Figure 2.9).

As the mean annual temperature varies, the type of permafrost also changes. **Continuous permafrost** occurs in the higher latitudes of the Arctic climatic zone, where at least 80 per cent of the ground is permanently frozen, although it also extends into northern Québec. Continuous permafrost is associated with very low mean annual air temperatures of −15° C or less. **Discontinuous permafrost** occurs when 30 to 80 per cent of the ground is permanently frozen. It is found in the Subarctic climatic zone where mean annual air temperature ranges from −5° C in the south to −15° C in the north. **Sporadic permafrost** is found mainly in the northern parts of the provinces, where less than 30 per cent of the area is permanently frozen. Sporadic permafrost is associated with mean annual temperatures of zero to −5° C.

Think About It

If our climate is warming, would you expect that permafrost would melt?

Major Drainage Basins

Facing three oceans, Canada is clearly a maritime nation (Vignette 2.8). With four distinct slopes to the land, four major drainage basins exist (Figure 2.10 and Table 2.6): the Atlantic Basin, the Hudson Bay Basin, the Arctic Basin, and the Pacific Basin. In addition, a small portion of southern Alberta and Saskatchewan drains southward to the Missouri River, which forms part of the Mississippi River system, which drains into the Gulf of Mexico. A **drainage basin** is land that slopes towards the sea and is separated from other lands by topographic ridges. These ridges form drainage divides.

The Continental or Great Divide of the Rocky Mountains, for example, separates those streams flowing to the Pacific Ocean from those flowing to the Arctic and Atlantic oceans. The Northern Divide extends from Labrador/Québec to the Rocky Mountains near the forty-ninth parallel and separates waters flowing into the Atlantic Ocean and the Hudson Bay. The Arctic Divide stretches across the middle of Canada from Baffin Island to the Rocky Mountains just to the northwest of Edmonton. Its waters, including the Athabasca and Mackenzie rivers, flow into the Arctic Ocean. The St Lawrence Divide marks the southern edge of the drainage basin of the St Lawrence River and Great Lakes.

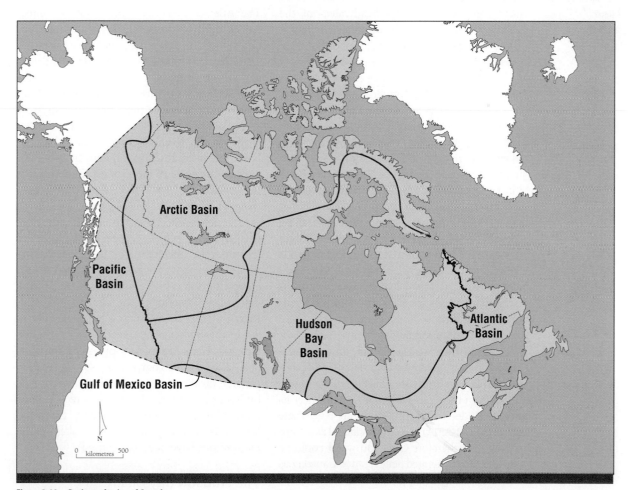

Figure 2.10   Drainage basins of Canada.
The four divides determine Canada's drainage basins. They are the Continental or Great Divide, the Northern Divide, the Arctic Divide, and the St Lawrence Divide. The Hudson Bay Basin lies between three divides—the Continental Divide, the Arctic Divide, and the Northern Divide—and is by far the largest of the five basins in Canada. It also serves as a boundary between southern Alberta and British Columbia, and between northern Québec and Labrador. (Further resources: *Atlas of Canada*, 2004, 'Drainage Basins', at: <atlas.nrcan.gc.ca/site/english/maps/environment/hydrology/drainagebasins>.)

Think About It

In the seventeenth and eighteenth centuries, why did drainage basins play a role in the setting of political boundaries in North America? Two examples are Rupert's Land (Hudson Bay Basin) and the Louisiana Territory (western section of the Mississippi River Basin).

A few rivers cross the US–Canada border and part of the Great Lakes lies in the United States. The Columbia River leaves British Columbia and continues its journey to the Pacific Ocean through the US states of Idaho, Washington, and Oregon. Two small rivers, the Milk and the Poplar, flow from southern Alberta and Saskatchewan into the Missouri River, which drains much of the northern half of the US Great Plains. The Red River flows from the US states of North Dakota and Minnesota into Manitoba and beyond to Lake Winnipeg and eventually to Hudson Bay by means of the Nelson River. The Richelieu River flows from Lake Champlain which lies mainly in New York, and drains into the St Lawrence River near Sorel, Québec. The headwaters of the Saint John River partially originate in Maine, and this river empties into the Bay of Fundy at the city of Saint John, New Brunswick.

Historically, the rivers in each basin have played major roles in providing access to the interior of Canada and in the development of the country. For example, Aboriginal peoples and Europeans used the St Lawrence and Mackenzie rivers as transportation routes during the fur trade. Today, the St Lawrence River plays a key role in Canada's internal and foreign shipments of goods to and from Montréal and other cities along the St Lawrence and along the shores of the Great Lakes. River barges bring food, building materials, and other goods to settlements along the Mackenzie River and along the coast of the Beaufort Sea. Oil and gas equipment also is barged to the Norman Wells oil fields and those in the Beaufort Sea. These rivers remain important waterways today.

Water is a scarce commodity, particularly in the dry Southwest of the United States. California, for instance, has had to divert water from the Colorado River to help meet its needs. Under the North American Free Trade Agreement (NAFTA), water was classed as a commodity and therefore Canadian water could be exported to the United States. Since Canada has the world's largest supply of fresh water, large-scale transfer of water from Canada has appeal to water-short American states. Prior to NAFTA, several inter-basin proposals appeared in the media but none were commercially feasible. Two such schemes were the Great Recycling and Northern Development Canal, which called for diversion of water from James Bay to the Great Lakes and then, by pipelines, to the water-short American Southwest. Another continental water diversion scheme, the North American Water and Power Alliance project, proposed diverting the Yukon River and the two major tributaries of the Mackenzie River, the Liard and the Peace, to the US along the Rocky Mountain Trench. Continental diversion of fresh water from one basin to another, however, is not a practical matter because Canadians rivers do not flow south but rather east, west, and north.

The Atlantic Basin

The Atlantic Basin is centred on the Great Lakes and the St Lawrence River and its tributaries, but the basin also includes Labrador. The Atlantic Basin has the third largest drainage area and also the third greatest streamflow. As seen in Figure 2.6, the Atlantic Basin receives considerable precipitation, making it second only to the Pacific Basin. The largest hydroelectric development in this drainage basin is located at Churchill Falls in Labrador, with the promise of more development in the Lower Churchill River. In January 2009, a five-member federal–provincial Joint Review Panel for the proposed Lower Churchill hydroelectric generation project was established. Earlier hydroelectric developments took place along the St Lawrence River in southern Québec and along its tributary rivers that flow out of the Laurentide Upland of the Canadian Shield. Rivers such as the Manicouagan River originate in the higher elevation of the Laurentide Upland. Here, abundant precipitation, natural lakes, and a sharp increase in elevation provide ideal conditions for the generation of hydroelectric power. Because there is a large market for electrical power in the St Lawrence Lowlands, virtually all potential sites in the Laurentides have been developed.

The Hudson Bay Basin

The Hudson Bay Basin is the largest drainage basin in Canada (Table 2.6), covering about 3.8 million km². Precipitation varies greatly across this basin. In the West, precipitation is

Table 2.6	Canada's Drainage Basins	
Drainage Basin	Area (million km²)	Stream-flow (m³/second)
Hudson Bay	3.8	30,594
Arctic	3.6	20,491
Atlantic	1.6	29,087
Pacific	1.0	24,951
Gulf of Mexico	<0.1	12
Total	10.0	105,135

Sources: Laycock (1987: 32); Dearden and Mitchell (2005: 124).

low while it is greater in the East (Figure 2.6), where the headwaters of its rivers in the uplands of northern Québec flow westward into James Bay. In northern Ontario and Manitoba, rivers drain into James and Hudson bays.

The large rivers and sudden drops in elevation that occur in the Canadian Shield make this part of the basin ideal for developing hydroelectric power stations. In fact, most of Canada's hydroelectric power is generated in the Canadian Shield area of the Hudson Bay Basin—the largest installations are on La Grande Rivière in northern Québec and on the Nelson River in northern Manitoba. La Grande Rivière's hydroelectric developments are the first stage in the James Bay Project. The Great Whale River Project was to follow the completion of the hydroelectric projects on La Grande Rivière, but a variety of circumstances (low energy demand, low prices in New England, and strong opposition from environmental groups and the Cree Indians of northern Québec) stalled its development.

The Arctic Basin

The Arctic Basin is Canada's second-largest drainage basin. The Mackenzie River dominates the drainage system in this basin. Along with its major tributaries (the Athabasca, Liard, and Peace rivers), the Mackenzie River is the second-longest river in North America. However, because of low precipitation in the Arctic, this basin has only the fourth-largest stream-flow. There are few hydroelectric projects in the Arctic Basin because of the long distance to markets, with the exception of the hydroelectric development on the Peace River in British Columbia. Here, power from the Gordon M. Shrum generating facility is transmitted to the population centres in southern British Columbia and to the United States, primarily to the states of Washington, Oregon, and California.

The Pacific Basin

The Pacific Basin is the smallest basin. However, it has the second-highest volume of water draining into the sea. Heavy precipitation along the coastal mountains of British Columbia accounts for this unusually high stream-flow. As a result, the Pacific Basin is the site of one of Canada's largest hydroelectric projects. Located at Kemano, this facility is owned and operated by Alcan, which uses the electrical generating station to supply power to its aluminum smelter at Kitimat. The ice-free, deep-water harbour at Kitimat and low-cost electric power generated at Kemano make Kitimat an ideal location for an aluminum smelter.

Environmental Challenges

In our contemporary world, humans are the most active and dangerous agents of environmental change. All human activities affect the environment. Cultivation of the land, building of cities, exploitation of renewable and non-renewable resources, and processing of primary products have forever changed our natural environment. Some engineering projects were necessary for the unity of the country. The most outstanding example was the construction of the Canadian Pacific Railway in the nineteenth century, while another major engineering effort saw the completion of the Confederation Bridge linking Prince Edward Island to the mainland in the last years of the twentieth century. Often, however, industrial activities, but particularly those associated with the burning of fossil fuels, have contributed to air pollution and, in turn, to acid rain, smog, and even global warming. Smog and acid rain are two products of air pollution, while another form of air pollution— the addition of ever-increasing quantities of greenhouse gases to the atmosphere—is triggering global warming, the environmental challenge of the twenty-first century.

Think About It

Ottawa, our capital city, lies in which drainage basin?

Canadians once believed that their country was so vast that human activities could never harm their environment. By the 1950s, Canadians began to suspect that this myth was incorrect. Industry has damaged the environment, sometimes without regard for the consequences. Clear-cut logging, for instance, has reduced the habitat for many plants and animals, caused erosion on steep slopes, and, in doing so, polluted stream and rivers. But industry alone is not to blame. Society as a whole is responsible for two major sources of pollution: the discharge of raw sewage into our oceans, rivers, and lakes, and car exhaust, a major source of air pollution in cities and a leading contributor of carbon dioxide to the atmosphere. Victoria, one of the most affluent places in Canada, continues to discharge raw sewage into the Pacific Ocean. Clearly, the extent and pace of human impacts on the environment have reached and, in some locations, exceeded the danger point. Walkerton, Ontario, was one such location. In May 2000, the town's water supply was contaminated with E. coli. Hundreds became ill and seven died (CBC News, 2004).

While the list of serious environmental problems is long, automobile pollution, coal-burning plants, and oil sands development in Alberta rank at or near the top. The Alberta oil sands supply both Canada and the United States with a large amount of crude oil. However, extraction of oil from these tar sands is a gamble that seems to have tossed the environmentalist's precautionary principle into the tailing ponds. It is highly expensive to extract the oil and extremely damaging to the land and to the atmosphere. Open-pit mining takes place at several sites, including the Suncor mine shown in photo 2.12.

Since environmental issues are the result of human actions, solutions are possible (Dearden and Mitchell, 2009: ch. 1). One solution is for more protected areas and parks (Slocombe and Dearden, 2002: ch. 12).

Todd Korol/Aurora Photos/GetStock.com

Photo 2.12

Tailing ponds at the Suncor oil sands operation near Fort McMurray, Alberta. While the oil sands produce the bulk of Canada's oil, the environmental cost is high.

Another is for more stringent regulations that will reduce damage to the environment caused by both new and existing projects. But perhaps the most significant solution lies in 'going green', which includes recycling waste products, moving towards electric automobiles, and increasing the production of electricity from natural sources such as solar and wind.

Federal, provincial, and territorial governments are introducing green policies and programs, and Ottawa and some provinces are coordinating their efforts with their counterparts in the US. In 2009, Prime Minister Harper and President Obama announced the launch of the US–Canada Clean Energy Dialogue—a collaborative scientific effort to develop new technologies aimed at reducing greenhouse gas emissions and combatting global climate change. At a regional level, Montana Governor Brian Schweitzer and Saskatchewan Premier Brad Wall have joined forces to secure $250 million from Washington for a **carbon sequestration** demonstration project that could help solve the global carbon crisis. Saskatchewan has a head start in carbon sequestration with its Weyburn Project, which involves injecting carbon dioxide into Saskatchewan's Weyburn oil field. The injection of gas accomplishes two goals. First, a liquefied version of carbon dioxide is stored underground, thereby reducing the release of this gas into the atmosphere. The origin of the gas is an oil refinery in the United States. By means of a 320-km pipeline, the liquefied carbon dioxide is transported to the Weyburn oil field. Second, the injection of carbon dioxide greatly increases the recovery of oil. In conjunction with the Weyburn Project, the University of Regina began conducting research into CO_2 capture techniques. Such technology, though very expensive to redesign, could reduce emissions from the upgraders at Alberta tar sands plants.

Under the section 'Economic Structure' in Chapter 8, page 336, the plans of Shell Oil to reduce its greenhouse emissions from its upgrader at Fort Saskatchewan are discussed.

Canadians are all too familiar with smog and other forms of air pollution. Most air pollution results from industrial emissions and from automobile and truck exhaust. Coal-burning plants and oil sands production account for most industrial pollution. Nanticoke Station in Ontario and TransAlta Plant in Alberta are the two largest coal-burning electric power stations in Canada. As Table 2.7 reveals, the top three provinces by industrial emissions are Alberta, Ontario, and Saskatchewan.

Table 2.7	Industrial Emissions by Provinces and Territories, 2006		
Rank	Province/Territory	Million Tonnes CO_2	%
1	Alberta	115.9	42.4
2	Ontario	71.8	26.3
3	Saskatchewan	22.5	8.2
4	Québec	22.0	8.0
5	British Columbia	11.8	4.3
6	Nova Scotia	10.9	4.0
7	New Brunswick	10.3	3.8
8	Newfoundland and Labrador	5.0	1.8
9	Manitoba	2.4	0.9
10	Northwest Territories	0.4	0.2
11	Prince Edward Island	0.1	0.1
12	Yukon	<0.1	<0.1
13	Nunavut	<0.1	<0.1
	Total emissions	273.1	100.0

Source: <www.pollutionwatch.org>.

Photo 2.13

Glacial retreat of the Athabasca Glacier, 1992 to 2005.

and synthesized by the IPCC—is based on anthropogenic pollution, that is, the release of greenhouse gases to the atmosphere caused by the ever-increasing use of fossil fuels, and on the model called the **greenhouse effect**. The greenhouse effect is related to the process of the atmosphere absorbing energy radiation from the earth. Only a few atmospheric gases, such as carbon dioxide, have the capacity to absorb radiation from the earth. These gases are called **greenhouse gases**. Thus, as more carbon dioxide and other greenhouse gases are added to the atmosphere by the burning of fossil fuels, the greenhouse effect causes air temperatures to rise. The IPCC is concerned that the increasing amounts added to the atmosphere will cause global air temperatures to increase rapidly. While only a relatively small increase in world air temperature has taken place over the last 250 years, much of this increase has taken place in the last 20 years. The expectation is that this warming trend will accelerate in the coming decades, causing great environmental change to climates, zonal soil and vegetation patterns, and wildlife associated with these natural zones. Some evidence of such consequences already exists (McKibben, 2010). For instance, many glaciers have diminished in size (see photo 2.13 and Figure 2.11), indicating a global trend, and the length of open water

Is global warming underway? Scientists think so. The leading group of scientists dealing with climate change is associated with the Intergovernmental Panel on Climate Change (IPCC) (Vignette 2.9). The argument of these scientists—well over 1,000 researchers from around the world whose work is evaluated

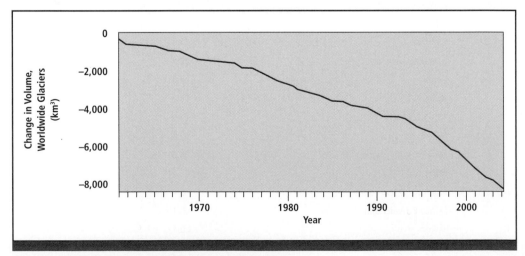

Figure 2.11   Worldwide change in volume (km³) of glaciers, 1960–2004.

Source: NASA, Earth Today, at: <www.nasa.gov/vision/earth/features/index.html>.

Photo 2.14

By September 2008, Arctic sea ice coverage had reached its lowest extent for the year and the second-lowest amount recorded since satellite observations began some 30 years ago. At this time, two routes through the Northwest Passage were ice-free, but only for a few days. Since the portion of the Arctic Ocean from the Beaufort Sea to Bering Strait is ice-free for a longer time and the ice-free zone is much larger than in the central and eastern Arctic, the possibility of ocean-going vessels plying these waters is more likely to occur before ships begin to use the entire Northwest Passage.

Photo: NASA/Goddard Space Flight Center Scientific Visualization Studio; Blue Marble Next Generation data courtesy Reto Stockli (NASA/GSFC).

in the short summer for the Arctic Ocean has reached record levels (see photo 2.14). While the disappearance of glaciers in the western Rockies spells trouble for the South and North Saskatchewan rivers, for farmers who irrigated with these river waters, for potash mines that need these waters, and for urban dwellers drinking these waters, a much longer ice-free Northwest Passage could revolutionize energy and mineral developments by reducing transportation costs to world markets.

Yet, while hard evidence supporting a general rise in global temperatures, as reflected by the continuing melting of glaciers and sea ice, is irrefutable, the rate of temperature change is less certain. The public's acceptance that the IPCC assumption is correct, namely, that global air temperatures will increase rapidly in the twenty-first century, has been challenged by three recent events. First, annual global temperature increases slowed down in 2008 and 2009, although this certainly is too short a period to mean a great deal. Second, **climategate**—when

e-mails from a British climate research institute were made public after a hacker obtained them in late 2009—damaged the reputation of the IPCC (see BBC News, 2010). Third, the extreme claims made by environmental advocates at the 2009 **Copenhagen Climate Conference** have muddied the scientific waters and thus strengthened the hand of climate change skeptics.

Climates have changed in the geological past (Vignette 2.10). However, the current climate change is caused by human actions. Hence, we have the power to prevent it or at least slow the rate of temperature increase.

How is Canada affected by global warming? Although no one can predict the precise impacts, higher temperatures seem assured, which would translate into longer, hotter summers and shorter, milder winters. Changes to Canada's precipitation patterns are more difficult to predict. On the one hand, the interior of Canada could become drier, but with more open water in Hudson Bay and the Arctic Ocean, Arctic marine air masses might bring

Think About It

Would you advocate the immediate construction of a natural gas pipeline along the Mackenzie River to supply the US Midwest or wait, assuming that the Arctic Ocean's ice-free season will increase, thus allowing LNG tankers to transport liquefied natural gas to markets along the US west coast and to Asia?

Vignette 2.10   The Intergovernmental Panel on Climate Change

Concerned about global warming, the World Meteorological Organization (WMO) and the United Nations Environment Program (UNEP) established the Intergovernmental Panel on Climate Change (IPCC) in 1988. This body provides information on climate change, its potential environmental and socio-economic consequences, and adaptation and mitigation options. However, the IPCC does not conduct any research, nor does it monitor climate-related data or parameters; rather, it assesses and then reports on the latest scientific, technical, and socio-economic literature every four years.

In its *Fourth Assessment Report* in 2007, the IPCC reported that the 100-year linear trend of global surface temperatures (1906 to 2005) indicated an increase in annual temperatures. The IPCC also stated that 11 of the last 12 years (1995–2006) rank among the 12 warmest years in the instrumental record of global surface temperature (since 1850). The IPCC expects to publish its *Fifth Assessment Report* in 2014.

Source: IPCC in Pachauri and Reisinger (2007) and IPCC (2010).

more precipitation to those parts of Canada. Still, six consequences for Canada of global warming seem likely:

1. Countries in high latitudes are expected to experience greater increases in temperature. In fact, Canada and other northern countries have already experienced greater temperature increases than the global average, a result of reduced snow and ice cover and, consequently, a lesser **albedo effect**. The greatest increases in annual temperatures would take place in formerly snow-covered areas of Canada, especially the Arctic, because solar energy can now warm the ground more effectively, as opposed to being reflected back into space (the albedo effect).[5] Similarly, as the ice cover on the Arctic Ocean disappears, more solar energy would be absorbed by the open water, thereby increasing evaporation and making Arctic air masses more moist.

2. Global warming, if IPCC predictions are correct for the end of the twenty-first century, could alter Canada's natural zones (Bouchard, 2001). If world temperatures rose sufficiently, Canada's climatic zones would shift northward, followed slowly but inevitably by natural vegetation, soil,

and wildlife zones. Eventually, the Arctic zone would be greatly reduced in its spatial extent.

3. Similarly, global warming could thaw the permafrost. Melting of the ice in permafrost could cause massive ground **subsidence**, resulting in an irregular relief referred to by physical geographers as 'thermokarst topography'. This would disrupt transportation and pipeline systems, and play havoc with foundations for buildings, bridges, and other human-made structures (Bone et al., 1997: 265–74). As a possible sign of things to come, in mid-March 2010, First Nations communities in northern Manitoba were thrust into crisis and isolation when the province had to close the network of ice roads they depend on for supplies and for inter-community contact because of an unusually early spring thaw.

4. Canada's agriculture would be affected by global warming. Agricultural activities could take place further north and less hardy crops could be grown, but, on the negative side, grain agriculture in the Canadian Prairies might be subject to greater risk of drought.

5. Water transportation at the end of the twenty-first century would

greatly benefit from longer navigation seasons. The Great Lakes and the St Lawrence River, for instance, could be navigable year-round.

6. The predicted rise in sea levels would have a devastating impact on some coastal areas, especially in the Lower Mainland of British Columbia and, in Atlantic Canada, on Prince Edward Island. Indeed, if the rise in sea levels continues, some countries of the world, such as the Maldives, a group of islands in the Indian Ocean, will cease to exist—although the Maldives government has begun to explore the possibility of buying land in Australia to become, in effect, a 'nation in exile'.

While the physics of global warming in the greenhouse model are elementary, the actual process of climate change is extremely complex and remains unclear. While the IPCC has established a warming trend over the past 100 years, what natural factors might alter that trend in the future? One possibility is a reduction in the amount of solar energy reaching the earth's surface as a result of a massive volcanic eruption that would release huge amounts of dust into the air, thus reflecting significant amounts of incoming solar energy back into outer space and thereby chilling the world's climate. But we can hardly count on such a natural event to reverse the current warming trend, and so efforts—such as more efficient automobile engines—are necessary to reduce the release of carbon dioxide into the atmosphere.

Regional Scale: Acid Rain

Acid rain—precipitation that has an unusually acidic chemical composition—affects the forests and lakes of Ontario, Québec, and Atlantic Canada. Acidity levels are described in terms of the pH factor, which measures the acidity or alkalinity of a substance on a scale of 0 to 14, with 0 being extremely acidic and 14 being extremely alkaline. The average pH of normal rainfall is 5.6, making it slightly acidic. Acid rain, however, has a pH of 2.4 or less. At this pH level, acid rain is a serious environmental problem in many parts of the world, including eastern Canada.

Most acid rain results from the chemicals derived from the burning of fossil fuels in industrial plants and from car and truck exhaust. Sulphur dioxide and nitrogen oxides are emitted in the smoke from coal-burning plants, while automobile engines are particularly prolific producers of nitrogen oxides. Acid rain is created when oxides of sulphur and nitrogen change chemically as they dissolve in water vapour in the atmosphere and then return to earth as tiny droplets with high concentrations of sulphuric acid and nitric acid.

The most visible impact of acid rain is in urban areas, where it slowly corrodes limestone buildings and defaces marble sculptures. Less visible but just as destructive are the effects of acid rain on forests, lakes, and soils. In eastern Canada, acid rain has damaged many forests and caused a sharp decline in fish stocks in many lakes. The problem of acid rain is compounded by the fact that pollutants emitted from coal-burning electrical power plants in the Great Lakes area of the United States fall as acid rain in Ontario and Québec. Canada and the United States have turned their attention to this problem. In 1991, the two countries signed a bilateral Air Quality Agreement, and a few years later, New England governors and eastern Canadian premiers agreed on an Acid Rain Action Plan to co-ordinate efforts to reduce acid rain. Progress has been made by reducing sulphur emissions at coal-burning power plants, substituting natural gas for coal at power plants, and introducing 'scrubbing' in existing coal-burning thermal facilities.

Local Scale: Smog

Air pollution is the number-one problem facing urban residents of major cities. The source of air pollution (smog) comes from emissions from automobiles and coal-burning plants. Smog, a mixture of smoke, sulphur dioxide, and other contaminants, takes the form of a brownish haze over cities. Most contaminants are produced by automobiles (carbon dioxide) and coal-fired thermal plants (sulphur dioxide). Vehicles account for over half of urban air

Think About It

Since China and India need more coal-burning thermal electricity plants to keep their economies growing, is it fair to ask those countries to reduce their greenhouse emissions?

Vignette 2.11   Fluctuations in World Temperatures

While it appears that our climate is relatively constant, wide fluctuations in world temperatures are part of our geological history. The last 10,000 years—the **Holocene epoch**—represents a relatively warm period in the earth's history. However, within that epoch, warmer and colder periods have occurred. Each has lasted around 500 years or so. For instance, a warm period known as the Medieval Warming Period took place between the tenth and fourteenth centuries. During this time, the Vikings established a settlement on the southern tip of Greenland and made voyages to neighbouring parts of Arctic Canada and southward to the northern tip of Newfoundland. At the same time, the Thule occupied much of the Canadian Arctic and were able to hunt the huge bowhead whales in the open summer waters. This warm spell was replaced by a cooler period known as the Little Ice Age, which lasted from about 1450 to around 1850.

The Little Ice Age had a dramatic impact on human beings. The Vikings were no longer able to sustain themselves in Greenland and the Thule had to make a drastic adjustment to their hunting economy. With a much more extensive and long-lasting ice cover over the Arctic Ocean, the bowhead whales no longer entered these waters for long periods of time. The consequences for the Thule inhabitants were devastating. They were forced to hunt smaller game—seals and caribou. The results were twofold: the new hunting system could not support as many people and it required smaller, more mobile hunting groups. Archaeologists believe that the Inuit, who were established in the Arctic by the mid-sixteenth century, are the descendants of the Thule people.

pollution. Coal-burning plants are the second-most serious source of urban air pollution. Densely populated areas like southern Ontario contain millions of automobiles and trucks. Furthermore, southern Ontario is an energy-deficient area. For years, coal thermal plants have produced much of the region's electricity at a relatively low cost. Now that urban air pollution has become a health problem, Ontario has plans to limit sulphur dioxide emissions by closings its coal-burning plants, replacing its electrical production with natural gas and nuclear power, and importing electricity from Québec, Michigan, New York, and Manitoba.

SUMMARY

Physical geography varies across Canada. This variation is critical in understanding Canada's regional character. At the macro level, physiographic regions represent large areas with similar landforms and geological structures. Climate creates a similar zonal arrangement of soils and natural vegetation. Climate therefore determines to a large extent the type of soil and native vegetation in a given region and hence influences land use. Together with topography, climate partly determines the land's ability to support a population. This link between the physical and human worlds identifies those regions having a more favourable mix of physical characteristics for economic development, and translates the abstract core/periphery model into a geographic reality. The Great Lakes–St Lawrence Lowlands region is the most favoured physical region in Canada. Physical barriers, such as the Rocky Mountains, and extreme climatic conditions, such as the very long and cold winters in northern Canada, have also affected the historical settlement of the country and continue to influence contemporary economic activities.

Canada has several physiographic regions and climatic zones. The seven physiographic regions are: the Canadian Shield, the Cordillera, the Interior Plains, the Hudson Bay Lowland, the Arctic Lands, the Appalachian Uplands, and the Great Lakes–St Lawrence Lowlands. The seven climatic zones are: the Pacific, Cordillera, Prairie, Great Lakes–St Lawrence, Atlantic, Subarctic, and Arctic.

Global warming represents a major environmental challenge. Canada's greenhouse gas emissions contribute to global warming. Alberta's tar sands developments, but especially its upgraders, contribute to the country's high emission levels. Carbon capture and storage represents one potential but very expensive solution.

CHALLENGE QUESTIONS

1. Which physiographic region has the most favourable natural condition for agricultural settlement?

2. What does the term 'friction of distance' mean? How has this friction altered accessibility between Canadian regions?

3. Arctic navigation has focused on global warming melting sea ice, but should isostatic rebound also be considered?

4. How does airborne pollution from Chinese coal-burning plants reach the Canadian Arctic?

5. Québec has the second-largest population as well as the second-largest industrial base, but, according to Table 2.7, it ranks fourth in industrial emissions. Why is this region not ranked second?

FURTHER READING

Dearden, Philip, and Bruce Mitchell. 2009. *Environmental Change and Challenge: A Canadian Perspective*, 4th edn. Toronto: Oxford University Press.

Canada's vastness is often equated with unspoiled wilderness and unlimited resources. Perhaps there was some validity to these images in the distant past when our numbers were smaller and our technology more limiting, but not now. False and misleading images are firmly put to rest in *Environmental Change and Challenge*. Our growing population and technological advances have placed our fragile environment at risk. Already, we have stepped over the line by decimating northern cod stocks and adding toxic chemicals to the Arctic food chain so that the Inuit must limit their consumption of seals and other sea mammals.

CANADA'S HISTORICAL GEOGRAPHY

INTRODUCTION

In one sense, Canada is a young country. Its formal history began in 1867 with the passing of the British North America Act by the British Parliament. In another sense, Canada is an old country whose human history goes back perhaps as far as 30,000 years. As an old country, its history has followed many twists and turns, but three events stand out because they continue to have a profound impact on the nature of Canadian society. These events are the arrival of the first people in North America and, most importantly, the capacity of their descendants to survive and endure the colonization and assimilation efforts of European settlers; the colonization of North America by France and England and the establishment of British institutions, laws, and values; and, in the early twentieth century, the influx of people from Central Europe and czarist Russia, many of whom settled the prairie lands. In later years, these 'outsiders' provided the political push behind government-backed multicultural programs and policies and, at the same time, challenged the previous vision of 'two Canadas'.

Given these circumstances, tensions between regions and groups were inevitable. Disagreements arose. Four faultlines stand out and they are based on fundamental differences between English- and French-speakers, centralists and decentralists, 'new' and 'old' Canadians, and Aboriginal and non-Aboriginal Canadians. With the exception of first-generation 'new' Canadians, all have their roots in Canada's history and geography. The search for solutions to these tensions is a dominant feature of modern Canadian society and, to a large degree, this process of seeking a middle ground accounts for Canadians' high degree of acceptance of various religious and ethnic groups as well as their desire to resolve regional disagreements. However, this tolerance did not magically appear; rather, it was 'learned' over time—often after reconsidering past intolerant acts towards minority groups, especially towards Aboriginal peoples and visible minorities. Using John Ralston Saul's metaphor, an underlying theme in this chapter is that Canada evolved from a 'hard' country to a 'soft' one.

CHAPTER OVERVIEW

In this chapter we will consider the following topics:

- The arrival of Canada's first people.
- The colonization of Canada by the French and the British.
- The settlement of Canada's West by peoples from Central Europe and czarist Russia.
- The territorial evolution of Canada.

- The four faultlines as they have developed in the context of Canada's geography and history.
- The notion of 'One Country, Two Visions'.
- Two power struggles: economic and political.
- Possible 'solutions' to complex problems.

A totem pole stands on the front lawn of the Legislature in Victoria, British Columbia. Photo: Josh McCulloch/All Canada Photos.

The First People

Around 30,000 years ago, according to some archaeological estimates, the first people to set foot on North American soil were Old World hunters who crossed a land bridge (known as Beringia) into Alaska and Yukon. Beringia was a product of the last ice advance when so much water was contained in the continental glaciers that the sea level dropped, thus exposing the ocean bottom between Siberia and Alaska. Scientists estimate that the sea level dropped by at least 100 metres, thus creating a land bridge between Asia and North America.

Beginning around 15,000 years ago, the climate warmed, the ice sheets retreated, and the sea level rose. Soon Beringia was again well below the surface of the Pacific Ocean. The Old World hunters were now in North America, but those remaining in Alaska were blocked from proceeding south by an ice sheet that was perhaps 4 km thick. When an ice-free corridor was formed just east of the Rocky Mountains, these ancient hunting tribes were able to migrate southward into the heart of North America (Figure 3.1). Archaeologists have estimated that this ice-free corridor appeared about 12,000 to 13,000 years ago. Just when the first people travelled along this narrow corridor is unknown, but it probably occurred after tundra vegetation had taken hold, supplying food

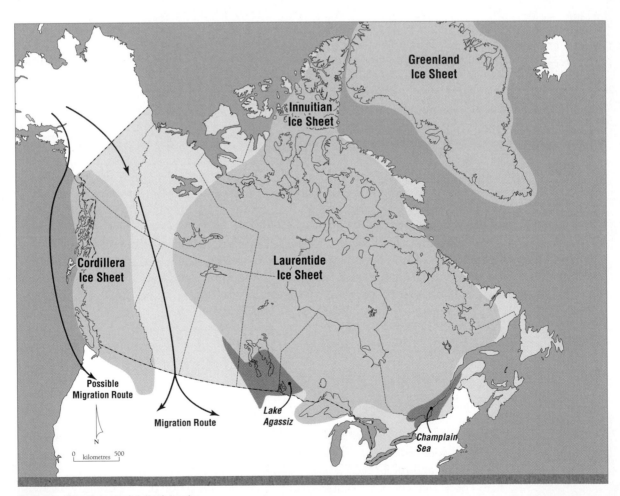

Figure 3.1   Migration routes into North America.
The Wisconsin Ice Sheet had retreated by 12,000 BP (before present), leaving an ice-free corridor between its two remaining parts (the Cordillera and Laurentide ice sheets). In places where the meltwater could not flow to the sea, water collected in low-lying areas and created huge glacial lakes such as Lake Agassiz. As explained in Vignette 2.6, however, the Champlain Sea was not a glacial lake but an extension of the Atlantic Ocean into the isostatic depressed valley of the St Lawrence River.

for the animals that also migrated into the heart of North America. With Beringia once more below sea level, further migration from northern Asia to North America was thought to be virtually impossible until the marine technology of the Paleo-Eskimos emerged about 5,000 years ago (Table 3.1). However, some researchers have posited a west coast sea migration of Old World hunters from much earlier times.

Some archaeologists even speculate that Old World hunters may have penetrated into the lands south of the ice sheet much earlier. According to one theory, these early hunters made a coastal migration along the edge of the Cordillera ice sheet that covered all the mountains and valleys of British Columbia. By slowly moving southward along the un-glaciated islands adjacent to the ice-covered British Columbia coast, these early people could have circumvented the Cordillera ice

sheet by island-hopping, thereby reaching the unglaciated area of the US Pacific coast much earlier than 12,000 years ago. However, archaeological evidence of such a route is lacking because these ancient campsites, if they existed, are now well below the current sea level.

Although the question of when, precisely, humans first reached North America is still unsettled, archaeological evidence indicates that '[b]y about 11,000 BP humans were inhabiting the length and breadth of the Americas, with the greatest concentrations being along the Pacific coast of the two continents' (Dickason with McNab, 2009: 15). Archaeologists have assigned the name Paleo-Indian to these first people of North America because they shared a common hunting culture, which was characterized by its uniquely designed fluted-point stone spearhead. In time, the Paleo-Indians developed more

Table 3.1   Timeline: Old World Hunters to Contact with Europeans

Date	Event
(BP = before present)	
30,000–25,000 BP	Old World hunters of the woolly mammoth cross the Beringia land bridge into the unglaciated areas of Alaska and southern Yukon.
18,000 BP	The Wisconsin Ice Sheet reaches its maximum geographic extent, covering virtually all of Canada.
15,000 BP	The Wisconsin Ice Sheet retreats rapidly in western Canada, exposing a narrow ice-free area along the foothills of the Rocky Mountains. Ice persists in northern Québec until about 6,000 years ago.
13,000–12,000 BP	Descendants of the Old World hunters migrate southward through the ice-free corridor east of the Rocky Mountains.
9,000 BP	Mammoths and mastodons become extinct, forcing early inhabitants of North America to adjust their hunting practices and thereby become more mobile and less numerous.
5,000 BP	As the ice sheet disintegrated across the northern limits of Canada, Paleo-Eskimos (known as the Denbigh) were the first peoples to cross the Bering Strait to the Arctic Coast of Alaska. Later they move eastward along the Canadian section of the Arctic coast.
3,000 BP	The Dorset people, who were descendants of the Denbigh, develop a more advanced technology suited for an Arctic marine environment.
1,000 BP	A third wave of marine hunters, known as the Thule, migrate across the Arctic, eventually reaching the coast of Labrador. At the same time Vikings reach Greenland and North America, where they establish a settlement on the north coast of Newfoundland (L'Anse aux Meadows).
1497	John Cabot lands on the east coast (Newfoundland or Nova Scotia).
1534	Jacques Cartier plants the flag of France near Baie de Chaleur.

Think About It

Some archaeologists believe that Old World hunters arrived in North and South America well before 12,000 years ago. If so, which migration route—land or sea—seems more plausible?

distinct hunting cultures, which Thomas (1999: 10) divides into three groups:

- Clovis culture from 9500 to 8500 BC;
- Folsom culture from 9000 to 8200 BC;
- Plano culture from 8000 to 6000 BC.

Paleo-Indians

How did these Old World hunting cultures evolve into a New World hunting system? The Paleo-Indians, the people who devised the fluted points, were descendants of the Old World hunters. The oldest fluted points found in North America are about 11,500 years old. These spearheads, along with the bones of woolly mammoths, have been discovered in the southern part of the Canadian Prairies. By 9,000 years ago, many of the large species, such as the woolly mammoths and the mastodons, had become extinct, possibly as a result of excessive hunting and/or climatic change. After the extinction of the woolly mammoths and mastodons, later Paleo-Indian cultures, which archaeologists refer to collectively as Folsom and Plano, developed a variety of unfluted stone points with stems for attachment to spear shafts, which made weapons more suitable for hunting buffalo and caribou. About 8,000 years ago, hunters in the grasslands of the interior of Canada pursued the buffalo, while those in the tundra and forest lands of northern and eastern Canada depended on caribou for most of their food. However, these smaller prey species could not support large numbers of people, so the Paleo-Indians had to develop new survival strategies. These strategies involved remaining in one area (and presumably keeping other peoples out of that area), developing effective hunting techniques for the local game, and making extensive use of fish and plants to supplement their principal diet of game.

This link between geographic territory and hunting societies marked the development of Paleo-Indian **culture areas**, which were geographic regions with the following two characteristics: (1) a common set of natural conditions that resulted in similar plants and animals, and (2) inhabitants who used a common set of hunting, fishing, and food-gathering techniques and tools. Under these conditions, Paleo-Indians formed more enduring social units that became the forerunners of the numerous North American Indian tribes at the time of contact with Europeans.

Indians

Most archaeologists support the idea that Algonquians (e.g., Cree, Ojibwa) are direct descendants of Paleo-Indians, but they are less certain about Athapaskans (e.g., Dene, Chipewyan, Gwich'in), whose ancestors may have arrived from Asia some 10,000 years ago. Since the Paleo-Indian culture had emerged some 11,500 years ago, the Athapaskans represent a distinct Indian culture. But how did the early Athapaskans cross the Bering Strait? Archaeologists suggest that the ancient ancestors of the Athapaskans either walked across the frozen Bering Strait or crossed it in small, primitive boats.

Around 10,000 years ago, the Laurentide Ice Sheet had retreated just to the north of the Great Lakes. At that time, climatic zones ranged from a tropical climate in Mexico to an arctic climate just south of the Laurentide Ice Sheet. These climatic zones provided different agricultural opportunities for Indian tribes in North America. About 5,000 years ago, Indians living in the tropical climate of Mexico began to domesticate plants and animals. This agricultural system and its people gradually spread northward into areas with more restrictive growing conditions. These climatic differences required Indians to adapt their agricultural system accordingly. About 3,000 years ago, Indians in what is today the eastern United States planted corn, beans, and squash (known as 'the three sisters'), which supplemented their diet of game and fish.

Agriculture was not possible north of the Great Lakes–St Lawrence Lowlands because of its shorter growing season for corn and other crops. Algonquian-speaking Indians who lived north of the Great Lakes had to hunt big-game animals, particularly caribou, for sustenance. They also traded with those more sedentary Indians, such as the Huron and Iroquois, who practised a form of slash and burn agriculture in the Great Lakes–St Lawrence Lowlands and in the Ohio Valley. By the sixteenth century, the Huron controlled the agriculture lands between Lake Simcoe and the southeastern corner of Georgian Bay, where about 7,000 acres were under cultivation and where Indian villages, with populations as large as 5,000 persons, were commonplace (Dickason with McNab, 2009: 46–7). In western North America, agriculture spread northward along the valleys of the Mississippi River and its major tributary, the Missouri River. Tribes on the Canadian prairies engaged in trade for agriculture products with tribes along the upper reaches of the Missouri River. In northwest Canada, Athapaskan-speaking Indians, whose ancestors probably came from Asia much later (perhaps between 7,000 and 10,000 years ago), continued to practise a nomadic lifestyle of hunting and gathering. They moved about in the forest lands stalking big-game animals and made summer hunting trips to the tundra where the caribou had their calving grounds.

Arctic Migration

Arctic Canada was settled much later than the forested lands of the Subarctic. Before people could occupy the Arctic Lands, two developments were necessary. The first was the melting of the ice sheets that covered Arctic Canada. About 8,000 years ago, only small remnants of the great Laurentide Ice Sheet remained in northeastern Canada, leaving the western Arctic free of the ice sheet. The second development was the emergence of a hunting technique that would enable people to live in an Arctic environment. About 5,000 years ago, the Paleo-Eskimos, who were living along the northeast coast of Asia and in Alaska, had developed an Arctic sea-based hunting technique, and they began to move eastward along the coast of Alaska and then along the Arctic coast of Canada. This Paleo-Eskimoan hunting culture is known as the Denbigh. Unlike previous marine hunting societies, these people invented a harpoon and other tools that enabled them to hunt seals and other marine mammals, though they also relied heavily on terrestrial game such as the caribou. About 3,000 years ago, a second migration from Alaska took place. Known as the Dorset culture, this culture replaced the Denbigh. The third and final Arctic migration took place roughly 1,000 years ago, when the

Photo 3.1

This tent ring representing the remains of a Thule house is located near Igloolik, Nunavut. Tent rings mark the sites where Thule families located their tents. The stone ring indicates the edges of the tent's skin walls, which were weighted down with rocks.

Alec Aitken

Thule people, who had developed a sophisticated sea-hunting culture, spread eastward from Alaska and gradually succeeded their predecessors. The Thule, the ancestors of the Inuit, hunted the bowhead whale and the walrus. However, the climate began to cool (known as the Little Ice Age) and whales no longer entered the Arctic Ocean in large numbers because the ocean was covered by ice for most of the year. The Thule were forced to hunt smaller game such as the seal and the caribou.

A fuller discussion of the Little Ice Age and its impact on the Thule is found in Vignette 2.11 page 68.

Initial Contacts

Long before the time of European contact, the descendants of Old World hunters occupied all of North and South America. Yet, Europeans consider the New World **terra nullius** or empty lands. North American Indian and Inuit tribes met the European explorers searching for a trade route to the Orient. While the total population of these tribes can only be estimated, many scholars now believe there may have

Think About It

In all societies, food security is a critical element. Which of the two groups—the hunters or the agriculturalists—would have better food security and thus be less likely to face starvation?

been as many as 500,000 Indians and Inuit living in Canada at the time of first contact. The greatest concentrations were found along the Pacific coast, where marine resources provide abundant food, and in the Great Lakes–St Lawrence Lowlands, where a primitive form of agriculture supported relatively large sedentary populations. Following contact, their numbers dropped sharply, perhaps declining to 100,000. By 1871, the Census of Canada reported an Aboriginal population of 122,700 (Romaniuc, 2000: 136). Loss of hunting grounds to European settlers and the spread of new diseases by explorers, fur traders, and missionaries greatly contributed to this depopulation. In the early seventeenth century, the sudden collapse of Huronia, the most powerful Indian group in the region, took place shortly after contact with the French. Its numbers dropped from 21,000 to less than half this number within a decade. This demographic catastrophe was not unique to Huronia, but it does provide one example of the deadly impact of European diseases and colonial alliances on the Aboriginal peoples. French missionaries brought diseases to the Huron villages, while the Iroquois, who opposed the French–Huron alliance, attacked the Huron villages and eventually destroyed the Huron Confederacy in 1649. In little more than a generation, Huron villages were abandoned, the cornfields of Huronia reverted to forest, and its people were greatly reduced in numbers and scattered across the land, some finding their way to the north shore of the St Lawrence in Québec, others being captured and assimilated by the Iroquois, and another remnant joining related tribes in what are today Michigan and Ohio.

John Cabot, the first European explorer to land in Canada after the brief settlement of the Norse at the tip of Newfoundland around AD 1000, reached the east coast in 1497. Cabot was followed by others, including Jacques Cartier and Martin Frobisher. In 1534 Cartier made contact with two Indian tribes along the Gaspé coast, and in 1576 Frobisher encountered an Inuit encampment along the Arctic coast of southern Baffin Island. Both explorers were searching for a route to Asia, but instead discovered a new continent and peoples. In both instances, contact with North Americans ended badly. Lives were lost on both sides, and the ore that the explorers took back to Paris and London, respectively, proved worthless (Vignette 3.1). Instead of gold, they had found fool's gold (iron pyrites).

Vignette 3.1   Contact between Cartier and Donnacona

In 1534, Jacques Cartier made contact at the Gaspé coast with Donnacona, the leader of the St Lawrence Iroquois. The Iroquois, whose village (Stadacona) was located at the site of Québec City, came to the Gaspé to fish for mackerel. The next year Cartier returned. After reaching the Iroquois village of Hochelaga, which later became the site of Montréal, Cartier realized that the St Lawrence did not provide a sea passage to China. Like the Spanish in Mexico and Peru who discovered gold and silver, Cartier hoped to find wealth in the New World. In 1541, on his third voyage to Stadacona, Cartier hoped to find diamonds and gold. He left France with several hundred colonists. By then, Donnacona and all but one of the Iroquois captives that Cartier had taken with him when he returned to France in 1536 were dead. The French authorities, who wished to conceal this fact, decided not to allow the surviving Indian woman to return to Stadacona with Cartier. The French arrived at Cap Rouge where they established a settlement just west of Stadacona. When the St Lawrence Iroquoians realized that Cartier was not going to return their chief and the other Iroquois captives, relationships quickly soured. Over the winter, the Iroquois killed at least 35 of the French settlers. Cartier and his surviving party left for France as soon as the river ice melted. He took along rocks that he believed contained gold and diamonds, but, like the ore mined on Baffin Island by Frobisher's men, Cartier brought back fool's gold and quartz. As there seemed to be no prospect for mineral wealth, the King of France lost interest in the New World. Sixty years went by before the French made another attempt at establishing a settlement at Stadacona. In 1608, Samuel de Champlain founded Québec City.

Source: Adapted from Ramsay Cook, *The Voyages of Jacques Cartier* (Toronto: University of Toronto Press, 1993).

Culture Regions

At the time of contact with Europeans, Aboriginal peoples occupied specific territories (cultural regions). Within each cultural region, these First Peoples developed distinct techniques suitable for the local environment and wildlife. Seven culture regions are the Eastern Woodlands, Eastern Subarctic, Western Subarctic, Arctic, Plains, Plateau, and Northwest Coast (Figure 3.2). The Inuit occupied the Arctic cultural region. In the Eastern Subarctic, the Cree were the principal Algonquian tribe. The Cree in this region had developed a technology—snowshoes—to hunt moose in deep snow. In the Western Subarctic, the Athapaskans, including a number of Dene tribes, hunted caribou and other big-game animals. Northwest Coast Indians harvested the rich marine life found along the Pacific coast. Tribes such as the Haida, Nootka (Nuu-chah-nulth), and Salish comprised the Northwest Coast cultural region. In the southern interior of British Columbia, the Plateau Indians—including the Carrier, Lillooet, Okanagan, and Shuswap—occupied the valleys of the Cordillera, forming the Plateau cultural region. Across the grasslands of the Canadian West, Plains Indians such as the Assiniboine, Blackfoot, Sarcee, and Plains Cree hunted bison. The Iroquois and Huron lived in the Eastern Woodlands of southern Ontario and Québec. Both groups combined agriculture with hunting. In the Maritimes, the Mi'kmaq and Maliseet also occupied the Eastern Woodlands, where they hunted

Think About It

Imagine yourself to be Captain Martin Frobisher. How would you communicate with the Baffin Island Inuit?

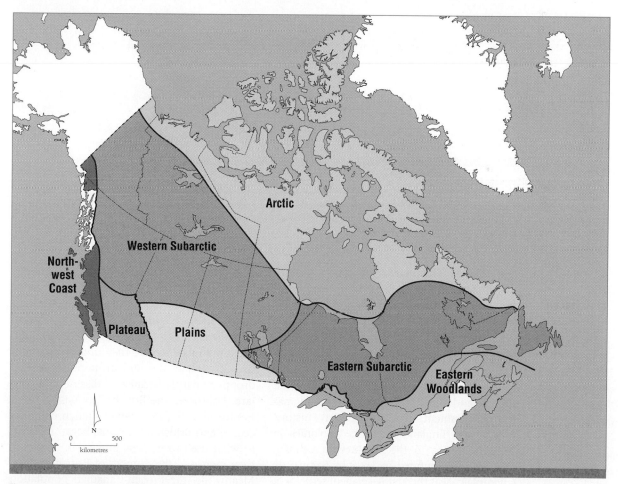

Figure 3.2 Culture regions of Aboriginal peoples.
Resources and natural conditions found in culture regions provided the foundation for the formation of unique spatial versions of Aboriginal hunting systems and social organizations.

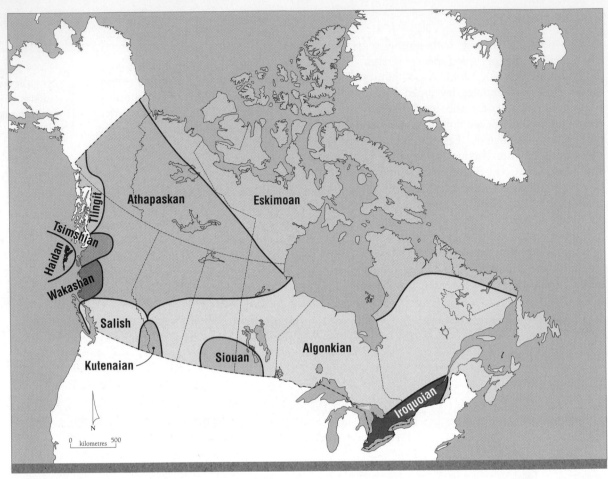

Figure 3.3   Aboriginal language families.
Aboriginal peoples in Canada form a very diverse population. At the time of contact, there were over 50 distinct Aboriginal languages spoken. These languages formed 11 language families, five of which were in one natural cultural region, the Northwest Coast. Following contact, language loss was swift. By the end of the twentieth century, only three Aboriginal languages, Cree, Inuktitut, and Ojibwa, had over 20,000 speakers. (Further resources: *Atlas of Canada*, 1996, 'Aboriginal Languages by Community, 1996' at: <atlas.nrcan.gc.ca/site/english/maps/peopleandsociety/lang/aboriginallanguages/bycommunity/1>.)

Think About It

Why do you think so many Indians taken to France by Cartier died?

Think About It

European countries claimed various parts of North America. What was the basis of their claim of sovereignty over lands they 'discovered'?

and fished. The complexity and diversity of Aboriginal peoples can be gleaned from the spatial arrangement of their languages (Figure 3.3).

The Second People

The colonization of North America by the French and the British was the second major development in Canada's early history. France and England established colonies in North America in the seventeenth century. Québec City, founded in 1608 by Samuel de Champlain, was the first permanent settlement in Canada. By 1663, the French population in New France was about 3,000 compared to a population of about 10,000 Indians (mainly Huron and Iroquois), who lived in the same area of the St Lawrence Valley, the Great Lakes, and the Ohio Basin. By 1750, the French Canadians constituted most of the population in New France, where they numbered about 60,000, while the Indian population had dropped sharply because of disease and warfare. Following the British Conquest of New France in 1759, the flow of French colonists ceased and British immigrants began to move to what used to be New France. In doing so, they increased the number of British settlers. Meanwhile, the French-Canadian population now had to depend entirely on natural population increase to expand their numbers.

The first large wave of British immigrants consisted of refugees from the United States. These Loyalists had supported Britain during the American War of Independence (1775–83). After the defeat of the British army, they sought refuge in other parts of the British Empire, including its North American possessions. In North America, most Loyalists settled in Nova Scotia, while a smaller number moved to the Eastern Townships of Québec and to Montréal. Others settled in what is now southern Ontario.

The second wave of immigrants came from the British Isles. From 1790 to 1860, almost a million people migrated from the British Isles to British North America, mostly because of deteriorating economic conditions in Great Britain and because of job opportunities in the merchant cod fishery. In the 1840s, the potato famines in Ireland added to the problems in the Old World, causing terrible hardships for the Irish people. Thousands fled the countryside and many left for the New World to settle in the towns and cities of British North America.

These two waves of migration greatly changed Canada. At the time of Confederation in 1867, the population of British North America had reached 3 million. Approximately 80 per cent of these British subjects lived in the Great Lakes–St Lawrence Lowlands, while about 20 per cent inhabited Atlantic Canada. Across the rest of British North America, Aboriginal peoples made up most of the population. In the Red River Settlement, a new Aboriginal people, the Métis, who were of Native and European descent, had emerged. By 1869, the Métis, who were split between French- and English-speaking, greatly outnumbered white settlers and fur traders in the settlement, which had a population of nearly 12,000. The Métis formed over 80 per cent of this population (Table 3.2).

But more important, the old balance of demographic power between the French and English had been reversed. British migration to Canada had, over the course of 100 years, changed the demographic balance between French-speaking and English-speaking

Photo 3.2

Relationships between explorers and Aboriginal peoples were not always peaceful. In this painting by John White, the artist, a member of Frobisher's second expedition in 1577, records a fight between Frobisher's crew and Baffin Island Inuit.

Canadians. Approximately 60 per cent of the European population was English-speaking. The new demographics also resulted in large English-speaking populations in the principal cities of Québec. Migration had also changed the ethnic composition of the English-speaking population from almost entirely English to a mixture of English, Irish, Scottish, and Welsh. Moreover, by the 1860s, Canada's ethnic character varied by region. In Atlantic Canada, the Scottish and Irish outnumbered the English. In Québec, the English and Irish formed a sizable minority in the towns and cities, though rural Québec remained solidly French-speaking except in the Eastern Townships. Ontario, like Atlantic Canada, was decidedly British.

North Wind Picture Archives

Photo 3.3

The American Revolution (1775–83) divided the residents of the Thirteen British Colonies. With the defeat of British forces, those British subjects who did not support the revolutionary cause were forced to leave, losing their property and sometimes their lives. Considered traitors by Americans, Loyalists were often subjected to mob violence.

Canada began as a collection of four small British colonies—Upper and Lower Canada, New Brunswick, and Nova Scotia. From Great Britain's perspective, union of its colonies was essential to withstand annexation by the United States. Yet, the challenge was how to convince each colony. For Québec, the issue was language, religion, and culture. The Maritimes economy was oriented to the sea, not linked by land to Central Canada. While the Maritimes harboured ill feelings towards Confederation for many decades because its economy suffered, the separation of the British and French in different geographic parts of British North America remains the greatest challenge to national unity and it ensures the

Table 3.2 Population of the Red River Settlement, 1869		
Ethnic Group	**Population Size**	**Population Percentage**
Whites born in Canada	294	2.5
Whites born in Britain or a foreign country	524	4.4
Indians	558	4.7
Whites born in Red River	747	6.2
English-speaking Métis	4,083	34.1
French-speaking Métis	5,757	48.1
Total Population	11,963	100.0

Source: Adapted from Lower (1983: 96).

continued existence of two visions of Canada. The greatest threat to Canadian unity stems from these two irreconcilable visions.

Further discussion of two visions of Canada is found later in this chapter in the section 'The French/English Faultline', page 114.

In 1867, 92 per cent of the population was either British or French. Each linguistic group developed its own vision of Canada: French Canadians saw Canada as two founding peoples, while English-speaking Canadians favoured the notion of equality among provinces. In English Canada, the notion of equal provinces grew out of the following factors:

- The nature of Confederation was such that provincial powers were shared equally.
- The British formed the majority of the population in three of the four provinces, thereby dominating the political affairs of those provinces.
- The British, while a minority within Québec, were the dominant business group.

The Third People

The hegemony of the British and French began to weaken in the early part of the twentieth century when the fertile lands of Western Canada were settled in large part by neither English-Canadian nor French-Canadian farmers. In 1870, Ottawa obtained the vast land of the Hudson's Bay Company and was faced with the question of settling this territory. Little progress was made until after the signing of treaties with various Plains Indian groups and the completion of the Canadian Pacific Railway (CPR) in November 1885. Now the time was right for massive immigration to settle these lands. For Ottawa, there were two key advantages in encouraging settlement. First, the threat of American settlers moving into the Canadian West and annexing these lands would be diminished. Second, the creation of a grain economy would provide freight for the Canadian Pacific Railway, thereby helping

Vignette 3.2   The Land Survey System and the Settling of the Prairies

In 1872, the federal government passed the Dominion Lands Act. This legislation established a survey system that divided the land into square townships made up of 36 sections, each measuring 1 mile by 1 mile, with allowances for roads. Each section was further subdivided into four quarters, each quarter section measuring one-half mile by one-half mile and comprising 160 acres. This survey system gave the Canadian Prairies a distinctive 'checkerboard' pattern.

Ottawa sought to populate the arable lands of Western Canada by offering 'free' land to farmers. The free land took the form of a homestead (a quarter section or 160 acres). A free grant of one quarter section was available to persons 21 years or older with the payment of a $10 fee. Upon fulfilling cultivation and residency requirements within three years of acquiring the property, the homesteader would receive title to the land. Since Ottawa had made substantial land grants to the Canadian Pacific Railway and to the Hudson's Bay Company (HBC), not all the land was free. Both the CPR and the HBC sold their land at market prices, thus making a considerable profit.

turn it into a viable operation. But where to find the people? Some came from Ontario, Québec, and Atlantic Canada to claim their 160 acres as homesteaders while others came from Britain and the United States (Vignette 3.2).

Still, much of Western Canada was unoccupied. Clifford Sifton, the Minister of the Interior, accepted the challenge to settle the West. By the beginning of the twentieth century, Sifton had launched an aggressive and innovative advertising campaign to lure people from Britain and the United States to 'The Last Best West', but this effort failed to bring sufficient immigrants. At that point, he recognized the need to go beyond these two countries. In a break with past immigration policy, Sifton turned his attention to the people of Central Europe, Scandinavia, and czarist Russia. Land-hungry peasants from Ukraine formed the largest single group of immigrants but Doukhobors and German-speaking Mennonites also came from czarist Russia, giving Western Canada a distinct and different mix of ethnic groups and landholdings from the rest of the homesteaders, with communal rather than individual landholdings the norm among some groups. Sifton recognized that Ukrainian peasants from czarist Russia came from a cold grassland environment and were therefore well suited for settling the Canadian Prairies. Sifton was even willing to accept Russian-speaking Doukhobors whose religious beliefs and social customs, including communal farming practices, separated them from other homesteaders. From 1901 to 1921, Western Canada's population increased from 400,000 to 2 million, with Saskatchewan becoming the third most populous province by 1921 (Table 3.3). As these ethnic groups and individuals spread across the Prairies, their impact was enormous on a landscape that only recently was populated largely by vast herds of buffalo and semi-nomadic Indian tribes (see Kerr and Holdsworth, 1990: Plate 17).

By opening the door for immigration from European countries without a French or British background, Sifton's immigration policy changed the face of Canada. His goal of settling the West was accomplished, and a new dimension had been added to Canada's social fabric—people with neither a French nor a British background. While the majority of these immigrants were homesteaders, some took jobs in Canada's booming mining and logging industries while others settled in Canada's major cities, especial Montréal and Toronto. Although most of these new settlers soon learned English, this cultural/linguistic difference from the two founding peoples surfaced later in the twentieth century as a powerful force for multiculturalism and pluralism in English-speaking Canada. The immigration to the Canadian Prairies from 1896 to 1914 forms the topic for more detailed discussion later in this chapter under the subheading 'The Immigration Faultline'.

The Territorial Evolution of Canada

Canada's formal history as a nation began with the proclamation of the British North America Act on 1 July 1867. This Act of the British Parliament united the colonies of New Brunswick, Nova Scotia, and the Province of Canada (formerly Upper Canada and Lower Canada) into the Dominion of Canada.[1] Canada soon acquired more territory. In 1870, the Deed of Surrender transferred Rupert's Land and the North-Western Territory to the federal government, at which time this large expanse that had been under HBC control was renamed the North-West Territories. In 1871, British

Think About It

Homesteaders far from the railway had to use local resources for building materials. What resource was readily available in the Prairie landscape?

Table 3.3	Canada's Population by Provinces and Territories, 1901 and 1921			
Political Unit	**Population 1901**	**%**	**Population 1921**	**%**
Ontario	2,182,947	40.6	2,933,662	33.4
Québec	1,648,898	30.7	2,360,510	26.9
Nova Scotia	459,574	8.6	523,837	6.0
New Brunswick	331,120	6.2	387,876	4.4
Manitoba	255,211	4.8	610,118	6.9
Northwest Territories*	20,129	0.4	8,143	0.1
Prince Edward Island	103,259	1.9	88,615	1.0
British Columbia	178,657	3.3	524,582	6.0
Yukon	27,219	0.5	4,147	>0.1
Saskatchewan	91,279	1.7	757,510	8.6
Alberta	73,022	1.3	588,454	6.7
Canada	5,371,315	100.0	8,787,949**	100.0

*Saskatchewan and Alberta did not become provinces until 1905 and were officially included in the population of the Northwest Territories.
**Includes 485 members of the armed forces.
Source: Adapted from Statistics Canada (2003).

Columbia joined Confederation, and Prince Edward Island followed two years later. In 1880, Ottawa acquired the Arctic Archipelago from Great Britain. Thus, while Canada began as a small country, consisting of what is now known as southern Ontario, southern Québec, New Brunswick, and Nova Scotia, it quickly became one of the largest in the world (Figure 3.4).

What was behind these real estate deals? For Britain, the union of its North American colonies had three advantages: (1) a better chance for their political survival against the growing economic and military strength of the United States, (2) an improved environment for British investment, especially for the proposed trans-Canada railway; and (3) a reduction in

British expenditures for the defence of its North American colonies. The British colonies perceived unification differently. The Province of Canada, led by John A. Macdonald, pushed hard for a united British North America because it would have a larger domestic market for its growing manufacturing industries and a stronger defensive position against a feared American invasion. The Atlantic colonies showed little interest in such a union. As they were part of the British Empire, the attraction of joining the Province of Canada had little appeal. Furthermore, unlike the Canadians, Maritimers continued to base their prosperity on a flourishing transatlantic trading economy with Caribbean countries and Great Britain. From 1840 to 1870, the backbone

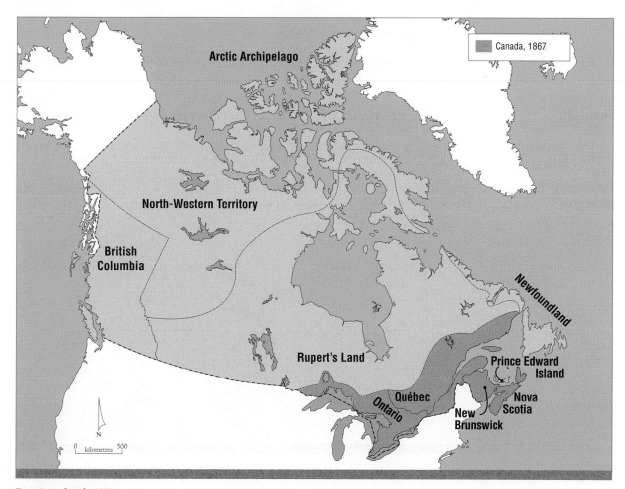

Figure 3.4 Canada, 1867.
At Confederation, Canada was only a fraction of its current territorial extent. The Hudson's Bay Company controlled most of British North America, including Rupert's Land and the North-Western Territory. (Further resources: *Atlas of Canada*, 2003, 'Territorial Evolution, 1867', at: <atlas.nrcan.gc.ca/site/english/maps/historical/territorialevolution/1867>.)

of the Maritime economy was the construction of wooden sailing-ships. Shipbuilding was so important that this period was known as the 'Golden Age of Sail' in the Maritimes. Even diplomatic pressure from Britain to join Confederation had little effect on Maritime politicians. But the Fenian raid into New Brunswick in 1866 and the termination of the Reciprocity Treaty with the United States quickly changed public opinion in the Maritimes (Vignette 3.3).[2] Shortly after the Fenian raids, the legislatures of both New Brunswick and Nova Scotia voted to join Confederation.

As we have seen, within a decade, the territorial extent of Canada expanded from four British colonies to the northern half of North America. The new Dominion grew in size with the addition of other British colonies and territories, and the creation of new political jurisdictions (Figure 3.5), while the British government transferred its claim to the Arctic Archipelago to Canada in 1880. Negotiations between Ottawa and the British colonies in British Columbia and Prince Edward Island soon brought them into the fold of Confederation, in 1871 and 1873, respectively, and the 'numbered treaties' were signed with Indian tribes of the West beginning in 1871. No such negotiations took place with the Red River Métis. The result was the Red River Rebellion of 1869–70, the formation of the Métis Provisional Government,

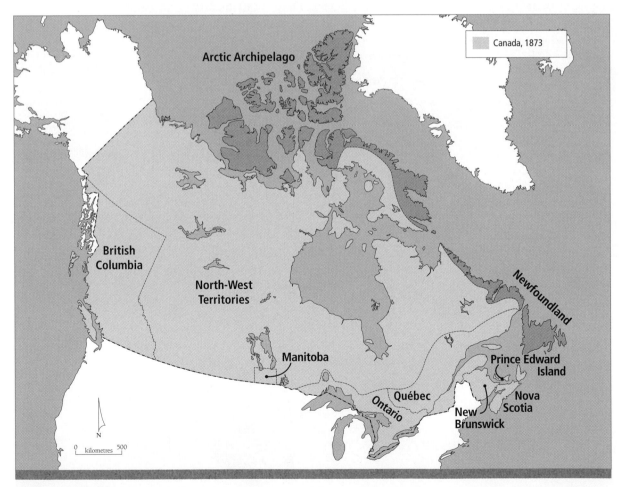

Figure 3.5   Canada, 1873.
Within seven years of Confederation, Canada had obtained the Hudson's Bay Company lands (including the Red River Settlement) and two British colonies (British Columbia and Prince Edward Island) had joined the new Dominion. For the North-Western Territory and Rupert's Land, the Crown paid the HBC £300,000, granted the Company one-twentieth of the lands in the Canadian prairies, and allowed it to keep its 120 trading posts and adjoining land. In 1870, these lands were renamed the North-West Territories. (Further resources: *Atlas of Canada*, 2003, 'Territorial Evolution, 1873', at: <atlas.nrcan.gc.ca/site/english/maps/historical/territorialevolution/1873>.)

Vignette 3.3   America's Manifest Destiny

The doctrine of Manifest Destiny was based on the belief that the United States would eventually expand to all parts of North America, thus incorporating Canada into the American republic. From its beginnings along the Atlantic seaboard, the United States had greatly increased its territory by a combination of force, negotiation, and purchase. In 1803, the United States purchased the Louisiana Territory (a vast land west of the Mississippi and east of the Rocky Mountains) from France, in 1846, the US gained the Oregon Territory, and in 1867 the country bought Alaska from Russia. To Americans, such expansion was an expression of their right to North America. As well, it would rid North America of the much-hated European colonial powers.

Not surprisingly, the Fathers of Confederation were concerned about American designs on British North America. First, in 1866, the Fenians raided Upper Canada, Lower Canada, and New Brunswick with the intention of seizing British North America and holding it for ransom until Ireland was free of British rule. Second, in 1867, the US purchase of Alaska left British Columbia wedged between American territory to its north and south, and the exact boundary along the coastline south of 60° N was uncertain. Canada and the United States settled this final border dispute in 1903.

and then negotiations with Ottawa that led to the entry of Manitoba into Confederation. The Manitoba Act of 1870 resulted in the Red River Settlement and a small surrounding area becoming the province of Manitoba.

For further discussion of the first Riel-led resistance, see 'The First Clash: Red River Rebellion of 1869–70, page 107.

By 1880, Canada stretched from the Atlantic to the Pacific and north to the Arctic. While still part of the British Empire, Canada had begun the slow journey to independence and nationhood. In 1905 the provinces of Alberta and Saskatchewan were created, and much later, in 1949, Newfoundland joined Canada, completing the union of British North America into a single political entity. (The territorial evolution of Canada is illustrated in Figures 3.4 to 3.8 and Table 3.4.)

Within a decade after Confederation, Canada's territorial extent had burgeoned, making it the second largest country in the world, which posed a serious problem for the nation's leaders, who had to somehow transform this vast territory into a nation. Sir John A. Macdonald, Canada's first Prime Minister, resolved this in part by authorizing the construction of a transcontinental railway, the Canadian Pacific Railway. The government had to heavily subsidize the building of the railway but, to Macdonald, the high cost was worth it.

National Boundaries

Well before Confederation in 1867, wars and treaties between Britain and the United States shaped many of Canada's boundaries. The southern boundary of New Brunswick, Québec, and Ontario was settled in 1783 when Britain and the United States signed

Think About It

Is it more accurate historically to refer to the Red River Rebellion as the Red River Resistance?

Table 3.4 Timeline: Territorial Evolution of Canada	
Date	**Event**
1867	Ontario, Québec, New Brunswick, and Nova Scotia unite to form the Dominion of Canada.
1870	The Hudson's Bay Company's lands are transferred by Britain to Canada. The Red River colony enters Confederation as the province of Manitoba.
1871	British Columbia joins Canada.
1873	Prince Edward Island becomes the seventh province of Canada.
1880	Great Britain transfers its claim to the Arctic Archipelago to Canada.
1949	Newfoundland joins Canada to become the tenth province.

the Treaty of Paris. Under this treaty, the United States gained control of the Indian lands of the Ohio Basin, with Britain controlling Québec lands draining into the St Lawrence River. Earlier, Britain had formally recognized the rights of Indians to the lands of the Ohio Basin, and to the north and west, in the Royal Proclamation of 1763, which provided the constitutional framework for negotiating treaties with Aboriginal peoples. This recognition was the basis of Aboriginal rights in Canada (see 'Aboriginal Rights' in this chapter). Based on the fur trade route to the western interior, the boundary of 1783 passed through the Great Lakes to the Lake of the Woods. In 1818, Canada's southern boundary was adjusted; it was set at 49° N from the Lake of the Woods to the Rocky Mountains. As for the northwestern boundary, there was some question as to where British territory ended and Russian territory began. British and Russians had come into contact through the fur trade. In the eighteenth century, Russian fur traders established trading posts along the Alaskan coast. Indians travelled along the Yukon River through the interior of Alaska to trade at these Russian forts. By the early nineteenth century, the HBC had reached the tributaries of the Yukon River. In 1825, Britain and Russia set the northern boundary at 141° W (the Treaty of St Petersburg). Russia also agreed to relinquish to Britain its claims to the coastal regions south of 54°40' N to 42° N. In Atlantic Canada, the boundary between Maine and New Brunswick had not been precisely defined in 1783. Minor adjustments took place in 1842 when Britain and the United States concluded the Webster-Ashburton Treaty.

In the early nineteenth century, the border separating Canada's western territories from the United States was not well defined. In fact, it depended on the natural boundary between Rupert's Land and the Louisiana Territory. Rupert's Land was defined as the lands whose waters flow into Hudson Bay while the lands of the Louisiana Territory were determined by the rivers draining into the Mississippi River system. Rupert's Land included the Red River Valley while the

Louisiana Territory had the Milk and Poplar rivers flowing from what is now Alberta and Saskatchewan into the Missouri River. In 1818, Britain and the United States decided on a compromise of the forty-ninth parallel because it was easier to delineate. Accordingly, the two countries established the North American Boundary Commission to establish and mark the western boundary along the forty-ninth parallel dividing the United States and Canada (see photo 3.4). The boundary survey began in 1857 and was completed in 1861. In 1869, their report was signed in Washington.

The last major territorial dispute between Britain and the United States took place over the **Oregon Territory**. The Oregon Territory stretched along the Pacific Coast from Alaska to Mexico—Mexico at that time extended northward to 42° N. Britain's claim to the Oregon Territory hinged on exploration and the fur trade. David Thompson of the North West Company reached this area in 1807, and after the 1821 merger of the North West Company with the Hudson's Bay Company, the HBC established Fort Vancouver at the mouth of the Columbia River in 1825, where it conducted trade with the local Indian tribes. In the 1830s, American settlers crossed the Rocky Mountains into the Columbia Valley, where they proceeded to cultivate the fertile soils of the Willamette Valley. Because both Britain and the United States had a foothold in the Oregon Territory, sovereignty was uncertain. Great Britain's claim was based on two factors: that (1) the HBC had established the first European settlement in the region, at Fort Vancouver, and (2) the Company exerted control over the land where the Indians trapped fur-bearing animals. The American claim was centred on the fact that the vast majority of settlers were Americans. In the final outcome, there was no doubt that occupancy was a more powerful claim to disputed lands than claims based on exploration and the presence of a fur-trading economy. Too late, the British urged the HBC to bring settlers from Fort Garry to the Oregon Territory (**Red River Migration**). With the Oregon Boundary Treaty of 1846, the boundary between British and American territory

Photo 3.4

A crew from the North American Boundary Commission building a mound marking the border between Canada and the United States, August or September 1873.

from the Rockies to the Pacific coast was set at 49° N with the exception of Vancouver Island, which extended south of this parallel.

Internal Boundaries

Since Confederation, the internal boundaries of Canada have changed (Figures 3.4–3.8). These changes have created new provinces (Alberta and Saskatchewan) and territories (Yukon and Nunavut). As well, the boundaries of Manitoba, Ontario, and Québec were extended. In all cases, these political changes took land away from the Northwest Territories.

In 1870, the boundary of Manitoba formed a tiny rectangle comprising little more than the Red River Settlement. The province's western boundary, while extended in 1881

and 1884, did not reach its present limit until 1912. By this time, Manitoba spread north to the boundary with the Northwest Territories (now Nunavut) and east to the Lake of the Woods.

Québec, too, received northern territories. In 1898, its boundary was extended northward to the Eastmain River and then eastward to Labrador. In 1912, Ottawa assigned Québec more territory that extended its lands to Hudson Strait. Canada also believed that the province of Québec should extend to the narrow coastal strip along the Labrador coast, while the colony of Newfoundland contended that Newfoundland owned all the land draining into the Atlantic Ocean. In 1927, this dispute between two British dominions (Canada and Newfoundland) went to London. The British government ruled in favour of

Think About It

Why did Canada and the US set the boundary from the Lake of the Woods to the Pacific coast at the forty-ninth parallel rather than use the natural boundary between the Mississippi/Missouri River Basin and the Hudson Bay drainage basin? As a result, the upper reaches of the Red River went to the United States and the headwaters of the Milk and Poplar rivers were assigned to British North America.

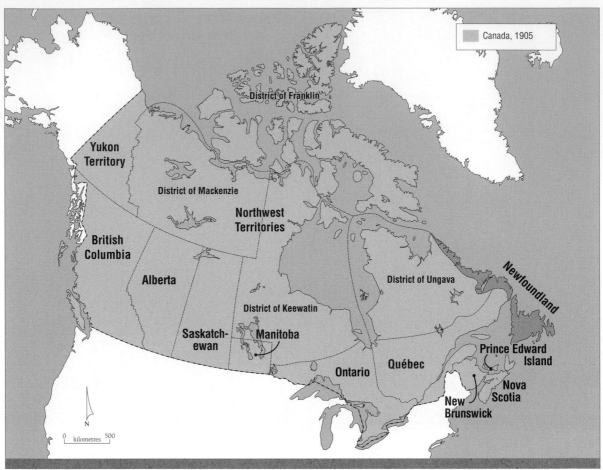

Figure 3.6   Canada, 1905.
By 1905, two new provinces (Alberta and Saskatchewan) and one territory (Yukon) were created out of the North-West Territories. In 1905, what remained was renamed the Northwest Territories, and by this time the Arctic Archipelago (District of Franklin) had been incorporated into the NWT. Ontario, Québec, and Manitoba expanded their boundaries. (Further resources: *Atlas of Canada*, 2003, 'Territorial Evolution, 1905', at: <atlas.nrcan.gc.ca/site/english/maps/historical/territorialevolution/1905>.)

Newfoundland (Figure 3.7). The Québec government has never formally accepted this ruling, though Québec has respected the boundary. However, the long-term contract for Québec to purchase electric power at 1960 prices from the Churchill Falls hydro-electric station located in Labrador remains a sore point for Newfoundland and Labrador, and plans to produce more hydroelectric power in Labrador, sell it to Hydro-Québec, and then transmit the power through Québec to markets in New England remain contentious. This controversial agreement is currently being challenged in the Québec court system by the Newfoundland and Labrador government on the basis of 'fairness'.

Vignette 9.11, 'Lower Churchill, Churchill Falls, and Hydro-Québec', page 393, discusses the hydroelectric relationship between Québec and Newfoundland and Labrador.

In the years following Confederation, Ontario gained two large areas. In 1899, its western boundary was set at the Lake of the Woods (previously this area belonged to Manitoba), while its northern boundary was extended to the Albany River and James Bay. In 1912, Ontario obtained its vast northern lands, which stretch to Hudson Bay.

In 1905, Canada formed two new provinces, Alberta and Saskatchewan. The final adjustment to Canada's internal boundaries

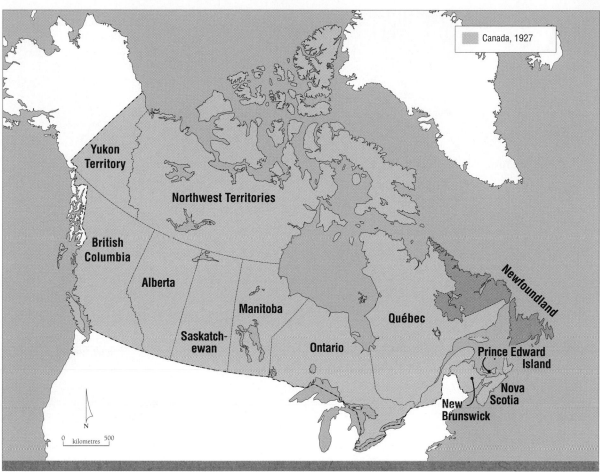

Figure 3.7　Canada, 1927.
The complicated history of Lower Canada and Newfoundland provided ample justification for both parties to claim the land between the Northern Divide and the coastal strip associated with the fisheries. In 1927, the Privy Council of the British Parliament ruled in favour of Newfoundland by selecting the watershed boundary, a decision that dated back to 1670 when King Charles II created Rupert's Land. In 1912 Ontario, Québec, and Manitoba gained additional northern lands to reach their current geographic size. (Further resources: *Atlas of Canada*, 2003, 'Territorial Evolution, 1927', at: <atlas.nrcan.gc.ca/site/english/maps/historical/territorialevolution/1927>.)

occurred in 1999 with the establishment of the government of the territory of Nunavut (Figure 3.8 and Table 3.5).

Faultlines

Canada's regional geography is partly defined by its faultlines, a notion introduced in Chapter 1. Tensions within Canadian society often begin as regional issues but soon turn into national issues. These disagreements are between one or more regions and Ottawa. Invariably, if one region wants something, another region or regions either will want the same or similar treatment or will object to special treatment that threatens the federation. The federal government, because it is charged with establishing national policies and programs, has the power to settle such issues. The challenge facing the federal government is to seek a balance between the regional factions, but such efforts rarely satisfy all parts of Canada. Nevertheless, the process of searching for solutions to these tensions informs Canadians of different views and, in doing so, lessens the tension—perhaps not initially but usually over time. Even so, Ottawa is most aware of the political power of Ontario and Québec and, whether real or not, federal policies seem to favour these two largest provinces.

Think About It

Examine Figures 3.6 and 3.7. Did the decision of King Charles in 1670 affect the determination by the British Privy Council of the 1927 border between Québec and Labrador?

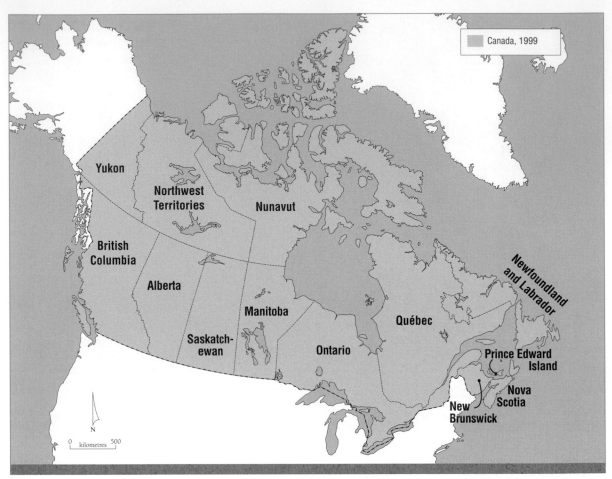

Figure 3.8   Canada, 1999.
On 1 April 1999, Nunavut became a territory. (Further resources: *Atlas of Canada*, 2003, 'Nunavut', at: <atlas.nrcan.gc.ca/site/english/maps/historical/territorialevolution/1999>.)

Table 3.5   Timeline: Evolution of Canada's Internal Boundaries

Date	Event
1881	Ottawa enlarges the boundaries of Manitoba.
1898	Ottawa approves extension of Québec's northern limit to the Eastmain River.
1899	Ottawa decides to set Ontario's western boundary at the Lake of the Woods and extend its northern boundary to the Albany River and James Bay.
1905	Ottawa announces the creation of two new provinces, Alberta and Saskatchewan.
1912	Ottawa redefines the boundaries of Manitoba, Ontario, and Québec, which extend their provincial boundaries to their present position.
1927	Great Britain sets the boundary between Québec and Labrador as the Northern Divide. Québec has never accepted this decision.
1999	A new territory, Nunavut, is hived off from the Northwest Territories in the eastern Arctic.

The Centralist/Decentralist Faultline

Geography poses powerful challenges to Canada's national unity. These challenges stem from Canada's vast space, economic competition between regions, regional trade patterns with the United States, and internal political differences between the provinces and Ottawa. In combination, these challenges to national unity are manifest as regional self-interest that often results in regional tensions. Sometimes such tensions are expressed by a group of individuals, such as fishers in Atlantic Canada, or through common interests among provinces, such as the oil-producing provinces. Whatever their form or geographic extent, these internal forces produce regional pressures that can be divisive.

Regional self-interest and resulting tensions (whether cultural, economic, or political) are a natural outcome of Canada's physical and human geography. For better or worse, regionalism is a fact of life in Canada and it may well be the most telling characteristic of Canada's national character. The regional challenge to Canada's east–west alignment, and therefore to Canada's national unity, may be summed up in four ways:

- Canadian regions are separated from each other by great distances, making trade and commerce between those regions more difficult (geographic forces).
- Regions compete with each other and provinces have trade barriers (forces derived from regional self-interest).
- Provinces compete over federal funding, since the division of political powers in the Canadian Constitution assigned costly services such as health care, education, and social services to provinces, and only the federal government has tax revenues large enough to pay for much of these services.
- Geography encourages Canadian regions to fall into the economic orbit of the United States.

Trade always has been a central issue between Canada and the US. For Canada, access to the American market is essential for its manufacturing and resource sectors. Canada's regions have their specific trade aspirations. As a consequence, the natural markets of North America have a north–south orientation. The Maritimes are geographically, historically, and economically tied to New England, while southern Ontario is linked to Michigan, Ohio, Pennsylvania, and New York. British Columbia is part of the Pacific Northwest and indeed has felt part of this international region referred to as Cascadia for many years.

Regional Tensions

Political solutions have attempted to overcome geography by creating transnational transportation systems, by fostering an industrial core, and by ameliorating regional disparities through equalization payments. The first challenge facing the government of Sir John A. Macdonald was clear: a railway must span the country from the Atlantic Ocean to the Pacific Ocean. Without a more substantial presence, Macdonald's Conservative government feared that the US might annex the unoccupied parts of Western Canada (Vignette 3.3). From Ottawa's perspective, the railway would exert effective sovereignty over its western territories. Though the magnitude of this project was nearly beyond its capacity, Ottawa proceeded with major financing from British investors. The government's goals for this transcontinental railway were:

- to link the West with the rest of Canada;
- to settle the Canadian Prairies;
- to provide an export route for prairie grain;
- to create a market in the West for eastern industries.

At the same time as Ottawa moved towards securing its western lands, it launched a new economic initiative in 1879, the National Policy, which reinforced the centralist/decentralist faultline. The National Policy and other federal policies stimulated manufacturing in Ontario and Québec. Rightly or wrongly, Canadians living outside Central Canada believe that the concentration of political power in Ontario and Québec has had an unfair influence over national policies and therefore favoured economic development in Central Canada over that in the rest of the country. In the following pages, these two themes—economic and political power struggles—are explored from the perspective of Central Canada (the core) and the rest of Canada (the periphery). In the case of economic power, Central Canada refers to the industrial core in southern Ontario and southern Québec. In the case of political power, Central Canada refers to Ottawa and, indirectly, to the influence of Ontario and Québec in Ottawa because of their larger populations and greater representation in the House of Commons and, ultimately, in the federal cabinet. Together, these two power struggles are the basis of the centralist/decentralist conflict or faultline.

Centralists advocate a strong central government, national policies that exert a political dominance over provinces, and a strong national economy (which means an expanding industrial core). Decentralists seek to strengthen the powers allocated to provinces. In particular, decentralists call for a devolution of federal powers to the provincial governments and the expansion and diversification of regional economies. Economic diversification is deemed necessary to generate more jobs and thereby encourage population growth. Eventually, it is thought that economic diversification will lead to larger populations in hinterland regions and thus more political representation, and therefore power, in Ottawa.

The National Policy and Regional Tensions

The National Policy protected Canada's manufacturing industries from foreign goods by means of high tariffs. Ottawa's goal was to create a national industrial base. In fact, this economic policy ensured that the country was divided into an economic core surrounded by resource hinterlands. For Central Canada, the benefits from the policy were considerable as it ensured the region's role as Canada's industrial heartland. It also resulted in the concentration of financial power in Montréal and Toronto. Hinterland regions were not so fortunate because they were compelled to sell their raw materials and foodstuffs on the world market and buy their manufactured goods in the domestic market. For the hinterland, this arrangement translated into 'selling low' and 'buying high'. People in the hinterland regarded the National Policy as an arrangement that benefited Central Canada at the expense of the rest of the country. They became suspicious of other national programs announced by Ottawa because they appeared to be designed to favour Central Canada and thereby to exploit other regions.

The National Policy accentuated the economic differences in various parts of the country. In doing so, it increased regional tensions, which resulted in the centralist/decentralist (or heartland/hinterland) divide. While each hinterland region faced a different set of economic relations with the industrial heartland of Canada, the basic economic relationship was the same—Central Canada produced the manufactured goods, while hinterlands produced raw materials and foodstuffs. But what prevented hinterlands from developing a manufacturing base under the National Policy?

Geography prevented manufacturers in the Maritimes from reaching markets in Central Canada and the growing market in the West. While the steel mill in Cape Breton did supply some of the steel rails for the construction of the Canadian Pacific Railway, most Maritime firms could not overcome the friction of distance to reach Central Canada and the West. Maritime industrialists also were no longer able to sell their products in the nearby market of New England because of high American tariffs. The combination of a small local market, great distance to the continental markets in Canada, and high American tariffs stalled economic development in the Maritimes. Before 1867, most Maritimers were uncertain about the benefit of Confederation; afterwards, many became convinced that union with Canada was a 'bad deal'.

At the same time, unrest in Western Canada was widespread in rural areas. Under the national economic policy, western farmers had to sell their grain on the world market where prices were low, but they had to purchase their farm machinery from manufacturers in Central Canada where prices were high. By the 1920s, this unrest took the form of political protest when the Progressive Party argued for free trade to allow farmers access to lower-priced American farm machinery. Geography compounded the problems facing grain farmers, who were located in the heart of North America. The cost of transporting their grain to foreign markets was very high, thereby making their returns even lower. How could farmers overcome the great distance to their grain markets, particularly those in Europe? They thought that the solution to this transportation problem lay in the construction of a railway to the nearest tidewater point—Hudson Bay! The Hudson Bay Railway, completed in 1929, never fulfilled the high expectations of western farmers because the savings derived from the shorter rail distance were offset by higher marine insurance costs due to the danger of icebergs. Farmers, however, believed the higher insurance charges were simply another trick by Central Canada to hold onto the grain trade at the expense of westerners.

British Columbia, though linked to the rest of Canada by the Canadian Pacific Railway, remained beyond the economic pull of Central Canada. British Columbia's economic order was driven by its geographic position on the west coast. The region's natural markets were overseas. At first, its raw materials, such as fish, timber, and minerals, were shipped to markets in the western United States, the British Isles, and the Far East. As well, because of British Columbia's geographic proximity to Western Canada, many of the Prairies' products were transported to the port of Vancouver for transshipment overseas. In 1914 the opening of the Panama Canal had a great impact on the forest industry in British Columbia because it provided a 'shorter' shipping route to markets in the British Isles, Western Europe, and the east coast of the United States. Like Canadians in Western Canada, British Columbians resented having to pay for higher-priced manufactured goods from Central Canada compared to those available just across the border.

The Territorial North suffered a different fate. Until World War II, Ottawa simply ignored it. Regarded by the federal government as a remote wilderness inhabited by Indians and Eskimos (later known as Inuit), the Territorial North remained a fur economy and did not participate in the events transforming Canada's emerging industrial society until well into the twentieth century. Ottawa's laissez-faire approach limited federal expenditures to Aboriginal peoples in the Territorial North. The few northern Royal North West Mounted Police posts (later the Royal Canadian Mounted Police) served as Ottawa's main expression of administration. Only when resource developments occurred, such as the Klondike gold rush, did Ottawa hurry to ensure a broader federal presence. After World War II, the federal government's laissez-faire policy was replaced by state involvement in northern development. This change was sparked by Prime Minister John Diefenbaker's 'northern vision', which translated into investments in northern infrastructure, especially highways leading to resources. Known as 'Roads to Resources' in the provinces and the 'Development Road Program' in the territories, the funding differed: for roads in provinces, the federal government shared costs with the provincial governments; in the territories, Ottawa paid all the road-building costs (Bone, 2009: 81).

Political Power Struggle

While the core/periphery model emphasizes the core's economic dominance over the periphery, the core also exerts other forms of dominance, including political dominance. Within the context of Canada, Ottawa not only represents the political core but, by favouring Central Canada (Ontario and Québec), reinforces their economic power. This leads to a sense of alienation and frustration among the remaining regions.

While the federal government employs equalization payments to even the economic discrepancies between regions, the uneven distribution of political and economic power across Canada makes it a 'troublesome country to govern'. Each Prime Minister from Sir John A. Macdonald to Stephen Harper has had to balance national economic interests against regional ones. Federal interventions in the marketplace are attempts to protect national interests, but they sometimes result in regional alienation.

But does Ottawa really favour Central Canada or is it just a problem of balancing national against regional interests? The reality lies in the division of powers between the two levels of government (see Vignette 3.4). Can such a power imbalance exist in a federal state? How much power does a province need? How much power can the central government give up? These questions go to the heart of a federal state and the fragile relations between the central and regional governments. This tug of war also is played by the separatist movement in Québec, but for an end (independence) that is different from that sought by other provinces and premiers (more powers). In 1968, Jacques Parizeau likened the Canadian federation to a chicken

being plucked by provincial governments led, of course, by Québec. When the last feather is plucked, the chicken will perish. Voilà, Québec has achieved its goal of independence (Nemni, 1994: 175). Québec is not the only province with a desire for more power, but such power in the rest of Canada focuses on financial power.

Just how are regions represented in Parliament? Canada's political geography is shaped by the election of representatives from various political parties to the House of Commons.[3] The electoral system is designed to reflect changes in Canada's population based on census figures each 10 years. In other words, the current apportionment of seats in the federal House of Commons is based on the 2001 census until 2011. In the 2008 federal election, the total number of seats was 308 (Table 3.6). The number elected and their electoral districts are determined by the size of Canada's population, while the assignment of electoral districts across the country is determined by Canada's population distribution. There are two exceptions. First, provinces must have at least the same number of seats in the House of Commons as in the Senate. Prince Edward Island has four seats in the Senate and therefore has four seats in the House of Commons, though its population only warrants one or possibly two seats. Second, each territory is allocated one seat irrespective of its population. Since Ontario and Québec have the largest populations, they also have the largest number of seats and, therefore, hold the political balance of power. The 2008 federal election results support this proposition, with the majority of seats in the House of Commons held by Ontario with 106 and Québec with 75—58.8 per cent of the total number of seats. This concentration of electoral power is further intensified by a concentration of political power within the governing party, which is naturally concerned with national issues and priorities, not regional ones. From the standpoint of those outside Central Canada, national policies appear to favour Ontario and Québec. The Senate is supposed to provide a regional balance in Canada's political system by giving a political voice

Vignette 3.4   Federal/Provincial Powers

Under Canada's federal system, the powers of government are shared between the federal government and the 10 provincial governments. All provinces have the same powers. At the time of Confederation, the powers of the provinces were carefully enumerated (and thus limited), while those of the federal government were not limited. Provincial governments are responsible for education, health and welfare, highways, civil law (property and civil rights), local government, and natural resources, while Ottawa has a much wider mandate, including authority over defence and external affairs, criminal law, money and banking, trade, transportation, citizenship, and Indian affairs. The two levels of government were assigned joint jurisdiction over agriculture, immigration, and taxation. Territorial governments are assigned their powers from Ottawa, while municipal governments obtain theirs from the provincial governments. First Nations are seeking self-government. While First Nations' status is not yet defined by the federal government, Indian leaders often refer to a 'third level of government', meaning a political level somewhere between municipal and provincial.

Table 3.6   Members of the House of Commons by Geographic Region, 2008

Geographic Region	Members (no.)	Members (%)	Population (%)	Difference	Seats per Population (000s)
Territorial North	3	0.9	0.3	0.6	1:33
Atlantic Canada	32	10.4	7.2	3.2	1:71
Western Canada	56	18.2	17.1	1.1	1:97
Québec	75	24.4	23.9	0.5	1:101
British Columbia	36	11.7	13.0	(1.3)	1:114
Ontario	106	34.4	38.5	(4.1)	1:115
Total	308	100	100		

Source: House of Commons, 2009. Party Standings: 40th Parliament. At <www.parl.gc.ca/information/about/process/house/partystandings/standings-e.htm>.

to 'regions'. However, since its members are appointed by the Prime Minister largely on the basis of long-time service to the political party currently in power, senators are more concerned about party loyalty than regional interests. In short, the Senate fails to provide a regional counterweight to the House of Commons. Instead, this role falls to the provincial governments.

Western Alienation

Western alienation represents a serious problem for Ottawa. It takes many forms but is based on western provinces' lack of power, whether real or perceived, to control their destiny. For Alberta, the fear is that Ottawa will 'steal' its oil wealth to benefit Ontario and Québec. Both Western Canada (especially Alberta) and British Columbia complain about their lack of political power in Ottawa, which often translates into complaints that Ontario and Québec receive favoured treatment from the federal government. Since these two central provinces contain almost 63 per cent of Canada's population and 59 per cent of the seats in the House of Commons, federal parties must have a strong foothold in one or both of these provinces to form the federal government. This demographic/electoral fact gives Ontario and Québec considerable political leverage in Ottawa. The National Energy Program was deeply resented by Alberta and while the program ended in 1984, the resentment and fear of a new version remains in the minds of Albertans.

Resource development falls under provincial powers. Yet, in 1980, the federal government imposed the National Energy Program on oil-producing provinces (Vignette 3.6). Ottawa had three goals: energy security, greater Canadian ownership of the oil industry, and a greater amount of wealth produced by the oil industry for Ottawa. A fourth but unspoken goal was to keep oil prices low in Ontario and Québec. The National Energy Program lasted only four years (1980–4), but its chief legacy was the western oil-producing provinces' deep distrust of the Liberal federal government. While more than 25 years have passed, the collective regional memory continues to reject Liberal candidates. For example, only seven out of 92 seats elected Liberal members in the 2008 election—five in

Vignette 3.5   Reshaping the House?

The next redistribution of seats in the House of Commons is scheduled to take place after the 2011 Canadian census. This detailed enumeration of all Canadians is set for 10 May 2011. Based on anticipated population figures, Ontario, Alberta, and British Columbia would gain seats in Parliament, and legislation introduced in April 2010 proposed 18 new seats for Ontario, 5 for Alberta, and 7 for BC. However, while seats are distributed among the provinces and territories in proportion to population, as determined by each decennial census, two other factors come into play. First, each province will have at least as many MPs as it previously had, which explains why Prince Edward Island has four seats. Second, each province has at least as many members of Parliament now as it had in 1976, which explains why Québec has 75 seats. Thus, if some provinces stand to gain representatives in the House of Commons following the 2011 census, no province or territory will lose members.

Source: Adapted from 'Canadian House of Commons', at: <www.absoluteastronomy.com/topics/Canadian_House_of_Commons>.

Vignette 3.6   Background to the National Energy Program

In the 1970s, the price of world oil rose rapidly because of the Organization of Petroleum Exporting Countries (OPEC) strategy: by curtailing the global supply of oil, the world price would rise. From 1972 to 1980, the world price of oil increased from US$2 a barrel to over US$20 a barrel. From 1973 to 1978, agreements were reached between the oil-producing provinces and the federal government to match the domestic price with the world price. Rising oil prices generated a huge profit for the oil companies and greatly increased oil revenues for Alberta and, to a much lesser degree, Saskatchewan and British Columbia. In 1979, however, Ottawa refused to match domestic oil prices with world prices, thereby creating a lower domestic price. For Central Canada, such a national policy gave its industrial firms a decided advantage compared to those across the US border. With world prices continuing to rise, subsidization of oil refineries in Atlantic Canada and Québec was necessary to meet the lower domestic price. The federal government needed a new energy strategy that addressed the question, 'How could Ottawa pay for this price differential and thus protect the manufacturing base in Ontario and Québec?' The answer was to extract part of this 'windfall' from oil-producing provinces, particularly the Alberta government. Peter Smith, a professor of geography at the University of Alberta, denounced the National Energy Program as 'one more manifestation of the customary heartland outlook; what is good for Ontario and Québec has to be best for the rest'.

Source: Smith, 1982: 301–2.

BC, none in Alberta, one in Saskatchewan, and one in Manitoba.

Why is it likely that this alienation will continue? Without a manufacturing base, the Canadian hinterland has had to depend on its natural resources as an extremely important source of regional development and provincial tax revenue. Thanks to geology, Alberta, British Columbia, and Saskatchewan have abundant petroleum deposits. These deposits are major sources of revenue for these three provinces and have turned first Alberta and later Saskatchewan from 'have-not' provinces to 'have' provinces. Since natural resources fall under the jurisdiction of the provincial governments, the federal government has no authority to 'tax' natural resources, yet these provinces suspect that Ottawa, but particularly a Liberal government, might try to find a way to 'grab' a portion of the revenues generated from natural resources. A major beneficiary of such a federal policy would be Ontario.

Equalization and Transfer Payments

Equalization and transfer payments represent a sharing of federal tax revenue with provinces and territories. Since some provinces are richer than others, the purpose is to ensure relatively equal access to public services across Canada (Table 3.7). Yet, these payments often cause tensions between Ottawa and the provinces. From a provincial perspective, equalization and transfer payments needed to be increased to meet the demands for public services in their jurisdictions. This debate has its roots in Confederation, when taxing and programs were divided between the two levels of government. Over the years, the original balance between taxing powers and program delivery changed radically. The federal government, because of its greater taxing powers, generates much more revenue than do the provinces. Also, provinces are responsible for delivering what turned out to be very costly programs such as health care and post-secondary education. This disparity between the two levels of government in regard to revenues and expenditures has caused the so-called fiscal imbalance. Efforts to adjust this imbalance take place through equalization and transfer payments. Equalization payments are designed to ensure that the citizens in 'have-not' provinces enjoy a standard of public services comparable to those in 'have' provinces. This simple concept involves the collection of taxes by Ottawa and then the redistribution of some of those taxes to 'have-not' provinces. While the equalization formula is complex, the federal government gives money to each province whose

Think About It

As Prime Minister, could you make a case that a portion of the revenue generated by natural resources should flow to Ottawa?

Table 3.7	Total Equalization Payments ($ millions), 2000–1 to 2009–10									
Year	PEI	NB	NL	NS	Man.	Que.	Sask.	BC	Ont.	Total
2000–1	269	1,260	1,112	1,404	1,314	5,380	208	0	0	10,947
2001–2	256	1,202	1,055	1,315	1,362	4,679	200	240	0	10,309
2002–3	235	1,143	875	1,122	1,303	4,004	106	71	0	8,859
2003–4	232	1,142	766	1,130	1,336	3,764	0	320	0	8,690
2004–5	277	1,326	762	1,313	1,607	4,155	652	682	0	10,774
2005–6	277	1,348	861	1,344	1,601	4,798	82	590	0	10,901
2006–7	291	1,386	632	1,451	1,709	5,539	13	260	0	11,281
2007–8	294	1,308	477	1,477	1,826	7,160	226	0	0	12,768
2008–9	322	1,584	197	1,571	2,063	8,028	0	0	0	13,765
2009–10	340	1,689	0	1,571	2,063	8,355	0	0	347	14,365

Note: Alberta received no equalization payments over this 10-year period.
Source: Canada, Department of Finance (2009a).

fiscal capacity is below a national standard. Fiscal capacity refers to a province's ability to raise revenue. For 2009–10, three provinces, Alberta, Newfoundland and Labrador, and Saskatchewan, were declared 'have' provinces (Table 3.7). The remaining seven provinces, including Ontario for the first time, received equalization payments. Québec continues to receive the largest payment—in 2009–10, Québec obtained close to $8.4 billion or 58 per cent of the total payments.

Besides making equalization payments to the 'have-not' provinces and the Territorial Formula Financing (TFF) program, the government of Canada provides significant financial support to provincial and territorial governments on an ongoing basis to assist them in the provision of programs and services. There are two main transfer programs: the Canada Health Transfer (CHT) and the Canada Social Transfer (CST). The CHT and CST support specific policy areas such as health care, post-secondary education, social assistance and social services, early childhood development, and child care. TFF provides territorial governments with funding to support public services, in recognition of the higher cost of providing programs and services in the North. In 2009–10, provinces and territories received over $60 billion through all of the major transfers (CHT, CST, equalization, and TFF), an increase of $6.7 billion from the previous year.

The Aboriginal/ Non-Aboriginal Faultline

The roots of the Aboriginal/non-Aboriginal faultline go back to the late fifteenth and sixteenth centuries with the voyages of John Cabot, Martin Frobisher, and Jacques Cartier. The relationship has evolved over time—first, resistance on the part of Aboriginal peoples to early explorers is recorded; then, the First Peoples are known in many instances to have assisted settlers and missionaries; and next, a partnership developed in the fur trade between Europeans and Aboriginal groups. But somewhere along this continuum of interaction relations changed: one group became stronger and the other weaker. The Europeans' presumptions of superiority and entitlement, the European technology in weaponry and tools, and the spread of European diseases among Native groups soon enough led to dependency. Such Aboriginal dependency became a fact of life and is at the hurtful rubbing edge of this faultline.

In 1867, the British North America Act transferred the responsibility for the Indian tribes from Great Britain to Canada. Much later, Ottawa's responsibility was extended to all Aboriginal peoples, including the Inuit and Métis. For that reason, the Aboriginal/non-Aboriginal faultline is cast into a framework of Ottawa versus Aboriginal peoples.

Think About It

If equalization payments are designed to provide similar public service in all provinces, why are there always complaints about the size of the payments?

In this model, Ottawa acts like a core, while Aboriginal Canadians living on the edge of Canadian society serve as the periphery.

Given that marginal position within Canadian society, what can the early contacts and the later, more continuous relationships tell us? By 1873, Canada consisted of the original four colonies, British Columbia, Prince Edward Island, and what had been the vast Hudson's Bay Company lands. Within that Canada, Aboriginal peoples occupied most of the territory. The issue that continued to face the country's leaders was what to do with the Aboriginal population. The principal solution was assimilation. The second solution was to settle the land. An example of the first solution was formulated in 1857 by the Province of Canada, which allowed Indians to become regular members of society. One requirement of The Act to Encourage the Gradual Civilization of Indian Tribes in this Province was the capacity to speak either English or French. An example of the second solution was the federal government's aggressive immigration program that went beyond earlier immigration efforts focused on British 'stock'. In the pre–World War I years, Ottawa did what was unthinkable before, eagerly inviting people from Scandinavia, Central Europe, and czarist Russia to help settle the prairie lands.

The Indians, Métis, and Inuit who lived in the so-called wilderness of the Subarctic and Arctic were beyond the pale of 'civilization', where they continued to practise their quasi-nomadic hunting-trapping-fishing-trading lifestyle. Life was good when game was readily available; but times were bad when food was in short supply. Starvation, while not widespread, was not uncommon. Only the fur traders and missionaries (and later the federal mounted police force) came into contact with them, affecting their way of life. However, geographic distance from most Anglo-Canadian 'newcomers' provided a certain protection from Euro-Canadian ways for those in the Subarctic and Arctic. The situation in the grasslands of Western Canada was a different matter. First, the principal game, the buffalo, had been decimated by hunters and by the east–west railways that cut across their natural migration routes in the Great Plains. Second,

Think About It

What factors had changed, resulting in a different outcome in the Northwest Rebellion of 1885 from that of the Red River Rebellion of 1869–70?

Ottawa wanted to secure the prairie land for settlers, hence the need for treaties with the Indians of the prairies. Third, the almost completed CPR line allowed the Canadian army to quickly reach the western prairies and suppress the Northwest Rebellion of 1885. The Métis and Plains Indians faced the full wrath of Ottawa after their brief struggle against unfair treatment and dispossession failed. Louis Riel, the spiritual leader of the Métis, and eight other rebels were hanged, Plains Cree chiefs **Poundmaker** and **Big Bear** were imprisoned, and the rest of the Plains Indians, many on the verge of starvation, were forced back onto their recently established reserves. Strict government regulations, such as **pass laws**, controlled their movements.

From the beginning, Ottawa's objective was the assimilation of Indian peoples into Canadian society (Milloy, 1999). Education was an important tool in federal efforts to 'civilize' Aboriginal peoples. One such effort, Indian residential schools, stands out. Spread across Canada but concentrated in Western Canada, the residential schools were operated by the major religious groups but especially the Roman Catholic Church. Without a doubt, residential schools were the most painful experience for many Indian children and their parents, and this learning experience has had long-term effects (Vignette 3.7). After a detailed examination of Indian–white relations, J.R. Miller (2000: 269) concludes that

> While some students of these residential schools were thoroughly converted by the experience, many more absorbed only enough schooling to resist still more effectively. It would be from the ranks of former residential school pupils that most of the leaders of Indian political movements would come in the twentieth century. By any reasonable standard of evaluation, the residential school program from the 1880s to the 1960s failed dismally.

In October 2001, Canada made an offer to pay 70 per cent of compensation in respect of joint government and church liabilities to victims of sexual and physical abuse at Indian residential

Vignette 3.7 Indian Residential Schools

In 1892, the federal government entered a formal arrangement with several Christian churches—Roman Catholic, Anglican, Methodist, and Presbyterian—to provide a boarding school education for young Aboriginal children. The churches ran the schools; Ottawa paid the bills. The plan was to quickly assimilate these young children into society by removing them from their families and home communities and by insisting that they not use their native languages. The effect was to destroy their culture and leave them between two worlds without a toehold in either one. While some parents wanted their children to attend these schools, many others were forced to send their children. From 1931 to 1996, about 150,000 Indian, Inuit, and Métis children attended boarding schools.

The federal government (and society in general) believed that Aboriginal children could be successful in modern society if they abandoned their culture and language and adopted Christianity, learned English or French, and had a basic education. Attendance was mandatory and this rule was enforced by Indian agents and other federal officers as well as by missionaries. By the 1980s, the failure of this assimilation program was self-evident. The last school was closed in 1996.

schools. On 23 November 2005, the Canadian government announced a $1.9 billion compensation package to benefit tens of thousands of survivors of abuse at Native residential schools. The settlement provides for a lump-sum payment to former students: $10,000 for the first school year plus $3,000 for each school year after that. The average payout is in the neighbourhood of $28,000. Those who suffered sexual or serious physical abuses, or other abuses that caused serious psychological effects, could apply for additional compensation or seek redress through the courts. Finally, on 11 June 2008 the Prime Minister made a formal apology in the House of Commons for the harm done to individuals, families, and cultures by the residential schools.

Since the 1970s, Ottawa has adopted a more enlightened policy towards resolving issues related to Canada's First Peoples, stressing three elements: settling of outstanding land claims, recognizing Aboriginal right to self-government, and accepting that the concerns of each category of Aboriginal peoples (Indians, Métis, and Inuit) are different and that such concerns require specific solutions. Aboriginal peoples' struggle for power is rooted in two questions:

- Who are the Aboriginal peoples of Canada?
- What are Aboriginal rights?

Aboriginal Peoples

The Constitution Act, 1982 refers to Indians, Métis, and Inuit under the umbrella term **Aboriginal peoples**, that is, those now living in Canada who can trace their ancestry to the original inhabitants who were in North America before the time of contact with Europeans in the fifteenth century. Indians are further distinguished between status, non-status, and treaty Indians. People legally defined as **status (registered) Indians** are registered or entitled to be recorded as Indians, according to the Indian Act as amended in June 1985, and have certain rights acknowledged by the federal government, such as tax exemption for income generated on a reserve. According to the registered Indian population compiled by Indian and Northern Affairs, the number of status Indians had grown to 778,000 by 31 December 2007. These registered Indians held membership in one of 615 bands or First Nations (INAC, 2009).

Non-status Indians are people of Indian ancestry who are not registered as Indians and therefore have no rights under the Indian Act. **Treaty Indians** are status or registered Indians who are members of (or can prove descent from) a band that signed a treaty. They have a legal right to live on a reserve and participate in band affairs. Less than half live on reserves. The **Métis** are people of European and North American Indian ancestry. The **Inuit** are Aboriginal people located mainly in the Arctic.

Statistics Canada records Aboriginal peoples by their identity as declared by those individuals on census day. In 2006, the census recorded 1,172,785 Aboriginal peoples: 698,025 North American Indians, 389,780 Métis, and 50,480 Inuit (Statistics Canada, 2008). The principal reason for the difference in population size of registered Indians and North American Indians is due to the method of recording. The registry kept by Indian and Northern Affairs Canada (INAC) is based on the list of Indians compiled by each of the 615 bands while the census records people based on an enumeration of the day of the census. Some people are missed and a few bands refused to allow a census enumeration.

The Indians, Inuit, and Métis are a highly diversified population. One indication of their cultural diversity is linguistic classification. As noted earlier, there were approximately 55 distinct Aboriginal languages (of 11 language families) spoken in Canada at the time of original contact (Figure 3.3). The largest language family is Algonkian. There are 15 distinct Algonkian-based languages, the most common of which are Cree and Ojibwa. Inuktitut, the Inuit language, has regional dialects and is spoken across the Canadian Arctic.

Another measure of Aboriginal diversity is self-identification. Many Indians prefer to identify themselves with the name of their tribal group, such as Cree or Iroquois, while others use the name of their First Nation (band) for a more precise identification. For example, the Cree occupy a vast territory that stretches from northern Québec to Alberta. There are many Cree tribes within that territory. A Cree living in northern Saskatchewan might identify himself or herself as a member of a Cree band, such as the Lac La Ronge band.

In 2007, INAC (2009) recorded the population of each band. The six largest First Nations are southern Ontario's Six Nations of the Grand River (Iroquois) with a population of 22,969; the Mohawk of Akwesasne (10,607) at St Regis on the Ontario–Quebec border near Cornwall; the Blood (10,253) in southern Alberta; Kahnawake (Mohawk) (9,570) near Montréal; the Saddle Lake Cree reserve (8,770) outside Edmonton; and the Lac La Ronge Cree in northern Saskatchewan (8,462). A majority of **First Nations people** live on reserves (53 per cent) while the rest live off-reserve, mainly in cities. Many of those who live in Canada's cities, however, periodically return to their reserves.

Aboriginal peoples are reclaiming their identity and place names. Some bands are relinquishing the names given to them by Europeans in favour of their original names, such as Anishinabe (for Ojibwa) and Gwich'in (for Kutchin). The landscape is also being reclaimed. For example, the Arctic community of Frobisher Bay, named after the English explorer, Martin Frobisher, is now Iqaluit, which means 'the place where the fish are', the capital city of Nunavut ('our land').

Aboriginal Rights

Aboriginal rights are group or collective rights that stem from Aboriginal peoples' occupation of the land before contact. Such rights apply most readily to status Indians and Inuit, while Métis are less well served. Aboriginal peoples' traditional attitudes and values towards land and wildlife are strikingly different from those held by most non-Aboriginal Canadians. For Canada's First Peoples, the land has not only economic value but also cultural, political, and spiritual value. Aboriginal attitudes and values are based on their former economic and social systems, that is, the largely subsistence system of hunting, fishing, and gathering that was in place before contact with Europeans.

Land rights are the most fundamental Aboriginal rights. Indeed, it is from land rights that most other collective rights flow, such as self-determination and self-government. Aboriginal peoples first began to secure land rights or treaty rights before Confederation and this process continues today. Some Aboriginal peoples, including the Métis, are still negotiating with Ottawa.[4] Treaties set aside **reserve** land, held collectively by and for the benefit of the band. Until the late 1950s, all Indian reserves were under the absolute control and governance of the federal Department of Indian Affairs through local Indian agents, federal administrators who oversaw the perceived education, health, legal, and financial needs and business dealings of local band members.

The reasons for signing treaties varied depending on the historical context. Authorities acting on behalf of the Crown often signed treaties to secure Aboriginal peoples as allies during times of turmoil, such as the War of 1812, or simply to procure land for the growing numbers of settlers coming to Canada. During the expansion to the West, treaties were signed to provide a place for Indian tribes so that the Indian Wars between the US military and various Indian tribes, which were common south of the border, would not erupt in Canada. For Aboriginal peoples, treaties often promised reserves that would not be available to settlers, as well as support during the transition from semi-nomadic hunting to sedentary farming. The numbered treaties for the Plains Indians therefore offered protection from the anticipated flood of settlers and some guarantee that the federal government would care for them now that their principal source of food, the buffalo, was gone. However, treaty assurances of federal assistance were often not met (see Carter, 2004; Brownlie, 2003).

The terms of each treaty varied, although they generally included cash gratuities and presents at the signing of the treaty, annual payments in perpetuity, the promise of educational and agricultural assistance, the right to hunt and fish on Crown land until such land was required for other purposes, as well as land reserves to be held by the Crown in trust for the Indians. In Treaty No. 6 of 1876, for example, which covered much of central Saskatchewan and Alberta, the amount of land assigned to each tribe was determined by its population size, i.e., each family of five received one square mile. Reserves are collectively owned by First Nations bands, though legally they are owned by the Crown in trust for them.

Conflicting ideas as to the significance of treaties between the signing parties largely shaped Aboriginal and non-Aboriginal relations in Canada during the twentieth century. When treaties were signed, Crown authorities viewed them as vehicles for extinguishing Aboriginal rights and titles to land and thus for opening the land to agricultural settlement. Aboriginal peoples, however, understood them as agreements between 'sovereign' powers to share land and resources. With such diverse perceptions, disagreements were inevitable. Added to the issue of perception, some Aboriginal groups never signed treaties, while other treaties were never fulfilled. The latter part of the twentieth century witnessed various movements to repair injustices related to Aboriginal land rights (for example, through land claims) and to recognize Aboriginal peoples' inherent right to self-determination (for example, through constitutional reform). Through these attempts and through the signing of modern treaties, Aboriginal peoples are seeking a new place in Canadian society in the hope of participating in the larger economy while still retaining their culture.

Think About It

Which Aboriginal group has the most 'rights'?

From Hunting Rights to Modern Treaties

History sometimes makes strange allies. Shortly after Pontiac, chief of the Odawa, led a successful uprising against the British in 1763, Britain decided to form an alliance with him and other Indian leaders.[5] For strategic reasons, George III issued the Royal Proclamation of 1763, which identified a part of British territory west of the Appalachian Mountains as Indian lands. At that time, Britain believed that it could claim 'uninhabited land', which the British defined as land without permanent occupation (that is, no cultivated land or permanent settlements). However, the British also believed that Indians had a limited ownership over the lands they inhabited, and that therefore such lands must be purchased from the Indian owners. This somewhat ambiguous concept remains the basis of land claims by Canadian Aboriginal peoples. The first major land grant, the Haldimand Grant of 1784, took place in the early days of British North America. The purpose was to reward the Iroquois who had served on the British side during the American Revolution. In his proclamation, the Governor of Québec, Lord Haldimand, prohibited the leasing or sale of land to anyone but the government in the tract extending from the source of the Grand River in present-day southwestern Ontario to the point where the river feeds into Lake Erie. However, Joseph Brant, the leader of the Iroquois, insisted that they had the same rights as the colony's Loyalist settlers, that is, freehold land tenure. And so the waters were muddied by the early sale and lease of plots of land in the original Haldimand Grant. This issue has been part of the contemporary conflict between non-Aboriginal residents and Six Nations Iroquois at Caledonia, Ontario.

 See Vignette 5.7, 'Historical Timeline of the Caledonia Dispute', page 194.

The legal meaning of Aboriginal title to land has evolved over time. Until the 1970s, Ottawa recognized two forms of land rights. Reserve lands were one type of right or ownership, which the Canadian government held for Indian people. The second type was a usufructuary right to use Crown land for hunting and trapping. At that time, Crown lands (both provincial and federal) included most of Canada's unsettled areas. Indian, Inuit, and Métis families lived on Crown lands, continuing to hunt, trap, and fish. However, federal and provincial governments could sell such lands to individuals and corporations or grant them a lease to use the land for a specific purpose, such as mineral exploration or logging, without compensating the Aboriginal users of those lands. By the 1960s, many Aboriginal groups still did not have treaties with the Canadian government. Atlantic Canada, Québec, the Territorial North, and British Columbia contained huge areas where treaties had not yet been concluded. As a consequence, Aboriginal peoples had no control over developments on these lands.

A combination of events radically changed this situation. One factor was the emergence of Native leaders who understood the political and legal systems. They used the courts to force the federal and provincial governments to address the issue of Aboriginal rights and land claims. The first major event took place in 1969, when Ottawa proposed reforms to the Indian Act in its White Paper on Indian Policy. This galvanized treaty Indians into action. The White Paper proposed to treat all Canadians equally. For Indians, it meant the abolition of their treaty rights and the reserve land system. About the same time, the Nisga'a in northern British Columbia took their land claim, known as the *Calder* case, to court. In 1973, the Supreme Court of Canada narrowly ruled (by a vote of four to three) against the Nisga'a argument that the tribe still had a land claim to territory in northern British Columbia. However, in their ruling, six of the seven judges agreed that Aboriginal title to the land had existed in British Columbia at the time of Confederation. Furthermore, three judges said that Aboriginal title still existed in British Columbia because it had not been extinguished by the British Columbia government, while three other judges stated that Aboriginal title had been extinguished by the various laws passed by the British Columbia government since 1871. The seventh judge ruled against the Nisga'a claim on a legal technicality. The Supreme Court's narrow verdict and the legal opinion of three judges that Aboriginal title still existed changed the course of Aboriginal land claims in Canada. Now the federal government agreed that Aboriginal peoples who had not signed a treaty may very well have a legal claim to Crown lands.

In the early and mid-1970s the James Bay Project in northern Québec and the proposed Mackenzie Valley Pipeline Project in the Northwest Territories added fuel to the political fire over Aboriginal rights. The possible impact of these industrial projects on Aboriginal peoples was made clear through the Mackenzie Valley Pipeline Inquiry of 1974–7 (the Berger Inquiry) into possible environmental and socio-economic impacts and in the media. It was obvious that Aboriginal organizations were prepared

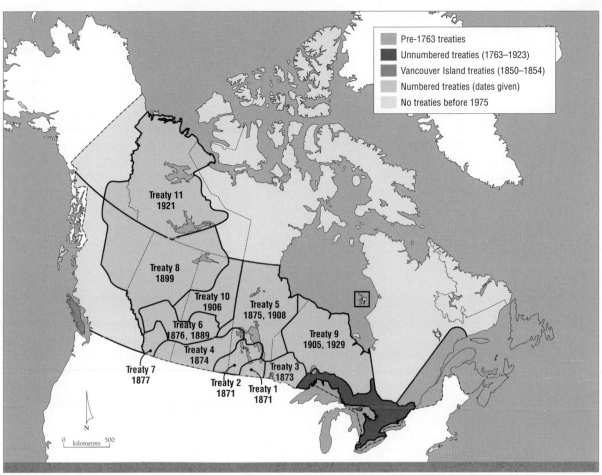

Pre-1763 treaties
Unnumbered treaties (1763–1923)
Vancouver Island treaties (1850–1854)
Numbered treaties (dates given)
No treaties before 1975

Treaty 11
1921

Treaty 8
1899

Treaty 10
1906

Treaty 5
1875, 1908

Treaty 6
1876, 1889

Treaty 9
1905, 1929

Treaty 4
1874

Treaty 7
1877

Treaty 3
1873

Treaty 2
1871

Treaty 1
1871

N

0   kilometres   500

Figure 3.9   Historic treaties.
The first treaties, made between the British government and Indian tribes, were 'friendship' agreements. In Upper Canada the Robinson treaties of 1850 set aside reserve lands in exchange for the title to the remaining lands. With the settlement of lands in the Canadian West, Indians became concerned about their future, so many of the 11 numbered treaties, which spanned a half-century from 1871 to 1921, included provisions for agricultural supplies. When the last numbered treaty was signed, many Aboriginal peoples in Atlantic Canada, Québec, and British Columbia were without treaties. (Further resources: *Atlas of Canada*, 'Historic Indian Treaties', at: <atlas.nrcan.gc.ca/site/english/maps/historical/indiantreaties/historicaltreaties>.)

to take action to defend their land claims. Their position in the 1970s was 'no development without land-claims settlements'. All these events changed both the public's views of Aboriginal rights and the government's position. At first grudgingly and then more willingly, governments, corporations, and Canadian society recognized the validity of Aboriginal land claims. The James Bay and Northern Québec Agreement in 1975 was the first modern land-claim agreement in Canada. Since then, all treaties are either specific or comprehensive agreements between an Aboriginal group and the federal government.

A **comprehensive land-claim agreement** is sought when a group of Aboriginal people who have not yet signed a treaty can demonstrate a claim to land through past occupancy. In 1984, the Inuvialuit of the Western Arctic became the first Aboriginal people to settle a comprehensive land claim with the federal government. Since then, 20 comprehensive claims have been finalized in northern Canada, involving over 90 Aboriginal communities with over 70,000 members. Negotiations are ongoing in approximately 60 processes across the country at various stages of negotiation and levels of activity. Still, many claims remain

Think About It

From an Indian perspective, what was wrong with the basic concept behind the White Paper?

unsettled, especially in British Columbia. Until they are concluded, relations between those pursuing agreements and the federal government will remain strained. For example, virtually the entire province of British Columbia, except for Vancouver Island, is claimed by First Nations. In British Columbia, progress has been slow. Until 1992, the provincial government claimed that British occupancy had extinguished Aboriginal title. However, in 1992 the British Columbia government accepted the principle of Aboriginal land claims. The following year, Ottawa and Victoria agreed to a formula for paying outstanding land claims.

The federal government would pay 90 per cent of the money needed to settle outstanding claims and the province would provide the land. In 2000, the Nisga'a Agreement was finalized, while other First Nations are negotiating with the British Columbia Treaty Commission. Progress has been slow. In March 2007, members of the Lheidli T'enneh First Nation voted to reject their Final Agreement. The 600-strong Tsawwassen First Nation signed their final agreement in 2007 and it became law in 2008. The Maa-nulth First Nation signed their final agreement in 2008 and expect it to be approved by Parliament in 2011. Four other

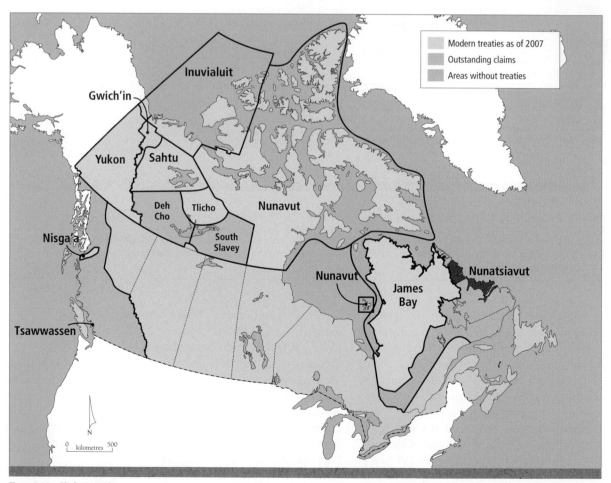

Figure 3.10   Modern treaties.
The first modern treaty, the James Bay and Northern Québec Agreement, was signed in 1975. By 2010, the main areas without treaties were much of British Columbia, Labrador, and lands in Québec. (The original inhabitants of Newfoundland, the Beothuk, had perished from disease, encroachment, and slaughter by the early nineteenth century.)

First Nations—Sliammon, Yale, Yekooche, and In-SHUCK-ch—are currently negotiating towards final agreements.

For more on BC land claims, see Vignette 7.2, 'Who Owns BC?', page 285.

Those Aboriginal groups who have concluded modern treaties are moving forward. They are able to focus on economic and cultural developments rather than expending their energies on land-claim negotiations. In 1993, the Nunavut agreement broke new ground by effectively establishing self-government over an entire territory. Since then, modern land-claim agreements, such as the Nisga'a Agreement, have included arrangements for self-government. As a result, a gap is emerging within the Aboriginal community between those who have a modern treaty and those who do not, as well as between those on reserve and those who live in urban areas among Canada's increasingly pluralistic majority society. Also, as with countries and with regions, some Aboriginal groups reside on lands that are rich in natural resources, resource developments, and development potential (e.g., oil and gas deposits, oil sands, pipelines, prime timber land) that provide a base for economic growth and considerable wealth, while many other groups live in areas with little resource potential where even subsistence from the land is marginal if not impossible.

Bridging the Aboriginal/ Non-Aboriginal Faultline

Aboriginal peoples are taking the control of their affairs away from Ottawa. Some Indian and Inuit peoples have made substantial advances in economic development, while others have moved into the area of self-government. Unfortunately, some Aboriginal peoples, including the Métis, have not yet begun this self-government process and remain on the political margins of Canadian society. For most, fortunately, the process of change has started.

Vignette 3.8   The Origin of the Métis Nation

The fur trade and the Métis are part of the historical fabric of Western Canada. With their command of English/French and Indian languages, the Métis were logical intermediaries in the fur trade. Over the centuries, the fur trade absorbed the mixed-blood offspring of Cree, Ojibway, or Saulteaux women with French fur traders from the North West Company or Scottish and English fur traders from the Hudson's Bay Company. In the early nineteenth century, the settlement near the confluence of the Red River and the Assiniboine River consisted mainly of French and Scottish 'half-breeds' and Scottish settlers brought from Scotland to fulfill Lord Selkirk's dream of an agricultural community. By 1821, the two fur-trading empires had amalgamated, throwing many English, French, and Scottish half-breeds out of work. Many gathered in the Red River Settlement, which provided the cultural melting pot for the formation of Métis Nation. The Métis culture was neither European nor Indian, but a fusion of the two.

In 1996, the *Report* of the Royal Commission on Aboriginal Peoples identified two major goals: Aboriginal economic development and self-government. The gap between Aboriginal and non-Aboriginal societies will not be bridged until these goals are achieved. The principal factor is transferring power from Ottawa (the political power core) to the various Aboriginal communities (the politically weak periphery). The economic and social well-being of Aboriginal reserves varies widely. Some, such as the Whitecap First Nation near Saskatoon with its casino and world-class golf course, have become tourist destinations. The Labrador Inuit provide another example when they gained a share of the royalties from the Voisey's Bay nickel mine within their land-claim agreement. Many others remain trapped in poverty and, without an economic base, breaking the dependency on the federal government is impossible. One consequence is the migration of Aboriginal peoples to cities. Developing an economic base on a reserve or remote community is not an easy task but it will result in a new and more positive relationship between Aboriginal and non-Aboriginal Canadians.

Vignette 3.9   From a Colonial Straitjacket to Aboriginal Power

Until 1969, Canada's Aboriginal peoples were largely invisible to other Canadians. Most were geographically separated, as many status Indians lived on reserves. Métis were in isolated communities, and Inuit still roamed the Arctic. They were outside the political process and thus denied access to political decision-making. In fact, status Indians did not receive the right to vote in federal elections until 1960. Shunned by Canadian society, these marginalized peoples had been subjected to assimilation policies for many years. In 1969, Ottawa made one last attempt to assimilate the Indian people of Canada through its 'Statement of the Government of Canada on Indian Policy', more popularly known as the White Paper. The federal government proposed to eliminate the legal distinction between Indians and other Canadians by repealing the Indian Act and amending the British North America Act to remove those parts of the Act that called for separate treatment for Indians, and to abolish the Department of Indian Affairs. Ottawa believed that the separation of Indians from other Canadians was not only divisive but also made the Indians dependent on government and thereby held them back. The remedy was 'equality'. In the context of the 1960s, when oppressed people in other countries, including blacks in the United States and in South Africa, fiercely sought equality and the American Indian Movement railed against colonial suppression, Prime Minister Trudeau believed that the White Paper was the answer to Canada's Indian problem. And some Indian leaders supported this solution. As it turned out, many others did not. Reaction was swift. In the same year, Harold Cardinal published *The Unjust Society* and the following year, under his leadership, the Alberta chiefs published a formal rebuttal to the White Paper, commonly known as the 'Red Paper' and titled *Citizens Plus: A Presentation by the Indian Chiefs of Alberta to the Right Honourable P.E. Trudeau*.

Over the next decade, the debate over the place of Indians in Canadian society took several different directions. First, there was legal support for the Indian position, beginning with the *Calder* case in 1973 when the Supreme Court held that the Nisga'a had Aboriginal rights. Second, the election of the Parti Québécois in 1976 called for 'nation-to-nation' discussions between the province and the federal government. Aboriginal leaders seized the opportunity to present their demands in the same constitutional language. Third, recognition of the Aboriginal peoples and their rights in the 1982 Constitution Act dramatically enhanced their status and bargaining power. Fourth, the Constitution Act did not define Aboriginal rights, leaving that task to negotiations or the courts. The courts have been active in defining Aboriginal rights. In 1997, the Supreme Court's landmark decision in the *Delgamuukw* case overturned the earlier decision denying that Indians in British Columbia had Aboriginal title. Furthermore, the Court ruled that Aboriginal title means that Indians have the right to the resources on those lands.

The Immigration Faultline

The history of non-British immigration to Canada is a complex and sometimes controversial topic. Most importantly, immigration has been a continuous stream of people coming to Canada, each having a distinct impact on the land and society. Before 1867, immigration was often an instrument of colonial power. After the British Conquest of New France, for example, the British government set the immigration policy and the French-speaking majority in Canada did not have a say in the shaping of this policy. This power lay exclusively with the colonial power. The British government's objective was to offset the large French-speaking population by encouraging large-scale immigration from the British Isles and outlawing immigration from France. In the case of the Acadians, the British, beginning in 1755, deported many of these people to England and other English colonies and many others fled to Québec or found their way to the Louisiana Territory. At the same time, the British sought to resettle the area with British subjects. Colonial-style immigration, therefore, not only generated tensions between the existing population and the newcomers, but it also imposed a way of life and a set of institutions on the existing population and often marginalized these people.

After 1867, the Dominion of Canada remained closely tied to the British Empire and its immigration policies continued to reflect the 'imperialist' attitude displayed in London, namely that Europeans but especially the British were superior to non-European peoples. This attitude, common to ancient as well as modern empires, flows from the ease with which Britain and other European nations carved out colonies in the rest of the world. Their power and capacity to form colonies stemmed from their industrialized, market-oriented economy and their efficient military system. After the end of slavery, local or imported non-white workers provided the labour force for the plantations and other economic endeavours in European colonies. Rudyard Kipling's concept of the White Man's Burden provides a more benevolent interpretation of this racist attitude: that the white man had the burden of extending his superior civilization to the rest of the world. It was an imperialist variant of the conservative, class-based notion of noblesse oblige. In any case, the building of the CPR line across the Cordillera provides a 'Canadian' variation of the use of non-white labour. In the sparsely populated western half of Canada, the CPR faced a severe shortage of workers. The solution was to import Chinese labourers, who worked for half the wages of white labourers and often were assigned the more dangerous construction work.

While the existing population viewed newcomers by asking how the new immigrants would benefit them and their society, the colonial power took the opposite position, i.e., how will the existing population benefit us? The economic, military, and social relationship between New France and the Huron Confederacy illustrates this point. In 1609, the Huron chiefs met with Samuel de Champlain to discuss both trade and a military alliance. The Huron had three objectives: (1) to gain access to European goods, including firearms, by supplying the French with beaver pelts; (2) to improve their material well-being with the trade goods and, in turn, trade these goods to more distant Indian tribes for profit; and (3) to strengthen their military position against their traditional enemies, the Iroquois, who were allied with the Dutch and later the English traders based in New York. The French had two objectives: (1) to secure a supply of furs; and (2) to convert the Huron to Christianity. At the height of the fur trade in the seventeenth century, New France greatly prospered and the Huron accounted for around half of the furs shipped to France (Dickason with McNab, 2009: 101). Trade was so important to the Huron tribes that when the French insisted that the Huron allow Jesuit missionaries to live among them as a condition for continued trade, the Huron reluctantly agreed. Unfortunately, the missionaries brought with them smallpox and other diseases that quickly swept through the Huron tribes, causing a sharp decline in their population.

Here, our focus is on the impact of immigration on the settling of Western Canada. The story begins with the purchase of Hudson's Bay lands by Ottawa, the reaction of the Métis in the Red River Settlement, the making of treaties, and then the subsequent settling of the Canadian Prairies by many people who were not of British ancestry. This historic period stretches from 1870 to 1914. During this time, although the face of British colonialism had changed from London to Ottawa, it had not softened. Immigrants and those being incorporated into the expanding Canada had to conform to Canadian society. The experiences of the original occupants of Western Canada—the Plains Indians and the Métis, and then of the Doukhobors, who were very clearly not British immigrants—ended badly. In all three cases, the pressure to conform to the majority society was both overt and covert; and in each case, the outcome pushed these three peoples to the margins of Western Canada and its society.

The First Clash: Red River Rebellion of 1869–70

With the transfer of the vast lands administered by the Hudson's Bay Company, Canada changed from a small territory to a truly continental country. While the boundary between Western Canada and the United States had been determined earlier, the survey of

Think About It

Despite the unfairness of two levels of payment for CPR workers, what do you think motivated these Chinese labourers to leave their homeland to work in the mountainous and dangerous terrain of British Columbia?

Ron Garnett/AirScapes.ca

Photo 3.5

The confluence of the Red and Assiniboine rivers is known as the Forks. Today, the Forks lies in the heart of Winnipeg. In times past, the strategic location of the Forks provided Plains Indians with ready access by canoe to the lands south of the forty-ninth parallel and to the vast western interior. In 1738, the French explorer La Vérendrye established Fort Rouge at the Forks. With the founding of the Red River Settlement in 1812, the Forks became its focal point.

lands for agricultural settlement took place in the 1880s. The survey system, based on a township and range model, stamped a rectangular-shaped grid on the cultural landscape, thus determining the shape and placement of farms, roads, and towns (Vignette 3.2). As Moffat (2002: 204) points out, this survey system 'enabled the division of western lands among the HBC, the Canadian Pacific Railway (CPR) and homesteaders, and set aside two sections in each township for the future of local education.' However, Ottawa failed to recognize the landholdings of the Métis and relegated Indians to reserves. Ottawa did not inform the residents of the Red River Settlement of its plans nor did the government recognize local landholdings. Events quickly spun out of control, resulting in the Red River Rebellion.

The Red River Rebellion pitted the existing population of the Red River Settlement against Ottawa, whose land survey and agricultural plans posed a deadly threat to the existing Métis settlement and its hunting economy. Even before the arrival of settlers, surveyors sent by Ottawa ignored the long-lot holdings of the Métis along the Red and Assiniboine rivers. In 1869, the Red River Colony was the only settled area of any size in the North-Western Territory, with a population of nearly 12,000 evenly divided between French- and English-speaking residents (Table 3.2). Most consisted of mixed-blood people, born of French and British fur traders and Indian mothers, who had settled in long lots along the banks of the two major rivers, and whose economy was based on the buffalo hunt and subsistence farming. By early 1869, news of the pending transfer of Hudson's Bay Company lands to Ottawa had reached the colony, and the arrival of land surveyors resulted in open hostility. When Canadian land surveyors began to survey lands occupied by the Métis, the Métis feared for their rights to those lands and even their place in the new society. Matters came to a boil when, in October 1869, Louis Riel put his foot on the surveyor's chain and told them to leave. Thus, the Red River Rebellion began, during which the Métis took control of Upper Fort Garry, the HBC headquarters, and William McDougall, who had been appointed lieutenant-governor of the HBC lands soon to be passed over to Canada, was turned back at the border in his attempt to claim Canadian sovereignty over the territory.

Two months later, the Métis under Riel formed a Provisional Government and soon began to negotiate with Canada over the terms of entry into Confederation. The three-man delegation sent to Ottawa by Riel's Provisional Government gained much of what they sought, including agreement to the establishment of a new province, but Orange Order elements from Ontario who had come to the Red River area were not pleased that the French-speaking Métis 'half-breeds' were in charge, and one man, Thomas Scott, who had been arrested by the Métis but persisted in being belligerent and unruly, was summarily executed after a brief trial. This inevitably led to further difficulties.

One advantage Riel had had in his negotiations with the Canadian government was 'remoteness'. Without rail connection to the settlement, Ottawa could not rush troops to quell the resistance, which, with the execution of Scott, seemed on the verge of full-scale warfare. Although a rail line reached St Paul in Minnesota, the US government refused to allow Canadian troops to cross the border. In April 1870, Macdonald authorized a military force of 1,000 troops—the Wolseley Expedition—to advance on Red River and assert Canada's sovereignty over the colony. The Canadian troops followed an old fur trade route and took four months to finally reach the Red River in August 1870. Fearing for their lives, Riel and his followers fled to the United States. On 15 July 1870, Manitoba became a tiny province of Canada with an area of about 2,600 square kilometres (1,000 square miles). The Métis had obtained most of their demands (the use of English and French languages within the government and a dual system of Protestant and Roman Catholic schools); at the same time, Prime Minister Macdonald had begun to ensure Canadian control over Western Canada.

Second Clash: Making Treaty

After Manitoba became part of the Dominion, Ottawa sought to expand its control into the empty Prairie lands. Making treaty with the Indians of this 'empty land' was essential before these potential farmlands were filled with homesteaders from Canada, the United States, and the Great Britain. From 1871 to 1877, seven treaties—the so-called numbered treaties 1 to 7—were negotiated to open the West to settlement.

The objective of Ottawa was to extinguish Indian rights to the land, as it had in Ontario with the **Robinson treaties** in 1850, and to promote the assimilation of Plains Indians into Canadian society. The formula was simple—cash, a small annual payment, and land for the exclusive use of Indians (now known as reserves). The assimilation goal soon enough took the form of Indian residential schools. But what were the goals of the various Indian tribes? While they did not speak with one voice, they were aware of the Robinson treaties, the Indian Wars in the United States, and the threat of agricultural settlement on their way of life. More importantly, their main source of food, the buffalo, was disappearing. Word from their cousins in the United States made them very aware of the impact of settlement and railways on the Indian way of life. Few options remained and their main goal was to survive as a people, but the path was not clear. However, treaty negotiations provided a small window of opportunity to improve the terms over the Robinson treaties. Of course, Indian tribes were acutely aware of earlier settlements, and by Treaty 3 they knew all the cards played by the federal negotiators and were able to use this information to wring additional concessions.

Think About It

Why do you think the Métis chose to negotiate with Ottawa rather than declaring their independence from Canada?

In this way, the Indians forced the federal government to consider issues far beyond the Robinson treaties. For instance, some Indian leaders hoped that agriculture might provide the basis for a new economy and they were able to have training in farming/ranching plus supplies and tools included in the treaties. Then, too, Indians leaders were able to include 'the medicine chest' in Treaty 6, which became the basis for subsequent free health care for First Nations peoples.

While both Canada and the Prairie Indians agreed to these seven treaties, the federal government and the First Nations saw treaties as necessary elements in achieving their very different goals. Ottawa, for instance, gained ownership of the land but it was not happy with the cost of the 'unanticipated concessions' granted to the Indian tribes by federal negotiators. Indians felt the fulfillment of their treaty rights, especially with regard to help in developing an agricultural base on reserve lands, was not forthcoming. Matters turned from bad to worse, culminating in the 1885 Northwest Rebellion.

The Third Clash: The Northwest Rebellion of 1885

While treaties had been signed, Indians faced desperate conditions, and those bands that were not docile in the face of drought and starvation found their meagre treaty provisions cut by federal officials. At the same time, the many Métis from Red River who had migrated north and west into present-day Saskatchewan in the years following the 1869–70 rebellion felt threatened once more by the advancing wave of settlers and by difficult conditions. A delegation went to Montana in 1884 and convinced Louis Riel, in exile as a schoolteacher and an American citizen, to return to Canada to lead their quest for their rights. Late in 1884, Riel sent a petition to Ottawa with various demands for all the inhabitants of the North-West—Indians, Métis, and whites—effectively asking that they be treated with the dignity deserving of loyal British subjects. Eventually, when no remotely supportive government response was forthcoming, Métis soldiers, with a few Indian

warriors, ambushed a NWMP contingent at Duck Lake on 26 March 1885, killing 12 men and losing six of their own. Big Bear, a Plains Cree chief, was seeking a peaceful solution to the plight of his people, but a few of his warriors, too, went on the warpath. On 2 April 1885 Cree warriors led by Wandering Spirit rode to Frog Lake to demand food. When the local Indian agent refused them, he was shot. The warriors then looted the settlement and left nine dead.

The Métis, under the leadership of Riel but led militarily by Gabriel Dumont, were prepared to fight the advancing Canadian army, which had arrived quickly from Ontario by means of the Canadian Pacific Railway. Attempts to unite with the Cree failed. Still, the Métis and a few Indians from nearby reserves were successful in surprising the Canadian troops, led by Major-General Frederick D. Middleton, at Fish Creek, but the larger and well-equipped Canadian army eventually wore down the smaller and less well-equipped Métis and Indian forces at Batoche. From Ottawa's perspective, the Northwest Rebellion was crushed. Louis Riel and eight Indian leaders were hung while Big Bear and Poundmaker were sent to prison. But the prairie uprising had an enduring effect on a nation, the Mètis people, and Ottawa relations with Québec.

The Flood of Newcomers

The settling of Western Canada represents one of the greatest world immigrations. By the late nineteenth century, this settlement pattern had placed a 'British/Canadian' stamp on the landscape and, with Indian treaties, relegated the earlier occupants to the margins. The emerging cultural landscape of Western Canada took three forms—the rectangular appearance of its rural landholdings, the orientation of villages and cities to the railways, and the symbols of ethnic diversity as expressed by farm buildings and churches. By 1895, Western Canada had a predominantly British population that had established its own land survey and ownership system, local governments and police to ensure 'law and order', and a variety of social institutions.

Theater of Operations - North West Rebellion - 1885

Canadian Pacific Railway Line	Modern Provincial Boundaries (after 1905)
Saskatchewan River System	Routes of the Three Main Military Columns
Territorial Boundaries - 1885	West - Gen. T. Bland Strange (Alberta Field Force) CENTER - Lt-Col. William Otter EAST - Gen. Frederick D. Middleton (Main Column)

Figure 3.11   Western Canada and the Northwest Rebellion of 1885.

The Canadian Pacific Railway played a key role in the Northwest Rebellion by transporting the Canadian forces quickly from Ontario to Qu'Appelle, Saskatchewan. In 1885, the political boundaries in Western Canada (except for Manitoba) still were part of the North-West Territories. The provinces of Alberta and Saskatchewan were formed in 1905 while Manitoba reached its current size in 1912.

Source: Based on Rattlesnake Jack's Old West Clip Art Parlour, Font Gallery, North West Rebellion Emporium and Rocky Mountain Ranger Patrol, at: <members.memlane.com/gromboug/P5NWReb.htm>.

As evident from the experience of the Métis pattern of landownership, elements of the landscape that did not conform ran into serious problems. During a 10-year span from 1870 to 1880, the Métis lost their majority due to an influx of immigrants from Ontario, many of whom either belonged to or supported the views of the Orange Order, a Protestant fraternal organization with strong anti-Catholic beliefs. Some newcomers saw no place for the Métis and Indians in the emerging society, thus creating tensions between the existing population and the newcomers. From 1871 to 1881, Manitoba's population increased from 25,228 to 62,260, with most immigrants coming from Ontario, the British Isles, and the United States (Table 3.8). At this time, those of British ancestry formed 54 per cent of the population, other Europeans made up 17 per cent, Métis, 17 per cent, and Indians, 11 per cent (Canada, 1882: Table III). The newly formed English-speaking majority

focused their attention on the dual school system. By 1891, Manitoba's population exceeded 150,000 (Table 3.8). In an example of the tyranny of the majority, the English-speaking population argued that with so few French-speaking students, funding for the Catholic school was not warranted. In 1890, the government of Manitoba abolished public funding for Catholic schools. This decision took on national significance by becoming a critical issue between Québec and the rest of the country (see below).

What caused this initial influx of settlers? One reason was that Ontario no longer had a surplus of agricultural land and sons of farmers looked to the unsettled lands on the Great Plains of the United States and to Manitoba. A second reason was that the promise of a railway would make farming in Manitoba more viable. With completion of the CPR line from Fort William on Lake Superior to Selkirk, Manitoba, in 1882, grain

Table 3.8	Population in Western Canada by Provinces, 1871–1911		
Year	Manitoba	Saskatchewan	Alberta
1871	25,228		
1881	62,260	21,652	9,875
1891	152,506	40,206	26,593
1901	255,211	91,279	73,022
1911	461,394	492,432	374,295

Note: The boundaries of Manitoba did not reach their present limits until 1912, and Saskatchewan and Alberta became provinces in 1905. Their populations for 1881, 1891, and 1901 have been calculated from the censuses of Canada for 1881 and 1891.
Sources: Canada (1882: 93–6; 1892: 112–13); Statistics Canada (2003).

could be transported by rail and ship to eastern Canada and Great Britain rather than by the more circuitous steamship route to St Paul and then by rail to New York. Wheat farming in Manitoba had become a profitable business because of advances in agricultural machinery and farming techniques, and rising prices for grain. Equally important, new strains of wheat, first Red Fife and then Marquis, both of which ripened more quickly than previous varieties, lessened the danger of crop loss due to frost. Marquis wheat, which matured seven days earlier than Red Fife, allowed wheat cultivation to take place in the parkland belt of Saskatchewan and Alberta where the frost-free period was shorter than in southern Manitoba.

The Doukhobors

By the end of the nineteenth century, Canada's West still needed more settlers. Clifford Sifton of Manitoba, the federal minister responsible for finding settlers, soon realized that he had to expand his recruitment area beyond Great Britain and into Europe and Russia. Even though this ran against a solely British-populated Western Canada, Sifton (1922) took a pragamatic approach, which he summed up in later years: 'I think a stalwart peasant in a sheep-skin coat, born on the soil, whose forefathers have been farmers for ten generations, with a stout wife and a half-dozen children is good quality.' Under Sifton, the pattern of immigration took a sharp turn from the main sources of immigrants to the West, namely Canada, the British Isles, and the

United States. Within two decades of entering Confederation, Manitoba's population had increased by just over 600 per cent (Table 3.8). Most were of British stock, but substantial numbers of Mennonites and Icelanders had also come to Manitoba. At the same time, few settlers had reached Saskatchewan and Alberta, though the Métis had relocated in Saskatchewan, primarily around the settlement of Batoche on the South Saskatchewan River just north of Saskatoon. In the next decade, the volume of immigrants from Central Europe, Scandinavia, and Russia increased substantially. As peasants, they were prepared for the harsh physical conditions associated with breaking the virgin prairie land and were willing to deal with the psychological stress of living on isolated farmsteads in a foreign country where their native tongue was not accepted. As their numbers grew, the anglophone majority became concerned about these newcomers and their possible effect on the existing social structure. The demographic impact of the non-British migration to Western Canada is shown in the 1916 census (Table 3.9).

This wave of Central Europeans had tremendous implications for Western Canada. While most newcomers assimilated into the larger society, a few did not. Often these ethnic groups settled in one area where they were somewhat insulated from the larger society and where they attempted to maintain their traditional customs, language, and religion. The federal government, by providing land reserves for ethnic groups such as the Mennonites and Doukhobors, reinforced this tendency.

Table 3.9   Population of Western Canada by Ethnic Group, 1916

Ethnic Group	W. Canada Population	% of W. Canada Population	% of Manitoba Population	% of Sask. Population	% of Alberta Population
British	971,830	57.2	57.7	54.5	60.2
German	136,968	8.1	4.7	11.9	6.8
Austro-Hungarian	136,250	8.0	8.2	9.1	6.4
French	89,987	5.3	6.1	4.9	4.9
Russian	63,735	3.7	2.9	4.5	3.8
Norwegian	47,449	2.8	0.6	4.2	3.4
Indian (Aboriginal)	39,147	2.3	2.5	1.7	2.9
Ukrainian	39,103	2.3	4.1	0.7	1.8
Swedish	37,220	2.2	1.4	2.5	2.7
Polish	27,790	1.6	3.0	1.0	0.9
Jewish	23,381	1.4	3.0	0.6	0.6
Dutch	22,353	1.3	1.3	1.4	1.3
Icelandic	15,800	0.9	2.2	0.5	0.1
Danish	9,556	0.6	0.3	0.5	0.9
Belgian	9,084	0.5	0.8	0.4	0.4
Italian	5,348	0.3	0.3	1.0	0.9
Other	26,219	1.5	0.9	1.5	2.3
Total	1,701,220	100	100	100	100

Source: Census of Prairie Provinces, 1916, Table 7. Data adapted from Statistics Canada, at: <www12.statcan.ca/English/census01/products/analytic/companion/age/provpymds.cfm>.

While they were successful farmers, the cultural differences between the more conservative Doukhobors and Canadian society were too great for the majority society to accept. Some Doukhobors were able to integrate into local society, but the Community Doukhobors simply were unable to adapt. They remained faithful to their religious beliefs that emphasized communal living and minimal dealings with the state, which included not informing the state about births and deaths of their members.

These people were deeply religious Russian peasants who rejected both the practices and beliefs of the Russian Orthodox Church and the secular authority. They were communalists and pacifists, and by refusing to serve in the army of czarist Russia they were considered by the state as outlaws and therefore needed to be punished. Consequently, they were persecuted by both the Church and the state. Seeking to be left alone, the Doukhobors sought a place far from the forces of authority where they could practise their religion and communal lifestyle. In choosing to settle in Canada they were granted blocks of land and exemption from military service.

In 1899, the Doukhobors—7,500 in total—arrived in Canada and took possession of lands in Saskatchewan, and they soon built 57 villages. These lands were selected by representatives of the Doukhobors before the peasant settlers arrived. Through negotiations with the Canadian government, they had obtained four large blocks of land totalling 750,000 acres. The four colonies were located just west of Swan River, Manitoba (North Colony), and at Prince Albert (Saskatchewan Colony) and Yorkton, Saskatchewan (South Colony and Good Spirit Lake Annex). Ottawa had allowed them to receive blocks of land rather than individual homesteads, thus facilitating their communal society. Farming was not only an economic activity, but it was also central to their religious beliefs, which emphasized the value of a simple, communal life. For example, Doukhobors shared in the returns from farming, and no one person owned the land or the tools. In a land of individual landholdings and the pursuit of profit, the Doukhobors were seen as 'out of step' with the surrounding community.

As the land around their villages and allotments was settled by other newcomers, the Doukhobors came into closer contact with their neighbours, who often coveted the uncultivated areas found on the edges of the land reserves of the Doukhobors. Two factors were at play in their not cultivating all of their land. First, the grant of land reserves was quite generous and went far beyond their immediate needs. Second, their compact communal settlements and intensive land cultivation near their villages resulted in a particular land-use pattern that left lands far from the village underutilized. In sharp contrast, homesteaders were required to cultivate land on their quarter-sections and to erect farmhouses, thus creating the checkerboard pattern of rural settlement. Neighbours also were puzzled by and perplexed with their communal way of life that separated them from the rest of the population. As public resentment increased, the federal government took action. In 1905, Frank Oliver succeeded Clifford Sifton as Minister of the Interior. Oliver decided to enforce the Dominion Lands Act, so when the Doukhobors refused to swear an oath of allegiance to the Queen, Oliver had his excuse.

Failure to take such an oath had two implications. First, it suggested that these people were disloyal to the Queen. Second, it meant that the Doukhobors could not obtain title to their homestead lands. Under this pretext, Oliver used the Dominion Lands Act to cancel their right to land. A hard core of Doukhobors remained committed to the communal way of life, most of whom eventually moved to British Columbia; others abandoned the village life and took title to homesteads. The villages gradually lost members and lands. The South Colony just north of Yorkton, Saskatchewan, was the last holdout, but by 1918 it ceased to exist on Crown land. It persisted in a much reduced area on purchased land until 1938, as did other communal settlements established in the Kylemore and Kelvington areas.

One explanation for the ultimate failure of the Doukhobor experiment was that Canada's model of individual settlement was simply too rigid to accept a communal one. Yet with ownership of land, the Doukhobors could establish agricultural villages. Another explanation focused on their unwillingness to swear allegiance to the Queen. In reality, the Doukhobors were successful farmers and their villages were working. However, they could not fit into the existing culture, which required conformity to the laws—and the informal values and lifeways—of the country. Primarily for that reason, the Community Doukhobors were unable to find a place in Western Canada. They represent a classic example of a people being too different—too 'other'—from the majority to be allowed a comfortable space within the cultural landscape. Ironically, the village model of settlement was perhaps the most effective way of settling the Prairies in the late nineteenth and early twentieth centuries. Professor Carl Tracie (1996: xii) puts it this way:

> At the very time when the individual homesteader was struggling with the very real problems of isolation and loneliness, the Doukhobor settlements, whose compact form allayed these problems, were being dismantled by forces which could not accommodate the communal aspects of the group. Also, although the initial government concern was the survival of the Doukhobors, their very prosperity, based as it was on communal effort, may have worked against them since it illustrated the success of a system diametrically opposed to the individualistic system dictated by government policy and assumed by mainstream society.

The French/English Faultline

Although the ancestors of Aboriginal peoples were the first to occupy North America, the colonization of the continent pushed them into the margins, leaving the two European powers—the French and the British—to place

Think About It

Does the open discrimination against Aboriginal peoples and non-British newcomers that was common in the nineteenth century and much of the twentieth century reflect Saul's concept of a 'hard' country? If so, why?

their mark on the land. Following the final British military victory, the Treaty of Paris (1763) confirmed British hegemony over a French-Canadian majority and its mastery over lands of New France. This historic fact underscores the dominant position of the British and their impact on later Canadian institutions and governments. Also, the British way of life was established across Canada, though rural Québec retained its French character, the seigneurial agricultural system, and Roman Catholic religion. Differences between these two cultures have come to represent a major faultline in Canadian society. However, the union of Lower Canada and Upper Canada (Ontario) in 1841 meant that French and English had to work together in a single parliament, which made each dependent on the other and was instrumental in the 1867 Confederation. This interaction of French and English has done much to shape the cultural and political nature of Canada. Over the years, they have accomplished much together. Nevertheless, significant differences between the two communities exist, and from time to time these differences flare into serious misunderstandings. Without a doubt, Canadian unity depends on the continuation of this relationship and the need for compromise, which has become a feature of modern Canadian political life and is a basic aspect of Canadian tolerance, between the two official language groups and to newcomers.

The serious nature of the French/English rift has profound geopolitical consequences for Canada. It is therefore crucial that we understand the origins and nature of this faultline. A well-known Canadian political columnist, Jeffrey Simpson, wrote:

> We can also hope that, in the 1980s, Canadians gained a deeper understanding of the faultlines running through their society, and that they will avoid measures that widen them, thereby concentrating on making new arrangements and reforming old ones, so that what the rest of the world rightly believes to be a successful experiment in managing diversity will endure and prosper. (Simpson, 1993: 368)

The beginning of formal French–English relations in Canada stretches back to the British Conquest of the French on the Plains of Abraham in 1759, an event that remains a dark page in French-Canadian history. In 1760, the French Canadians watched the remnants of the French army and the French elite board ships to return to France. The French Canadians had no thought of leaving as they were born in the New World, but what would happen to them under British military rule? Would they, like their Acadian brethren, be deported to other British colonies? Britain did not need to take such drastic action by this time—the Acadian deportations had occurred in the 1750s, prior to the Plains of Abraham and the Treaty of Paris—as there was no French threat to Britain's North American possessions. In the Treaty of Paris, France ceded New France to Britain, which placed the French-Canadian majority under the British monarchy. While the English lived in cities in Québec and dominated the Québec economy, French Canadians lived mostly in rural areas where they successfully maintained their culture within a British North America, an achievement made possible because of two factors. First, the French Canadians were a large homogeneous population that occupied a contiguous geographic area. Second, Britain forged a close relationship with the former elites of New France (the Roman Catholic clergy and the landed gentry) to ensure the French Canadians' loyalty because Britain wanted to secure its northern colony against its restless American colonies to the south. This relationship between Britain and the French Canadians would be strengthened with the Québec Act of 1774.

The Québec Act, 1774

With the Québec Act of 1774, the unique nature and separateness of Québec were recognized, thus affirming its place in British North America. This Act is sometimes described as the Magna Carta for French Canada.[6] Its main provisions ensured the continuation of the aristocratic seigneurial landholding system and guaranteed religious freedom for the colony's Roman Catholic majority and, by implication, their right to retain their native language.[7] This gave the most powerful people in New France a good reason to support the new rulers. The Roman Catholic Church was placed in a particularly strong position. Not only was the Church allowed to collect tithes and dues but its role as the protector of French culture went unchallenged. Therefore, the clergy played an extremely important role in directing and maintaining a rural French-Canadian society, a role further enhanced by the Church's control of the education system. The habitants (farmers) were at the bottom of French-Canadian society's hierarchy. They formed the vast majority of the population and continued to cultivate their land on seigneuries, paying their dues to their lord (seigneur) and faithfully obeying the local priest and bishop. The British granted another important concession, namely, that civil suits would be tried under French law. Criminal cases, however, fell under English law.

The seigneurial system formed the basis of rural life in New France and, later, in Québec. In 1774, there were about 200 seigneuries in the St Lawrence Lowland. This type of land settlement left its mark on the landscape (the long, narrow landholdings and the vast estate of the seigneur) and on the mentality of rural French Canadians (close family ties, a strong sense of togetherness with neighbouring rural families, and staunch support for the Church). A habitant's landholding, though small, was the key to his family's prosperity, and by bequeathing the farm to his eldest son the habitant ensured the continuation of this rural way of life. In 1854, the habitant was allowed to purchase his small plot of land from his seigneur, but the last vestiges of this seigneurial system did not disappear until a century later. Even today, the landscape along the St Lawrence shows many signs of this type of landholding.

While the heart of this new British territory was the settled land of the St Lawrence Lowland, its full geographic extent was immense. Essentially, the Québec Act of 1774 recognized the geographic area of former French territories in North America. Québec's territory in 1774 was extended from the Labrador coast to the St Lawrence Lowland and beyond to the sparsely settled Great Lakes Lowland and the Indian lands of the Ohio Basin.

The Loyalists

The American War of Independence changed the political landscape of North America. Within the newly formed United States, a number of Americans, known as the **Loyalists**, remained loyal to Britain. Like the French-speaking people in North America, most of these Loyalists were born and raised in the New World. For them, North America was their homeland. During the revolution, they had sided with the British. They were hated by the American revolutionaries and lost their homes and property. As they were not welcome in the new republic, many Loyalists resettled in the remaining British colonies in North America, where Britain offered them land. The majority (about 40,000 Loyalists) settled in the Maritimes, particularly in Nova Scotia. About 5,000 relocated in the forested Appalachian Uplands of the Eastern Townships of Québec. A few thousand, including Indians led by Joseph Brant, took up land in the Great Lakes Lowland in present-day Ontario. The Six Nations of the Iroquois Confederacy, who had been loyal to the British, settled along the Grand River in southwestern Ontario on a vast tract, known as the Haldimand Grant, that was given to them in 1784 by the lieutenant-governor of Québec, Lord Haldimand.

In 2006, the Six Nations became tangled in a nasty 40-acre land development dispute. See Vignette 5.7, 'Historical Timeline of the Caledonia Dispute', page 194, and Figure 5.6, 'The Haldimand Tract', page 193, for more details about the Haldimand Grant.

Within a few decades, the English-speaking settlers in the Great Lakes region grew in number. Soon they sought to control their own affairs so they could have a more British government with British civil law, British institutions, and an elected assembly. In the Constitutional Act of 1791, Québec was split into Upper and Lower Canada.

The Constitutional Act, 1791

The Constitutional Act of 1791 represented an attempt by the British Parliament to satisfy the political needs of the French- and English-speaking inhabitants of Québec. These were the main provisions of the Act: (1) the British colony of Québec was divided into the provinces of Upper and Lower Canada, with the Ottawa River as the dividing line, except for two seigneuries located just southwest of the Ottawa River; and (2) each province was governed by a British lieutenant-governor appointed by Britain. From time to time, the lieutenant-governor would consult with his executive council and acknowledge legislation passed by an elected legislative assembly.

In 1791, Lower Canada had a much larger population than Upper Canada. At that time, about 15,000 colonists lived in Upper Canada, most of whom were of Loyalist extraction, plus about 10,000 Indians, some of whom had fled northward after the American Revolution. Lower Canada's population consisted of about 140,000 French Canadians, 10,000 English Canadians, and perhaps as many as 5,000 Indians.

Following the Constitutional Act, Upper and Lower Canada each had an elected assembly, but the real power remained in the hands of the British-appointed lieutenant-governors. In Lower Canada the lieutenant-governor had the support of the Roman Catholic Church, the seigneurs, and the Château Clique. The **Château Clique**, a group consisting mostly of anglophone merchants, controlled most business enterprises and, as they were favoured by the lieutenant-governor, wielded much political power. In Upper Canada the **Family Compact**—a small group of officials who dominated senior bureaucratic positions, the executive and legislative councils, and the judiciary—held similar positions in commercial and political circles. While these two elite groups promoted their own political and financial well-being, the rest of the population grew more and more dissatisfied with blatant political abuses, which included patronage and unpopular policies that favoured these two groups. Attempts to obtain political reforms leading to a more democratic political system failed. Under these circumstances, social unrest was widespread.

In 1837 rebellions broke out. In Lower Canada Louis-Joseph Papineau led the rebels, while William Lyon Mackenzie headed the rebels in Upper Canada. Both uprisings were ruthlessly suppressed by British troops. The goal of both insurrections was to take control by wresting power from the colonial governments in Toronto and Québec and putting government in the hands of the popularly elected assemblies. In Lower Canada the rebellion was also an expression of Anglo-French animosity. While both uprisings were unsuccessful, the British government nevertheless sent Lord Durham to Canada as Governor General to investigate the rebels' grievances. He recommended a form of responsible government and the union of the two Canadas. Once the two colonies were unified, the next step, according to Durham, would be the assimilation of the French Canadians into British culture. When Durham left in 1838, a second rebellion broke out in Lower Canada, but it was as unsuccessful as the first.

The Act of Union, 1841

In response to Durham's report, in 1841 the two largest colonies in British North America, Upper and Lower Canada, were united into the Province of Canada. This Act of Union gave substance to the geographic and political realities of British North America. The geographic reality was that a large French-speaking population existed in Lower Canada, while an English-speaking population was concentrated in Upper Canada. The political reality was twofold. Both groups had to work together to accomplish their political goals and neither group could achieve all its goals without some form of compromise. When the two cultures were forced to work together in a single legislative assembly, a new beginning to the French/English fault-line surfaced.

The new governor, Sir Charles Bagot, was appointed by and reported to the Colonial Office in London. The governor had the authority to appoint members to a Legislative Council and an Executive Council. The only representative body was a Legislative Assembly. Even though Lower Canada (after union known as Canada East) was somewhat larger in economic strength and population size (670,000) than Upper Canada (Canada West, with a population of 480,000), each elected 42 members to the Legislative Assembly. The vast majority of inhabitants in Canada East and Canada West lived in the rural countryside. For example, in 1841, the principal towns of Montréal and Toronto had populations of about 40,000 and 15,000, respectively.

Demographic Shifts

By the time of the Constitutional Act in 1791, the balance of French- and English-speaking inhabitants of British North America had begun to tilt in favour of the English. This demographic shift began after the American Revolution when thousands of Loyalists from the former American colonies relocated in British North America. In 1791, the European population of British North America was about 225,000 (mostly French Canadians). Some 162,000 (72 per cent) lived along the St Lawrence River in what is now the province of Québec. Perhaps as many as 50,000 (22 per cent) lived in Atlantic Canada. The remaining 15,000 (6 per cent) were scattered along the north shores of Lake Erie and Lake Ontario in what is now part of Ontario. By this time, the Aboriginal population of Upper Canada, Lower Canada, and Atlantic Canada had declined substantially to about 25,000.

Aboriginal population is discussed in Vignette 4.6, 'Early Estimates of Aboriginal Population', page 161. Also see Figure 4.8, 'Aboriginal population by ancestry, 1901–2006', page 159.

Within 50 years, not only had this geographic pattern changed but the balance of demographic power had shifted. While their numbers increased due to high fertility rates, French Canadians were no longer the majority in British North America because of the flood of British immigrants. In 1841, British North America had about 1.5 million inhabitants, 45 per cent of them located in Lower Canada, about 33 per cent in Upper Canada, and 12 per cent in the Atlantic colonies. Over the next 30 years, the balance continued to swing in favour of English-speaking regions, thanks to massive immigration from the British Isles. By 1871, Québec's population was only 34 per cent of Canada's population (Table 3.10).

As the country expanded its boundaries and more land was settled, Canada's English-speaking population grew, while Québec's French-speaking population diminished in proportion. Manitoba joined Confederation in 1870 with a population of almost 12,000, which was made up largely of French- and English-speaking Métis. In 1871, British Columbia, with an estimated population of 28,000 British subjects, joined Canada. Beyond these provinces, Indian and Inuit peoples inhabited the land. The total Aboriginal population of all territory that would eventually become Canada was about 100,000 at the time

Colony/Province	1851	1861	1871
Ontario	41.1	45.2	46.5
Québec	38.5	36.0	34.2
Nova Scotia	12.0	10.7	11.1
New Brunswick	8.4	8.1	8.2
Manitoba			< 0.1
British Columbia			< 0.8
Total per cent	100.0	100.0	100.0

Table 3.10   Population by Colony or Province, 1851–71 (percentages)

Source: McVey and Kalbach (1995: 38). © 1995 Nelson Education Ltd. Reproduced by permission.

of Confederation. In the subsequent decades, these new lands would be settled by Canadians, Europeans, and Americans. With few exceptions, English became the adopted language of these settlers. For a while, Manitoba was an exception, but when the English-speaking majority gained control of the government and the public institutions, the Métis found it difficult to maintain their culture and language.

Strained Relations

During these formative years, several events seriously strained relations between the Dominion's two founding peoples:

- the Red River Rebellion, 1869–70;
- the Northwest Rebellion, 1885;
- the Québec Jesuits' Estates Act, 1888;
- the Manitoba Schools Question, 1890.

The Red River Rebellion, 1869–70

The Métis rebellion, led by Louis Riel, soon became a national issue, reopening differences between English, Protestant Ontario and French, Roman Catholic Québec.[8] Québec considered Riel a French-Canadian hero who was defending the Métis, a people of mixed blood who spoke French and followed the Catholic religion. Protestant Ontario, on the other hand, considered Riel a traitor and a murderer. For Canada, the larger issue was the place of French Canadians in the West. A compromise was achieved in the Manitoba Act of 1870. Accordingly, the District of Assiniboia became the province of Manitoba. Under this Act, land was set aside for the Métis, although a number of them sold their entitlements (scrip) to land allotments for a cheap price to incoming settlers and moved further west or sought to continue their former hunting lifestyle in Manitoba. The elected legislative assembly of Manitoba provided a balance between the two ethnic groups with 12 English and 12 French electoral districts. Equally important, Manitoba had two official languages (French and English) and two religious school systems (Catholic and Protestant) financed by public funds.

The background to the Red River Rebellion is presented earlier in 'The First Clash: Rebellion of 1869–70', page 107.

The Northwest Rebellion, 1885

During the 1870s, many Ontarians settled in Manitoba while some Métis sought a new home on the open prairie. Seeking to remain hunters, one group settled along the South Saskatchewan River where they established a Métis colony around Batoche, about 60 km northeast of present-day Saskatoon. Batoche became the new centre of the French-speaking Métis in Western Canada. As settlers spread into Saskatchewan, the Métis again feared for their future. In 1884, when a party of Métis went to Montana to plead with Louis Riel to return to Batoche and lead them again, Riel, convinced of his destiny, accepted this challenge. As we have seen, this uprising ended in failure for the Métis and their Indian allies. For Québec, the defeat of the Métis and the subsequent hanging of their leader not only represented a defeat for a French presence in the West but also widened the gulf between French and English Canadians. Riel's link to Québec and the Roman Catholic Church made him a powerful symbol of language, religious, and racial divisions for over 100 years. Indeed, to this day historians remain divided about Louis Riel's legacy, his place in the story of Canada, and even his sanity as the messianic leader of a doomed rebellion.

The background to the Northwest Rebellion is presented earlier in 'The Third Clash: The Northwest Rebellion of 1885', page 110.

The Québec Jesuits' Estates Act, 1888

In the nineteenth century, religious and linguistic intolerance was widespread. For example, Protestant extremists in Ontario were ready to pounce on any perceived injustice to their cause. The Jesuits' efforts to obtain financial compensation for lands that the British took from them in 1763 and later transferred to Lower Canada proved to be such a case.

The Jesuit estates, which were granted under the French regime and used for schools and missions, were appropriated by Britain after the Conquest and given to Lower Canada in 1831. In 1838, Catholic bishops petitioned unsuccessfully for the return of the Jesuit estates. After Confederation, the ownership of the estates passed to the Québec government, with which the Jesuits began negotiating in 1871 for financial compensation. However, the archbishop of Québec argued that the money should be divided among Catholic schools rather than given in its entirety to the Jesuits, who wanted to establish a university in Montréal that would compete with Québec's Université Laval. Québec Premier Honoré Mercier asked Pope Leo XIII to act as arbiter in the dispute among the Roman Catholic hierarchy. In 1888, Québec's Legislative Assembly passed the Jesuits' Estates Act, which determined the division of the financial compensation: $160,000 went to the Jesuits, $140,000 went to the Université Laval, and $100,000 went to selected Catholic dioceses.

Ontario's Orange Order, a Protestant fraternal society, opposed the settlement, arguing that the arbiter, Pope Leo XIII, was an intruder into Canadian affairs and that public funds should not be used to support Catholic schools. In March 1889, the House of Commons debated the motion to disallow the Québec Jesuits' Estates Act and eventually voted against this motion. Similar anglophone, Protestant, anti-Catholic sentiment surfaced in Manitoba regarding the Manitoba Schools Question.

The Manitoba Schools Question, 1890

The British North America Act of 1867 established English and French as legislative and judicial languages in federal and Québec institutions. The remaining three provinces (New Brunswick, Nova Scotia, and Ontario) had only English as the official language. The question of French language and religious rights in acquired western territories first arose in Manitoba.

The French/English issue became the focal point for the entry of the Red River Settlement (now Manitoba) into Confederation. Local inhabitants—mostly French-speaking, Roman Catholic Métis and the less numerous English-speaking Métis—were determined to have some influence over the terms that would include their community as part of Canada. One of their concerns was language rights, an issue that was ultimately resolved when a list of rights drafted by the provisional government became the basis of federal legislation. When the settlement and surrounding territory of Red River entered Confederation in 1870 as the province of Manitoba, it did so with the assurance that English- and French-language rights, as well as the right to be educated in Protestant or Roman Catholic schools, were protected by provincial legislation.

During the 1870s and 1880s, with the influx of a large number of Anglo-Protestant settlers from Ontario, the proportion of Anglo-Protestants in the Manitoba population increased and the proportion of French and Roman Catholic inhabitants decreased. This demographic change created a stronger Anglo-Protestant culture in Manitoba. In 1890, the provincial government ended public funding of Catholic schools. From Québec's perspective, this legislation shook the very foundations of Confederation. Sir Wilfrid Laurier became Prime Minister in 1896 and, in the following year, Laurier negotiated a compromise agreement with the government of Manitoba. The compromise allowed for the teaching of Catholic religion in a public school when there were sufficient Catholic students. Similarly, if there were sufficient French-speaking students, classes could be taught in French.

One Country, Two Visions

The greatest challenge to Canadian unity comes from the cultural divide that separates French- and English-speaking Canadians and their respective visions of the country. In the early years of Confederation, events such as those outlined above widened the French/English faultline. For French Canadians these events demonstrated the 'power' of the English-speaking majority and their unwillingness to accept a vision of Canada as a partnership between the two founding peoples. The root of each vision lies in the history of Canada.

One vision of Canada is based on the principle of two founding peoples. This vision originated in French-Canadian historical experiences and compromises that were necessary for the sharing of political power between the two partners. This vision began with the Conquest of New France, but its true foundation lies in the formation of the Province of Canada in 1841. From 1841 onward, the experience of working together resulted in a Canadian version of cultural dualism.

Henri Bourassa, an outstanding French-Canadian thinker (and Canadian nationalist) in the early twentieth century, was a strong advocate of cultural dualism. He wrote, 'My native land is all of Canada, a federation of separate races and autonomous provinces. The nation I wish to see grow up is the Canadian nation, made up of French Canadians and English Canadians' (quoted in Bumsted, 2007: 307). Bourassa argued that a 'double contract' existed within Confederation. Even today, Bourassa's 'double contract' is an essential element in the two founding peoples concept. He based the notion of a double contract on a liberal interpretation of section 93 of the BNA Act, which guarantees denominational schools. Bourassa expanded the interpretation of the religious rights to include cultural rights for French- and English-speaking Canadians. In more practical terms, Bourassa regarded Confederation as a moral contract that guaranteed French/English duality, the preservation of French-speaking Québec, and the protection of the language and religious rights of French-speaking Canadians in other provinces.

From a geopolitical perspective, Canada is a bicultural country. In one part the majority of Canadians speak English, and in another part French is the majority language. Thus, French culture predominates in Québec and has a strong position in New Brunswick. In addition to provincial control over culture, two other geopolitical factors ensure the dynamism of French in those provinces. One factor is the large size of Québec's population—the vitality of Québécois culture is one indication of its success. The second factor is the geographic concentration of French-speaking Canadians in Québec and adjacent parts of Ontario and New Brunswick. In New Brunswick the French-speaking residents, known as Acadians, constitute over one-third of the population.

Federal bilingualism policies instituted in the late 1960s and 1970s also helped rejuvenate francophone minorities. Before these policies, assimilation into the much larger English-speaking culture had seriously weakened the position of francophones in all the provinces, except Québec, and in the two territories. While the attraction of joining the dominant anglophone culture remains, financial support from Ottawa for French educational and cultural facilities in the English-speaking provinces has ensured a place for bilingualism in all provinces and territories.

The Royal Commission on Bilingualism and Biculturalism was an attempt to bridge the gap between English and French Canadians. This Commission, set up in 1963, examined the issue of cultural dualism, that is, an equal partnership between the two cultural groups. But by the 1960s, Canada's demographics revealed a third ethnic force and the concept of duality no longer reflected reality. English-speaking Canada had changed. English-speaking Canada had evolved from a predominantly British population to a more diverse one with several large minority groups who also spoke other languages besides English, especially German and Ukrainian. Ottawa, in searching for a compromise, established two policies, bilingualism (1969) and multiculturalism (1971).

In the second vision, Canada consists of 10 equal provinces—yet this, too, is complex. On the one hand, this vision represents the simple notion based on provincial powers granted under the British North America Act, which ensured that Canada consists of a union of equal provinces, all of which have the same powers of government. Nonetheless, by assigning provinces, including Québec, powers over education, language, and other cultural matters within their provincial jurisdictions, Québec's French culture was secure from political tampering by the anglophone majority in the rest of Canada. Confederation then provided a form of collective rights for French culture within Québec. Under Canada's federal system, the powers of government are shared between the federal government and 10 provincial governments. But are all provinces really equal? As noted earlier, population size, geographic extent, and financial strength vary considerably, which is reflected in the need for equalization payments.

The vision of 10 equal provinces may reflect an English-Canadian nationalism. For some time, English-speaking Canadians have been searching for their cultural identity. Before World War I, Canadians saw themselves as part of the British Empire. By the end of World War II, this perspective began to change. The maple leaf flag, adopted by Parliament in 1964, and 'O Canada', the new national anthem approved by Parliament in 1967 and officially adopted in 1980, were signs of this cultural change. While the Québécois culture was flourishing, thanks in part to generous provincial funding for the arts, English-speaking Canadians continued to lean heavily on American culture. Some looked with envy at the cultural accomplishments of the Québécois and wondered aloud if similar achievements in English-speaking Canada were possible. The answer was yes, providing the provincial governments offered similar financial support for the arts.

Compromise

Given the incompatibility of the two visions— two founding peoples versus 10 equal provinces—and the historical development of the country, Canadian politicians have had the unenviable task of trying to accommodate demands from different groups—especially French Canadians, new immigrants, and Aboriginal peoples—and from different regions without offending other groups or regions. As in the past, politicians have continued to struggle with this Canadian dilemma, but in reality there is no perfect solution, only compromise. With this object in mind the federal government has made many efforts in search of the elusive middle ground.[9] It seems the search for an acceptable compromise between the two opposing visions of Canada will never end, and perhaps that is a good thing because the process is more important than the end result. To understand the current struggle for compromise, it is important to understand the political, economic, and cultural developments that have taken place in Québec over the past five decades.

Resurgence of Québec Nationalism

After World War II, Québec broke with its past. A rise of Québec nationalism had begun much earlier but gained political momentum during the **Quiet Revolution** of the early 1960s. This development was the result of four major events. The most important was the resurgence of ethnic nationalism, that is, a pride in being a Québécois. The second was Québec's joining the urban/industrial world of North America and the subsequent expansion in the size of its industrial labour force and business class. The third was the removal of the old elite. This reform movement was profoundly anticlerical in its opposition to the entrenched role of the Church in Québec society, particularly the Church's control over education. In many ways, this reform was based on the aspirations of the working and middle classes in the new Québec economy. The fourth was the state's aggressive role in the province's affairs.

With the election of Jean Lesage's Liberal government in 1960, which held power until 1966, the province moved forcefully in a new direction. It created a more powerful civil service that allowed francophones access to middle and senior positions often denied them in the private sector of the Québec economy, which was controlled by English-speaking Quebecers and American companies. It nationalized the province's electric system, thereby creating the industrial giant known as Hydro-Québec, now a powerful symbol of Québec's revitalized economy and society. In turn, Hydro-Québec built a number of huge energy projects that demonstrated the province's industrial strength. By 1968, this Crown corporation had constructed one of the largest dams in the world on the Manicouagan River. Called Manic 5, this dam demonstrated Hydro-Québec's engineering and construction capabilities. To Quebecers, Hydro-Québec was a symbol of Québec's economic liberation from the years of suffocation associated with Maurice Duplessis and his Union Nationale government, which had been closely tied to big businesses owned by English-speaking Canadians and Americans. Clearly, Lesage's political goal of becoming 'maîtres chez nous' (masters in our own house) had materialized with the success of Hydro-Québec, thus sparking a growth in Québec nationalism. Québec's desire for more autonomy in its own affairs intensified with increased confidence. In short, a new society had arisen in Québec, a society that wanted to chart its destiny. Charles Taylor (1993: 4) summed up this new feeling as 'a French Canada which, after a couple of centuries of enforced incubation [under London and then Ottawa], was ready to take control once more of its history.' The political question Taylor raised is a simple one: Would this 'control' take place within the framework of Canada's political system or outside it?

Photo 3.6

French President Charles de Gaulle during his incendiary 'Vive le Québec libre' speech in Montreal, 24 July 1967.

Separatism

Separatism grew out of the Quiet Revolution. It is a form of ethnic nationalism that is popular with francophones but unpopular with anglophones and allophones (those whose first language is neither French nor English). In 1967, French President Charles de Gaulle visited Québec at the time of Expo '67 in Montréal and ignited the forces of French-Canadian nationalism with his now famous words, 'Vive le Québec. Vive le Québec libre' (photo 3.6). From that moment on, separatism gained support and took on a mainstream political form. By the time of the first referendum on independence in 1980, the separatists formed a substantial minority within Québec's population, with perhaps as many as 20 per cent dedicated separatists and another 20 per cent strongly dissatisfied with their place within Canada.

Separatism has had two distinct branches, represented by the Front de libération du Québec (FLQ) and the Parti Québécois (PQ). The FLQ was a small fringe group within the separatist movement. It sought political change through revolutionary means, including bombing, kidnapping, and murder. However, the vast majority of separatists sought change through democratic means. The PQ, first elected to government in November 1976, was committed to a democratic solution by means of a referendum followed by negotiations with the rest of Canada. Referendums, the process of referring a political question to the electorate for a direct decision by general vote, are notoriously tricky political instruments, but Québec Premier René Lévesque, who had been the architect of Hydro-Québec as a minister in the Lesage government and in the late 1960s left the Liberals to form the PQ, offered Quebecers what he thought was a clear choice—the unpopular status quo or a bold new beginning under sovereignty-association. **Sovereignty-association** meant political separation but a new economic association with the rest of Canada. Lévesque recognized that the integrated nature of Canada and its east–west economic axis made continued economic ties with Canada essential for the survival of Québec.

Prior to the May 1980 referendum, the PQ vigorously tackled challenging economic problems and critical cultural issues, all of which had three purposes:

- to accelerate the modernization processes that began with the Quiet Revolution;
- to promote the Québécois culture;
- to demonstrate that a PQ government could run the affairs of an independent state.

The provincial government was involved in the marketplace, often through Crown corporations and government assistance for francophone business operations. The provincial government also promoted Québécois culture in a variety of ways. Under the Liberal government of Robert Bourassa, the French language had been declared the sole official language in 1974, but the PQ government went much further with Bill 101 in 1977.[10] This bill made it necessary for most Québec children, regardless of background or preference, to be educated in French-language schools, and was a key measure in ensuring the supremacy of the French language in the province. Among its many goals, Bill 101 was designed to ensure that the children of new immigrants went to French schools and thus to guarantee that the French-speaking population of Québec would continue to grow.

Québec voters rejected the sovereignty-association referendum, with almost 60 per cent voting to remain in Canada, which suggests that just over half of the francophone voters stood with the 'Non' side, along with almost all the English-speaking residents. The rest of Canada responded with a collective sigh of relief, but separatism was far from dead. Several political events renewed separatist sentiment. One was the Constitution Act of 1982, which patriated the Constitution and gave Canadians the Charter of Rights and Freedoms. The Trudeau government accomplished this political feat at the cost of poisoning relations with Québec City by including the Charter of Rights and Freedoms in the Constitution. The Charter curtailed the power of the Québec government, and the Constitution was patriated without the approval of the Québec government. There were also the failed attempts of Brian Mulroney's Conservative government to achieve provincial unanimity for constitutional reform. The first attempt was the Meech Lake Accord, a package of constitutional revisions incorporating Québec's 'minimum' demands for political reform. This Accord was agreed to in principle by Ottawa and the 10 provinces in 1987, but two provinces, Manitoba and Newfoundland, failed to pass it within the required three years. Quebecers felt humiliated and rejected by the rest of Canada. In 1991, the Mulroney government attempted a second round of constitutional negotiations culminating in the Charlottetown Accord, which would give Québec distinct society status, the provinces more power and input on the selection of judges, and Aboriginal peoples the entrenched right to self-government, as well as provide for reform of the Supreme Court and the Senate to make these institutions more representative of provincial interests and power. In 1992, the Charlottetown Accord was roundly rejected in a national referendum and only narrowly approved in four provinces (but not Québec).

These last two political misadventures, as well as lingering bitterness from the patriation process, revived the spirits of the separatists, led by Jacques Parizeau and Lucien Bouchard. Parizeau's party, the PQ, returned to power in 1994, promising a referendum on sovereignty. While the referendum question referred to a new partnership with the rest of Canada, Parizeau believed that such an arrangement was impossible and saw only one solution—an independent Québec. In the previous year, the Bloc Québécois (a new federal party representing Québec interests), led by Lucien Bouchard (a former cabinet minister in the Mulroney government), formed the official opposition in the House of Commons. The Québec public expressed their dissatisfaction with Ottawa by rejecting traditional political parties. This left Québec federalists in a vulnerable position where they grew steadily weaker and more disorganized. The federal Liberals could offer little help. In fact, Jean Chrétien, himself a Quebecer, was extremely disliked by many in Québec for a variety of reasons, including his role in the patriation of the Constitution in 1982. He had become a 'tête de turc' (a scapegoat for federal policies), a symbol of those Québecers who, as federal ministers, put Québec 'in its place'.

For historical background on the French/English faultline in Québec, see Chapter 6, 'British Colony, 1760–1867', page 238.

The results of the 1995 referendum vote in Québec were extremely close. 'No—by a Whisker!' screamed the headline of the *Globe and Mail* on the morning after the referendum of 30 October 1995. Québec came within 40,000 votes of approving the separatist dream of becoming an independent state (Vignette 3.10).

Moving Forward

The 1995 referendum was a low point in French/English relations, and its after-effects were many and varied. English Canada, dazed by the outcome, attempted to respond. Ottawa reacted almost immediately after the October referendum by passing a unilateral declaration that recognized Québec as a distinct society. The leader of the separatist forces, Jacques Parizeau, had shocked all Quebecers on the night of the referendum when he blamed the 'Oui' side's loss on 'money and the ethnic vote'. In an electrifying moment, the dark side of ethnic nationalism had been revealed. Other centrifugal forces were released, too, including the partitionists, who argued that 'if Canada is divisible, so is Québec.' The Cree in northern Québec threatened secession, and partitionists pressured dozens of municipalities around Montréal and Hull (since renamed Gatineau) to declare their allegiance to Canada.

By 1996, the federal government had decided to take a hard line with Québec, which included having the Supreme Court of Canada determine the conditions of separation. In the same year, the premiers attempted to address the unity issue. In a much more conciliatory manner, they announced the Calgary Declaration: 'the unique character of Québec society with its French-speaking majority, its culture and its tradition of civil law is fundamental to the well-being of Canada.' In the typical fashion of Canadian provincial leaders, the premiers

added to their Declaration that 'any power conferred to one province in the future must be available to all.' This Declaration was the third attempt at reconciliation with Québec since the patriation of the Constitution in 1982.[11] The next step was for each provincial government to pass the appropriate legislation, giving this declaration legal status. By July 1998, all provinces (except Québec) and territories had passed this resolution in their legislatures.

Changes are taking place in Québec, too. Separatism, while not gone, has lost its spark for the time being. Then, too, the threat of the English is more a thing of the past. Quebecers, confident in their language and culture, are more comfortable and secure than ever before. Equally important, Ottawa is more comfortable with the idea of Québécois being recognized as a 'distinct cultural group' or nation within Canada. In November 2006, the House of Commons overwhelmingly passed a motion by Prime Minister Harper that recognized Québécois as a nation within Canada.

Vignette 3.10   The Results of the 30 October 1995 Referendum

The Question: 'Do you agree that Québec should become sovereign, after having made a formal offer to Canada for a new Economic and Political Partnership within the scope of the Bill respecting the future of Québec and of the agreement signed on June 12, 1995?'

The Answer (at 10:30 p.m. Eastern Time, 21,907 of 22,427 polls):

	Number	Per cent
No	2,294,162	49.5
Yes	2,254,496	48.7
Rejected	83,340	1.8
Total	4,631,998	100.0

Source: *Globe and Mail* (1995).

Think About It

Why won't separatism go away?

SUMMARY

History and geography explain the nature and complexity of contemporary Canada. Canada is both a young and an old country. Complexities are reflected in its four faultlines. The newcomer/old-timer faultline hinges on the 'accommodation' issue. The centralist/decentralist argument has taken a twist with Ontario now declared a 'have-not' province. Aboriginal peoples are settling their outstanding issues with the Crown—the land-claim settlement process continues to function and the deep sores caused by Indian residential schools, one can hope, have begun to heal. History teaches Canadians that differences will continue to emerge but compromises are necessary for national unity, regional harmony, and social justice.

As for national unity, the French/English accommodation goes to the heart of the nation. Geography compels Canadians to recognize that Québec represents a distinct region of Canada in which a different language and culture dominate. Canada's history reflects over 200 years of French/English interactions. Through conflicts and compromises, these innumerable interactions, both large and small, have shaped the essential components of the national character, namely, the capacity and willingness to find solutions to complex questions. Both Québec and the rest of Canada are different places and those changes have shaped the French/English faultline and the attempts at compromise in this relationship.

On the surface, reconciliation seems an impossible task, but political realities demand some form of compromise or at least a willingness to search for a solution. Perhaps Paul Villeneuve (1993: 104) was correct when he observed that 'for Canada, survival lies in the travelling toward an identity.' For most Canadians, this implies recognition of the deep divide between French and English Canada; between Aboriginal and non-Aboriginal peoples; between newcomers to Canada and native-born Canadians; and between Ottawa and the various regions of Canada. But Canadians also recognize that underlying each faultline is a particular vision of Canada. Canadians have learned that the political gains achieved by having the dominant English-Canadian society impose its will over the minority French-Canadian society, Aboriginal peoples, and newcomers are of short duration and eventually will weaken national unity. Perhaps the House of Commons recognition of Québécois as a nation within Canada, the healing associated with the Indian residential schools debacle, and the apology to Japanese Canadians for their internment reinforce the image of a tolerant and understanding Canada. Over the course of its short history as a nation of nations, Canada, as a collective entity, has learned 'tolerance' the hard way, and, it would seem, has chosen a 'soft' path into the twenty-first century.

CHALLENGE QUESTIONS

1. Why did the doctrine 'terra nullius' allow Europeans to consider North America 'unoccupied' and therefore open to European ownership and settlement?
2. How has Canada become a 'soft' nation and thus learned the benefit of compromise and tolerance?
3. Do you believe that political reality forces federal governments to favour Ontario and Québec over other parts of the country? Can you supply an example?
4. Why does Québec support the concept of Canada as 'two founding peoples' rather than the concept of Canada as '10 equal provinces'?
5. The rural-to-urban migration is a worldwide phenomenon. While a parallel between Indian peoples and rural folk moving to cities exists, why are more First Nations peoples likely to return to their reserves?

FURTHER READING

Harris, R. Cole, ed. 1987. *Historical Atlas of Canada, Volume I: From the Beginning to 1800.* Toronto: University of Toronto Press.

Gentilcore, R. Louis, ed. 1993. *Historical Atlas of Canada: Volume II: The Land Transformed 1800–1891.* Toronto: University of Toronto Press.

Kerr, Donald, and Deryck W. Holdsworth, eds. 1990. *Historical Atlas of Canada, Volume III: Addressing the Twentieth Century 1891–1961.* Toronto: University of Toronto Press.

The historical geography of Canada recalls past events. Maps play a large role in this rediscovery of Canada's past. In 1970, several geographers and historians explored the idea of preparing a major Canadian historical atlas focused on social and economic themes. These three volumes, which parallel the discussion in this chapter, are the successful outcome. The editors weave together the various historic strands that constitute Canada's historical geography and provide a rich legacy for Canadian scholars and students. Four more recent major events, however, are not covered—the rise of Aboriginal political power; the threat of separation of Québec from the rest of Canada; the influx of non-European immigrants; and the Canada–US Free Trade Agreement. These issues are discussed in Chapter 4.

CANADA'S HUMAN FACE

INTRODUCTION

Canada is home to over 33 million people. The country's population continues to grow, thanks in large measure to the flow of newcomers to its shores. Canada is now a pluralistic society and accommodation of newcomers' distinctive cultural traditions within Canadian society remains an ongoing challenge. Without a doubt, Canada is undergoing profound demographic, economic, and social changes. One such change is related to demographic and social changes that are largely driven by a declining rate of natural increase (except for Aboriginal peoples) and the cultural gifts brought to Canada by immigrants, especially those from non-European countries. Second, the process of urbanization continues to march forward with more and more people living large cities. Third, Canada's centre of population geography continues to shift westward, particularly to Alberta and British Columbia but even to formerly slow-growing Manitoba and Saskatchewan. Until the global economic crisis that began in late 2008, relatively few immigrants located outside of the three major cities of Toronto, Montréal, and Vancouver. Since then, a growing number have settled in Western Canada where the economy has remained relatively strong compared to that of Ontario. Similarly, more and more Canadians are attracted to Western Canada because of its more buoyant economy. Before 2008, the population shift was driven by economic factors, including the strong demand and record high prices for energy and mineral products, while Canada's manufacturing sector faced ever-increasing competition from low-wage foreign competitors, causing serious restructuring, relocation to offshore sites, and, in too many cases, loss of jobs, especially higher-paying jobs.

Then, in late 2008, the global economy collapsed, sending Canada's economy on a downward spiral. Resources faced lower prices while manufacturing was badly buffeted by a sharp drop in exports, especially to its number-one customer, the United States, and by the near collapse of its automobile industry. The impact of these economic difficulties has varied across Canada, with Western Canada and British Columbia dealing with lower prices for commodity exports while the manufacturing sector in Ontario and Québec has been in full retreat. Ontario was confronted with the bankruptcy of two of its major automobile manufacturers, Chrysler and General Motors. The federal and Ontario governments came to their rescue with financial support. By 2010, the worst appeared to have passed, economically, but a full recovery will require Canada's major market, the United States, to regain its economic strength. Until then, uncertainty is the operative word for the Canadian economy, especially its manufacturing sector.

CHAPTER OVERVIEW

Issues examined in this chapter are:

- The factors causing Canada's population to increase.
- The implications of immigration for Canadian society.
- Multiculturalism and the 'accommodation' of newcomers.
- The new reality of Canada's North American and global economy.
- The impact of the global economy on Canada's economy.
- The emergence of a north–south trade axis.
- The impact of demography on Canada's four faultlines.

Jean Talon Market, one of the most popular markets in Montréal, Québec. Photo: Chris Cheadle/All Canada Photos.

Canada's Population

Canada's population continues to grow and age, but also to move in new geographic, social, and religious directions. For the first time, the population exceeded 33 million in 2009. From 2001 to 2006, Canada's population increased by 0.5 per cent annually. Over the last decade, most growth has taken place in three geographic regions—Western Canada, Ontario, and British Columbia. Equally important, Canada's **demography** (Vignette 4.1) has a new look—ethnic composition and cultural diversity now reflect fresh elements in our society. Canada has a substantial number of adherents to Islam, Sikhism, and Hinduism, its larger cities are home to many so-called visible minorities, and Chinese forms the third-most commonly spoken language in Canada. Clearly, immigration plays a major role in these demographic and social changes. Other forces of demographic and social change include the rapid increase in Canada's Aboriginal population, the place of the French language within the nation, and Canada's economic role within the North American community. These factors have sparked an economic and social revolution that has yet to run its course. The transformation of the Canadian transportation system, especially its major trucking and railway routes, from an east–west orientation to a North American one marks a visible sign of the shift from a national economy to a continental one.[1]

Population Size

Since Confederation, Canada's population has increased by nearly 10 times. In 1871, Canada had 3.4 million people compared to 33.7 million in 2009. Three primary factors accounting for this growth over the last 143 years are natural increase, population gained from territorial expansion, and immigration. For over a hundred years, natural increase was the primary factor accounting for Canada's population growth. In the latter part of the twentieth century, birth rates dropped significantly, although these have recovered somewhat in the

---

Vignette 4.1 Demography, Census-taking, and Statistics Canada

Demography—the statistical study of human populations, including their size, age and gender structure, distribution, density, growth, and related socio-economic characteristics—is central to an understanding of regional geography. Statistics Canada is responsible for enumerating the population of Canada every five years. Between census years, Statistics Canada prepares estimates of Canada's population. Enumeration figures are usually lower than the computer-estimated figures because some Canadians are not enumerated in the census. This 'gap' is referred to as an **undercounting**. Statistics Canada prepares both annual and quarterly estimates (Statistics Canada, 2009c, 2009d). Table 4.1 provides an example of estimated population figures for Canada. Statistics Canada is also required to 'collect, compile, analyse, abstract and publish statistical information relating to the commercial, industrial, financial, social, economic and general activities and conditions of the people of Canada' (Statistics Canada, 2008e).

Table 4.1 Canada's Annual Population Estimates, 1 July 2001–2009

Year	Population
2001	31,019,020
2002	31,353,656
2003	31,639,670
2004	31,940,676
2005	32,245,209
2006	32,576,074
2007	32,927,372
2008	33,311,389
2009	33,739,859 (estimate revised in Dec. 2009)

Source: Statistics Canada, Population Estimates, various years. See vignette text for references.

twenty-first century. Immigration, which has always played a role, became the dominant factor in population growth. Of course, Canada gained population as its territory expanded. The last territorial expansion took place in 1949 when Newfoundland, with a population of 360,000 at that time, joined Confederation. With its current population, Canada can be considered a medium-sized country on the world stage—only one-quarter of the nations of the world have larger populations than Canada.

Population Density

As the second-largest country in the world by geographic area, Canada's population density is one of the lowest in the world. **Population density** is determined by dividing the number of people by the land area. Canada has a population density of 3.5 persons per square kilometre, which means the country has an extremely low population density (but not the lowest in the world). Australia, another country with much of its dry lands unsuitable for settlement, has only 2.6 persons per km^2. Mongolia, another large country with little land suitable for settlement, has an even lower density of 1.7. All other countries of the world have higher population densities. The United States, for example, has 30 persons per km^2, while Bangladesh is one of the most densely populated countries with a density of 1,000 people per km^2.

The explanation for these variations is simple. Land varies greatly in its capacity to support human settlement. Most land in Canada lies beyond the northern limits of agriculture, and land in the Territorial North has an exceptionally low capacity to support human life. The Territorial North's population density is only 0.03 people per km^2 (Table 4.2). In comparison, Ontario has a population density of 13.4 people per km^2.

Tony Tremblay/iStockphoto.com

Photo 4.1

Founded in 1642, Montréal, Québec, is one of Canada's oldest cities. Located on the St Lawrence River, Montréal is a transportation hub for international shipping. The city represents the largest francophone urban population in North America and the second largest in the world.

Michael Caven/iStockphoto.com

Photo 4.2

Toronto, Ontario, with a population approaching 5 million, is Canada's most populous city and serves as the economic engine for Ontario and as the financial capital for Canada. Toronto has also become the nation's most culturally diverse city because of its capacity to attract new Canadians.

Population density figures are more meaningful if they are expressed as the amount of arable land per person. This measure is called physiological density. By this measure, Canada's physiological density is similar to that found in the United States.

Population Distribution

Population distribution is the dispersal of people within a geographic area. Canada's population is extremely unevenly distributed across the country (Figure 4.1 and Table 4.2). In fact, few nations have so much of their population concentrated in such a relatively small area of the country, while the rest of the country is almost vacant. So pronounced is this uneven population distribution that an American geography text described Canada's population as if it were 'drawn by a magnet toward the giant neighbor on the south, for

they [Canada's inhabitants] are strikingly concentrated along the United States border' (Trewartha et al., 1967: 542). Canadian scholars often view this same distribution as consisting of a national population core surrounded by a sparsely populated hinterland. The population core is sometimes described as Canada's national **ecumene** (inhabited area). Within this ecumene lies the core of Canada's highway system (see Figure 4.1).

As Table 4.2 illustrates, Ontario has the highest density at 13.4 persons per km² followed by Québec at 5.6 persons per km². These two geographic regions combine to form a demographic core with 62 per cent of Canada's population. Combined or individually, these geographic regions exert consider political force within the Canadian federation and, as a result, they are a source of much regional alienation. Their dominant demographic position is enhanced by their

Think About It

Why is the 2006 census population figure shown in Table 4.2 different from Statistics Canada's estimated 2006 figure in Table 4.1? Which do you think is more accurate?

Photo 4.3

Vancouver, British Columbia, is Canada's leading ocean port, with most goods coming and going to China and other Asian countries. The third-largest Canadian city, Vancouver was home to the 2010 Winter Olympic Games. In the foreground, the Cambie Street Bridge spans False Creek and leads to BC Place; to the west of the bridge, at the bottom of the photo, is Granville Island with its Public Market.

geographic situation in the Great Lakes–St Lawrence Lowlands, proximity to the manufacturing heartland of the United States, and their economic/financial strength.

Population zones provide a more exact geographic picture of Canada's population distribution. As shown in Figure 4.1 and Table 4.3, four population zones vary in population size from very large (60 per cent of Canada's population) to very small (less than 1 per cent of Canada's population). Similarly, the four zones vary considerably in population density. The overall spatial pattern reinforces the image of a highly concentrated population core surrounded by more thinly populated zones.

Table 4.2 Population Size, Percentage, and Density, Canada and Regions, 2006

Geographic Region	Population	Population (%)	Land Area (000s km²)	Population Density (per km²)
Territorial North	101,310	0.3	3,778	0.03
Atlantic Canada	2,284,779	7.2	502	4.7
British Columbia	4,113,487	13.0	893	4.4
Western Canada	5,406,908	17.1	1,756	3.2
Québec	7,546,131	23.9	1,358	5.6
Ontario	12,160,282	38.5	917	13.4
Canada	31,612,897	100.0	9,203	3.5

Source: Adapted from Statistics Canada (2007a).

laughingmango/iStockphoto.com

Photo 4.4

With the Parliament Buildings (left) and the Château Laurier in the background, the Rideau Canal provides a winter skating experience in Ottawa, Canada's capital. The canal, completed in 1832, was originally built as a military supply route between and Kingston and Ottawa.

its agriculture lands contain the most fertile farmlands in Canada.

The secondary core zone extends in a narrow band across southern Canada. In general, its northern boundary corresponds with the polar edge of arable land. As the second-most favoured zone, it occupies the more southerly portions of the Appalachian Uplands, the Canadian Shield, the Interior Plains, and the Cordillera. About 12 million Canadians (over one-third of the country's total population) live in this moderately populated zone. Canada's remaining major cities are located within this zone, including Vancouver, Edmonton, Calgary, Winnipeg, and Halifax. Within the secondary zone, some areas, such as southern Alberta and British Columbia, are growing quickly while other areas have experience much slower growth and even population losses, such as Newfoundland and Labrador. As a result, the population of the secondary zone is increasing slowly and unevenly.

The third zone, characterized by sparse population, is associated with a narrow band of the boreal forest that stretches across mid-Canada. Three physiographic regions are found here—the Canadian Shield, the Interior Plains, and the Cordillera. As the third-most populous zone, less than 1 per cent of all Canadians (about 300,000) live here. Only one of Canada's major cities, Fort McMurray, is in this zone. Fort McMurray, Alberta, is an outstanding example of a booming **resource town**. As the hub of northern Alberta's oil sands extraction and exploration, Fort McMurray (which is within the Wood Buffalo Regional Municipality) is the largest city in the tertiary zone with a population exceeding 51,000. Other larger urban centres range in size from 10,000 to 20,000. Whitehorse and Yellowknife, as the capital cities of Yukon and the Northwest Territories, are administrative centres and **regional service centres**, since they also provide most of the service functions for their areas (Figure 4.2). These two cities, with populations of 20,461 and 18,700, respectively, in 2006, also anchor the poleward edge of zone 3.

Most of Canada's expansive territory is found in the last, almost uninhabited lands of the North. Here, because of the extremely

Canada's core population zone lies in the Great Lakes–St Lawrence Lowlands where the bulk of Canada's population is found. As the most naturally favoured physiographic region, the Great Lakes–St Lawrence Lowlands contains 19 million people and almost three-quarters of Canada's major cities. This population core includes Toronto, Montréal, Ottawa–Gatineau, Québec City, Hamilton, Oshawa, London, and Windsor, to name only some of the largest cities in the region. As Canada's most densely populated area, its economy is based on manufacturing and

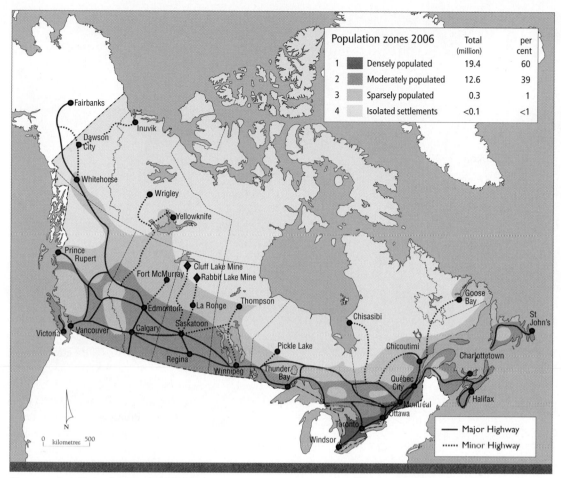

Figure 4.1 Canada's population zones and highway system.

Canada's population is heavily concentrated in southern Ontario and southern Québec, where a favourable physical geography and an advantageous geographic location have resulted in a dense population. A secondary belt of population spans a southern strip of Canada. Together, the densely and moderately populated zones account for 99 per cent of Canada's population. The core highway system connects cities and towns in Canada's ecumene, which consists of population zones 1 and 2. Outliers of the highway system extend into population zones 3 and 4. Nunavut is the only political territory not connected to the national highway system. (Further resources: *Atlas of Canada*, 'Population Distribution 2001', at: <atlas.nrcan.gc.ca/site/english/maps/peopleandsociety/population/population2001/distribution2001>).

Table 4.3 Population Zones, 2006

Zone	Population (millions)	Percentage of Canada's Population	Major City	Population of Major City
1. Core zone: densely populated	19.0	60	Toronto	5,113,149
2. Secondary zone: moderately populated	12.3	39	Vancouver	2,116,581
3. Sparsely populated zone	0.3	1	Fort McMurray*	51,496
4. Almost uninhabited zone with isolated centres	<0.1	<1	Labrador City	7,240

*Wood Buffalo Regional Municipality.
Sources: Statistics Canada (2007a, 2007b, 2007g).

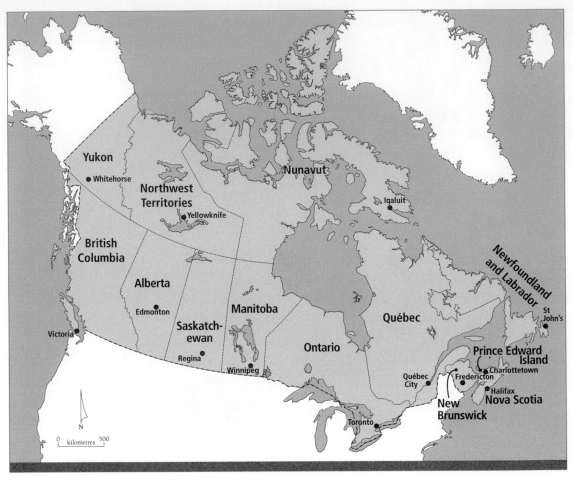

Figure 4.2   Capital cities.
With five exceptions, the capital cities of the 10 provinces and three territories are the largest urban centres in each political jurisdiction. The exceptions are: New Brunswick, Saint John; Québec, Montréal; Saskatchewan, Saskatoon; Alberta, Calgary; and British Columbia, Vancouver.

cold climate, the presence of permafrost, and the polar ice pack, the land is inhospitable for settlement. This zone includes fewer than 100,000 inhabitants. Most reside in small, isolated **Native settlements**. With a small population and a large geographic area, this population zone has the lowest population density. From a resource perspective, it is the least productive area in the country. Most commercial activities centre on exploitation of non-renewable resources, including mineral bodies and petroleum deposits. Unlike in the other zones in Canada, Aboriginal peoples are the majority in this zone. In spite of a high rate of natural increase among the Aboriginal population, the quaternary zone is affected by a net out-migration. Urban centres are small, most with populations under 5,000. In 2006,

the iron-mining town of Labrador City had the largest population in the fourth zone with 7,200 inhabitants, followed by the rapidly growing Iqaluit, the capital city of Nunavut, with 6,184. Unlike that of the resource town of Fort McMurray, Labrador City's population is decreasing. From 2001 to 2006, Labrador City saw its population decline by 6.5 per cent while Iqaluit, with its increasing public sector, had a remarkable jump in population of 18.1 per cent.

Urban Population

Canada is an urban country with 80 per cent of its population living in cities and towns. Statistics Canada (2007e) defines an urban area as having a population of at least 1,000 and no fewer than 400 persons per square

Percentage

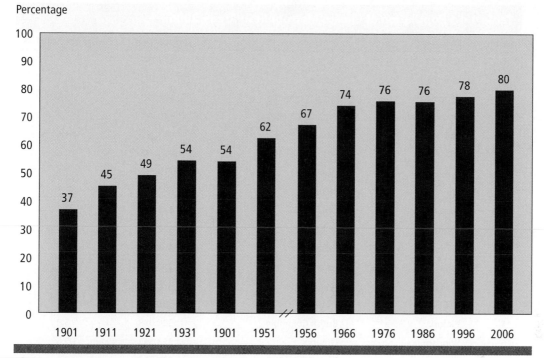

Figure 4.3   Percentage of Canadian population in urban regions, 1901–2006.
Source: Adapted from Statistics Canada (2007c).

kilometre. Canada's urban population has grown remarkably. For over a century, a shift from rural to urban places has been at work. From 1901 to 2006, the country's urban population increased steadily from 37 per cent to 80 per cent (Figure 4.3).

Census Metropolitan Areas

The emergence of large cities across Canada is the latest outcome of urbanization. These cities serve as the economic and cultural anchors of their hinterlands. Statistics Canada defines **census metropolitan area** (CMA) as an urban area (known as the urban core) together with adjacent urban and rural areas that have a high degree of social and economic integration with the urban core. The urban core population of a CMA must be at least 100,000 based on the previous census.

The proportion of Canada's population residing in census metropolitan areas has increased from 30.3 per cent in 1931 to 68 per cent, or 21.5 million Canadians, in 2006. In 1931, there were 10 CMAs: Halifax, Hamilton,

Montréal, Ottawa, Québec, Saint John, Toronto, Vancouver, Windsor, and Winnipeg. By 2006, Canada had 33 census metropolitan areas (Table 4.4). Of the 21.5 million people residing in these CMAs, 14.1 million lived in one of the six metropolitan areas with a population of more than 1 million: Toronto, Montréal, Vancouver, Ottawa–Gatineau, and, for the first time topping the million mark, Calgary and Edmonton.

Since 1951, the population growth of the five largest CMAs has clearly outstripped the national rate of increase. Over the period 1951–2006, Toronto has gained the greatest number of people and has consistently ranked in the top five cities by growth rate. By 1971, Toronto surpassed Montréal as Canada's largest metropolitan centre, and Toronto, with a higher growth rate, continues to outdistance Montréal in population size. From 2001 to 2006, for instance, Toronto's growth rate was 9.2 per cent while that of Montréal was 5.3 per cent. Yet, the greatest rate of increase among the census metropolitan areas—a phenomenal 19.2 per cent—took place in Barrie, Ontario. The five leading

Table 4.4   Population of Census Metropolitan Areas, 2006

Order	CMAs	Province	Population	% Change, 2001–6
1	Toronto	Ontario	5,113,149	9.2
2	Montréal	Québec	3,635,571	5.3
3	Vancouver	British Columbia	2,116,581	6.5
4	Ottawa–Gatineau	Ontario–Québec	1,130,761	5.9
5	Calgary	Alberta	1,079,310	13.4
6	Edmonton	Alberta	1,034,945	10.4
7	Québec	Québec	715,515	4.2
8	Winnipeg	Manitoba	694,668	2.7
9	Hamilton	Ontario	692,911	4.6
10	London	Ontario	457,720	5.1
11	Kitchener	Ontario	451,235	8.9
12	St Catharines–Niagara	Ontario	390,317	3.5
13	Halifax	Nova Scotia	372,858	3.8
14	Oshawa	Ontario	330,594	11.6
15	Victoria	British Columbia	330,088	5.8
16	Windsor	Ontario	323,342	5.0
17	Saskatoon	Saskatchewan	233,923	3.5
18	Regina	Saskatchewan	194,971	1.1
19	Sherbrooke	Québec	186,952	6.3
20	St John's	Newfoundland and Labrador	181,113	4.7
21	Barrie	Ontario	177,061	19.2
22	Kelowna	British Columbia	162,276	9.8
23	Abbotsford	British Columbia	159,020	7.9
24	Greater Sudbury	Ontario	158,258	1.7
25	Kingston	Ontario	152,358	3.8
26	Saguenay	Québec	151,643	−2.1
27	Trois-Rivières	Québec	141,529	2.9
28	Guelph	Ontario	127,009	8.2
29	Moncton	New Brunswick	126,424	6.5
30	Brantford	Ontario	124,607	5.5
31	Thunder Bay	Ontario	122,907	0.8
32	Saint John	New Brunswick	122,389	−0.2
33	Peterborough	Ontario	116,570	5.1

Sources: Statistics Canada (2002c, 2007d).

census metropolitan areas by growth rates for the last five years were Barrie, Calgary, Oshawa, Edmonton, and Kelowna. Only two CMAs suffered a decline—Saguenay (–2.1 per cent) and Saint John (–0.2 per cent).

What is the attraction of cities? First and of greatest importance, most business and employment opportunities are found in cities, especially large cities. As well, Canadians prefer to live in an urban setting where amenities are readily available. Cities are also important for other reasons. Major urban centres are at the cutting edge of technological innovation and capital accumulation. In the new world of the knowledge economy, manufacturing does not determine a city's prosperity; rather, the creativity of its business and university communities is the determining factor.

Despite the steady and in some instances remarkable growth of Canadian cities, all is not well in metropolitan Canada. Competition from malls and big-box stores in the suburbs has hurt downtown retail areas. Also, automobiles are ill-suited for the downtown while generously accommodated in shopping areas on the urban perimeter. As a result, downtown cores have lost their focus of commercial and social activities and some even seem deserted at night. In the twenty-first century, cities face the daunting task of finding solutions to these problems that are associated with the growth of suburbia and the shift of commercial activities from downtown to the outlying areas. Among the leading challenges for cities are smog, traffic congestion, shortage of parking spaces, the homeless, water contamination, and garbage disposal. Efforts to make downtowns more pedestrian- and bicycle-friendly as well as to launch numerous 'green' programs signal a new direction.

Variation in Urban Population by Geographic Region

Urbanization is associated with economic development. For that reason, the rate of urbanization has varied across Canada. Not surprisingly, Ontario and British Columbia

Joan Vicent Cantó Roig/iStockphoto.com

Photo 4.5

Calgary's downtown is dominated by skyscrapers, many of which are associated with the petroleum industry.

have the highest percentages of their populations classified as urban, while Atlantic Canada and the Territorial North have the lowest. For years, however, Ontario led all other geographic regions (Table 4.5).

By 2006, as shown in Table 4.5, British Columbia had surpassed Ontario. The 2006 census figures continue to reflect an increase

Table 4.5	Percentage of Urban Population by Region, 1901–2006						
Region	**1901**	**1921**	**1941**	**1961**	**1981**	**2001**	**2006**
Ontario	40.3	58.8	67.5	77.3	81.7	84.7	85.1
British Columbia	46.4	50.9	64.0	72.6	78.0	84.7	85.4
Québec	36.1	51.8	61.2	74.3	77.6	80.4	80.2
Western Canada	19.3	28.7	32.4	57.6	71.4	75.7	76.7
Atlantic Canada*	24.5	38.8	44.1	50.1	54.9	53.9	54.1
Canada	34.9	47.4	55.7	70.2	76.2	79.7	80.0

*Newfoundland is not included in Atlantic Canada's figures until 1961.

Note: While comparable statistics are not available for the Territorial North, two observations are possible. (1) Prior to the 1950s, few people in this region lived in settlements. (2) By 2006, Statistics Canada (2007f) classified nine centres in the Territorial North as urban areas: Whitehorse in Yukon; Hay River, Inuvik, and Yellowknife in the Northwest Territories; and Iqaluit, Pangnirtung, Rankin Inlet, Cambridge Bay, and Arviat in Nunavut. Their total population in 2006 was 55,137, resulting in 54 per cent of the Territorial North's population defined as urban.

Sources: McVey and Kalbach (1995: 149). © 1995 Nelson Education Ltd. Reproduced by permission. Statistics Canada (1997b, 2002a, 2007f, 2008a).

Photo 4.6

The North Saskatchewan River frames Edmonton's downtown and provincial legislative buildings.

in the urban population. This trend is most obvious in Western Canada where, over the past 100 years or so, the rural nature of the region has declined sharply. In 1901, Western Canada had less than 20 per cent of its population living in urban places. By 2006, the figure had jumped to almost 77 per cent. The shift from rural to urban communities, although most obvious in Western Canada, is a national phenomenon and is associated with two factors: the declining numbers involved in agriculture due to mechanization, and the increase in job opportunities in urban places.

Population Change

Canada's population increased by 1.6 million (5.4 per cent) from 2001 to 2006. Population change has three components: births, deaths, and migration. **Population increase** is the sum of natural increase and net migration over a given period. The term **population growth** is used when this increase is expressed as a rate, that is, as a percentage change over time. The **rate of natural increase** is the difference between the **crude birth rate** (CBR) and the **crude death rate** (CDR). CBR is the number of live births per 1,000 people in a given year. CDR is the number of deaths per 1,000 people in a given year. **Net migration** is the difference between in- and out-migration. One theoretical explanation lies in the **push-pull model**, which, in its most simple form, sees adverse factors at home 'pushing' people out, making them want to emigrate. Attractive factors abroad can 'pull' people in.

Canada has a large influx of migrants, but a number of Canadians also leave the country, often for the United States. While reasons for out-migration vary, many Canadians move to the United States because of job opportunities and/or because of a more temperate climate.

Since 1851, Canada has enjoyed continuous population growth. At first, high rates of

Table 4.6 Population by Regions, 2001–6

Region	2006	2001	Change 2001–6	% Change 2001–6
Territorial North	101,310	92,779	8,531	9.2
Atlantic Canada	2,284,779	2,285,729	–950	–0.1
British Columbia	4,113,487	3,907,738	205,749	5.3
Western Canada	5,406,908	5,073,323	333,585	6.6
Québec	7,546,131	7,237,479	308,662	4.3
Ontario	12,160,282	11,410,046	750,236	6.6
Canada	31,612,897	30,007,094	1,605,803	5.4

Source: Statistics Canada (2007j).

natural increase and high levels of immigration propelled this growth. The highest rate of population growth occurred over the 1901–11 period, with a remarkable increase over 10 years of 34 per cent, spurred by the large influx of immigration to the Canadian Prairies (see Table 4.7). As more Canadians moved to cities, parents opted for smaller families, causing the rate of natural increase to decline. During the 1950s and 1960s, however, the baby boom pushed the average annual rate of population increase to nearly 3 per cent (Vignette 4.2). After that short burst, the rate fell. From 2001 to 2006, the annual average rate was 1.1 per cent. Most population increase is now related to immigration rather than natural increase. Since most immigrants relocated to Ontario, Québec, and British Columbia, much of the population increase in these regions is attributed to new Canadians. The bulk of increase in Western Canada results from Alberta drawing Canadians from other parts of the country to job opportunities.

For the thousands of Newfoundlanders working in the Alberta oil fields, see Vignette 9.12, 'The Big Commute and Transfer Payment', page 396.

Think About It

While Québec's population is increasing, its rate of increase is below the national average. Is this what you would expect of a 'core' region?

Vignette 4.2 The Baby Boom Effect

During the Great Depression and World War II, economic and social conditions did not favour large families. With the return of Canada's servicemen and women, a stable political world, and improving economic conditions, couples' attitude towards family formation took a positive turn. In 1946, the birth rate suddenly increased, reaching nearly 30 births per 1,000 people in 1951. The crude birth rate continued at this level until 1966. Demographers refer to this aberration as the baby boom. By the late 1960s, however, fertility rates decreased sharply. The end of the baby boom is sometimes described as the baby bust.

The baby boom lasted for 20 years. During that time of high fertility, there were almost 10 million births, which created a bulge in the age structure of Canadian society that continues to have both economic and social implications. Thus, while this demographic phenomenon was short-lived, it has left its mark on Canadian society (Foot with Stoffman, 1996). As consumers of goods and services, baby boomers have had a decided impact on the economy as they move through their life cycle. Companies have geared their products to meet the strong demand for goods and services created by baby boomers. In the early 1950s, the emphasis was on baby products and larger houses. In the 1960s, a similar age-related pressure was exerted on school facilities, creating a demand for more schools and teachers. As the baby boomers approach old age, the demand for health-care services is expected to rise. Companies are already targeting their advertisements at the growing number of retirees from the baby boom generation. Governments, on the other hand, are concerned about rising health-care costs associated with the expected increase in senior citizens.

Natural Increase

Natural increase is determined by the number of births minus the number of deaths. Until 1986, the greater portion of Canada's population growth (Table 4.7) was due to natural increase. At that time, the number of immigrants began to exceed the number due to natural increase, i.e., the net of births minus deaths (Statistics Canada, 2007g). Until 1921, the majority of Canadians lived in rural settings where large families were the rule. In the 1870s, for example, natural increase exceeded 2 per cent per year. At that rate, Canada's population would double every 30 to 35 years. Birth rates were extremely high, while death rates were low, allowing for a high rate of natural increase. As most available farmland became occupied, however, children of farm parents had no choice but to seek their fortunes elsewhere. Many went to the cities in search of work. As Canada became an industrial country, fertility rates declined. Over the period 1951–2006, except for a short time after World War II, the crude birth rates declined steadily from about 45 births per 1,000 people per year to approximately 11 births per 1,000 people. Over the same period, the crude death rate dropped from about 20 deaths per 1,000 people per year to about seven. While there is no single explanation for these changes, public health measures such as water purification, improved nutrition, medical advances, and the development of public health-care systems across the country account for most of the sharp reduction in the **mortality rate**. In recent years, mortality rates have inched upward, signalling the impact of

Table 4.7	Population Increase, 1851–2006		
Year	**Population (000s)**	**Percentage Change**	**Average Annual Rate**
1851	2,436.3	—	—
1861	3,229.6	32.6	2.9
1871	3,689.3	14.2	1.3
1881	4,324.8	17.2	1.6
1891	4,833.2	11.8	1.1
1901	5,371.3	11.1	1.1
1911	7,206.6	34.2	3.0
1921	8,787.9	21.9	2.0
1931	10,376.8	18.1	1.7
1941	11,506.7	10.9	1.0
1951	14,009.4	21.8	1.7
1956	16,080.8	14.8	2.8
1961	18,238.2	13.4	2.5
1966	20,014.9	9.7	1.9
1971	21,568.3	7.8	1.5
1976	22,992.6	6.6	1.3
1981	24,343.2	5.9	1.2
1986	25,309.3	4.0	0.8
1991	27,296.9	7.9	1.6
1996	28,846.8	5.7	1.1
2001	30,007.1	4.0	0.8
2006	31,612.9	5.4	1.1

Sources: McVey and Kalbach (1995: 42); Statistics Canada (1997b, 2002a, 2007a).

Table 4.8   Canada's Rate of Natural Increase, 1851–2006

Year	Crude Birth Rate	Crude Death Rate	Natural Increase (%)	Natural Increase (000s)
1851	45	20	2.5	61
1861	45	20	2.5	81
1871	42	20	2.2	81
1881	40	19	2.1	90
1891	38	18	2.0	97
1901	35	16	1.9	97
1911	32	14	1.8	129
1921	29.3	11.6	1.8	160
1931	23.2	10.2	1.3	138
1941	22.4	10.1	1.2	145
1951	27.2	9.0	1.6	255
1961	26.1	7.7	1.8	335
1971	16.8	7.3	1.0	205
1981	15.2	7.0	0.8	200
1991	14.3	7.0	0.7	207
1995–6	12.6	7.1	0.6	163
2001–2	10.5	7.1	0.3	108
2004–5	10.5	7.3	0.3	103
2005–6	10.6	7.3	0.3	108

Note: Data on births and deaths are now recorded from 1 July to 30 June.

Sources: Adapted from Statistics Canada (1997c, 2003b, 2006b, 2007h, 2007i); McVey and Kalbach (1995: 268, 270).

Canada's aging population. The decline in the **fertility rate** is more complex. A series of social and economic changes caused parents to opt for smaller families. Among these changes, three stand out: the shift of people from rural areas to towns and cities; the sharp increase in the number of women in the labour force; and the widespread acceptance of family planning. Family planning was greatly helped by the birth control pill. Introduced in 1960, the birth control pill has played an important role in reducing the birth rate and was a factor in ending the baby boom. In 2007, however, the crude birth rate jumped to 11.2 per thousand persons, a sharp increase from the figure of 10.6 in 2006 but still far short of the crude birth rates associated with the baby boom (Statistics Canada; 2009k; Table 4.8).

By the twenty-first century, Canada's crude birth and death rates were extremely low. These vital rates are similar to those found in other industrial countries. Consequently, Canada has a low rate of natural increase. By 2005–6, the natural rate was 0.3 per cent. Will this rate continue to decline? Based on the experience of other industrialized countries, Canada's rate of natural increase could reach zero.

The Demographic Transition Theory

The **demographic transition theory** is the most widely accepted theory that describes population change in industrial societies. This theory is based on the assumption that changes in birth and death rates occur as a society moves from a pre-industrial to an industrial economy. These demographic changes occur in five phases, each of which has a distinct set of vital rates that coincide with the phases in the process of industrialization (Table 4.9).

Canada did not experience a late **pre-industrial phase**, which was associated with a feudal society. A cursory examination

Table 4.9 Phases in the Demographic Transition Theory

Phase	Birth and Death Rates	Rate of Natural Increase
Late pre-industrial	High birth and death rates	Little or no natural increase but possible fluctuations because of variations in the death rate
Early industrial	Falling death rates	Extremely high rates of natural increase
Late industrial	Falling birth rates	High but declining rates of natural increase
Early post-industrial	Low birth and death rates	Little or no natural increase; stable population
Late post-industrial	Falling birth rates	Declining population

of Canada's birth and death rates over the last 150 years reveals strong similarities to the early industrial, late industrial, and post-industrial phases. In the nineteenth century high birth rates and lower death rates were common. These vital rates are similar to those in the second phase of the demographic transition theory. Then Canada entered the late **industrial phase**, characterized by low death rates and a declining birth rate. By the 1980s, birth rates had declined significantly, resulting in little natural increase. Such rates characterize the early **post-industrial phase** of the demographic transition theory. Based on a more precise measure of fertility, demographers argue that Canada's rate of natural increase has fallen below its replacement level (Vignette 4.3). A few countries, such as Spain, Italy, and Japan, have seen their populations decline, which places them in the late post-industrial phase. Within Canada, the Aboriginal population continues to have significantly higher birth rates and higher rates of natural increase.

Age Structure

Canada's demographic picture clearly indicates a 'greying population'. In 2006, the median age was 38.8 years, but Statistics Canada projects that by 2056 the median age will be 46.9 years (Table 4.10). The demographic implications for Canada are most significant, and include a smaller proportion of children (under 15 years of age), a smaller proportion of the population in the workforce (ages 15–64), and a larger percentage over 64 years of age. Added to this demographic picture, Canadians are living longer. The economic implications also are significant because a smaller percentage of workers means a growing tax burden on these people and ever-increasing medical costs and demands by Canada's senior citizens.

Vignette 4.3 The Concept of Replacement Fertility, Total Fertility, and Canada's Sex Ratio

The concept of replacement fertility refers to the level of fertility at which women have enough daughters to replace themselves. If women have an average of 2.1 births in their lifetime, then each woman, on average, will have given birth to a daughter and a son. The number 2.1 was determined to represent the minimum level of replacement fertility because, on average, slightly more boys than girls are born. In 1961, the Canadian total fertility rate was 3.8 births per woman of child-bearing age (15–49). Since then, the fertility rate has declined. The estimated 2007 total fertility rate was 1.6 births (CIA, 2007).

Nature has ensured that slightly more male than female babies are born. However, male mortality rates are higher than those for females. The 2006 **sex ratio**, defined as the number of males per 1,000 females in Canada's population, was 959, meaning that there are slightly more females than males.

Table 4.10 Canada's Aging Population as Defined by Median Age					
1946	1966	1986	2006	2011	2056
27.7	25.4	31.4	38.8	40.11	46.9

Source: Statistics Canada (2006d).

Age Dependency

Age dependency ratio is the ratio of persons in the 'dependent' age groups (under 15 and over 64 years) to those in the 'economically productive' age group (between 15 and 64 years). The assumption is that productive members of society are those between the ages of 15 and 64, while unproductive members are either too young (under 15) or too old (over 64) to make an economic contribution. The purpose of the age dependency ratio is to compare the number of dependants with the number of economically productive members of society, thus giving a rough measure of the economic burden on those in the economically productive age group.

In Canada, the age dependency ratio declined substantially from 1961 to 2001. In 1961, the ratio recorded 70 dependants per 100 people of working age. By 2006, it had declined to about 44 per 100. Why is the age dependency ratio declining? The answer lies in the baby boom sector of the population, which still remains in the economically productive age group (Vignette 4.2). However, by 2010, the leading edge of this bulge in the age structure of Canadian society will be joining Canada's senior citizens, causing the age dependency ratio to rise. From 2010 onward, the age dependency ratio will increase each year, leaving fewer taxpayers to cover social costs, such as education and health care.

Immigration

Canada is a country of immigrants. Figure 4.4 illustrates the variation in annual immigration from 1901 to 2006. Ottawa encourages immigration for three reasons: (1) newcomers keep Canada's population increasing;

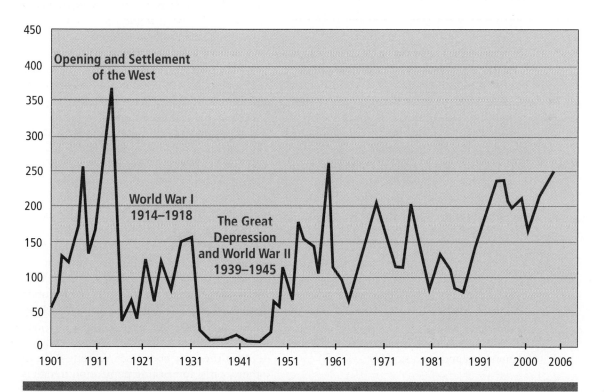

Figure 4.4   Annual number of immigrants admitted to Canada, 1901–2006.
Source: Statistics Canada (2003a: 2; 2009b).

Percentage

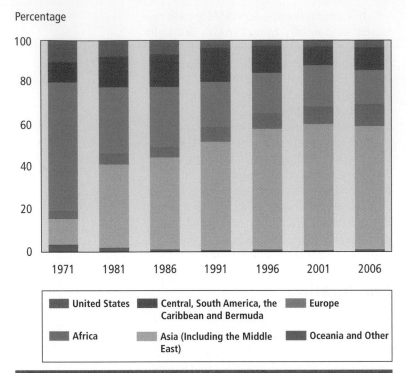

Figure 4.5   Region of birth of immigrants to Canada, 1971–2006.
Source: Statistics Canada (2009j).

The proportion of Canada's population who were born outside the country has reached nearly 20 per cent, the highest level in 70 years. From 1901 to 2004–5, the annual number of people settling in Canada has varied: the greatest annual numbers of newcomers arrived in 1912 and 1913 (settling of the West) and the lowest numbers came in the years of the Great Depression and World War II.

The changing proportion of immigrants born in Europe and Asia is shown in Figure 4.5. Table 4.12 provides a broader picture of the country of origin of newcomers. Most immigrants in recent years were born in Asia and the Middle East (58 per cent) with smaller numbers coming from Europe (20 per cent), the Caribbean and Central and South America (11 per cent), Africa (8 per cent), and the United States (3 per cent).

Why Immigration?

Canada has always felt the need for immigrants—to populate unsettled lands, to do the dirty, dangerous, and low-paying jobs that the native-born would not do, to supplement the trades and professional workforce, and to maintain population and economic growth. Since the 1970s, the principal engine of Canada's population increase has been immigration. In the future, immigration is expected to play an even more significant role. Today, immigration represents close to 70 per cent of the population growth. Since natural increase is expected to continue to decline, Statistics Canada projects that the proportion of Canada's population growth attributed to natural increase will drop to zero by 2030. At that time, deaths are expected to start outnumbering births. From that point forward, immigration would be the only growth factor for the Canadian population (Figure 4.6). The assumption underlying this projection is that the birth rate will continue to decline. While such a decline over the next 20 years seems reasonable, Statistics Canada did not forecast the baby boom (see Vignette 4.2). In other words, forecasting population trends is subject to error caused by unforeseen changes in human behaviour. In fact, the number of births has increased unexpectedly in

(2) newcomers add valuable members to Canada's workforce; (3) Canada takes in refugees who are fleeing oppressive socio-political conditions in their homelands. Since 1992, some 200,000 immigrants have come to Canada each year. By 2007–8, the figure had reached 250,000 (Statistics Canada, 2009b). A key social issue is not the number of immigrants but how well they enrich and add to the cohesiveness of Canadian society. Until the 1970s, most came from Western countries. Today, most come from Asia. As a result, the adjustment process has become more challenging for newly arrived immigrants and for other Canadians. For those belonging to non-Christian religions and those who are 'visible' Canadians, social barriers into the mainstream society often exist. These barriers include language and education as well as discrimination and exclusion.

The issue of cultural adjustment is discussed in Chapter 1 under 'Newcomers and Old-timers', page 12.

Numbers (in thousands)

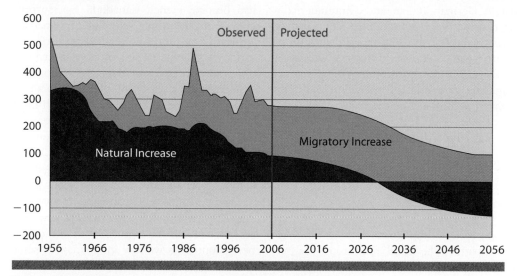

Figure 4.6   **Immigration: An increasingly important component of Canadian population growth.**
Source: Statistics Canada (2009k).

recent years, but fertility still has a way to go to reach levels achieved in the early 1990s (Table 4.11).

The demographic implications of immigration for Canada are significant, as the data in Table 4.11 suggest.

- Immigration ensures that Canada's population keeps increasing.
- Immigrants will form a larger and larger proportion of the national population, thus further increasing the cultural diversity of Canada's society.
- Canada's increasing diversity will call into question the validity of the vision of two founding peoples.
- Concentration of new Canadians in Canada's largest cities has changed the demographics of these cities, making them different from the rest of Canada.

Ethnicity

What is ethnicity? An **ethnic group** is made up of members of a population who share a culture that is distinct from that of other groups. Each group has a common identity, shared values, and cultural/linguistic/religious bonds and symbols. **Culture** is the learned collective behaviour of a group of people. Canadian society is composed of many ethnic groups,[2] with more than 200 different ethnic origins reported in the 2006 census question on ethnic ancestry compared to 25 in 1901.

According to Statistics Canada data from the 2006 census, the leading ethnic group is defined as 'Canadian'. But more interesting is that almost 50 per cent of Canadians selected an ethnic origin from the British Isles (English, Scottish, Irish)

Table 4.11	Demographic Impact of Immigration, Selected Years				
Year	Births	Deaths	Natural Increase	Immigration	Difference
1993–4	389,286	207,528	181,758	227,860	46,102
2000–1	327,187	231,232	95,955	255,999	160,044
2007–8	364,085	237,202	126,883	249,603	122,729

Sources: Adapted from Statistics Canada (2003b, 2006a, 2009a).

while only 15.8 per cent declared a French origin. German, Italian, and Ukrainian make up nearly 20 per cent (Table 4.12). Chinese form the largest non-European ethnic group, while North American Indians (4 per cent), Métis (1.3 per cent) and Inuit (0.2 per cent) together represent 5.5 per cent. Statistics Canada defines **ethnic origin** as:

> the ethnic or cultural origins of the respondent's ancestors. An ancestor is someone from whom a person is descended and is usually more distant than a grandparent.

In spite of immigrants coming from different parts of the world, time erodes the ethnic connection with the country of origin. For Canadians born and raised in this country, the connection with their original (foreign) ethnic origins is tenuous at best (Beaujot, 1991: 297). Place, as cultural geographers insist, plays a critical role in the development of a regional/national identity. This phenomenon is well known. Consider, for example, the attachment of the French born in New France who had no interest in returning to France in 1763 because their lives were centred on the

New World. They were no longer French but had become French Canadian. Not surprisingly, then, the ethnic selection of 'Canadian' by nearly one-third of Canadians is attributed to the geographic notion of place overriding the concept of ethnicity (Table 4.12). To put it differently, the resettlement of people in a new place causes their ethnicity to fade with time, especially if marriages take place with members from other groups. Loyalty to the 'old country' tends to fade with the second generation of immigrant families. This process of putting down cultural roots is a normal process as long as the newcomers are accepted into the mainstream society. Of course, there are cases of newcomers finding it difficult to put down roots in their new country. As we saw in Chapter 3, some 100 years ago the Doukhobors found that their collective way of farm life, as well as their strict adherence to their faith and Russian language, was not accepted by other Canadians living in Western Canada.

With the flood of non-British and non-French migrants settling the Prairies early in the twentieth century, and some of these people eventually moving to the towns and cities of industrial Canada, the emergence of a third cultural force broke the dominance of

Think About It

While ethnicity is a very fluid concept in Canada, is the Census of Canada correct in asking for the identity of 'North American Indians' rather than for traditional ethnicity, such as Cree?

Table 4.12   Ethnic Origins of Canadians

	1996			2006		
Ethnicity	Number	%	Ethnicity	Number	%
Total population	28,528,125	100.0	Total population	31,241,030	100.0
Canadian	8,806,275	30.9	Canadian	10,066,290	32.2
English	6,832,095	23.9	English	6,570,015	21.0
French	5,597,845	19.6	French	4,941,210	15.8
Scottish	4,260,840	14.9	Scottish	4,719,850	15.1
Irish	3,767,610	13.2	Irish	4,354,155	13.9
German	2,757,140	9.7	German	3,179,425	10.2
Italian	1,207,475	4.2	Italian	1,445,335	4.6
Ukrainian	1,026,475	3.6	Chinese	1,346,510	4.3
Chinese	921,585	3.2	North American Indian	1,253,615	4.0
Dutch (Netherlands)	916,215	3.2	Ukrainian	1,209,085	3.9

Notes: (1) Table shows total responses. Because some respondents reported more than one ethnic origin, the sum is greater than the total population or 100 per cent. (2) Figures referring to North American Indian are based on Aboriginal ancestry population., i.e., those persons who reported at least one Aboriginal ancestry (North American Indian, Métis, or Inuit) to the ethnic origin question. 'Ethnic origin' refers to the ethnic or cultural origins of a person's ancestors.

Sources: Statistics Canada (2003a, 2010).

the British and French ethnic groups. After World War II, immigrants from all over the world came to Canada in ever-increasing numbers. Canadian society now consisted of a number of ethnic groups whose members shared a sense of identity based on country of origin, language, religion, tradition, and other common experiences. Overarching these shared ethnic experiences, however, was a sense of belonging to Canada, and since 1971 Ottawa has promoted multiculturalism, the idea that a cultural identity and a national one are not mutually exclusive.

In Québec, a strong sense of cultural identity exists among the people of French origin, but it is tempered by centuries of life in North America. The term 'Québécois', if it refers to the descendants of the original French settlers, more accurately reflects their ethnic background and continues to drive that province's society and politics. History and geography are the causes of this sense of ethnic nationalism. Nationalism is nurtured by Québec's place, not in Canada but in North America. As a French enclave in the English-speaking North American continent, Québec naturally fears for its cultural existence. Its close proximity to the United States underscores its precarious position in North America much more than its 'sheltered' position within Canada. The United States exerts an enormous cultural impact on the world through the mass media and now, increasingly, through the Internet. Within North America, American culture poses the major threat to French- and English-Canadian cultures. This fear of assimilation is the basis of ethnic nationalism in Québec, a nationalism generally expressed by the Québécois in one of two ways: (1) Québec society is distinct within Canada; (2) only through independence will Québec be fully protected and emancipated. Not surprisingly, concern over change to the existing way of life by accommodating new cultural ways exists across Canada but was most openly expressed by Hérouxville's municipal council when they announced their 'code of standards' for newcomers (see Vignette 1.2).

Dual loyalties exist in all geographic regions, but they are strongest among Québécois when they look to their provincial government to protect their interests rather than to the federal government. Dual loyalties also exist among **recent immigrants** whose ties to their home country remain strong and who retain a passport from their former country. Dual passports have the advantage of making travel for pleasure or business more convenient, but dual citizenship is not legally recognized in all countries, which can lead to serious difficulties. Then, too, there is the question of loyalty. As the former Minister of Citizenship and Immigration, Judy Sgro, stated, 'We need to be loyal to one country as far as your citizenship. Your heart can be where you were born, but I think the commitment to Canada has to be strong and I think dual citizenship weakens that' (Woods, 2006).

Language

Language is a key component of ethnicity. Indeed, language represents the most durable link to the past and the tool for maintaining a culture. Canada's two official languages are English and French. The 2006 census (Statistics Canada, 2008b) reported that:

> In 2006, 98 per cent of the population can speak one or both official languages. In addition, English or French is spoken at least regularly at home by 94 per cent of Canadians and most often at home for 89 per cent of the population, sometimes in combination with a non-official language.
>
> Due to increased immigration since the mid-1980s, and the tendency of most immigrants to have a mother tongue other than English or French, the share of the allophone population has grown: from 18 per cent in 2001 to 20 per cent in 2006.

Language is a key issue for French-speaking Canadians and immigrants from non-English or French countries. From an immigrant perspective, 'fitting in and getting a job' relates closely to fluency in an official language, while the 'health' of the French language remains a sensitive matter. In fact, this topic is the basis of a population faultline discussed later in this chapter.

Think About It

Do immigration and the resulting multiculturalism pose a threat to Québec society?

Think About It

What is the implication if you declare yourself as 'Canadian' rather than as one of the more conventional ethnic groups when answering the census question on ethnicity?

A number of French organizations across Canada aim to protect and promote their language and culture. Such organizations have received federal funding to finance their cultural and educational activities. This public support, plus the presence of French-language radio and television programs, has resulted in a rejuvenation of these provincial organizations.

Many languages, other than French or English, are spoken across the country, especially in larger cities where new immigrants tend to settle. Without the official-language status enjoyed by French and English, however, other languages are often difficult to maintain. The survival of these other languages in Canada is bolstered somewhat by continued immigration and the efforts of non-official language groups to maintain their languages. However, Canadian-born members of ethnic groups—the second and subsequent generations of immigrants—usually lose the language of their parents and grandparents.

For Aboriginal peoples, the issue of language has two components, which, unfortunately, may work against each other. First, cultural viability means maintaining Aboriginal languages. Second, economic viability means finding a place within the Canadian economy. David Newhouse, a noted Aboriginal scholar (2000: 402–3), sees the emergence of English as the lingua franca among Aboriginal peoples. By 2006, only three of the 50 languages spoken by Canada's Aboriginal peoples had a secure future (Norris, 2007). They are Cree, with 87,285 speakers (up 7 per cent from 2001), Inuktitut, with 25,500 (down 3 per cent from 2001), and Ojibwa, with 30,256 (down 2 per cent from 2001).

Religion

Religion is another key element of culture. The changing nature of Canada's religious composition is indicative of the changing cultural mix of people. The recent shift in Canada's religious makeup is linked with the large number of immigrants arriving from non-Christian countries and the increasing number of Canadians who declare 'no religious beliefs' and who have no association—except perhaps at birth, marriage, and death—with the religious institution of their parents. On the other hand, the number of Canadians who subscribe to religions such as Islam, Hinduism, Sikhism, and Buddhism has increased substantially.

Culture is not only a durable link to the past but also provides the institutional organization to preserve ethnicity. Religious organizations provide an institutional structure that consolidates people of similar beliefs. One example is the Roman Catholic Church. In the early history of Canada, the Roman Catholic Church helped sustain Catholicism, the French language, and the French-Canadian way of life. While its role has diminished in recent decades, the Church was the dominant cultural force among francophones from the conquest of New France to World War II. Not only did it give spiritual direction to French Canadians within Québec but it also organized its parishes to sponsor group immigration to more isolated areas of Québec (such as the Clay Belt in northwestern Québec), to other provinces, and, after initial opposition, to New England. In these group migrations, priests provided the leadership for the move and for organizing the new settlement, including its educational, social, and religious structures.

Religious freedom is the right of all Canadians. This right was inscribed in the British North America Act and is contained in the Constitution Act, 1982. In fact, many immigrants have come to Canada seeking religious freedom. The Hutterite Brethren, for example, are a small Christian group who live in agricultural colonies. Like the Amish and Mennonites, they place pacifism at the core of their faith. In 1899, Hutterites fled Europe to the United States. The Hutterites immigrated to Canada in 1918 because the US government refused to exempt them from military service. Initially, Hutterite colonies were formed in Manitoba and Alberta. Later, colonies were established in Saskatchewan. Hutterites number about 30,000 and the majority live and work in one of 300 agricultural colonies where they remain committed to their religion, a communal lifestyle, and pacifism (Statistics Canada, 2003d).

Canada is thought of as a Christian country. This image was certainly true in 1867, but today Canada is much more religiously diverse. It is true that most Canadians are Christians. For example, in the 1960s nearly 90 per cent of Canadians declared themselves to be Christian (though some may not have been active church members). By 2001, this figure had dropped to 77 per cent, while 16 per cent claimed to have no religious beliefs (ibid.). Since Statistics Canada asks Canadians about their religion every 10 years, the next census (2011) will provide fresh data. In the meantime, the 2001 census data revealed that Roman Catholics remain the largest group of Christians, comprising 44 per cent of all Canadians, followed by Protestants at 29 per cent. Canadians with no religious affiliation make up 17 per cent of the population. At the same time, the numbers of Muslims, Hindus, Buddhists, and Sikhs had more than doubled from 1991 to 2001, making up 5 per cent of Canada's population (ibid.).

Multiculturalism

Multiculturalism, the cornerstone of Canada's new and distinctive demographic character, is a product of immigration and, strangely enough, biculturalism. The Royal Commission on Bilingualism and Biculturalism (1970) recommended that Ottawa recognize the multiplicity of Canada's population. The following year, in 1971, the federal Liberal government made multiculturalism official policy and in 1972 the cabinet position of Minister of State for Multiculturalism was created. Government funding to ethnic organizations and for publishing and other educational programs soon followed. In 1988, the federal government passed the Canadian Multiculturalism Act, which is designed to encourage greater human understanding and stronger bonds among Canadians of different cultural backgrounds and ethnic origins, and today official multiculturalism is under the purview of the Department of Canadian Heritage.

Photo 4.7

The Basilica of Notre Dame, opened in 1829, is the principal Roman Catholic Church in Montréal, and is a reminder of the Church's powerful role in the history of French-Canadian society.

Dunjava/Dreamstime/GetStock.com

Photo 4.8

The growing Muslim population in Alberta is reflected in Canada's largest mosque, which serves the Ahmadiyya Muslim community.

Think About It

What does Charles Taylor mean by the term 'life of diaspora'? Do you think he was a good choice to serve on the Québec Commission on Accommodation Practices Related to Cultural Differences?

For Charles Taylor, multiculturalism is a way for the Canadian government and society to recognize the worth of newcomers' distinctive cultural traditions without compromising Canada's basic political principles (Vignette 4.4). The outer boundary of multiculturalism is where it rubs against the edge of conventional values held by mainstream Canadians. An example was the rejection of shariah law to settle family disputes within the Muslim community in Ontario. On the other hand, subtle adjustments are occurring in a wide range of areas. For example, the Toronto Stock Exchange has sought to accommodate Muslim investors by launching a Canadian stock index (the S&P/TSX 60 Shariah) in May 2009. This index excludes banks, pork producers, and entertainment and gambling stocks, thus allowing Islamic investors to abide by Islamic law and still participate in equity investment strategy (www.theglobeandmail.com/report-on-business/sp-launches-canadian-sharia-compliant-index/article1155888/). While multiculturalism does not focus on Aboriginal peoples, who are hardly newcomers, Aboriginal peoples also seek respect for their traditional practices. For that reason, Aboriginal leaders across Canada applauded Governor General Michaëlle Jean in 2009 when she sliced off and ate a raw piece of a seal's heart to show solidarity with Inuit and their culture at a festival at Rankin Inlet, Nunavut.

For further discussion of Michaëlle Jean's 'seal meal' and the reaction it received, see Chapter 10, page 427, and photo 10.5.

In many ways, multiculturalism is the opposite of **ethnocentricity**. But tolerance and respect of others are not an automatic outcome of life in pluralistic societies. Canadian tolerance and respect were learned the hard way, going back centuries to often bitter (and sometimes violent) conflicts between French- and English-speaking Canadians. For Canada to survive as a nation, the two antagonists had no choice but to become partners. One product of this so-called partnership was biculturalism. The other path to nation-building—a

Vignette 4.4   Charles Taylor on Multiculturalism

One of Canada's leading philosophers, Charles Taylor (1994: 63), argues in *Multiculturalism: Examining the Politics of Recognition*, that:

> . . . all societies are becoming increasingly multicultural, while at the same time becoming more porous. Indeed, these two developments go together. Their porousness means that they are more open to multi-national migration; more of their members live the life of diaspora, whose centre is elsewhere. In these circumstances, there is something awkward about replying simply, 'This is how we do things here.' This reply must be made in cases like the Rushdie controversy, where 'how we do things' covers issues such as the right to life and to freedom of speech. The awkwardness arises from the fact that there are substantial numbers of people who are citizens and also belong to the culture that calls into question our philosophical boundaries. The challenge is to deal with their sense of marginalization without compromising our basic political principles.

classic European-style nation-state founded on a single common ethnicity and language—was not possible in the northern half of North America. Reaching an accord (not a solution) between the two founding peoples was not a simple task and disputes continue to emerge, as discussed in Chapter 3. Since dominance is not feasible in the long term, then compromise becomes a political necessity and eventually the search for compromise becomes ingrained as a national trait. In *Reflections of a Siamese Twin: Canada at the End of the Twentieth Century* (1997), John Ralston Saul described Canada as a 'soft' country to capture this flexible trait.

See the section 'Faultlines within Canada' in Chapter 1, page 10, for more on the subject of Canada as a 'soft' country.

Multiculturalism is a form of cultural pluralism; in its Canadian form multiculturalism simply reflects the demographic reality created by liberal immigration policies. Canada has a pluralistic society. Cultural pluralism means that Canadians are not of any one cultural background, race, or heritage. But Canadian identity, flexible as it is, includes core values that are defined by Canada's history and geography. Three examples are: (1) government is based on British parliamentary institutions; (2) two official languages ensure a place for French as well as English, which also means that other languages have no standing in the political and public affairs of Canadian society; (3) Aboriginal peoples have special rights, which flow out of historic treaties, modern land-claim agreements, recognition in Canada's Constitution, and numerous court cases, as we have seen in Chapter 3.

Multiculturalism is a distinctively Canadian approach to social equality in nation-building, encouraging respect for cultural diversity. Ideally, multiculturalism is a shared vision of people of diverse cultural backgrounds seeking to live together in harmony. In reality, tensions exist, and four of these are articulated by the faultlines we have discussed. Multiculturalism should not lead to barriers between citizens and thereby threaten the cohesion of Canadian society. Most Canadians, but particularly new Canadians, recognize that multiculturalism binds Canada's diverse populations together. Canadians as a whole see no conflict between citizens having three identities—a cultural (ethnic/religious/language) identity, a regional identity (province/city/First Nation), and a national one. The national identity is based on tradition and law, specifically the Constitution, which includes Canada's Charter of Rights.

Canada remains a coherent society with different ethnic, religious, and social groups sheltered under this broad concept of multiculturalism. But are there dangers to national unity? European Union countries are having second thoughts about multiculturalism: the Netherlands and Germany have hardened their immigration policies. Even some prominent Canadians are questioning the concept because of their fear of ethnic ghettos and parallel societies. In 2005, Bernard Ostry, who was responsible for drafting and implementing the federal multicultural policies in the 1970s, expressed concerns about multiculturalism:

> The multiculturalism policy of the 1970s is under increasing criticism. This policy, developed from grassroots, was intended to assure new citizens full participation in the nation's life. Some now claim that it has had the opposite effect, that it has encouraged minorities to retreat to their own corners. (Ostry, 2005)

It certainly is arguable whether minorities in Canada are retreating into ethnic enclaves (Walks and Bourne, 2006). Two factors may be at play. First, social and economic barriers such as racism and low incomes may keep visible minorities in certain neighbourhoods. Second, social cohesiveness may cause minorities to live in the same neighbourhoods for the comfort and security of living near others of the same socio-cultural background.

Over time, immigrant children and grandchildren are more likely to find a place in mainstream Canada. Upward mobility is associated with education and perhaps interracial marriages. But the question remains: 'Might multiculturalism increase ethnic group identification at the expense of Canadian social cohesion?' Without doubt, tensions have arisen from time to time in Canada as people of different cultures, languages, and racial origins have expressed themselves when they encounter what they perceive as barriers within Canadian society, but peaceful discord is not in itself a failure of multiculturalism but part of a process of social interplay necessary to expose and resolve differences. Canadian history is on the side of both multiculturalism and immigration because immigrants—particularly their children—have found a place within Canadian society. While this process has not always been easy in the past, the challenge for visible minorities is even greater due to racism.

In sum, the issue confronting the security of Canada is not multiculturalism, but rather the potential for violent behaviour on the part of a handful of extremists or from those who have brought too much unsorted cultural baggage from their countries of origin. One example of 'extremists' cultural baggage' is the 1985 bombing of an Air India flight from Montréal to London en route to Bombay (Mumbai), which killed 329 passengers and crew, most of whom were Canadians. A more recent example, in this case a much more benign form of cultural baggage, was the agitated but peaceful protests of thousands of Tamils in Ottawa and Toronto who called for an end to the civil war in Sri Lanka (photo 4.9). Then, too, ongoing disputes with First Nations over treaty lands keep resurfacing, with the confrontation at Caledonia, Ontario, that flared up in 2006 and remains unresolved and the 2007 blocking of the main rail line between Toronto and Montréal by the Mohawks of Tyendinaga in

CP/Darren Calabrese

Photo 4.9

Tamil supporters block the Gardiner Expressway in downtown Toronto, 10 May 2009, to call attention to the escalating civilian death toll in the Sri Lankan civil war. Reports that nearly 400 civilians were killed during an overnight artillery bombardment against the Tamil Tigers by the Sri Lankan army brought protestors in Toronto into the streets.

eastern Ontario only the most recent. There is no single answer to why citizens resort to protests or even attack, harm, or kill fellow citizens, but, rightly or wrongly, their motivation is based on deeply held feelings of injustice stemming from Canada's colonial past or from disputes in people's countries of origin.

Population Trends and Demographic Faultlines

The uneven geographies of population and social change are well known to Canadian geographers. Bourne and Rose (2001) have discussed such changes (see 'Further Reading' in Chapter 1). From 2006 to 2008, the internal shifting of Canada's population was responding to growing economic opportunities, especially those in Western Canada. Over this two-year period, all regions saw their population increase but the greatest percentage increases took place in Western Canada, Ontario, and British Columbia (Statistics Canada, 2009a). Within Western Canada, Saskatchewan has enjoyed an economic boom that seems to be immune to the current global crisis and has resulted in an increase in population, which now exceeds one million. Why? First, Saskatchewan's resources base—agriculture, natural gas, oil, and potash—has had several 'good' years. Second, knowledge-based industries have taken hold, especially in the bio-agricultural area. Third, with major urban centres expanding, the construction industry has benefited. Population trends, whether regional gains or losses, provide the basis for the exploration of the four demographic tensions within Canada. In our discussion, these four faultlines focus on the implications of immigration for Canada and its cities, Aboriginal population growth, the changing French/English balance, and shifts in regional population. As Bourne and Rose point out, these population trends began some time ago and are already having an impact on Canada and its regions. Atlantic Canada, but especially Newfoundland and Labrador, faces a massive out-migration with considerable implications

for the regional economy. Former federal cabinet minister John Crosbie put it this way:

Christ, everywhere you turn around there's somebody leaving for Alberta. That they've got somewhere to go to better themselves and get work is not looked upon negatively. But we're exporting the ones that have the most initiative and are the most adventuresome and this weakens us. (Martin, 2005: A18)

The rapidly expanding Aboriginal population is clearly outpacing the growth in the rest of Canada's population. The myth of the vanishing Indian has been replaced with the re-emerging Indian. While the population of reserves continues to increase, many have moved to cities (see Peters, 2010). Over the past 20 years, cities in Western Canada have experienced substantial increases in the number of Canadians of Indian, Métis, and even Inuit ancestry.

The opposite situation faces the French/English balance. While the percentage of French-speakers in Québec remains above 80 per cent, in the rest of the country it has declined from 29 per cent of the national population in 1951 to 22 per cent in 2006, and the French-Canadian population is growing very slowly.

Immigration is, without a doubt, a most complex and somewhat sensitive subject. Canada has always been a land of immigrants. The annual inflow of immigrants now outnumbers those born in Canada. With many immigrants coming from different cultural and religious backgrounds, the settlement of most immigrants into Canada's three major cities has had an enormous impact on city landscapes, especially with the emergence of ethnic enclaves. Sociologists such as Herberg (1989) and Li (2003) see immigration as a tension between maintaining the sense of a common identity and heritage and the need to incorporate some elements of the cultures of the immigrants into mainstream Canadian culture. Geographers such as Hiebert (2000; Hiebert and Ley, 2003) and Ley (1999; Ley and Hiebert, 2001) examine one aspect of immigrant culture, namely the

Think About It

Why do you think Québec has been cool to the idea of multiculturalism and, instead, has reframed it as 'interculturalism'?

Think About It

In the short time between the 2006 census and 2009, both Newfoundland and Labrador and Saskatchewan have become 'have' provinces. What changed?

effect of transplanting foreign landscapes into Canadian cities. Immigration, therefore, is one of the most significant elements of change in Canadian society. The Canada described by Kenneth Hare in 1968, which was on the cusp of major social, economic, and political change, is gone (see 'Further Reading' in Chapter 1). No one would recognize Toronto 'the Good' of the 1950s in the cosmopolitan Toronto of today. Toronto and other cities have undergone dramatic transformations of their cityscapes and their ethnic/racial makeup.

Newcomers and Old-timers

Everyone is a newcomer—at first. As shown in Figure 4.7, newcomers, defined as foreign-born, have always formed a significant proportion of Canada's population. The first key event in massive migration was the need to settle the Canadian Prairies. As a result, the percentage of foreign-born jumped from 13 per cent in 1901 to nearly 25 per cent by 1911. The percentage fell to 15 per cent in 1951 because of low immigration levels during the Great Depression and World War II. Since then, the percentage has

steadily increased to reach 20 per cent, or 6,186,950 foreign-born, in 2006.

Why such an emphasis on immigration? Canada has had need for newcomers and they, in turn, seek a more secure life in a more open and affluent society. Children of newcomers become old-timers. This pattern of inclusiveness—some would say integration—forms the basis of the evolving Canadian society. Inclusiveness still leaves geographic space for cultural differences, as expressed by the province of Québec, by the territory of Nunavut, and by cultural enclaves in Canadian cities. Aboriginal peoples are a special case. As Peter Li (2003: 9) pointed out, 'Over time, earlier immigrants and their descendants become old-timers of the land, and their charter status places them in a privileged position vis-à-vis the newcomers who must accept the conditions of entry and rules of accommodation laid down by those who came before them.'

Vignette 3.9 explains why the 'newcomer to old-timer' model is not applicable to Aboriginal peoples, who for the most part were excluded from participating in mainstream Canadian society until recently, and

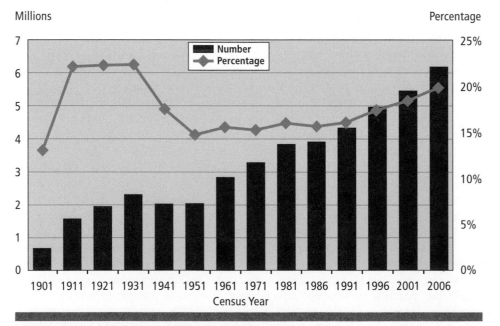

Figure 4.7   Number and share of the foreign-born population in Canada, 1901–2006.

Source: Chui, Tran, and Maheux (2008).

remain hindered by locational and institutional factors. For example, only after 1959 were status Indians allowed to vote in federal elections without losing their Aboriginal status, and it was not until 1958 that an Indian band—the Tyendinaga Reserve near Belleville, Ontario—was allowed any control over band funds (Dickason with McNab, 2009: 371). The Inuit and Métis, while not excluded legally, had little opportunity and encouragement to participate in political affairs, and, like the status Indians, sought some form of autonomous existence within Canada. The place of Aboriginal peoples within Canadian society has evolved over time.

For a more complete discussion of Aboriginal peoples in Canadian society, see 'Aboriginal Peoples' in Chapter 3, page 99.

Since newcomers often come from a different cultural world, they may bring with them different languages, customs, foods, and religious beliefs from those found in mainstream Canada. While these cultural imports often enrich Canadian society (Vignette 4.5), a few have caused a cultural clash between some members of the two groups. On top of these cultural differences, since many recent immigrants come from non-European countries, racism can add another element to these cultural tensions. Visible minorities make up a major and growing share of the populations of Canada's major cities. Bourne and Rose (2001: 115–16) argue that 'the challenge of overcoming exclusionary practices, including those based on racism in all its forms, and in particular the impact of these practices on labour market opportunities for youth, becomes all the more urgent.'

Immigration brings benefits, including maintaining Canada's population, supplying much needed skilled labour to Canada's workforce, and bringing investment capital. Yet, another side to immigration exists in the minds of the old-timers. The two primary concerns are that the newcomers will somehow lessen their economic position in Canadian society, and that, coming from a different cultural, linguistic, and racial background, they will remake Canadian society in their image. This latter fear is reinforced by the concentration of immigrants in Canada's three major cities, where they may create cultural enclaves that prevent their integration into Canadian society.

Since 11 September 2001, new concerns have emerged. First, as part of a campaign of violence against the West, a few Muslims who are **jihadists** may slip through Canada's immigration screening system. Some Muslims, perhaps with a certain degree of justification, believe that the West (and the United States in particular) is waging war against them, not just in the deserts and mountains of Iraq and Afghanistan but through the extent to

---

Vignette 4.5 Russell Peters, a Premier South Asian–Canadian Comic

David Johns

Russell Peters is a Canadian comedian who explores attitudes towards race (his and others) in a way that is challenging and cheeky. Born and raised in Ontario by parents who moved to Canada from India, Peters combines his experiences of growing up 'as the only brown child on the block' with insightful observations on race relations today. Now a North American star who has toured globally, Peters delights in exposing the humorous side of clashing cultures in Canada and the United States.

which Western cultural, political, and economic values are exported to and imposed on people throughout the world. For this reason, some followers of fundamentalist Islam have sought to fight this expansionist American hegemony by all available means, including sporadic and persistent acts of terrorism against their own people and against targets in the West or targets that represent what they revile of Western and American influence.

The history of immigration has shown two remarkable developments over time. First, Canadian society absorbs some of the cultural imports and multiculturalism fosters a social expression of the cultural contribution made by newcomers. Second, recent immigrants, especially their children, have shown a capacity not only to integrate into Canadian society, but also to reshape it. Charles Taylor would consider such reshaping as part of the porous nature of Canadian identity (Vignette 4.4). The cohesiveness of Canadian society depends on such integration and on an acceptance by all Canadians of cultural adaptation within the society to Old World cultures and religions. Such adaptation, of course, includes the ability to laugh at oneself and to appreciate the humour of other people and within other cultures. Russell Peters, a standup comedian who pokes fun at racism in Canada, exemplifies this positive side to immigration (Vignette 4.5).

Regional Patterns of Immigration

Immigrants are not evenly dispersed across Canada. Nearly 70 per cent locate in Toronto, Montréal, and Vancouver. In contrast, slightly over one-third (34.4 per cent) of Canada's total population live in these three census metropolitan areas. The first attraction of these cities to newcomers is economic because most jobs are found there, both in numbers and in range of opportunities. The second attraction is social, with newcomers drawn to places where fellow immigrants have already established and can help them to find housing and jobs. Meniy Zewde, for example, a parking lot attendant from Addis Ababa, was attracted to social factors found in Toronto but not in Saskatoon (Jiminez and Lunman, 2004):

> I went to Saskatoon when I first came here 12 years ago but I was so lonely and it was hard to find work. Here [Toronto], there is a large community of Ethiopians. I go to the Evangelical Church of Ethiopia and we have lots of Ethiopian restaurants.

The geographic result is that nearly 95 per cent of new Canadians reside in Canada's census metropolitan areas. The regional pattern of immigration, therefore, takes on an urban/rural dichotomy. Ottawa has expressed some concern about this dichotomy. In October 2002, Denis Coderre, at that time the federal Immigration Minister, proposed that prospective immigrants sign a social contract promising to live for three to five years in cities, towns, and rural communities outside of Canada's three major cities in return for Canadian citizenship. Such immigrants would be offered jobs and the right to bring their families to Canada. The underlying assumption is that the newcomers would fit into these communities and decide to stay after their contractual period expires (Fife, 2002). Coderre's proposal never took effect.

The Expanding Aboriginal Population

In 2006, the Aboriginal population, as measured by ethnicity, was around 1.3 million, marking a remarkable rate of population growth and demographic recovery from the low point of approximately 100,000 in the nineteenth century. In 1951, Aboriginal peoples comprised less than 2 per cent of Canada's population. By 2006, the proportion of Aboriginal identity population (as opposed to Aboriginal ancestry population, which is higher) had climbed to 4 per cent. Between 1996 and 2006, the Aboriginal population had increased 45 per cent, nearly six times faster than the 8 per cent rate of increase for the non-Aboriginal population (Statistics Canada, 2009h). Clearly, the early twentieth-century

Think About It

Can you think of an economic and/or demographic situation when Ottawa does not or would not encourage immigration?

Think About It

Why do religious institutions play such a prominent role for newcomers compared to other Canadians, many of whom have no religious affiliation and who are not active in an institution-based faith?

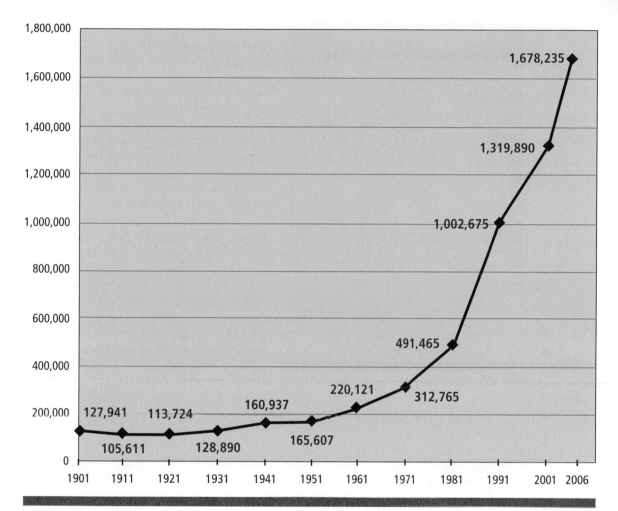

Figure 4.8   Aboriginal population by ancestry, 1901–2006.

Note: For discussion of the difference between 'Aboriginal identity' and 'Aboriginal ancestry', see Table 1.2.
Sources: Adapted from Statistics Canada (2003c, 2009h).

myth of the vanishing Indian has been put to rest (Figure 4.8).

The Aboriginal population boom began in the post World War II period and is associated with the relocation of Aboriginal peoples into settlements, food security, and access to public services. The next geographic shift began late in the twentieth century when Aboriginal peoples began to move to cities and towns. By 2006, over half lived in urban centres (ibid.). These demographic changes add to the challenge for Aboriginal individuals and organizations as they break away from dependency on Ottawa and try to adjust to the modern economy and society while still maintaining their cultural identity. This challenge is particularly powerful in Western Canada, where the ratio between Aboriginal and non-Aboriginal population is highest (Figure 4.9).

Aboriginal peoples are diverse. While the 2006 census (Statistics Canada, 2008c) categorizes Aboriginal peoples into three groups: North American Indians (about 62 per cent), Inuit (about 4 per cent), and Métis (about 34 per cent), First Nations alone represent many former tribes, different economies, and native languages. Added to the cultural and economic diversity, a divide exists between those living in their homelands such as Nunavut and those living in cities and towns.

Think About It

What does the urban/ homeland divide among Aboriginal peoples reflect?

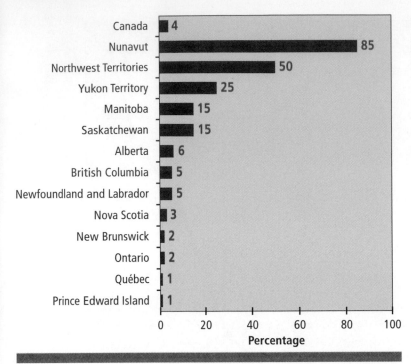

Figure 4.9   Percentage of Aboriginal people in the population, Canada, provinces, and territories, 2006.
Source: Statistics Canada (2009h).

Aboriginal peoples have gone through great changes since European contact. These demographic changes are classified into four phases in Table 4.13. Early contact was associated with population decline while population stabilized in late contact. The last phase shows a high rate of natural increase, though the birth rate has begun to decline. During the most recent phase, Aboriginal population has increased at a higher rate than Canada's population (see Table 4.13 and Figure 4.8). One measure of this expanding Aboriginal population is that Aboriginal people accounted for 4.0 per cent of the total population of Canada in 2006 compared to 2.8 per cent in 1996.

The economic and political implications of this high rate of natural increase are profound. Already the population increase plus the movement of Aboriginal Canadians to cities have made the presence Aboriginal people more visible in Canadian society. Across Canada, their proportion of the population by the six geographic regions varies widely. In the Territorial North, fully 50 per cent

Table 4.13 Major Phases for the Aboriginal Population in Canada	
Phase	**Characteristics**
Pre-contact	The Aboriginal population in Canada in the centuries preceding European discovery and settlement was at least 200,000 and possibly as large as 500,000. This population may have varied in size due to the carrying capacity of the land, which, in a hunting society, is controlled by the availability of game (food). For instance, natural conditions, especially weather, could affect the size and migration routes of animal populations.
Early Contact (1500–1940)	Aboriginal peoples who came into contact with Europeans were exposed to new diseases, and these new diseases often spread across the land prior to the arrival of Europeans in a particular place. Population losses were heavy. Loss of hunting lands also added to their demise. By the end of the nineteenth century, the Aboriginal population was just over 100,000.
Late Contact (1940–1960)	Rising fertility rates coupled with high mortality rates resulted in the stabilization of the Aboriginal population. Towards the end of this phase, fertility rates were high and mortality rates declined. The result was the start of rapid population increase.
Post-contact (1960 to present)	High fertility and low mortality account for remarkable population rebound. The net result was a population explosion. While the Aboriginal population is likely to increase at a rate well above the national average in the coming decades, its natural rate of increase is expected to diminish due to a declining fertility rate. The Aboriginal population now exceeds 1 million with a growing number living in cities.

Vignette 4.6 Early Estimates of Aboriginal Population

When Jacques Cartier sailed into Baie de Chaleur in 1541, the Indian and Inuit population in Canada may have been as high as 500,000. The exact figure will never be known. What we have are only estimates. By reconstructing the land's capacity to support wildlife and therefore also hunting societies, anthropologist James Mooney (1928: 7) estimated that about 220,000 Indians and Inuit lived in Canada at the time of contact. More recently, scholars have revised this figure upward. Dickason (Dickason with McNab, 2009: 40) and Denevan (1992: 370) estimate that the number of Aboriginal peoples living in Canada was closer to half a million. Whatever the exact figure, initial contact with Europeans resulted in a rapid depopulation of Aboriginal peoples. Factors include loss of hunting grounds and therefore food shortages, increased warfare, the spread of new diseases from Europe among the Indian tribes, and, in some instances, the

intentional slaughter of Native people by the European newcomers. Communicable diseases, such as smallpox, caused great suffering and many deaths among Indian tribes. Epidemics sometimes quickly reduced the size of tribes by half. Depopulation did not take place across British North America at once but in a series of regional depopulations associated with the arrival of British settlers, although European epidemic diseases, spread through Aboriginal trade networks, often preceded the actual appearance of Europeans. In 1857, the first comprehensive counting of the Aboriginal population for British North America, undertaken by the Hudson's Bay Company at the request of British House of Commons, totalled 139,000 (Bone, 2009: 62). By 1881, the Census of Canada recorded 108,000 Aboriginal peoples (Canada, 1884: Table 3.1). As shown in Figure 4.8, the low point was reached in 1911 when the Aboriginal population was recorded as 105,611.

of the population is of Aboriginal ancestry; Western Canada and British Columbia have the next highest, though much lower, percentages. Within the Territorial North, 85 per cent of Nunavut's population is classified as Aboriginal (Inuit in this case). Looking at Aboriginal population as it is distributed across the six regions, most reside in Western Canada (505,650 or 43 per cent), Ontario (242,495 or 21 per cent), and British Columbia (196,075 or 17 per cent). The Territorial North, on the other hand, has only 4 per cent of the total Aboriginal population.

With this rapidly growing population, the economic challenge—both for Aboriginal peoples and for governments—is to facilitate their participation in the economy. This challenge is especially evident in Manitoba and Saskatchewan. Aboriginal political power is, in part, linked to population size, so with the percentage of Aboriginal peoples in Canada increasing (exceeding 5 per cent by 2011), the federal, provincial, and territorial governments, but especially western provinces, will have to heed the political voice of Native leaders more carefully. Population increase within Aboriginal communities can

also have negative impacts, such as placing greater demands on scarce resources. For example, Aboriginal peoples are already facing increased housing demands and most live in public housing. Another implication

The Canadian Press/Nathan Denette

Photo 4.10

Elder Siipa Isullatak teaches children how to sew traditional Inuit clothing at Nakasuk Elementary School in Iqaluit, Nunavut, on 1 April 2009, the tenth anniversary of Nunavut becoming a separate territory.

Table 4.14   Aboriginal Population by Identity, Canada and Regions, 2001–6

Region	Aboriginal Population 2006	Aboriginal Population 2001	Change 2001–6	Change 2001–6 (%)	% by Region
Territorial North	53,135	47,990	5,145	10.7	4
Atlantic Canada	67,010	54,120	12,890	23.8	6
Québec	108,430	79,400	29,030	36.6	9
British Columbia	196,075	170,025	26,050	15.3	17
Ontario	242,495	188,315	54,180	28.8	21
Western Canada	505,650	436,455	69,195	15.9	43
Canada	1,172,795	976,305	196,490	20.1	100

Note: For more information on Aboriginal populations by identity and ancestry, see Table 1.2 and Figure 4.8.
Source: Statistics Canada (2008c).

of population increase has a regional component, with Western Canada, Ontario, and British Columbia facing the greatest increases in the next decade. As more and more reside off-reserve, the five provincial governments in these regions will face considerable pressure from 'urban' Indians who fall under the jurisdiction of the provinces and cities for social and health benefits, education and training, and public housing. Perhaps the most important federal program to assist First Nations students to enter post-secondary institutions and thereby equip themselves for a more productive role in society has been the Post-Secondary Student Support Program (PSSSP), which covers tuition, living expenses, and travel. Unfortunately, this program is unable to fund all students who apply. The federal government provides funding for status Indians and Inuit (as defined by the Indian Act) through the PSSSP. This funding is distributed by band councils under their eligibility criteria. For example, some bands fund more students at a portion of the total cost of their education, whereas other band councils give a grant covering all of a student's expenses.

French/English Language Imbalance

In the 2006 census, the French/English language numbers continue to favour English-speaking Canadians. However, the most striking gains occurred among allophones. In 2006, 18 million or 58 per cent of Canadians declared English as their mother tongue,

while French-speakers accounted for nearly 7 million or 22 per cent of all Canadians (Table 4.15). Allophones—those declaring neither English or French as their mother tongue—grew to just over 6 million to form 20 per cent of the total population. At the same time, Canadians who spoke both official languages increased to 17.4 per cent.

The historic trend indicates an ever-increasing percentage of English-speaking Canadians—until 2006, when a slight decline took place due to an increase in allophones, who comprised 20.1 per cent of the national population (Statistics Canada, 2008d). The percentage of francophones in the total population continued to drop, although their numbers increased. Thus, those who reported French as their mother tongue made up 22.1 per cent of the Canadian population in 2006, down from 22.9 per cent in 2001 and 29.0 per cent in 1951 (Table 4.15).

French- and English-speaking Canadians represent the traditional duality of Canadian society. With the establishment of the Province of Canada in 1841, the relationship flowered into a partnership with English-speaking Canada West and French-speaking Canada East sharing political power. In 1867, the British North America Act (section 133) established that both French and English 'may be used by any Person in the Debates of the Houses of Parliament of Canada and of the Houses of the Legislature of Quebec', as well as in federal courts, but it was not until the Official Languages Act of 1969 that French truly began to be entrenched in the

Table 4.15	Population with French Mother Tongue, 1951–2006					
Year	Canada (000s)	Canada (%)	Québec (000s)	Québec (%)	Rest of Canada (000s)	Rest of Canada (%)
1951	4,069	29.0	3,347	82.5	722	7.3
1961	5,123	28.1	4,270	81.2	854	6.6
1971	5,794	26.9	4,867	80.7	926	6.0
1981	6,178	25.7	5,254	82.5	924	5.2
1991	6,562	24.3	5,586	82.0	976	4.8
1996	6,637	23.5	5,747	81.5	970	4.5
2001	6,782	22.9	5,802	81.4	980	4.4
2006	6,892	22.1	5,917	79.6	975	4.1

Source: Adapted from Harrison and Marmen (1994: Table 2); Statistics Canada (1997a, 2002b, 2008d).

institutional fabric of Canadian society. The assignment of education to the provinces in the BNA Act (section 93), and the stipulation that Roman Catholic (i.e., francophone) schools had equal standing with Protestant (i.e., anglophone) schools ensured that French would retain a dominant position in Québec, at least for the time being.

Linguistic rights for both French and English as the official languages, as outlined in the Official Languages Act and in the Constitution Act, 1982 (sections 16, 23, and 29), give them a special place in Canadian society. One implication is that new Canadians whose language is neither French nor English gravitate to one linguistic group or the other. Within Canada, English is the dominant language. English is also the business language of North America and, to a large degree, the world. Immigrants therefore have a compelling reason to learn English. Also, most immigrants settle in English-speaking areas and so are naturally drawn into the anglophone linguistic group. Outside of Québec and New Brunswick, the French language is losing ground. Anne Gilbert (2001: 173) points to the decision of the government of Ontario to reject the concept of recognizing both English and French as official languages as a missed opportunity to showcase Canada's serious commitment to its dual languages.

French/English dualism is a fundamental aspect of the geopolitical nature of Canada. As Jacques Bernier (1991: 79) of Université Laval stated: 'Canada's duality is intrinsic, and as long as it is not clearly recognized and dealt with,

the issue of Canadian unity will remain.' This duality is a political concept embedded in the historical relationship between the two cultures. The main indicator of the stability of this French/English dualism is language; in other words, the stability of this concept depends on a relatively constant number of Canadians speaking each language. But how should we measure duality? Should mother tongue or household language hold the key? Or does the number of bilingual Canadians hold sway?

In 2006, 18.1 million Canadians declared that their mother tongue (the first language learned that is still understood) was English, while 6.8 million indicated French (Statistics Canada, 2008d). More significantly, the number of French-speaking Canadians is increasing at a rate slower than the national rate of population growth (Table 4.15). Over the 1951–2006 period, the proportion of Canadians indicating that French is their mother tongue has dropped from 29 per cent to 22 per cent, although the proportion of bilinguals has increased to 14.4 per cent.

The French/English duality remains strong at a national level, but the increasing number of Canadians whose mother tongue is neither French nor English reflects the arrival of immigrants. Within Québec, the low fertility level among francophones means that the adoption of French by newcomers is extremely critical for the long-term maintenance of the French language. To ensure the well-being of the French language in Québec, the provincial government passed a number of language laws, including Bill 101 in 1977.

Table 4.16 Changes in Regional Populations, 2001–6				
Geographic Region	2001	2006	Increase/Decrease	% Change
Atlantic Canada	2,285,729	2,284,779	–950	–0.1
Québec	7,237,479	7,546,131	308,652	4.3
Ontario	11,410,046	12,160,282	750,236	6.6
Western Canada	5,073,323	5,406,908	333,585	6.6
British Columbia	3,907,738	4,113,487	205,749	5.3
Territorial North	92,779	101,310	8,531	9.2
Canada	30,007,094	31,612,897	1,605,803	5.4

Source: Adapted from Statistics Canada (2007a).

Core/Periphery Populations

Canada's population is unevenly distributed across the country's regions. Population remains concentrated in Central Canada, though population growth in Western Canada and British Columbia has been well above the national average. According to the 2006 census, the hinterland regions have increased their share of the national population, but that increase has been geographically uneven (Table 4.16). Ontario has continued to grow rapidly while Québec's rate of population increase has slowed. More recent 2008 estimates from Statistics Canada (2009c, 2009d) indicate that the greatest gains continue to occur in Western Canada and British Columbia, with Ontario's rate slowing to below the national average.

Economics dominates regional shifts. Interprovincial migration follows. The global slowdown that began in 2008 has dramatically affected Canada's manufacturing regions, causing their population increases to fall below the national average. Resource-based economies in British Columbia and Western Canada have fared better and their population increases are well above the national average (Statistics Canada, 2009c, 2009d). Even Atlantic Canada has stemmed its high levels of out-migration. By the next census in 2011, these short-term demographic blips may turn out to be just that—blips—or a more permanent trend.

Canada's Economic Face

Canada is firmly caught in the global recession. Restructuring of the economy is taking place and the manufacturing base continues to contract (Figure 4.10). Already, the offshore relocation of the lower end of that base has taken place and more is likely to follow because of wage differentials between Canada and emerging industrial countries. This offshore relocation began when Canada reduced its tariffs on imported manufactured goods as a consequence of the creation of the FTA (1989), NAFTA (1994), and the World Trade Organization (1995). It made more business sense to produce products in countries with low labour costs and then export these products to Canada. The global recession has accelerated this process of the **hollowing-out** of Canada's manufacturing base. Even the high dollar (which approached parity with the US dollar by early 2010) has taken its toll in this process. Plants that close and relocate offshore because they are unprofitable at current exchange rates are highly unlikely to return even if the Canadian currency declines.

Until the global economic meltdown, the Canadian economy was growing at a pace well ahead of other G8 countries. The suddenness of the economic collapse is revealed in the unemployment figures for 2008 and 2009, which show a jump from 6.1 per cent to over 8 per cent (Statistics Canada, 2009e). The negative impact on Canada's workers has been significant, and recovery to former employment levels will take time. By the mid-point in 2009, however, the nadir of the recession had passed and economic recovery appeared to be in sight. But just when was not clear because the timing of the cyclical nature of the global economy is not well understood and certainly not predictable (Figure 4.11). Equally important for

Think About It

Unless Canadian wage rates decline significantly, is the process of hollowing-out eventually going to relocate most manufacturing, including automobiles, to China, India, and other low-wage countries?

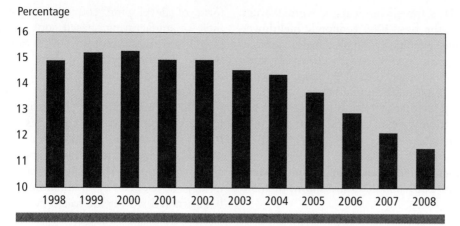

Figure 4.10   Manufacturing's declining share of employment.
Source: Bernard (2009: Chart B).

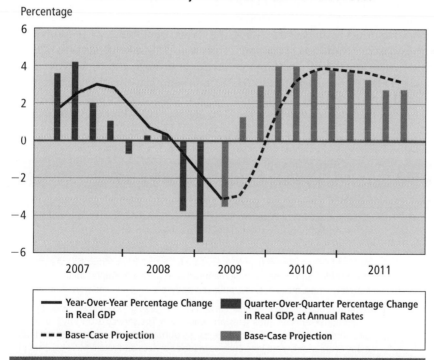

Figure 4.11   The cyclical nature of Canada's economy.
Canada's economy is strongly affected by outside forces, especially trade with the United States. This graph indicates the impact of the decline of the US economy and that of the rest of the world on Canada's trade and therefore its economy. In January 2010, the Bank of Canada looked at Canada's overall economic performance from 2007 to June 2009 and then projected its performance into the latter half of 2009 and beyond to 2011. The bank used a base-case projection, assuming no change in current public and private policies, such as a change in the bank rate, which was set extremely low in 2009 in order to stimulate the economy.

Source: Bank of Canada (2010).

Think About It

If Canada's automobile industry declines, will Ontario's economy follow suit?

Canada is the shape of its economic structure in the post-2008 economic meltdown. For the Canadian economy, a major caveat is the US economy. With such a heavy dependency on exports to the US, Canada's economy, but particularly its manufacturing and forestry sectors, will not hit stride until the US economy recovers—and that may take some time.

Provincial and Canadian unemployment data for 2007 and 2009 are shown in Table 5.1, page 195.

Economic Structure

Like other modern industrial countries, Canada's economic structure has evolved from an agrarian to a post-industrial economy. By structure, we mean the relative share of activity in the **primary**, **secondary**, and **tertiary/quaternary sectors** of the economy (Tables 4.17 and 4.18). With the shift of manufacturing activities to China and other low-wage countries, economic structures of developed and developing countries are changing. The implications for Canada and especially its secondary sector are profound. This global shift takes the

form of developing countries expanding their manufacturing based on low-cost labour and, with few trade restrictions between countries, exporting these products to wealthy industrialized countries, especially the United States. In turn, these exports have eroded much of the manufacturing base of industrialized countries, forcing them to turn to more sophisticated forms of manufacturing that rely heavily on research and innovations. High-speed train technology developed by Bombardier, which has garnered worldwide sales, including in China, is but one example.

More information on Bombardier's train technology can be found in Chapter 6, page 246.

Economists Peter Drucker (1969) and Daniel Bell (1976) and geographer David Harvey (1989) were among the first scholars to recognize this global-driven adjustment leading to an information society and the growing importance of a **knowledge-based economy**. Unlike Bell, who describes this change through the rose-coloured glasses of capitalism, Harvey looks at this historic process through darker-tinted glasses of Marxist analysis. The knowledge-based economy

Table 4.17	Economic Structure by Sectors
Economic Sector	**Characteristics**
Primary	Economic activities concerned with the extraction of natural resources of any kind, such as fishing, farming, forestry, mining, and trapping.
Secondary	Economic activities that process, transform, fabricate, or assemble the raw materials derived from primary activities, or that are high-tech knowledge-based activities that reassemble, refinish, or package manufactured goods. Examples include automobile manufacturing, meat-packing, pulp and papermaking, and software manufacturing.
Tertiary	Economic activities involve the sale and exchange of goods and services, including professional services, retail sales, and education. The service economy is extremely bifurcated with some high-end careers (accountants, bankers, doctors, and lawyers) and many more poorly paid jobs, such as in food services and retail (McJobs), that often offer only part-time employment at minimum wage.
Quaternary	Economic activities that deal with the handling and processing of knowledge and information that leads to decision-making by companies and governments. The leading edge of the knowledge-based economy is found at research centres located at universities and within governments and corporations. Innovative research clusters form around these centres with companies transforming the innovations into commercial production.

Table 4.18   Sectoral Changes in Canada's Labour Force, 1881–2006

Development Stage	Year	Primary	Secondary	Tertiary*
Agricultural	1881	51	29	19
Early industrial	1901	44	28	28
Late industrial	1961	14	32	54
Post-industrial	2006	4	20	76

*Because data for the quaternary sector are not available, percentages for the tertiary sector conventionally include both tertiary and quaternary jobs.
Sources: Adapted from McVey and Kalbach (1995: Table 10.3); Statistics Canada (2006c).

offers hope for advanced industrial countries, like Canada, to offset a decline in traditional manufacturing activities. Success depends on integrating scientific knowledge, in the form of technology and innovation, into new products and services that can successfully compete on the international stage. Investments by both the public and private sectors are essential. Each region of Canada has a different set of possibilities in the knowledge-based economy, and these possibilities are discussed in the regional chapters.

Vignette 4.7   Global Economic Collapse

In late 2008, the global economy froze due to the financial crisis in the United States, caused in large part by overly risky lending practices on the part of banks and investment firms. As David Parkins's cartoon demonstrates, the suddenness of the downturn caught everyone by surprise and the collapse of the US financial system sent shock waves around the world. The Governor of the Bank of Canada, Mark Carney (2009), summarized Canada's economic situation as:

> These are difficult economic times, with the Canadian economy being buffeted by an intense and synchronized global recession. In recent months, that global recession has been exacerbated by delays in implementing measures to restore financial stability around the world.

Courtesy of David Parkins

Photo 4.11

The global financial crisis that struck in late 2008 caught a lot of people by surprise, although many of the financial firms on Wall Street that were at the centre of crisis received a multi-billion dollar bailout from the US government.

Think About It

Does the city where you reside show any signs of research activity and, if so, is it associated with the local university and its research park?

What is an information society? First, such a society has a large portion of highly educated citizens concentrated in major cities. These workers function within the digitized world, and a significant number are employed in scientific research. Richard Florida's theory of the creative class fits into the concept of the information society, and his theory supports both the notion that creative people are the drivers of urban and regional growth and the concept of 'clustering', namely, that the creative class is attracted to cities with a cultural soul, where a wide variety of artistic and cultural events flourish and where ethnic diversity and tolerance to non-mainstream lifestyles are well established (Florida, 2002a, 2002b, 2005, 2008). Florida further claims that innovative firms, public research centres, and universities situated in culturally rich cities have an advantage in recruiting and retaining highly creative labour.

Second, an information society is supported by a knowledge-based economy that places a high priority on scientific research. Such research creates new products that transform society: the potential impact of an electric car on city design and greenhouse emissions; biotechnical innovations that have led to higher-yielding grains; application of green technology (e.g., green roofs) to city buildings, thus making large cities more livable places. The role of Canadian universities is crucial. Canada's most recognized technological cluster,

the Technology Triangle, consists of Waterloo, Kitchener, and Cambridge, Ontario, and in that cluster the University of Waterloo is an essential catalyst both in research and in training scientists. Universities, by establishing research parks, not only foster partnerships with federal research institutions and private firms, but also provide a link between pure and applied research. In this process, the **National Research Council of Canada** plays an important role both in conducting research and in providing funding through various programs. These innovative incubators create scientific research and innovation zones like the Waterloo region, thus demonstrating the *power of place*. Such scientific nodes are one expression of the *clustering approach* where the search for innovations in science and technology flourishes. Innovation Place on the grounds of the University of Saskatchewan is but one instance of a Canadian university fostering innovation, commercialization, and economic growth through university, industry, and government partnerships. In 2009, Saskatoon's Innovation Place received the '2009 Outstanding Research-Science Park Award' from the Association of University Research Parks, based in Tucson, Arizona (Association of University Research Parks, 2009).

A broader concept of the power of place is discussed in Chapter 1, page 14, while more on the Technology Triangle is found in Chapter 5, page 200.

Vignette 4.8   Public Support for Scientific Research and Knowledge-based Nodes

Scientific research is expensive but essential in post-industrial economies. Canada is no exception. Basically, scientific research to produce innovative products and services is a key factor in transforming Canada's economy from an industrial one to a post-industrial economy. Public support is essential and the federal Scientific Research and Experimental Development (SRED) program is one example. SRED provides tax rebates and credits for companies engaged in scientific research for product/service development. Provinces also support such research, with Québec providing the most generous support with a tax credit of 37.5 per cent compared to Ontario's similar credit of 10 per cent (PriceWaterhouseCoopers, 2009). The impact of federal support on one software company in Saskatoon is revealed in comments by its CEO, Amit Gupta of Solido Design: 'The big trend right now is to export R&D operations to China and India. As a result of this SRED program, it becomes cost effective for us not to export those jobs and keep those jobs here locally in Saskatoon'.

Source: PriceWaterhouseCoopers Canada (2009) and Kyle (2009: C9).

SUMMARY

Canada's population continues to grow thanks in large part to immigration, and population distribution remains concentrated along the southern border with the United States. The process of urbanization remains a dominant driver of Canada's population geography with more and more people living in urban settings. Given Canada's increasingly pluralistic society, multiculturalism remains a key plank in Ottawa's effort to accommodate newcomers. The search for cultural accommodation without jeopardizing Canada's traditions remains an ongoing process, reflecting the dynamic nature of Canadian culture.

Canada's economy is under pressure to find its place within the global economy. This search is made more challenging by the worldwide economic crisis that began late 2008. Given Canada's strong trade dependency on US markets, full recovery is closely tied to the United States economy.

Stresses within Canadian society take the form of four faultlines, namely Aboriginal/non-Aboriginal, core/periphery, French/English, and newcomers/old-timers. Key demographic changes are occurring along these divides, reflecting internal forces and revealing population and political shifts. The first faultline is characterized by a much higher rate of natural increase among the Aboriginal population. The second faultline involves changes in the sizes of Canada's regional populations. Since Confederation, population growth has occurred in all regions but at varying rates. The Ontario, Québec, and Atlantic Canada shares of the national population have declined, while those of British Columbia, Western Canada, and the Territorial North have increased. The third faultline has experienced a faster rate of population increase among English-speaking Canadians than French-speaking Canadians; the explanation lies not in fertility rates but in immigration. The fourth faultline—between new Canadians and those born in Canada—is reflected in the federal policy of multiculturalism and is played out daily in Canada's major urban centres, where the vast majority of new Canadians settle. Indeed, according to a recent Statistics Canada study, the European-origin ('white') population in the Toronto region will become the new 'visible minority' by 2017, and by 2031 the so-called visible minorities will comprise 63 per cent of the Toronto population. Vancouver is projected to be 59 per cent visible minority by 2031 (Javed, 2010).

The following chapters, 5 through 10, focus on the regional nature of Canada, with each chapter devoted to exploring one of Canada's six geographic regions. Friedmann's version of the core/periphery model provides a guide for the ordering of these six regions (Table 1.4). The regional discussion therefore begins with Ontario and Québec, which represent Canada's economic core. These two chapters are followed by chapters on British Columbia, Western Canada, Atlantic Canada, and the Territorial North. While many of the same topics and issues appear in each of the six geographic regions, each is examined from the perspective of the region under discussion. Each regional chapter will illustrate the physical diversity within the region and show how this diversity affects human activities and settlement. Each chapter also has a section on the region's historical development to enrich our sense of the present by providing a link with the past. One recurring theme is the relationship among the six regions. This relationship is cast in the core/periphery model, but it is changing due to continental economic integration (NAFTA), global trade (the WTO), the restructuring of the world economy and the North American/Canadian economies, plus the shift of power—demographically and economically—beyond the core regions of Ontario and Québec. Within this discussion, the shift of regional economies towards a greater participation of knowledge-based activities is highlighted.

Finally, Canada's regions are witnessing radical changes in their economic circumstances that challenge the core/periphery theory. This 'grand' theory assumes that manufacturing prices would always increase at a greater rate than foodstuffs and commodities; hence, the manufacturing centres of the world would gain more and more over time. Two watershed events—the global market and the industrialization of Asia—have tipped the world economy towards higher prices for foodstuffs and commodities. If this price shift is permanent, the economic and demographic consequences for Canada and its six regions will vary. The implications for each region and these differences are noted in each regional chapter.

CHALLENGE QUESTIONS

1. From 2001 to 2006, some geographic regions increased their share of Canada's population while others did not. Is this just a short-term 'blip' on the demographic radar screen or are the fundamentals in place for this demographic shift to continue for the next decade?

2. The former Minister of Citizenship and Immigration, Judy Sgro, stated that 'We need to be loyal to one country as far as your citizenship. Your heart can be where you were born, but I think the commitment to Canada has to be strong and I think dual citizenship weakens that.' Should Canada abolish the option to have dual citizenship for the reasons she suggested?

3. In recent years, the natural rate of population increase for Aboriginal peoples has been significantly above the national average. Does the demographic transitional theory offer anything to suggest that the rate may slow down?

4. What could slow and even reverse the process of hollowing-out where most manufacturing, including automobile assembly, is relocated to other countries, thus leaving Canada with a very weak secondary sector?

5. Is the paradigm of continuous population growth sustainable without reducing the quality of life and damaging our environment, or should Canadians think about an alternative paradigm?

FURTHER READING

Bunting, Trudi, Pierre Filion, and Ryan Walker, eds. 2010. *Canadian Cities in Transition: New Directions in the Twenty-First Century*, 4th edn. Toronto: Oxford University Press.

Canada is a highly urban society. Its cities are complex, dynamic, and ever-changing. In this most recent edition of *Canadian Cities in Transition*, the editors have brought together many of Canada's leading urban geographers and planners to discuss why and how Canadian cities work or don't work. The underlying theme of the 25 chapters is transformation, as cities today are in the midst of remarkable change and new adaptations brought about by such factors as environmental degradation, traffic congestion, immigration and internal migration, and the new economy.

The range of subjects covered in this volume is broad, and includes such new topics as obesity and the built environment, food systems and food 'deserts', Aboriginal peoples in urban Canada, gentrification, cities with declining populations, and social polarization and neighbourhood inequality. Of special interest for those concerned about the future of Canadian cities in a changing world is Chapter 5, by William E. Rees, 'Getting Serious about Urban Sustainability: Eco-Footprints and the Vulnerability of Twenty-First-Century Cities', in which the author introduces the startling concept of cities as 'human feedlots' and outlines why cities must become 'green' to survive and thrive. The final chapter, by Larry S. Bourne and R. Alan Walks, 'Challenges and Opportunities in the Twenty-First-Century City', examines the issue of homelessness and the tensions between the reality of urban sprawl and the goal of contemporary planners and politicians to achieve the 'compact city'. Chapters on the history of Canada's cities, as well as on transportation, housing, economic restructuring, inner cities and the creative class, and governance also are of special interest.

ONTARIO

INTRODUCTION

Ontario's prominent position within Canada and North America has been badly shaken by the sudden economic downturn that began in 2008. Manufacturing has been hit hard, especially the automobile industry. Yet, the basic factors that propelled Ontario to the lead position within Canada and its prominent role in North America have not changed. Ontario remains the heartland of Canada's manufacturing, continues to play a dominant in the financial industry, and serves as the cultural centre for English-speaking Canada. While the old economy is gone, the shape of the new economy is uncertain. For Ontario to prosper and progress into the next decade as well as to retain its title as the economic engine of Canada, bold and innovative measures will be necessary. Two key components of its economy—forestry and manufacturing—are stalled in the old economy and they need two things: (1) fresh directions and investments to get back on the economic highway; and (2) recovery of the US economy and the resumption of high levels of exports to the US.

CHAPTER OVERVIEW

The upheaval in the North American automobile industry makes our Key Topic the restructuring of the automobile industry. As well, this chapter will consider the following topics:

- Ontario's physical geography and historical background.
- The downsizing of the manufacturing sector.
- The basic characteristics of Ontario's population and resources.
- The economic and environmental challenges facing Ontario.
- The significance to Ontario of several important trade agreements.
- Ontario's dichotomy—an industrial core in the south and a resource hinterland in the north.
- Ontario's changing role within Canada and North America.

Queen's Park, one of Canada's oldest urban parks, is located in downtown Toronto. Within the park, the Ontario Legislative Building, built in 1893, represents a form of Romanesque style architecture while the surrounding landscape retains its original 'English' designed gardens. Photo: © All Canada Photos/Alamy.

Ontario within Canada

The first decade of the twenty-first century has not been kind to Ontario. Its economy took an economic jolt, especially its forest and manufacturing industries. While signs of an economic decline appeared earlier, the worst blow came in 2008 with the collapse of two of its principal automobile manufacturers, Chrysler and General Motors. The negative spinoff rippled through the manufacturing sector, causing massive layoffs. For the first time, Ontario received equalization payments in 2009, making it—at least for 2009—a 'have-not' province. Yet, in 2006, Ontario, by contributing 39 per cent of Canada's **gross domestic product**

Figure 5.1   The province of Ontario.
Ontario has the largest population of Canada's regions and has been the economic motor of the country, but a number of factors related to economic restructuring, the loss of manufacturing jobs, rising unemployment rates and the recent global economic crisis meant that in the 2009, for the first time, the province was classified as a 'have-not' province and received equalization payments from the federal government. (*Source: Atlas of Canada*, 2006, 'Ontario', at: <atlas.nrcan.gc.ca/site/English/maps/reference/provinceterritories/ontario>.)

(GDP), remained the foremost economic region in the nation (Figure 5.2). By 2008, this measure of economic power had dropped to 37 per cent (Statistics Canada, 2009a); and, in the same year, nearly 18,000 residents migrated to other provinces (Statistics Canada, 2009e). Ontario, instead of leading the country in population growth, has fallen well below the national average. Only the inflow of international migrants has allowed Ontario to have a positive population growth rate.

What happened to the strong man of Confederation? First, Ontario was caught in a Canada-wide decline in manufacturing and forestry. Second, the sudden and unexpected global economic collapse of 2008–9 put a serious dent in Ontario's economic performance, especially in its automobile manufacturing and parts industries. Exports to its major customer, the United States, fell precipitously, thus magnifying Ontario's economic downturn. Third, the days of an 'open border' are gone, making cross-border manufacturing more costly and less attractive for locating new plants in Ontario; in fact, with the hard line taken by the head of Homeland Security in the US, Janet Napolitano, border crossing could become even more difficult.

Next, the western provinces have outpaced the economic and population growth of Ontario, with the result being that Western Canada and British Columbia have increased their share of the national GDP and becoming the leading choices of destination as measured by interprovincial migration (Statistics Canada, 2009e). Finally, while cross-border trade in manufactured goods had been the norm, the US policy of '**Buy America**' left Ontario's manufacturers unable to participate in the massive US stimulus spending program of 2009–10.

Ontario's current economic woes are complex and interwoven. In the past, Ontario benefited greatly from a series of trade agreements with the United States, beginning with the 1965 Auto Pact and including the North American Free Trade Agreement (1994). With its economic wagon linked so closely to US markets, the roots of Ontario's economic troubles go beyond its borders. A report by Statistics Canada (2009a) put it this way:

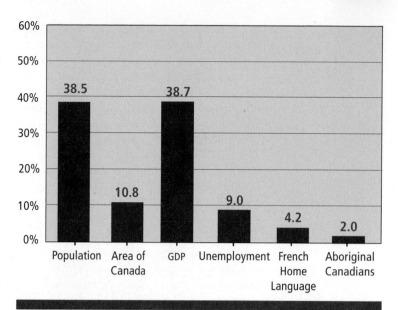

Figure 5.2   Ontario vital statistics.
Ontario's economic and political strength within Canada is revealed by its share of the nation's GDP and population. With 12.2 million people as of 2006, Ontario comprises the largest single market in Canada and holds the demographic key to political power in Ottawa, and its economy, by accounting for nearly 39 per cent of national GDP, exhibits a high level of productivity. Franco-Ontarians and Aboriginal peoples form small minorities while newcomers represent a growing force within Ontario, especially the Greater Toronto Area. In 2007, Ontario's unemployment rate of 6.4 per cent was above the national average of 6.0 per cent, but by 2009 Ontario's rate had soared to 9 per cent, well above the much lower rates found in the western provinces and even those of Québec and New Brunswick (Table 5.1). (*Note:* Unemployment percentage is for 2009; demographic data are based on the 2006 census.) *Sources:* See Tables 1.1, 1.2, 5.1.

GDP in Ontario fell 0.4 per cent in 2008 after a 2.3 per cent increase in 2007. Ontario's growth has generally been below the national growth rate since 2003. Automobile and auto parts production declined more than 20 per cent, while production of wood products dropped sharply. Weakened export demand contributed to a decline in 16 of 21 major manufacturing industry groups in Ontario. Construction activity fell back slightly as several engineering projects neared completion and residential investment declined 2.5 per cent. Personal expenditure grew but at a slower pace than in 2007, in tandem with a deceleration in labour income.

For more on international trade agreements, see Table 1.3, 'Key Steps Leading to a More Open Canadian Economy', page 18.

Can Ontario recover? While Ontario struggles with a host of economic challenges,

Think About It

Was the loss of an 'open border' one reason for Siemens to announce the closing of its Hamilton plant in 2011 (Marr, 2010) and then move its gas-turbine division to Charlotte, North Carolina?

including the downsizing of its automobile sector, the province's strength remains its central location within Canada and North America; its large population/market; its natural resources; and its vibrant cities and universities. With this market power plus ample support from Ottawa for its automobile industry, Ontario will continue to be the leading manufacturing area in Canada. Its products are sold across North America and, to a much lesser extent, in other parts of the world. Without a doubt, Ontario will remain the economic pulse for Canada's east–west economic/transportation axis as well as its north–south axis in North America. According to Thomas Courchene, one of Canada's leading economists, Ontario had evolved into one of North America's key economic regions before the end of the twentieth century (Courchene with Telmer, 1998). But that trade relationship, so pivotal to Ontario's economic well-being, has a downside. Perhaps Professor Glen Norcliffe's (1996: 44) warning some 14 years ago about too heavy a reliance on a 'continentalist [trade] option' still has merit and should lead to a widening of not only Ontario's but Canada's export strategy. The good news is that, by late 2009, the worst seemed to have passed and automobile manufacturing was showing signs of stabilization and recovery. The bad news is the return of the **Dutch disease**, which offers a theoretical explanation as to why a rapidly expanding oil resource in Alberta can push Canada's exchange rate to higher levels, thus making the price of its manufactured goods more expensive in the US. In 2002, Canada's dollar was valued at 64 cents on the US dollar. By April 2010, the Canadian dollar had reached parity with the US dollar and was predicted to remain at or near par for the foreseeable future.

For further understanding of the extent of Canada's export dependency on the US, see Vignette 1.4, 'Is Continentalism Inevitable?', page 24 and Table 1.6, 'Canadian Exports, 2006', page 22.

Historically, four natural resources—agriculture, forests, minerals, and water—have spurred the province's economic development, and the processing of these products created a strong industrial base, which in turn accounts for Ontario's rapid population growth. The Great Lakes provided low-cost water transportation and, at Niagara Falls, low-cost hydro-electric power. But today, Ontario faces three challenges. One challenge is energy. Not only does Ontario require more energy, but the cost of energy has risen rapidly. A second challenge confronts the manufacturing sector, which is coping not only with fierce competition in the Canadian market from countries with much lower labour costs but also, when the price of oil is high, with the higher value of the Canadian dollar—not to mention the difficulty in access to US markets, partly because of the 'Buy America' policy. A third challenge, which affects northern Ontario in particular, is the forest industry. Again, dependency on US markets has seen good times for this industry, but for the last 10 years at least the forest sector has seen only bad times.

Ontario's position within Canada goes beyond its economic role. Ontario remains the political linchpin in Confederation and has become the cultural sparkplug for English-speaking Canada. Its cultural and political role can be attributed largely to Ontario's large and affluent population. Ontario, with nearly 13 million people, has by far the largest population of the six regions. For that reason, Ontario sends more representatives to the House of Commons than any other province. Its large market has the capacity to support the arts, including the Toronto Symphony Orchestra, the Hamilton Symphony, the Canadian Opera Company, the National Ballet, the Stratford Shakespearean Festival, the Shaw Festival, many other venues for staging theatrical works and popular music concerts, the Art Gallery of Ontario, and the Royal Ontario Museum; an extensive network of post-secondary education institutions; and professional sports teams not found elsewhere in Canada—the Toronto Blue Jays (baseball), the Toronto Raptors (basketball), and Toronto FC (soccer).

Ontario's political power in Ottawa often draws the federal government to seek solutions, whether to ease access across the US border or to provide financial support for Ontario's automobile industry. Since Ontario's economic future lies in expanding trade with

Courtesy of Tim Griffith

Photo 5.1

Toronto's high-order cultural facilities includes the Four Seasons Centre designed specifically for opera and ballet. Opened in August 2006, the performance of Wagner's Ring Cycle attracted Wagner enthusiasts from around the world. *Source:* Website for the Canadian Opera Company: <www.coc.ca/house/house.html>.

the United States, Ontario is interested in having Ottawa pursue trade policies with the United States that support its industrial base, facilitate ease of border crossings for its goods, and promote the well-being of its manufacturing industry, especially the automotive sector. Both the federal and provincial governments have responded. Their investments demonstrate a strong commitment to manufacturing. While the federal and provincial governments have made substantial contributions to the Ontario Automotive Investment Strategy over the past 10 years, thus ensuring a partnership with automobile companies to strengthen the industry's competitiveness in the global marketplace, more recently Ottawa and Ontario have committed billions of dollars to keep Chrysler and General Motors afloat.

Regardless of the current economic difficulties, Ontario will remain Canada's core industrial region for the foreseeable future. Even recent **net interprovincial migration** figures, which strongly favour Western Canada and British Columbia over Ontario, do not signal an end to Ontario's dominant position within Canada; rather, they indicate a continuation of the shift of economic and demographic power westward due more to the relatively weaker demographic and economic performances of Québec and Atlantic Canada. As shown in the Figure 5.3, prepared by two of Canada's leading economists, Don Drummond and Derek Burleton (2008), Ontario's economy could shift either way. As they put it:

With much of Ontario's economic success driven by advantages that no longer exist, a new direction is required. We look to the provincial government to take leadership on this front by developing a vision on where it plans to take the economy down the road.

Think About It

In your opinion, is Ontario's becoming a 'have-not' province and receiving equalization payments a temporary or permanent change?

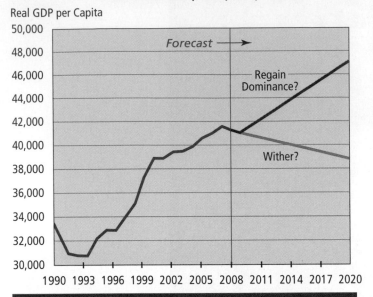

Ontario Real GDP per Capita up to 2020

Real GDP per Capita

Forecast →

Regain
Dominance?

Wither?

Figure 5.3   Ontario's economy to 2020: Which way?
Ontario's manufacturing sector could suffer from deindustrialization and become part of the adjacent US
Rust Belt; or, Ontario could reinvent itself.
Source: Drummond and Burleton (2008: 1).

Ontario's Physical Geography

Ontario is larger than most countries (Figure 5.1), stretching out over 1 million km². Extending over Ontario are three of Canada's physiographic regions (Great Lakes–St Lawrence Lowlands, Canadian Shield, and Hudson Bay Lowland), and three of the country's climatic zones (Arctic, Subarctic, and Great Lakes–St Lawrence) (Figures 2.1 and 2.6). Manitoba lies to its west, Hudson and James bays to its north, while Québec, on its eastern boundary, is bordered in part by the Ottawa River. This central location within Canada and its close proximity to the industrial heartland of the US have facilitated Ontario's economic development.

Ontario is not a homogeneous natural region. For that reason, Ontario is divided into two sub-regions (northern and southern Ontario). This division is based on its physiographic regions. Accordingly, northern Ontario consists of the Canadian Shield and Hudson Bay Lowland while southern Ontario matches the geographic extent of the Great Lakes–St Lawrence Lowlands that fall within

Ontario. In turn, each sub-region has a different economy. Northern Ontario has the characteristics of a resource hinterland, while southern Ontario is the epitome of an agricultural-industrial core (Vignette 5.2).

Northern Ontario stretches across 10 degrees of latitude—from 46° N to nearly 57° N. Fort Severn First Nation is located on the shores of Hudson Bay at 56° 37' N while Sudbury is at 46° 30' N. The Subarctic climate of northern Ontario has longer and colder winters as well as shorter and cooler summers than those occurring in southern Ontario. Even along its southern edge at 46° N, short summers make crop agriculture vulnerable to frost damage. In addition to a difficult climate for agriculture, the rocky Canadian Shield has very little agricultural land while the Hudson Bay Lowland has none. The rugged, rocky terrain of the Canadian Shield has only a few pockets of agriculture where former lakebeds provide the basis for soil development. Climate, soils, and physiography combine to limit agriculture in northern Ontario.

Unlike Québec, the Ontario section of the Canadian Shield has relatively low elevations, limiting the opportunities for hydroelectric power developments. However, this sub-region does have vast forests, superb scenery (which supports tourism), and extensive mineral wealth. The Hudson Bay Lowland consists of a poorly drained plain associated with muskeg and permafrost. Few opportunities for resource development exist and it remains an area for hunting and trapping inhabited mainly by Cree Indians. As a resource hinterland, northern Ontario is limited economically to the Canadian Shield almost entirely, and its development is dependent on forestry, tourism, and mining.

Southern Ontario, located in the southernmost part of Canada, has Canada's longest growing season. Windsor, for example, is at latitude 42° N and Toronto is close to 44° N. Southern Ontario has a moderate continental climate. This climate is noted for long, hot, and humid summers, warm autumns, short but cold winters, and cool springs. Annual precipitation is about 1,000 mm. The greatest amounts of precipitation occur in the lee of the Great Lakes, where winter snowfall is particularly heavy (see Vignette 5.3). The Great Lakes modify

Think About It

Distance provides a geographic context. For Ontario, the distance from Sudbury to Fort Severn is approximately 1,200 km. Using the scale in Figure 5.1, what is the distance from Sudbury to Toronto?

Vignette 5.1   The Burden of Confederation

One sign of Ontario's 'Burden of Confederation' is that Canadians living in provinces less prosperous than Ontario have benefited from equalization payments. Ontario and Alberta have been the major source of tax revenue for Ottawa and these revenues have provided much of the funds for equalization payments. In 2009–10, however, Ontario, for the first time, received an equalization payment while Saskatchewan and Newfoundland and Labrador joined Alberta in shouldering the 'Burden of Confederation'. While Ontario is expected to bounce back quickly, Danny Williams, the Premier of Newfoundland and Labrador, enjoyed the moment as shown in the cartoon below by flipping some cash to Dalton McGuinty, the Premier of Ontario.

Photo 5.2

For the first time, Ontario became a 'have-not' province in 2009.

temperatures and funnel winter storms into this region. During the winter, this region experiences a great variety of weather conditions.

Southern Ontario is the most favoured physical area in Canada. Not surprisingly, the vast majority of the province's population, nearly 12 million, or 93 per cent, live in southern Ontario. The area is underlain by slightly tilted sedimentary rocks, which are covered by a thin deposit of glacial till. Except for the Niagara Escarpment, there is little relief topography. Formed since deglaciation, the Niagara Escarpment is an erosional remnant of the more resistant sedimentary rock that extends from the Bruce Peninsula to Niagara Falls. The Escarpment provides the most spectacular scenery in southern Ontario and a nature trail known as the Bruce Trail extends some 725 kilometres along this limestone topographic feature. Prior to settlement, a

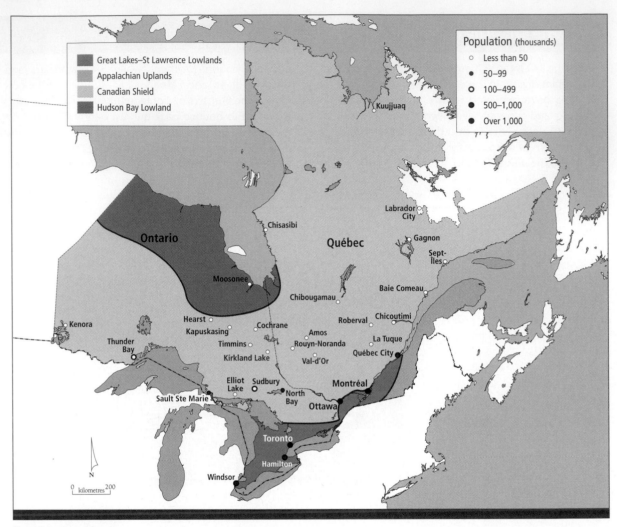

Figure 5.4   Central Canada.

Central Canada consists of Ontario and Québec. Three physiographic regions are found in Ontario and four in Québec. Canada's most productive agricultural lands, its manufacturing belt, and its core population zone all are located in one physiographic region—the Great Lakes–St Lawrence Lowlands.

Vignette 5.2   Contrasting Sub-Regions: Northern and Southern Ontario

Ontario is a geographic paradox. The contrasts between southern Ontario and northern Ontario are extreme. Each has very different physical conditions and geographic situations. Over time, two distinct economies have emerged, each with its own spatial pattern of population distribution. While southern Ontario forms the industrial and population heartland of the province, northern Ontario is an old resource hinterland where economic and population growth have stalled. Because so many young people are moving to more prosperous areas in Canada, the population of northern Ontario is losing a valuable segment of its labour force, leaving its cities and towns with disproportionately large numbers of older and very young people. By October 2009, northern Ontario had a population of just under 800,000, while southern Ontario had a population of 12.3 million (Statistics Canada, 2009b).

Vignette 5.3   Ontario's Snowbelts

Ontario's snowbelts are legendary. On the upland slopes facing Lakes Huron and Superior and Georgian Bay, huge snowfalls totalling in the 300–400-cm range occur each winter from November to late March. The uplands on the northeastern shores of Lake Superior receive the greatest total snowfall amounts of any area in Ontario, exceeding 400 cm annually. Much of the snowfall in snowbelt areas can be attributed to cold northwesterly to westerly winds blowing off the lakes and ascending the highlands. As the Arctic air travels across the relatively warmer Great Lakes, it is warmed and moistened. Snow clouds form over the lakes and, once onshore, intensify as the air is then forced to ascend the hills to the lee of the lakes, triggering heavy snowfalls. Areas on the downslope side of the higher ground to the lee of the lakes receive less than half the annual snow totals of the upslope snowbelt areas. For example, Toronto, Hamilton, and other places to the lee of the Niagara Escarpment are snow-shadow regions with winter amounts of 100 to 140 cm. In the snowbelt regions, snowfall accounts for about 32 per cent of the year's total precipitation; but in the snow-sparse area of southwestern Ontario around Windsor and Chatham, the snow contribution is only about 13 per cent.

mixed forest vegetation flourished in this temperate continental climate (Figure 2.7). With a long growing season, ample precipitation, and fertile soils, the southern Ontario lowland has the most productive agricultural lands in Canada.

The section on physiographic regions and climate in Chapter 2, especially the figures and tables, provide important background information for Ontario.

Darwin Wiggett/All Canada Photos

Photo 5.3

The Niagara Escarpment, stretching over 700 km from Queenston on the Niagara River to Tobermory on the Bruce Peninsula, is the most significant landform in southern Ontario. The Escarpment reaches over 350 metres above the surrounding land at various points.

Climate and Agriculture

Southern Ontario's climate is dominated by its long, warm summer that extends from May to September. Tropical air masses that originate in the Gulf of Mexico extend over this area, resulting in hot and humid weather. Winter, on the other hand, takes hold for three to four months from mid-November to March when occasional invasions of Arctic air masses bring exceptionally cold weather. Such weather is associated with cold, sunny days and frigid nights. During the early spring and late fall, unsettled, cloudy weather dominates, with minimum temperatures falling below the freezing point. Proximity to the Great Lakes affects the local weather by funnelling air masses into this region and by increasing local precipitation from air masses absorbing moisture from the surface of the Great Lakes. During the winter, high winds funnelled along the Great Lakes storm track often accompany cold spells, driving the wind chill to a dangerous level. Storms are short-lived followed by relatively mild weather. Annual precipitation ranges around 100 cm. Warm, moist air masses from the Gulf of Mexico provide most of the moisture for the region, which falls as convectional and frontal precipitation.

Because of its mild climate and fertile soils, southern Ontario accounts for much of Canada's agricultural output by value. Southern Ontario has over half of the highest-quality agricultural land (known as Class I) in Canada. With a combination of fertile soils and a warm summer, farmers in southern Ontario account for more than $7 billion of agricultural products each year. In 2006, Ontario had just over 57,000 farms, a 4.2 per cent decrease over the past five years compared to the national decline of 7.1 per cent (Statistics Canada, 2007b). This decline in the number of farms is related to increase in farm size from 92 hectares per farm in 2001 to 94 hectares per farm in 2006, which is necessary for farmers to achieve economies of scale (ibid.).

Located south of 44° N, southern Ontario has abundant moisture and a long, warm-to-hot summer permitting the production of highly specialized crops such as grain corn, soybeans, and sugar beets as well as a wide variety of fruits, grapes, and vegetables that cannot be grown in other parts of Canada. In the Niagara Fruit Belt, a unique set of growing conditions has provided the basis for grape production, resulting in high-quality wine production (Vignette 5.4). Equally important, the large urban population provides a stable local market for its dairy, livestock, soft-fruit, and vegetable products. Although the amount of cropland is relatively small, farming operations in southern Ontario are much more intense than in Western Canada, where extensive grain crops dominate farming operations. The difference between the two agricultural areas is revealed in the size of farms and the types of crops. In 2006, the average size of farms in Ontario, 94 ha, compares to an average of 587 ha for farms in Saskatchewan (ibid.). Grain farming and cattle ranching dominate in Western Canada, while a much more diversified use of agricultural land prevails in southern Ontario, where corn, barley, and winter wheat are important crops along with highly specialized ones such as tobacco and vegetables. Corn and grain often serve as fodder for the hog, dairy, and beef livestock farms that operate in Ontario.

Agricultural land use varies within southern Ontario due to subtle but important physical differences. For example, soils are less fertile and the growing season is shorter east of Toronto than southwest of Toronto. There are three highly specialized agriculture zones west of Toronto: the Essex–Kent vegetable area, the Norfolk Tobacco Belt, and the Niagara Fruit Belt. All lie in southwestern Ontario where the growing season is relatively long. These zones, located south of 43° N, are the southernmost lands in Canada.

Unfortunately, the expansion of cities and towns in southern Ontario has spread into these farmlands. Valuable farmland is now part of the urban landscape. Until the price of agricultural land is equal to or surpasses that of urban/industrial land prices, this process of losing irreplaceable arable land will continue. How to resolve this land-use conflict is beyond the market economy because the price of farmland is well below that of urban land. Ontario's government has recognized this dilemma but state intervention in the real estate market has been ineffective (Vignette 5.5).

Think About It

Would you, as Premier of Ontario, consider protecting Niagara Fruit Belt farmland as a priority? If so, how would you justify such a policy?

Vignette 5.4 The Niagara Fruit Belt

The Niagara Fruit Belt lies on the narrow Ontario Plain that extends from Hamilton to Niagara-on-the-Lake. The Ontario Plain has fertile, sandy soils suitable for most types of agriculture. Lake Ontario marks its northern limit while the Niagara Escarpment forms its southern edge. This small agricultural zone contains the best grape and soft-fruit growing lands in Canada, and is home to vineyards that account for most of Canada's quality wines, including the unique ice wine.

While the Niagara Fruit Belt occupies a northerly location for grape vines and soft-fruit trees, local factors have more than offset the threat of frost at 43° N latitude. These factors are:

- Air drainage from the Niagara Escarpment to Lake Ontario reduces the danger of both spring and fall frosts.
- The water of Lake Ontario is warm in autumn and its proximity to the Niagara Fruit Belt helps to moderate advancing cold air masses.
- In early spring, the cool waters of Lake Ontario keep air temperatures low, thereby delaying the opening of the fruit blossoms until late spring when the risk of frost is much lower.

Barrett & MacKay Photography

Photo 5.4

The Niagara Fruit Belt extends about 65 km between Hamilton and Niagara-on-the-Lake. The Niagara Fruit Belt is one of the major fruit and grape producing areas in Canada. Most vineyards are located on the slopes of, or below, the Niagara Escarpment. For many years, hardy vines that produced low-quality grapes resulted in a poor-quality (but inexpensive) wine. With the Free Trade Agreement, Canada had to remove its tariffs, making it difficult to compete with foreign wines. Since that time, farmers in the Niagara Fruit Belt have been successful in growing the finest varieties of grapes and have been able to make some of the finest wines in the world.

Environmental Challenges

Ontario faces two major environmental challenges—air pollution and water pollution. Solutions are costly and they require behaviour changes in lifestyles. Nevertheless, action is needed and without more robust action by governments, these two problems will only get worse, causing serious harm to the environment and to the health of citizens. Both types of pollution represent the hidden costs of our industrial world.

Ontario's cities present a special version of pollution's hidden costs. Two major costs

Vignette 5.5   The Tragedy of Urban Sprawl

Urban sprawl devours valuable but underpriced agriculture land. Since the economic value of urban/industrial classified land is higher than land designated for agricultural uses, urban/industrial encroachment continues to spread into the rural landscape. Although Ontario has over half of the best agriculture land in Canada, urban encroachment and other non-agricultural interests have cut deeply into the amount of farmland. For instance, over the three decades between 1976 and 2006, nearly 200,000 acres or 25 per cent of the province's Class I land was lost to agriculture. Even the extremely valuable lands in the Niagara Fruit Belt remain under pressure from residential and commercial land developers. An indirect impact of urban sprawl is the release of more greenhouse gases because of the longer drives from suburbia to places of work. Ontario has put into place measures like the Greenbelt Plan, the Places to Grow Act, and the Growth Plan for the Greater Golden Horseshoe Area; but these efforts fall far short of a comprehensive land-use plan to regulate urban sprawl and to convince the public to switch from the automobile to public transportation.

For more on the high cost of developing new technologies in Ontario's knowledge-based economy, see 'Ontario Today', page 195. For more on a knowledge-based economy, see 'Economic Structure' in Chapter 4, page 166.

Pollution has a health cost. The densely populated area of the Golden Horseshoe—which includes Toronto and Hamilton—has significant health problems caused by smog. Smog becomes a serious health hazard in the summer months when high temperatures and air inversions combine to keep the smog hanging over the city as a brownish-yellow haze. The word 'smog' is actually a combination of 'smoke' and 'fog'. Smog is the most visible form of air pollution, and is caused when heat and sunlight react with various pollutants emitted by industry, vehicle exhaust, pesticides, and oil-based home products. A study by the Ontario Medical Association estimated that illnesses caused by smog cost Ontario more than $1 billion a year in hospital admissions, emergency room visits, and absenteeism. Even more startling, the OMA study (2005) concluded that about 5,800 people die prematurely in Ontario each year because of smog-related respiratory problems.

In 2003, the Ontario government recognized the need to deal with the smog problem. Most air pollution comes from automobile exhaust and coal-burning thermoelectric plants. According to Environment Canada (2006), Ontario Power Generation is the largest air polluter in Canada because of its coal-burning generating plants. The Ontario government recognizes this problem and proposes to close its five coal-burning thermoelectric generating stations. While the coal-burning plants contribute 16 per cent of Ontario's electrical consumption, they are, after automobile pollution, the major source of air pollution. Progress has been slow. Only the Lakeview generating station near Toronto has closed. Plans to phase out the Thunder Bay and Atikokan plants in northwestern Ontario by 2007 were extended to 2012 because of the ongoing need for their generating capacity. Without coal, natural gas is one option while nuclear power is another. Yet both require time to construct new power plants and their cost of production is higher

involve disposing of waste products and the health cost of **air pollution** from vehicular traffic in cities. City governments have had limited success in converting their residents from automobiles to public transportation, and in reducing their waste products through recycling programs and composting. Toronto's solution has been to export its garbage to a dump site in Michigan, but that option ends in 2010. The 'greening' of Toronto and other major cities in Ontario will remain a vision until urbanites change their lifestyles. The solution to urban garbage disposal lies in research to find different means of converting garbage into energy. Gasification, a low-emission method of extracting energy from a variety of waste materials, may offer a solution. While the landscape is littered with garbage-to-energy technologies that promised breakthrough advances, none have cleared the commercial hurdles; the cost of moving from a demonstration plant to a commercial one is high. **Sustainable Development Technology Canada** has provided base funding for promising technologies such as the Plasco Energy Group's test facility in Ottawa that converts urban wastes (garbage) into reusable products (Plasco Energy, 2009).

than that of coal. For instance, in 2010 TransCanada completed a gas-fired plant at Halton (just north of Oakville) but its proposed plant for Oakville remains on hold due to local opposition. At the same time, gas and nuclear plants are not without hazards and drawbacks, especially the issue of the disposal of nuclear waste, and past cost overruns and problems with Ontario's nuclear generating facilities have saddled all of Ontario's electricity users with a massive debt: a pro-rated debt retirement charge is added to every customer's monthly bill.

With demand for electricity increasing each year, Ontario realized that closure of all its coal plants would result in energy shortages and the possibility of widespread brownouts. Ironically, it is the Nanticoke and Lambton coal-fired stations in the heart of southern Ontario that produce most of the smog-causing pollutants affecting the health of residents of southern Ontario. Yet, the province does not want to cease electric production from its remaining coal plants until clean-air energy production is in place—and their construction will take time. Here are some examples:

- Two natural-gas generating stations that will replace the electrical output

from the Lambton plant are not expected to be operative until after 2010.
- More hydroelectricity is coming by 2010 with the anticipated completion of the Niagara Tunnel Project.
- Even more time is required for increasing the capacity of existing nuclear generating plants.

Alternative energy sources, such as wind farms, have begun to feed into Ontario's electrical grid, but these provide a minuscule amount of total energy needs and at a higher cost than conventionally produced electricity.

Water pollution constitutes a serious problem across Ontario. Industrial effluents, farm chemicals, and livestock waste are the main culprits. At the local level, in May 2000, the contamination of drinking water with E. coli at Walkerton in southwestern Ontario provided a deadly example of farm-waste hazards with seven deaths and many illnesses. At the regional level, the integrity of the Great Lakes has been compromised by human activities. Over 35 million people, many of whom live in large urban cities such as Chicago, Detroit, Cleveland, Toronto, and

Photos courtesy of Ontario Power Generation

Photo 5.5

In 2007, the Ontario government planned to replace its Thunder Bay coal-burning generating station (left) with a gas-fired unit and to close its Atikokan plant (right). The Thunder Bay station has burned low-sulphur lignite coal from Western Canada and low-sulphur sub-bituminous coal from the United States.

Montréal, depend on the Great Lakes/St Lawrence for water supply and sewage disposal. The area surrounding Lake Michigan, Lake Erie, and Lake Ontario contains many industries and factories that, in the past, dumped their toxic wastes into the Great Lakes and buried their chemical wastes along the shores. Later, these chemical wastes began to leak into the lakes. Measures were taken to regulate these polluters, but pollution from the runoff from agricultural lands, the waste from cities, and toxic discharges from industry continue to affect the Great Lakes.

Canada and the United States have recognized the need to regulate the discharge of toxic wastes into the Great Lakes. The Great Lakes Water Quality Agreement, first signed in 1972 and renewed in 1978, expresses the commitment of the two countries to restore and maintain the chemical, physical, and biological integrity of the Great Lakes Basin (see Figure 5.5). By the beginning of the 1990s, cleaner water and increasing fish populations indicated that these joint efforts were making solid progress. However, industrial discharges of pollutants remain a problem. Growing levels of phosphorus, primarily from waste water containing detergents and from agricultural runoff containing high levels of chemical fertilizers, are

Figure 5.5   The Great Lakes Basin.

Except for the state of Michigan, much of the Great Lakes Basin is located in Ontario, as the watersheds of the Mississippi River and Ohio River in the United States, part of the Gulf of Mexico Basin, extend close to the southern and western perimeters of the Great Lakes. (Based on *Atlas of Canada*, 2004, 'Great Lakes Basin', at: <atlas.nrcan. gc.ca/site/English/maps/reference/provincesterrritories/gr_lks/referencemap_image_view>)

contributing to the creation of a 'green slime': a dead zone where only toxic organisms can survive. In addition, the careless introduction of exotic species is resulting in sea lampreys, Asian carp, and zebra mussels squeezing out native species and thereby radically changing Great Lakes ecosystems. Ocean-going ships have carried unwanted passengers such as zebra mussels and sea lampreys into the Great Lakes via the St Lawrence Seaway. Both have had a negative impact on domestic fisheries.

Under 'Environmental Challenges' in Chapter 6, page 235, the impact of the zebra mussel on the St Lawrence River and Great Lakes is presented.

Ontario's Historical Geography

When the American Revolution began in 1775, Ontario—except for the French settlement around Detroit (established 1701)—was a densely forested wilderness inhabited by a few fur traders and Indians. By 1750, Detroit had a French population of 600, of whom 150 lived across the Detroit River at Petite Côte. In 1760, British troops took control of Fort Detroit and the surrounding settled area.

After losing the American colonies, loyal British subjects returned to England or relocated in one of Britain's colonies. In 1782–3, Loyalists moved northward to Nova Scotia and New Brunswick, while others resettled in the Eastern Townships of Québec. A smaller number, perhaps as many as 10,000 Loyalists and members of the Six Nations who had fought alongside British troops, settled on land along the St Lawrence River that was upstream from French-speaking habitants, thus colonizing the wilderness area that later became known as Upper (relative to the mouth of the St Lawrence) Canada. In 1784, Britain rewarded its Indian military allies, the Six Nations, with the Haldimand Tract, which consisted of a strip of land on each side of the Grand River totalling 385,000 hectares (Darling, 2007). These Loyalists were followed by American, British, and European newcomers, who flooded into Upper Canada over the next half-century in search of land.

In less than 100 years, the natural forest landscape was transformed into a British agricultural colony.

Until the War of 1812, many settlers had come from the United States. They, like most other colonists, sought land in the fertile lowlands of the Great Lakes. When hostilities between Britain and the United States escalated, the Americans launched an attack on British North America. The War of 1812 effectively ended the influx of American settlers into Upper Canada. After the war ended with the signing in December 1814 of a treaty of stalemate at Ghent, an increasing number of settlers came from the British Isles, especially Ireland and Scotland. By 1851, Canada West (as Upper Canada had become with the Act of Union of 1841) had reached a population of 952,004 with 86 per cent living in rural settings (Statistics Canada, 2007a). By Confederation, virtually all the arable land in the Great Lakes Lowland had been cleared of forest by settlers who created an agricultural landscape. With a severe land shortage, some turned further north to try their luck in the few pockets of arable land within the Canadian Shield, but few were successful. Others migrated to towns and cities where they became absorbed into the urban workforce, and still others moved west to what would soon become Manitoba. Many more took their chances by participating in the last great American land rush that took place west of the Mississippi River. By 1867, the demographic balance of power remained in the rural community—urban areas, such as villages and towns, contained less than 20 per cent of the population of Ontario.

When Canada West joined Confederation in 1867, it was renamed Ontario (Figure 3.4). At that time, the geographic extent of Ontario was about 100,000 km²—a fraction of its present size—but as Canada acquired more territory from Great Britain, Ontario and Québec obtained some of these new lands (Figures 3.5–3.7). They had little immediate value for economic development and settlement, however, because they were carved from two physiographic regions (the Canadian Shield and the Hudson Bay Lowland) that were far from markets and had little or no agricultural potential. Since Confederation, the borders of Ontario have been extended three times, greatly increasing the geographic size of the province, but not its agricultural lands. The first expansion occurred in 1874 when Ontario's boundaries were pushed northward to about 51° N and westward towards the Lake of the Woods. Ontario's second expansion, in 1889, ended the bitter contest between Manitoba and Ontario for the land around the Lake of the Woods. At the same time, Ontario's northwest boundary was adjusted to the Albany River, which flows into James Bay, gaining Ontario access to James Bay. In 1912, the final boundary modification occurred when the District of Keewatin south of 60° N was assigned to Ontario and Manitoba. As a result, Ontario extended its political boundary to the northwest, stretching from Manitoba to Hudson Bay at the latitude of 56°51' N. Through these boundary adjustments, Ontario reached its present geographic extent of 1 million km².

The Birth of an Industrial Core

The economic essence of Ontario is its industrial base. Transportation routes have played a key role, and especially the Welland Canal, which has facilitated low-cost transportation (Vignette 5.6). Manufacturing in southern Ontario (then Upper Canada) began in the nineteenth century and received a boost in 1854, when Britain and the United States signed the Reciprocity Treaty, which allowed trade between the British North American colonies and the United States. By 1867, Ontario had the largest population of Canada's four founding provinces and a fledgling industrial base. Small manufacturing outfits in small villages often consisted of less than five employees (e.g., a village blacksmith; a miller). Larger manufacturing took place in towns and cities, where sawmill, gristmill, and distillery operations were commonplace. At that time, most manufacturing activities depended on water power, so most were located near a stream or river.

The Americans allowed the Reciprocity Treaty to lapse in 1866, cutting off access to the American market for Canada West. However, Confederation and the 1879 National Policy of Prime Minister Macdonald enabled what was now Ontario to secure the Canadian markets. Under the National Policy high tariffs were imposed on imported manufactured goods, which allowed manufacturing in southern Ontario to flourish. The consequences of Confederation and protective tariffs for Ontario were threefold: (1) the creation of a Canadian market for Ontario products; (2) an increase in the size of the more successful manufacturing companies; and (3) the growth of the industrial workforce in Ontario. In the rest of the country, however, prices for manufactured goods (made in Ontario) were generally higher than in the adjacent areas of the United States. The regional price variations reflected two factors: Canadian transportation costs and Ontario's more limited economies of scale compared to those of US manufacturers.

For more on the National Policy, see Chapter 3, page 91.

The National Policy of high tariffs provided the impetus for a national industrial core in southern Ontario, while the rest of the country—except for Québec—was relegated to a domestic market for these manufactured goods. This economic arrangement made necessary the east–west transportation axis and, in doing so, favoured Ontario and Québec, thus laying the groundwork for western alienation and Maritime dissatisfaction with Confederation. Not until the Auto Pact of 1965 and then the 1989 Free Trade Agreement did this national core/periphery relationship undergo significant change and a north–south transportation axis emerge as a prominent aspect of Canada's manufacturing trade.

See Chapter 1, page 18, for more on the Auto Pact and the FTA.

By the early twentieth century, four factors had led to successful manufacturing in southern Ontario, which can be illustrated by the automobile industry. The first factor was Ontario's geographic advantage: close proximity to America's manufacturing belt

Vignette 5.6   The Welland Canal

The Welland Canal connects Lake Ontario and Lake Erie, allowing ocean-going ships to enter the heart of North America. To avoid the Niagara River and its huge falls, the first canal-builders faced the daunting task of constructing a canal across the Niagara Peninsula, a distance of some 44 km. The first canal, opened in 1829, was dug by hand. A series of locks made from hand-hewn timbers connected a series of creeks and lakes. As the size of ships increased, the original canal proved inadequate and a new canal was built in 1845.

Within 40 years, even larger ships required a third renovation, which was opened in 1887. The present canal was completed in 1932. In 1973, to bypass the city of Welland, a new channel was constructed, for which a series of lift locks were needed to overcome a difference in elevation of nearly 100 metres between Lake Ontario and Lake Erie. The Welland Canal has been part of the St Lawrence Seaway since 1959 and is operated by the St Lawrence Seaway Management Corporation.

Al Harvey/Slide Farm

Photo 5.6

The Welland Canal is a strategic link between Lake Ontario and Lake Erie that provides a water route around Niagara Falls. To accommodate ever increasing traffic, the lock system was divided into two at several places, as shown in this aerial photograph. In 2006, approximately 40 million tonnes of goods passed through these locks. The three leading products were grain, iron ore, and coal.

led American industries to locate branch plants in Canada early in the twentieth century. This spillover effect first took place in the Windsor–Detroit area in 1904, when the Ford Motor Company established an automobile assembly plant in Windsor. The second factor was trade restriction on foreign manufactured goods (imposed by the National Policy in 1879). Automobile parts from Ford's plant in Detroit had to be ferried across the Detroit River and assembled in an old carriage factory in Windsor. The third advantage was access by American branch plants in Canada to the lower tariffs for Canadian-made products in

Library and Archives Canada C34334

Photo 5.7

In 1794, York became the capital of Upper Canada. Despite its political status, this frontier village remained on the western edge of British settlement that stretched westward from Lower Canada along the north shore of Lake Ontario. By 1812, York had only 700 residents. This painting (dated 1804) illustrates a group of houses strung along the shore of Toronto Bay. Beyond this narrow strip of cleared land lies the original forest of southern Ontario.

the British Empire. As a result, Canadian-built Ford cars were sold not only in Canada but in various places in the British Empire. General Motors followed the same strategy by opening a branch plant in Oshawa. The size of its domestic market provided southern Ontario with its fourth location advantage for the automobile industry: most GM cars were sold in southern Ontario, thus minimizing transportation costs.

With the liberalization of trade, manufacturing in Canada escalated. First came the Auto Pact of 1965. The Big Three automobile companies in the early 1960s had proposed a common market between Canada and the US for the production and marketing of their products. In the year Ottawa and Washington passed the appropriate legislation, Canada produced nearly 900,000 vehicles. By 2000, the figure had reached 2.5 million. Until 2008, annual production hovered around 2.5 million cars and trucks, most of which were exported to the US market. At this level of

production, Canada was accounting for 14 per cent of North America's vehicle output. Ontario's automotive assembly and parts firms remain the anchor of manufacturing in the province, employing about 150,000 people. These well-paid workers produce high-quality motor vehicles that account for 12 per cent of the country's gross domestic product.

The second step took place in 1989 when the Canada–US Free Trade Agreement brought other components of the economy under a single North American market, thus removing tariffs (some high) between the two countries. Since American firms had already benefited from economies of scale, their costs of production were considerably lower than those of similar Canadian firms. As well, the American firms were well established in the marketplace and dislodging them was a daunting task. Canadian-based manufacturing firms were forced to play catch-up and many did not survive. American firms under stress often closed their smaller, less efficient

Library and Archives Canada 1669

Photo 5.8

View of King Street [Toronto], Looking East (1835) by Thomas Young. At the time of this painting, York had just been renamed Toronto and had a population of nearly 10,000.

branch plants and served their Canadian customers from their American factories. Some Canadian firms merged to attain a size deemed necessary to compete in the North American and global markets and to avoid takeovers by large foreign companies. Mergers, however, resulted in the closing of redundant offices, thus reducing the number of employees.

Aboriginal Territory within Ontario

Ontario has 126 First Nations holding Aboriginal territory, known as reserves, that was obtained through treaty negotiations between government and individual tribes. Most were classified as 'unnumbered' and took place before 1923. Three numbered treaties—3, 5, and 9—cover northwestern and northern Ontario. The first land grants to Indians took place during the days of British North America, beginning with the Haldimand Proclamation of 1784, which assigned land along the Grand River to the Iroquois who fought alongside the British in the American Revolution (Figure 5.6). In 1850, the two Robinson treaties were completed; they covered large areas east and north of Lake Huron and north of Lake Superior.

Figure 3.9, 'Historic treaties', page 103, shows the location of the numbered and unnumbered treaties that blanket Ontario.

In Ontario, Indians were granted land hundreds of years ago. When disputes arise today, reaching an agreement is challenging because the 'facts' are buried in time (see Vignette 5.6). The slow pace of resolution is frustrating to Native Canadians and two land disputes (Ipperwash and Caledonia) have led to violent protests by Indians from the Kettle and Stony Point and Six Nations reserves. In the Ipperwash dispute, the facts are relatively clear: land from Stony Point was taken in 1942 to serve as a military training camp, named Camp Ipperwash. After the war, the land

Barrett & MacKay/All Canada Photos

Photo 5.9

With an ideal climate and proximity to large urban markets, dairy farming dominates the agricultural landscape of much of Ontario.

was supposed to have been returned but the Department of National Defence decided to keep the camp to train cadets. Promises from Ottawa to return the land were not fulfilled. By 1993, the Stony Point Indians were utterly frustrated with the repeated failure of Ottawa to act on its promises and in September 1995 Indian protestors moved into Ipperwash Provincial Park, where a confrontation with the Ontario Provincial Police took place, ending in the shooting of an unarmed Native protestor, Dudley George, by an Ontario police sniper. In 2003, the Ontario government asked Justice Sidney Linden to conduct a public inquiry into the circumstances surrounding the 1995 death of Dudley George, including the role of the then Premier, Mike Harris. In May 2007, Justice Linden issued the Ipperwash Inquiry Report, in which he concluded that Harris had not explicitly ordered the Ontario police into the Ipperwash Provincial Park to remove the Indian protestors. At the same time, Justice Linden called

for the immediate return of Camp Ipperwash to the Kettle and Stony Point First Nation. Jim Prentice, the federal Minister for Indian Affairs and Northern Development, responded by stating that, 'We'll do something immediately' (*National Post*, 2007), but, in spite of the commencement of 'fresh' negotiations, ownership of Camp Ipperwash remained in limbo as of early 2010.

Specific land claims by Native groups are not usually as straightforward as the Ipperwash claim—and even that claim is unresolved. The Caledonia dispute exemplifies the complexity of some claims. An outline of the historical evolution of the Six Nations claim to a 40-hectare parcel owned by a land developer at Caledonia, Ontario, near Hamilton, suggests why, in many instances, settlements have been achieved at such a slow rate (Vignette 5.7). The basis of the Six Nations claim goes back to the original Haldimand Grant of 1784 and land surrenders in the eighteenth and nineteenth

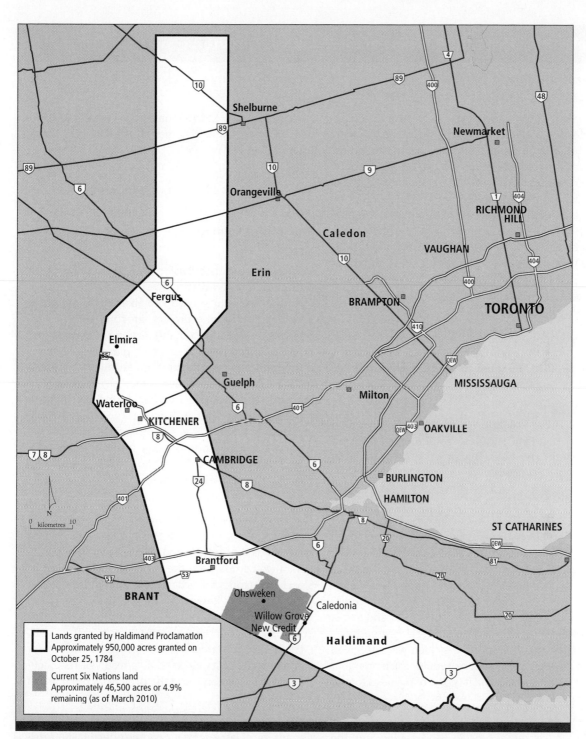

Figure 5.6   The Haldimand Tract.

Source: Six Nations Land Resources. Based on map at <www.sixnations.ca/LandsResources/HaldProc.htm>.

Vignette 5.7   Historical Timeline of the Caledonia Dispute

Eighteenth Century

1784

The British Crown allows the Six Nations (Iroquois Confederacy) to 'take possession of and settle' a strip of land nearly 20 kilometres wide along the Grand River, from its source to Lake Erie, totalling about 385,000 hectares; called the Haldimand Grant, it is named after the governor of Québec, Frederick Haldimand, who previously had negotiated the purchase from the Mississauga Indians of over 1.2 million hectares on the Niagara Peninsula, for a little over £1,000. Haldimand's intent in this purchase was to provide land for settlement to loyal Iroquois who had fought beside the British in the American Revolution.

1792

Upper Canada's lieutenant-governor, John Graves Simcoe, reduces the grant to the Six Nations by two-thirds, to 111,000 hectares.

1796

The Six Nations Confederacy grants its chief, Joseph Brant, the power of attorney to sell some of the land and invest the proceeds. The Crown opposes the sales but eventually concedes.

Nineteenth Century

1840

The government of Upper Canada recommends that a reserve of 8,000 hectares be established on the south side of the Grand River and the rest sold or leased. Six Nations council agrees to surrender for sale all lands outside those set aside for a reserve, on the agreement the government would sell the land and invest the money for them. A faction of Six Nations petitions against the surrender, saying the chiefs were deceived and intimidated. Six Nations would challenge that claim in a 1995 lawsuit and it is part of the basis for the current protest.

1850

The Crown passes a proclamation setting out the extent of reserve lands on both sides of the Grand River—about 19,000 hectares agreed to by the Six Nations chiefs.

Twentieth Century

1992

Henco Industries Ltd purchases 40 hectares of land near Caledonia and renames it the Douglas Creek Estates.

1995

The Six Nations disputes the ownership of this Crown land.

Twenty-First Century

2005–6

Henco Industries' subdivision plan for Douglas Creek Estates is registered with the province of Ontario. The following year, a group of Six Nations members occupies the housing project, erecting tents, a teepee, and a wooden building. Court challenges and non-Native counter-protests take place. Several hundred non-Native demonstrators gather in Caledonia to protest occupation and what they call police inaction. Dozens of police officers form lines between Native and non-Native protestors. Three people arrested. In a search for a compromise, the Ontario government buys out Henco's interest in the disputed property for $15.7 million, thus maintaining Crown ownership of this disputed land.

2007

The federal government enters negotiations with the Six Nations to resolve the historic and current land claims disputes. Canada makes an offer of $125 million to compensate the Six Nations for four outstanding historic claims based on nineteenth-century land surrenders known as Grand River Navigation Company investment; Block 5 (Moulton Township); Welland Canal flooding; and the Burtch Tract. As well, Ottawa compensates Ontario for $26.4 million for the province's costs incurred as a result of the occupation near Caledonia and the province's purchase of the land.

2008–present

Canada receives a formal counter-offer of $500 million from the Six Nations. The following year, Canada rejects the claim for $500 million and restates its offer of $125 million. Negotiations continue.

Sources: Adapted from CBC News (2006a, 2006b) and Indian and Northern Affairs Canada (2009). Chronology of Events at Caladonia at: <www.ainc-inac.gc.ca/ai/mr/is/eac-eng.asp>.

centuries. History is not clear on these issues. While the protest finally ended, negotiations between the Six Nations and the federal government continue. The most recent development took place in 2009, when the federal government rejected the $500 million claim of the Haudenosaunee/Six Nations, leaving its offer of $125 million on the table.

Ontario Today

Canada's centre of gravity—as measured by economic performance and population size—continues to shift westward. Yet, the sheer size of Ontario's economy and population keeps that centre anchored in Ontario. This westward shift is due more to the weak performance of Québec and Atlantic Canada, though Ontario's recent performance—as indicated by its 2009 record unemployment rate (Table 5.1)—also made a contribution to this shift. Yet, long-term statistical indicators of the powerful position of Ontario within Confederation include:

- Largest economy and population of the six regions.
- Average personal income well above the national average.
- Greatest cluster of major cities, universities, and technological centres of any region.

Table 5.1 Before and After the Crash: Unemployment Rates by Province, 2007 and 2009

Province	2007	2009
Saskatchewan	4.2	4.8
Manitoba	4.4	5.2
Alberta	3.5	6.6
British Columbia	4.2	7.6
Québec	7.2	8.5
New Brunswick	7.5	8.9
Ontario	6.4	9.0
Nova Scotia	8.0	9.2
Prince Edward Island	10.3	12.0
Newfoundland and Labrador	13.6	15.5
Canada	6.0	8.3

Source: Statistics Canada (2009c, 2010).

- Elects more members of Parliament than any other region.
- Central location within North America further facilitated by its hub position in the east–west and north–south transportation systems.

While Ontario enjoys its paramount position within the nation, the road has recently become bumpy and the future less certain. Ontario, like the other regions, needs a vision and investment strategy based on the next stage of economic development, the so-called knowledge-based economy. The hope is that innovation and technology will move developed regions, like Ontario, to a more sophisticated information society where cost of labour is less critical to determining the success of businesses and where exports can penetrate global markets. Labour is critical, however, because bright workers are the key to the knowledge-based industry. Indeed, if Florida (2002) is correct in thinking that the creative class (which includes the so-called knowledge-based workers) want to live in 'interesting' cities, then that relationship reinforces the concentration of these firms in such urban places. Because of the inventive nature of this type of economy, small as well as large firms can be engaged in the search for new and better ways of production. One example is the small Toronto-based firm Thoora, which has created a tool to monitor the blogosphere for both news and opinions. However, firms with a global reach, like Research In Motion (RIM), are the principal drivers of innovation and they have important spinoff effects for the surrounding scientific community. For example, because these global firms are at the cutting edge of world technology, they push local businesses and university research to higher levels of scientific investigations and the invention of new forms of technology. Unfortunately, the loss of Nortel has had the opposite effect. Nevertheless, Ontario, with a majority of Canada's creative wealth as reflected in its universities and provincial/federal research facilities, is well placed to lead; interestingly, Québec devotes more public funds to its research efforts.

Think About It

What is a fair price for land? Canada's offer of $125 million as a financial settlement for the four outstanding Six Nations claims falls short of the $500 million counter-offer by the Six Nations. Should the two sides simply split the difference and move on?

In 'Economic Structure', Chapter 4, page 166, Florida's theory is discussed within the broader context of the information society.

Think About It

Faced with this double whammy, how is the Ontario government going to square its desire to reduce greenhouse gas emissions by replacing low-cost coal-generated electricity and not increase its electrical rates?

While resource development and manufacturing will remain pillars of Ontario's economy, the spearhead for the future is the knowledge-based economy that can compete on the world stage. As part of the emerging information society, such an economic sector has already surfaced, but it requires vast public support within a coherent federal/provincial plan for innovative research institutions that provide the intellectual base for the knowledge-based economy. Canadian universities form the heart of such innovation, while research-connected companies supply the applied component. The Waterloo 'cluster' of knowledge-based companies, the University of Waterloo, and public research institutions provides one example. Anchored by RIM, Canada's globally successful high-technology company, the Waterloo Region has one of the most diverse economies in Canada, with strengths in advanced manufacturing and communications technology. This advanced technology cluster boasts a healthy mix of small-, medium-, and large-sized innovative firms that often leads to joint innovation efforts and commercial breakthroughs.

In the short run, Ontario must deal with several economic obstacles. One is an energy shortage. Another is the restructuring of its automotive industry. A third is the management of its forestry. A fourth is the thickening border with the US. Energy provides an example of the complexity of these challenges. Even though Ontario produces much power from hydroelectric installations at Niagara Falls and along the Ottawa River, from four thermal coal-fired plants using coal from Pennsylvania and West Virginia, and from nuclear generators in plants at Pickering and Darlington along Lake Ontario east of Toronto and at Tiverton on the Bruce Peninsula, the province faces the threat of an energy shortage. To meet that demand, additional electricity is imported from Québec and produced within Ontario by natural gas from Alberta. While Manitoba has the capacity to produce more low-cost power along its Nelson River, its cost delivered to southern Ontario is around 10 cents per kilowatt hour, about double the current price to residential consumers. Still, what is Ontario to do? With the record high prices of coal and uranium expected to continue into the next decade, Ontario faces both energy shortage and higher energy costs. Green energy provides one alternative, but an expensive one; natural gas has relatively low prices now, but will they stay low?

Added to these long-term challenges, the global economic crisis hit Ontario particularly hard. The first blow occurred in 2008. With the US economy in deep recession, exports to the US dropped dramatically, leaving Ontario's manufacturing industry without its major market. Recovery over the next five years will be difficult and the nature of that recovery will determine the future shape of Ontario's economy. As for a recovery in 2009, Royal Bank economists Paul Ferley and Robert Hogue (2009: 4) put a gloomy—perhaps too gloomy—spin on Ontario's future.

> Ontario's economy entered 2009 covered in bruises after it contracted the most in the fourth quarter [of 2008] in nearly 18 years. Far from regaining its footing early in 2009, however, evidence so far indicates that it likely sunk even deeper as the key auto sector almost completely shut down during January and only partly resumed operations subsequently amid free-falling motor vehicle sales in North America and deep troubles at GM and Chrysler.

Ferley and Hogue may have underestimated the resilience of the Canadian automobile industry. On 22 August 2009, for instance, Chrysler announced a bold plan to produce right-side-drive minivans for European markets at its Windsor plant, and to do so Chrysler would recall some 1,200 employees to operate the third shift in September (Keenan, 2009). Then, too, the US 'cash for clunkers' rebate program stimulated sales and caused General Motors and Ford to increase production and to recall workers in Canadian plants.

Vignette 5.8 Importance of Central Location to Ontario

The role of proximity cannot be underestimated. For example, trucks can deliver Ontario goods to firms in Michigan, Ohio, Pennsylvania, and New York in a matter of hours, allowing a just-in-time production system to flourish on both sides of the border. Backups at the major border crossings—especially between Windsor and Detroit—caused by heightened concerns about security in the US have become a major problem for 'just-in-time' delivery. With much of Canada's $1.7 billion worth of daily exports to the US coming from Ontario, hampering the flow of people and goods across the border to satisfy Washington's concerns about security and terrorism has a negative impact on Ontario—the only question is 'how much'. The answer to that question is still unknown, largely because of the difficulty of collecting information on the cost. A 2004 report to the Ontario Chamber of Commerce estimated the slowdown at the border cost Ontario $5.25 billion the previous year. More recently, Len Crispino, chief executive of the Ontario Chamber of Commerce, stated: 'That border now has become more of a choke point, rather than a conduit for trade' (French, 2007). He estimated that Canada loses as much as Cdn$8 billion every year to border delays.

Ontario Advantage: Trade with the United States

Ontario is geographically positioned to engage in trade, both domestically and internationally. Yet, over 80 per cent of Ontario's exports go to the United States. The importance of the US market to Ontario firms is critical. As Stephen Beatty (2009), managing director of Toyota Canada, stated:

> I want to emphasize, though, that we could not do this—the business case would not exist for assembly plants in Canada—if we did not have open access to the North American market. Even though our Ontario plants build such a high proportion of vehicles destined for Canadian driveways, the vehicles we export are crucial to supporting our investments here in Canada.

Production of forest, mineral, and other products in northern Ontario follows the same pattern: production is far greater than the Canadian market can absorb. The advantage has to do with costs of production. At a high level of output, economies of scale are achieved, keeping the cost per unit of output low. Forestry companies, like automakers, sell much of their lumber, pulp, paper, and other products to the US market. In southern Ontario, manufactured goods ranging from aircraft to steel products are produced for both these markets. **Restructuring** of Ontario's manufacturing industry took place after 1989 when the FTA was signed. Canadian firms, especially those with high labour costs in their manufacturing process, were faced with a highly competitive market that often resulted in plant closure and/or relocation to the US or Mexico.

The FTA and then NAFTA, along with Ontario's geographic location within North America, have allowed the region's business firms to penetrate the huge US market and that has led to greater integration into the North American economy. For example, Ontario's exports to the United States in the 1980s were of about the same value as those to the rest of Canada; but by 1998, Ontario's exports to the US soared to two-and-a-half times the value of exports to the rest of Canada (Courchene with Telmer, 1998: 276–7). Driven by automobile shipments to the US market, this trade gap continued to widen in the first years of the twenty-first century as the sheer size of the US market has tilted the province's trade in that direction.

The lopsided nature of Ontario's international trade is not an accident. Geography has played a role by providing easy access to US markets. Close proximity has been reinforced by the federal policy of 'free trade' with the United States. Manufacturing, especially automobile manufacturing, is Ontario's

Think About It

As the Premier of Ontario, what is your strategic plan for increasing exports of manufactured goods?

economic linchpin and sales to the United States remain crucial to the province's economic well-being. Given these natural and political advantages, what province would not want an automobile industrial base plus access to the richest market in the world?

Southern Ontario is, in effect, a northern extension of the American manufacturing belt. With US automobile plants just across the border in Detroit, automobile manufacturing spread across the Detroit River into Windsor. Initially, this northward diffusion of the automobile industry took the form of branch plants owned by the Big Three (General Motors, Ford, and Chrysler). Until the 1965 Auto Pact, Ottawa placed high tariffs on imported automobiles and their parts. In doing so, the federal government encouraged American and other foreign firms to open branch plants that produced similar goods for Canadians, but because of the size of the Canadian market these were smaller manufacturing plants than those in the US and in various European countries. The net effect for Canadian consumers was higher prices due to higher production costs, the high cost of shipping these goods long distances to regional markets in Canada, and the provincial and federal taxes on automobiles, which were higher than those in the United States. Nevertheless, Ontario benefited financially from the expansion of its industrial base.

The Canada–US Auto Pact was designed to integrate Canada's automobile industry into the North American market. At that time, over 90 per cent of Canadian automobile production was controlled by the Canadian branch plants of the Big Three automobile companies. These companies wanted to rationalize their production and marketing of automobiles within North America to be prepared for competition from foreign producers. While the Auto Pact ended in 2001 as a consequence of a WTO ruling, by then this Canada–US agreement not only had altered the nature and purpose of automobile manufacturing in Ontario, but it also had created a shift in federal policy towards more liberal trade arrangements and towards economic continentalism in North America.

Most of Ontario's exports to the United States cross the Detroit River either on the Ambassador Bridge or through the Detroit–Windsor Tunnel. Automobile trade between Windsor and Detroit is a critical factor in this international trade and accounts for 30 per cent of Canada's trade with the United States. Economic hard times can alter this trade significantly. Canadian auto sales (mainly to the US market) were $61 billion in 2007 and $47 billion in 2008, a drop of 22 per cent (Kowaluk and Larmour, 2009: Table 1).

Continentalism is discussed in Chapter 1; see especially Vignette 1.4, 'Is Continentalism Inevitable?', page 24.

Ontario's Economy

Economies vary from one region to the next. Ontario is seen as the manufacturing and financial centre of Canada. More precisely, an economy contains an industrial structure. Economists measure this structure in several ways, but the most common method is by the number of workers in an industrial sector. Each sector describes a particular set of economic activities, such as agriculture, forestry, manufacturing, education, and trade. To simplify matters, these sectors are grouped together into three broad categories: primary, secondary, and tertiary activities.

Economic structure is discussed in Chapter 4. In particular, see Table 4.17, 'Economic Structure by Sectors', page 166, and Table 4.18, 'Sectoral Changes in Canada's Labour Force, 1881–2006', page 167.

Until 2008, Ontario's economic structure was changing relatively slowly, with its tertiary sector[1] gaining ground and its primary and secondary sectors losing ground. However, this shift accelerated in a short period of time—three years. The catalyst was the global economic crisis, which hit Ontario's manufacturing sector particularly hard. While the total number of workers in Ontario increased from 6.4 million to 6.7 million from 2005 to 2008, only the tertiary sector, recorded an increase in the number of workers (Table 5.2). By 2008, the tertiary sector had just over 77 per cent of the workforce, up 2.8 percentage

Table 5.2   Ontario Employment by Industrial Sector, Number of Workers, 2005 and 2008

Industrial Sector	Workers 2005 (000s)	Workers 2008 (000s)	Difference 2005 to 2008
Primary	128	123	−5
Secondary	1,509	1,405	−104
Tertiary	4,761	5,160	399
Total	6,398	6,688	290

Source: Statistics Canada (2006b, 2009c).

points from 2005 (Table 5.3). Most job losses in the secondary sector were associated with manufacturing, especially in automobile manufacturing and parts production.

Manufacturing

Southern Ontario accounts for almost half of all manufacturing jobs in Canada. Several factors account for this concentration but paramount is the availability of a skilled and hardworking labour force, which is essential to producing quality products. Since the FTA, the face of manufacturing has changed, and so has southern Ontario's mix of manufacturing activities. Textile firms and other manufacturers that required semi-skilled labour working at or near the minimum wage have largely disappeared. New to the scene are high-technology companies that depend on highly skilled workers who can command high wages. Ontario is at the leading edge for Canada's high-technology industry (Vignette 5.9). Then, too, Ontario's economy received a boost with the recently announced provincial **Green Energy Act** designed to stimulate green construction projects and to produce equipment for these anticipated projects. Ontario hopes to become a global leader in

clean, renewable energy and conservation, and in so doing to create thousands of jobs, economic prosperity, energy security, and climate protection. The intent also, of course, is to export the resulting technology around the world. While the emphasis on a green economy could reduce greenhouse gas emissions and provide renewable energy sources, Ontario will have to provide large subsidies in the form of a 'feed-in tariff'; that is, green energy will be marketed at relatively high rates—and certainly much higher than for coal energy—for electricity sold to Ontario Power Authority.

Yet, not all is well in the manufacturing heartland, especially not in the auto industry. Ontario's manufacturing dropped by nearly 5 per cent between 2007 and 2008 and during the same period automobile production declined by 22 per cent (Kowaluk and Larmour, 2009: Tables 1 and 2). Ontario is facing the challenge of a global shift in manufacturing. Europe and the United States have seen a massive drop in their manufacturing while China and other low-wage countries have become the 'workshops' of the world. Until 2004, manufacturing was relatively healthy in Ontario, but from 2004 to 2008 the tables turned with a dramatic loss of

Table 5.3   Ontario Employment by Industrial Sector, 2005 and 2008 (%)

Industrial Sector	% of Workforce, 2005	% of Workforce, 2008	Difference 2005 to 2008
Primary	2.0	1.8	−0.2
Secondary	23.6	21.0	−2.6
Tertiary	74.4	77.2	2.8
Total	100.0	100.0	

Source: Statistics Canada (2006, 2009d).

Vignette 5.9   The Wave of the Future: Clusters and High Technology

High-technology companies are found in many places across Canada but particularly in Canada's major urban centres. In fact, Britton (1996: 266) states that 'Canadian urbanization is the key to understanding the location of technology-intensive activities.' Marc Busch takes this idea further. His answer is 'clusters' (Atkin, 2000: C1). A cluster is a place where institutions, companies, and individuals have a commitment and enthusiasm for innovative technological research along with the capital necessary to develop and market the product. High-tech clusters often are anchored around a university or a public research agency like the National Research Council. Most high-technology companies are found in Toronto, Montréal, and Ottawa rather than in Saskatoon, Halifax, or Victoria. The latter do have high-tech operations, but they are on the edge of the high-tech world and depend on a niche operation. Saskatoon's niche is agricultural biotechnology. For St John's, it is marine technology. Can technological innovations rejuvenate Ontario's struggling economy? Waterloo has already experienced the technological boom. The Waterloo region hosts the Technology Triangle of Canada and boasts 550 technology companies, including the region's anchors: Research In Motion, manufacturer of the BlackBerry; Open Text, the largest software company in the country; and Christie Digital Systems Inc., maker of a high-end projection system.

---

nearly 200,000 jobs, or one in five manufacturing jobs (Bernard, 2009: 10). Is this the end? With massive restructuring of the automobile industry promised for 2009–10, the answer is 'probably not' (see Figure 5.3). But the real question for Ontario is, can it position itself for a recovery in the next upturn of the **business cycle** and regain its export position, especially in the US, or will it become a manufacturing Rust Belt?

 Table 1.5, 'Economic Structure of Canada's Six Geographic Regions, 2006', page 20, provides an overview of each region's employment pattern.

Key Topic: The Automobile Industry

The automobile industry remains the key manufacturing activity in Ontario. By 2004, Ontario had become the leading motor vehicle producer in North America. Previously, Michigan produced the most vehicles among states and provinces. Eighty-five per cent of Canadian manufactured vehicles are exported to the United States. The impact of the 2008 economic crisis saw the demand for cars and other vehicles drop dramatically in the United States. Consequently, from 2007 to 2008, Canada's annual production dropped by 20 per cent to 2.1 million vehicles (1.2 million automobiles and 0.9 million commercial vehicles) (Table 5.4). While production figures for 2009 are not yet available, sales figures provide an indication of production. Carlos Gomes (2010), an economist with Scotiabank, noted that automobile sales in Canada hit a low point in 2009 with the sale of 1.46 million cars, but he forecasts a slight upturn in sales to 1.53 million vehicles in 2010.

Why is production expected to increase? By 2010, North American sales are anticipated to increase sharply because of the need to replace an aging fleet of vehicles and demand from those just entering the workforce. Just how that expected demand will affect Ontario's automobile production remains unclear, but it seems doubtful that the 2004 annual production level of 2.7 million vehicles will be regained anytime soon (Table 5.4).

Automobile Production: Industrial Core or Dead End?

Most industrial nations have an automobile industry that provides an anchor for manufacturing with spinoffs to 'numerous other industries', such as the steel industry. Historically, according to geographer Peter Dicken (1992: 268), '[t]he motor vehicle industry came to be regarded as a vital ingredient in national

Table 5.4   Canadian Motor Vehicle Production, 1999–2008

Year	Cars	Commercial Vehicles*	Total
1999	1,626,316	1,432,497	3,058,813
2000	1,550,500	1,411,136	2,961,636
2001	1,274,853	1,257,889	2,532,742
2002	1,369,042	1,260,395	2,629,437
2003	1,340,175	1,212,687	2,552,862
2004	1,335,516	1,376,020	2,711,536
2005	1,356,197	1,332,165	2,688,362
2006	1,389,536	1,182,756	2,572,292
2007	1,342,133	1,236,657	2,578,790
2008	1,195,426	882,153	2,077,579

*A commercial vehicle is a larger type of motor vehicle used for transporting goods or 10 or more passengers; thus, a taxi would be classified as an automobile while a city bus would be a commercial vehicle.

Source: OICA (2009).

economic development strategies.' Today, however, the North American auto industry and its Big Three automakers are in a state of crisis, with recent bankruptcy bailouts from governments to GM and Chrysler (Vignette 5.10), falling sales, and strong competition from Asian car manufacturers—notwithstanding the loss of consumer confidence in Toyota resulting from numerous recalls for defects at the end of 2009 and in early 2010. These global events raise a critical question for the future of Ontario's economy and its workers, namely: Is the automobile industry in Canada going the way of the textile industry? Low-cost labour and easy access to the Canadian market encouraged Canadian textile producers to move offshore where much lower production costs of labour more than offset higher transportation costs.

The Auto Pact

The Canada–US Automotive Products Agreement of 1965, more commonly called the Auto Pact, is an example of a successful single-industry production-sharing agreement between Canada and the United States. Before the Auto Pact, the auto industry in Canada had small, high-cost plants, each a smaller version of the much larger plants in the United States. Even though Canadian wages for auto workers were lower than those for American workers, Canadian automobile prices for the same models were considerably

Vignette 5.10   The Bailout of Chrysler and GM

Governments are loath to see the bellwethers of their manufacturing industries—motor vehicles—die, for then, it is asked, who can lead the manufacturing sector? In 2009, Canada and Ontario provided financial assistance to Chrysler Canada (US$3.8 billion) and General Motors Canada (US$10.6 billion) to keep the automobile industry in Canada at its historical share of North American production (roughly 20 per cent). In 2009, Fiat's successful buyout of Chrysler marked a new beginning and direction for Chrysler Canada, with Fiat's Sergio Marchionne taking over as Chrysler's CEO. Federal and Ontario governments secured a 2 per cent stake in Chrysler. In the US, a $25 billion government bailout of GM and Chrysler went through, although both of these companies were in bankruptcy protection by mid-2009 and many auto dealership franchises had been forced to close.

higher than in the United States. The Auto Pact revolutionized Canadian trade policy by reversing a century-old structure that sheltered Canadian automakers behind high tariff barriers. As such, it was a precursor to the FTA and NAFTA. From Canada's perspective, the Auto Pact served three objectives:

- It secured guarantees that Canadian automobile plants would not close.
- It brought to Canadian plants the advantage of economies of scale by allowing them to specialize in a few types of automobiles that would supply the North American market.
- It reduced the price of cars for Canadian customers.

The agreement called for Canada to eliminate the 15 per cent tariff on imported automobiles and US parts for Canadian manufacturers, and for the US to eliminate its corresponding tariff. The agreement included special protection for Canadian plants. Canada was guaranteed a minimum level of automobile production based on production levels for each type of automobile manufactured in Canada in 1964, that is, the ratio of vehicle production to sales in Canada for each class of vehicle had to be at least 75 per cent or the percentage attained in the 1964 model year, whichever was the highest. Furthermore, **value added** in vehicles produced in Canada had to be no less than the dollar amount achieved in 1964.

Under the terms of the Auto Pact, qualified motor vehicle manufacturers (Ford, General Motors, and Chrysler) were able to import both vehicles and automotive parts duty-free into Canada. Other motor vehicle manufacturers, however, had to pay a 6.1 per cent import duty. Japan and the European Union appealed to the WTO, arguing that the Auto Pact discriminated against imported vehicles. In July 2001, as a consequence of a WTO ruling in favour of Japan and the EU that it did indeed amount to an unfair trade practice, the Auto Pact, which had helped to build Canada's automobile assembly and parts industry, ceased to exist. Its demise helped Toyota and Honda expand their market share in North America, though Toyota's brand may have been damaged by the large number of

recalls of its automobiles in late 2009 and early 2010.

The Growth of the Auto Industry

The automobile industry drives the Ontario economy. With 150,000 workers employed by the assembly and parts firms, nearly one in seven Canadian manufacturing jobs depends directly on this industry. Wages in the assembly plants are relatively high for semi-skilled workers, but somewhat lower for workers in parts firms. Nevertheless, these incomes enable workers to make substantial purchases, thereby stimulating southern Ontario's retail sector. In addition, the auto industry accounts for nearly one-quarter of Canada's merchandise exports. (Most of Canada's remaining exports are primary products, such as energy, grain, and forest products.)

From 1995 to 2002, exports of automotive products (passenger autos and chassis, trucks and other motor vehicles, and motor vehicle parts) increased in value from $62.9 billion to $97 billion (Statistics Canada, 2003), but by 2005 the value of these exports dropped to $88 billion (Statistics Canada, 2006a). During this period, Ontario's strong hold on automobile production was largely due to:

- Canadian assembly plants had a higher productivity than comparable US plants. In 2001, Canadian workers had a 7 per cent advantage over American workers in the time required to produce an automobile (Brent, 2002).
- The low Canadian dollar allowed Canadian exports of cars and trucks to cost less than similar vehicles produced in the United States.
- Health-care costs are covered by Ontario while these costs are part of the wage package in the United States.

Both the federal and Ontario governments supported the automobile industry by supporting research into advanced technologies and by providing training in automotive manufacturing technologies at Ontario universities. For example, McMaster University and the University of Waterloo have established an Automotive Manufacturing Innovation

New Motor Vehicle Production in Ontario

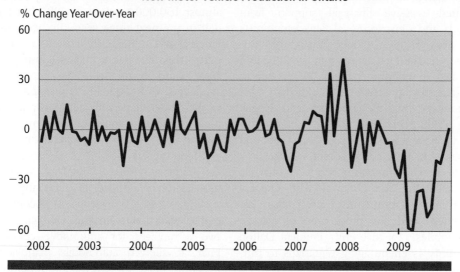

Figure 5.7   Ontario automobile production.
Source: Ferley and Hogue (2009).

initiative with 35 industry partners and support from the Ontario government. Another example is the launching of Auto21 at the University of Windsor with funding from the federal government.

After 2002, however, two problems surfaced. First, North America developed a mismatch between production and sales, forcing companies to close less productive plants. The North American automobile industry had the capacity to assemble 23 million units a year, but sales in 2002 fell below 18 million units and sales predictions of 17 million units for 2003 pointed to layoffs and plant closures (Keenan, 2003). New, more efficient, and flexible assembly plants were constructed and, by 2005, these additional plants added another 3 million units of capacity. Second, with the exchange rate on the Canadian dollar rising from 64 cents US in late 2002 to 96 cents in September 2007, the price of Canadian exports to the United States rose significantly. If the Canadian dollar retains this position or continues to rise, automobile exports to the United States may be affected, as will investments in new assembly plants and auto parts firms.

The Canadian automotive industry has seen a shift from the domination of the Big Three to the rising power of Japanese-based firms. A May 2007 report by Statistics Canada confirmed this changing nature of the industry:

> The downsizing of the traditional Big 3 firms has been widely-publicized, reflecting the traumatic impact on their employees and their suppliers. Less appreciated, however, is how rapidly overseas firms have ramped up production in Canada to offset many of the losses. (Statistics Canada, 2007d)

Japanese-based manufacturers have filled much of the gap left by the traditional North American automakers in production in Canada. In 2006, the Japanese Automobile Manufacturers Association of Canada (JAMA) reported that just over 901,000 automobiles were produced in Canada, double their level in 1998, but by 2008, their production had declined to 795,859 (JAMA Canada, 2010). In spite of the turmoil facing Toyota over recalls, its sales in Canada increased in 2009 while Honda's sales slumped sharply (ibid.). As a result, Japanese automobile manufacturers' share of the domestic market jumped from 16 per cent to 36 per cent during this period (Statistics Canada, 2007d).

By 2008, the automobile industry was in shock. Sales dropped to record lows, causing production to fall. Chrysler and General Motors required massive financial support from Canada and Ontario. Royal Bank economists Ferley and Hogue (2009) estimated that production reached its low point in late 2008 and expected a rebound in early 2009—but to nowhere near pre-2008 levels. But like most economists' predictions, Ferley and Hogue were off the mark, with the low point occurring in 2009 and signs of a modest rebound appearing in early 2010 (Gomes, 2010).

Is there another approach to automobile manufacturing for Ontario? The Ontario government is promoting both a green environment and the production of electric automobiles. To that end, Premier Dalton McGuinty is encouraging Ontarians to purchase electric cars:

Electric cars will become part of the Ontario government's fleet and consumers will get up to $10,000 in rebates to buy one of the experimental vehicles. This plan helps get more people behind the wheel of a green vehicle to create jobs, reduce smog and equip Ontario for the 21st century. (CBC News, 2009)

However, electric cars are expensive: General Motors' Chevrolet Volt is priced at over $40,000, so it remains to be seen how many people, even with the rebate, will buy an electric car.

Automobile Parts Firms

The automobile industry consists of two separate operations: the assembly of automobiles and trucks and the production of their parts. In addition, some manufacturing firms supply semi-processed materials. In southern Ontario, fabricating firms produce steel, rubber, plastics, aluminum, and glass parts for automobile assembly and parts plants in Canada and the United States. Finally, service firms, ranging from the advertisers and designers to the sales and service staff, manage the finished product. In short, the auto industry is a final-product type of manufacturing, and as such, its added value reaches a maximum.

The automobile parts industry employed almost 130,000 workers in 2005 but fell by half in the next three years. By being highly efficient and strategically located, automobile parts firms can operate on a just-in-time principle—auto components are produced in small batches and quickly delivered as needed to their customers. This allows the assembly plants to achieve considerable savings by reducing their inventories, warehousing space, and labour costs.

In the late 1980s, the Big Three automobile manufacturers decided to subcontract their parts business. This practice of subcontracting parts is called **outsourcing**. Outsourcing had two advantages for automobile companies. First, it allowed automobile manufacturers to concentrate on assembling automobiles, thereby reducing their costs and improving the quality of their product. Second, parts companies were not unionized and therefore had lower wages. This wage differential—and the savings it provides—is the main reason why General Motors, Ford, and DaimlerChrysler continue to divert work to parts firms. In Ontario, the Canadian-based Magna International has grown into the third-largest auto parts company in North America.

The 'Canadian advantage' encouraged the expansion of Canadian-based auto parts firms. This advantage was based on Canadian workers' higher productivity as compared to their American counterparts, Canada's healthcare program, and a weak Canadian dollar. Since then, the Canadian advantage has slipped for two reasons. First, the strengthening of the Canadian dollar caused the exchange advantage to largely disappear. Second, access to the US market became more complicated because of security concerns related to the 11 September 2001 attacks on New York and Washington. Delays at the border increased costs for truckers and compromised the 'just-in-time' system for the parts industry.

The border issue is a particularly critical matter for the parts industry because many items cross the border multiple times before they are transformed into parts for the assembly plants. Some appreciation of transborder movement of parts lies in knowing that new vehicles are composed of over 8,000 individual parts that must be worked into the components we think of as building the car.

Ottawa and Washington currently are seeking a solution to the traffic congestion that hampers trade between the two countries, with plans well underway for a new international bridge called the Detroit River International Crossing (DRIC) that would open in 2015. This new border crossing is expected to tie in directly with Highway 401 as it approaches Windsor and thereby avoid the present crawl through the city streets of Windsor to reach the Ambassador Bridge, which lies north of the proposed DRIC. In 2009, the project was approved by both Canadian and American environmental assessment agencies, and in April 2010 agreement was reached between Windsor and the province on a $1.8 billion project, involving 12,000 construction jobs, for building the freeway connection to the new bridge (CBC News, 2010). Funding for Canada's portion of the multi-billion dollar project will come from Ottawa's infrastructure funds, making Canada's section publicly owned, while the United States section may be privately owned. Of course, the financing for bridge construction is secured by future toll revenues, so Ottawa and the Canadian taxpayer may even make money in the long run (Detroit River International Crossing, 2009; Transport Canada, 2009; Potter, 2010).

The automotive crisis affected the Ontario economy directly and the Canadian economy indirectly. Ontario's parts companies were hurt by the downturn in automobile production. The near failures of General Motors and Chrysler delivered a harsh message—diversify or die. In 2009, for example, Magna nearly succeeded in a takeover bid for Opel, a German automaker, while Fiat took control of Chrysler. Successful parts companies have turned to a broader range of clients, ranging from solar and wind energy technology to consumer products, forestry/mining equipment, and aerospace components. While the automobile industry remains their primary

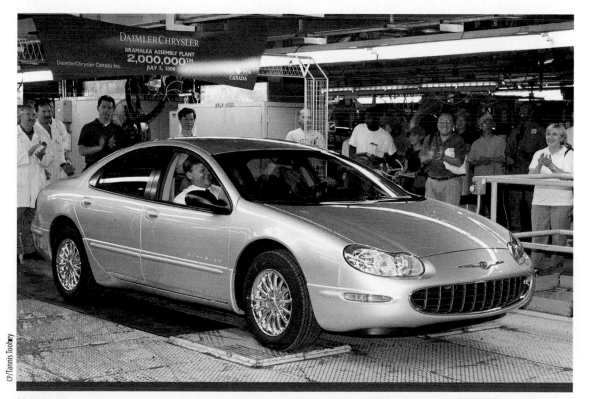

Photo 5.10

A new car rolls off the assembly line at the Chrysler plant in Brampton, Ontario, but will Chrysler survive?

customer, diversification provides greater stability. The magnitude of the automotive crisis to both the Canadian and US economies provided the rationale for the enormous bailout by Ottawa and Washington.

Automobile Assembly Plants

Automobile assembly plants are concentrated in southern Ontario where transportation links to the major markets of Canada and the United States are readily available and driving distances are short (Figure 5.8). All Canadian-based automobile companies sell most of their cars and trucks in the United States. In 2005, approximately 85 per cent of all vehicles produced in Ontario were sold in the US. Three companies, General Motors, Ford, and DaimlerChrysler, accounted for 60 per cent of North American vehicle production in 2005. Ten years earlier, the Big Three dominated the automobile market by controlling 90 per cent of car and truck sales. Competition from Japanese and Korean companies with assembly plants located in North America, including southern Ontario, is fierce. These Asian-based firms have captured 40 per cent of Canada's market and an equivalent part of the American market.

While Japanese auto factories are located in the United States, the Canadian advantages

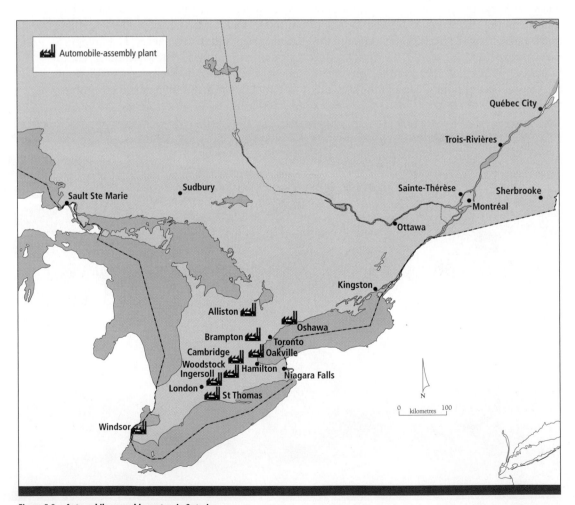

Figure 5.8   Automobile assembly centres in Ontario.
Canada's auto assembly plants are located in southern Ontario. The Sainte-Thérèse GM plant north of Montréal closed in 2002 and the Ford plant in St Thomas is scheduled to close in 2011.

mentioned earlier—a highly motivated work-force, the lower value of the Canadian dollar, and the public medical system—have attracted substantial Japanese investment in Ontario. In American-based assembly plants, the cost of providing medical insurance is often built into the wage agreement.

With the closing of the General Motors plant at Sainte-Thérèse near Montréal in 2002, all auto assembly plants are located in southern Ontario close to both the Canadian and American markets (Table 5.5). This closure affected auto-parts firms based in the Montréal area because these suppliers tend to locate their factories near assembly plants to reduce transportation costs. For example, in 1996 Magna International decided to build a truck frame plant in St Thomas, Ontario, from which truck frames could be delivered to any of three General Motors truck assembly plants within one day. One of these assembly plants is located at Oshawa, Ontario, while the other two are in Pontiac, Michigan, and Fort Wayne, Indiana.

Troubled Times: Lose Some, Win Some

A fierce fight for the highly prized North American market is underway. The closing of the General Motors plant at Sainte-Thérèse near Montréal in 2002 signalled the start of restructuring by the Big Three. By 2006, the Big Three's hold on this market had slipped while Honda and Toyota, the two leading Japanese automobile manufacturing companies, had grabbed more of the market. The Big Three have dominated in the sale of light trucks, minivans, and SUVs. For the Big Three, the recent drop in their sales is due to (1) a shift from the larger vehicles to smaller, more fuel-efficient cars because of higher gasoline prices; (2) customer satisfaction with the performance and quality of Japanese vehicles, which results in both new sales and return sales; and (3) the new demographics, i.e., the baby boomers are becoming empty-nesters and no longer require large vehicles such as minivans and SUVs.

The Big Three automobile companies are in trouble. Their new strategy is to reduce production and cut costs by closing older, less efficient assembly plants and replace them with more efficient and flexible factories. However, in 2003, DaimlerChrysler, troubled by overproduction, abandoned its plans to build a state-of-the-art assembly plant at Windsor. At the same time, the company announced that it would close its Dodge Ram van assembly plant in Windsor, thus eliminating 1,000 jobs (Brent, 2003). Since 2003,

Table 5.5	Automobile Assembly Plants in Southern Ontario, 2009
Location	**Products**
Ingersoll	Chevrolet Equinox; Pontiac Torrent; Suzuki XL-7* (a GM–Suzuki joint venture)
Brampton	Chrysler 300; Dodge Magnum; Dodge Charger
Windsor	Dodge Caravan; Chrysler Town and Country; Pacifica
Oakville	Ford Freestar; Ford Edge; Lincoln MKX
St Thomas (to close in 2011)	Ford Crown Victoria; Mercury Grand Marquis
Oshawa	Chevrolet Monte Carlo; Impala
Oshawa	Buick LaCrosse (Allure In Canada); Pontiac Grand Prix
Oshawa	GMC Sierra; Chevrolet Silverado (pickups)
Alliston	Honda Civic; Acura CSX; Acura MDX; Pilot; Ridgeline
Woodstock	Toyota RAV4
Cambridge	Toyota Corolla and Matrix
Cambridge	Lexus RX350

*Production of the Suzuki XL-7 ceased in May 2009.

Source: Assembly Plants in Canada, June 30, 2010, Industry Canada. Reproduced with the permission of the Minister of Public Works and Government Services, 2010.

the bad news has continued. Ford announced in 2006 that it was closing its Essex engine plant in Windsor while General Motors plans to shut its power train plant in St Catharines (CBC News, 2006b).

In February 2007, DaimlerChrysler declared that the company would eliminate as many as 2,000 jobs at its plants in Brampton (Keenan, 2007). In spite of these cutbacks, most plant closures are occurring in the United States, and Ontario is making some gains. For instance, Ford will transfer production of its Lincoln Town Car from St Thomas to Oakville in 2008 and begin assembling a new vehicle, the Ford Edge. Another example is that General Motors, in a surprise move, plans to produce the Camaro, a 'muscle' car, at Oshawa. The reductions by the Big Three have left parts plants feeling the pinch and layoffs have been announced for firms in Stratford, St Mary's, Newmarket, and Toronto (Van Praet, 2006). By late 2009, signs of a recovery had appeared. While sales in Canada remained strong, US demand for new automobiles finally showed an upswing, due in part to the US rebate program worth $4,500 for older cars. Then, Chrysler's decision to reopen its Windsor plant to produce right-side-drive minivans for markets in Britain, Japan, and India signals a new direction and a break from such heavy reliance on the US market.

Asian automakers are capturing more and more of the North American market. While Honda and Toyota are scouting for sites to build manufacturing facilities in North America, GM and Ford are closing plants and slashing jobs after losing a steady stream of customers to foreign-based rivals. In Ontario, Honda and Toyota are expanding their production capacity, thus creating more jobs and more demand for parts. Toyota, in 2008, built an assembly and parts plants at Woodstock with production of the RAV4 sports utility. Honda's new engine plant near Alliston began production in 2008. Honda and Toyota have expanded their operations in Ontario primarily because of strong demand for their automobiles in the US. Ontario is a preferred production site in North America for the Japanese manufacturers as a consequence of support from both the federal and provincial governments and what is recognized as a strong work ethic among small-town workers. Then there is the issue of the massive recall of Toyota cars. Could the negative reaction of the US media and the US government affect Toyota's decision in selecting future sites?

Northern Ontario

The physical base of northern Ontario consists of the Canadian Shield and the Hudson Bay Lowland. While both have disadvantages for settlement, development has taken place along the two transportation corridors between Ontario and Manitoba. One corridor follows the Trans-Canada Highway (Ontario Highway 17) and the Canadian Pacific rail

Vignette 5.11   The Threat of 'Hollowing-Out'

In the coming decade, Ontario's principal challenge will be its manufacturing base. How can Ontario avoid the threat of 'hollowing-out'? China, now the manufacturing centre of the world, presents fierce competition: its firms can produce manufactured goods at lower cost because of the wage differential between China and Ontario. Many Ontario firms have responded by relocating offshore where wages are much lower and, even with the cost of shipping the products back to Canada, production and transportation costs still are lower than they would be for domestically manufactured goods. This geographic shift of manufacturing from Canada to other countries began with the textile industry and is now affecting more sophisticated forms of manufacturing. With China now producing automobiles for its domestic customers, how long will it be before China is exporting surplus production to North America? This ongoing erosion of Canada's industrial base is a result of the open world markets. For Canada, its vulnerability has serious consequences for its two core manufacturing regions—Ontario and Québec.

line while the more northern corridor parallels the Canadian National rail line and the northern link of the Trans-Canada Highway (Ontario Highway 11).

Figure 2.1, 'Physiographic Regions and Continental Shelves in Canada', page 34, and Figure 5.4, 'Central Canada', page 180, delineate Ontario's physiographic regions. Also see 'Physiographic Regions' in Chapter 2, page 34.

As an old resource hinterland, northern Ontario is troubled by a sluggish economy, a declining population base, and high unemployment rates. With the housing crisis in the US, exports for lumber have dropped while pulp and paper products are facing an ever-diminishing demand: to some extent, computers have replaced newspapers, which have gone out of business or shrunk in size, and also have largely replaced paper in such business applications as billing, accounting, and banking; retailers use less paper; packaging of food and other products tends to use plastic (or consumer-purchased cloth bags) in preference to paper; recycling programs have replaced a small percentage of new paper products. Even northern Ontario's trans-shipment role between Ontario and Western Canada has weakened as more products, such as grain and potash, are shipped to Vancouver for sale in Asian countries. With a faltering economy, northern Ontario's population has taken on four demographic characteristics that are strikingly different from southern Ontario:

- an aging population;
- a net out-migration, especially of younger members of its population;
- few immigrants;
- a small but rapidly expanding Cree and Métis population.

The physical base of northern Ontario is considerable but its population consists of fewer than 750,000 people. In percentage terms, northern Ontario comprises 87 per cent of the geographic area of Ontario but has less than 7 per cent of Ontario's population. With virtually no agricultural base, most people live in towns and cities along the two transportation routes that link Montréal and Toronto with Winnipeg. Both of these trans-Canada routes cross the rugged Canadian Shield. The southern route includes the Canadian Pacific Railway and the Trans-Canada Highway. The Trans-Canada Highway connects North Bay with Sudbury, Sault Ste Marie, Nipigon, Thunder Bay, Dryden, and Kenora. The northern route includes the Canadian National Railway and a major highway. This northern highway connects North Bay, Timmins, Kirkland Lake, Cochrane, Kapuskasing, and Nipigon to Thunder Bay.

Mining, forestry, and tourism are the major economic activities, although many people are employed in the public sector. Like other resource hinterlands, the development of northern Ontario's economy was linked to external markets. Resource exploitation in northern Ontario began after the construction of railways in the late nineteenth century. Soon thereafter, resource development took place along the lines of the Canadian Pacific Railway (CPR) and the Canadian National Railway (CN). The same pattern of resource development took place along the Temiskaming and Northern Ontario Railway (also known as the Ontario Northland Railway).[2] The railway extends northward from North Bay to New Liskeard, Kirkland Lake, and Cochrane, which is on the CN main line. Northern Ontario exports minerals and forest products, including gold, nickel, newsprint, and lumber. These resource products contribute less than 10 per cent of the value of Ontario's exports to foreign countries.

Unlike the northern hinterland of Québec, the natural conditions in northern Ontario are not conducive to major hydroelectric developments. While a number of rivers flow across northern Ontario to James Bay and Hudson Bay, the gentle slope of the land, especially in the Hudson Bay Lowland, does not make the construction of hydroelectric dams feasible. A few established sites for hydroelectric production, such as along the Ottawa River and at Abitibi Canyon about 100 km north of Cochrane, are important, but they cannot meet the energy demands in Ontario. The much higher elevations in the Canadian Shield area of Québec provide the necessary natural drop to drive the turbines that produce hydroelectric power. Since Québec has a surplus of electrical power, some of it is transmitted to southern Ontario.

Northern Ontario has a strikingly different settlement pattern from that of southern Ontario. Long distances separate the four major cities. For example, the distance from North Bay to Sudbury is over 100 km; from Sudbury to Sault Ste Marie is about 300 km; and from Sault Ste Marie to Thunder Bay is about 500 km. The explanation for this oasis-like settlement pattern is due to northern Ontario's physical geography. The rocky terrain of the Canadian Shield discourages continuous settlement; a series of isolated settlements are located at mining sites, pulp plants, and key transportation junctions. Similar to other old hinterlands, the urban centres of northern Ontario are declining and the populations of single-industry towns like Cobalt, Kirkland Lake, and Porcupine rose and fell as the cycle of resource exploitation ran its course.

Forest Industry

While forestry was by far the most important primary industry in northern Ontario, the forest industry has faced tough times. By 2008, at least a dozen of communities in northern Ontario that rely on the forest industry have been especially vulnerable to recent economic conditions. Shutdowns have put thousands out of work. In Kenora, Red Rock, Dryden, Thunder Bay, Terrace Bay, Kapuskasing, and other centres in Ontario's north, workers in the forest industry are facing layoffs, closures, and limited prospects. The negative spinoff effects reverberate through these communities and across northern Ontario. While plant closures have taken place across the country, the high cost of energy is an added factor affecting mills in Ontario. Electricity prices have skyrocketed in the province, to the point where energy makes up 30 to 40 per cent of the cost of getting wood from the forest to the mill and then to the market. For many companies in northern Ontario, energy has become the make-or-break number on the balance sheet.

Ontario has 57 million hectares of productive forest area with most classified as softwood. The main species are black spruce, poplar, and Jack pine. From this resource, the forest industry of northern Ontario normally produces approximately $15 billion of products each year, with 60 per cent exported to markets in the US. Ontario is one of the leading exporting provinces of softwood lumber. Access to the American market is critical. Access is controlled by agreements between Canada and the US. The most recent agreement, the 2006 Softwood Lumber Agreement (SLA), limits Canadian softwood exports. Under this agreement, Canadian lumber firms are allocated 34 per cent of softwood lumber sales in the US market. Each province is allocated a share of those exports based on 2004–5 exports, so that Ontario gets around 9 per cent of the Canadian softwood lumber exports to the US (CTV, 2006).

For a fuller discussion of the SLA, see Chapter 7, 'North American Free Trade Agreement', page 306.

Across Ontario, the northern coniferous forest is classified into two regions: the boreal barrens and the boreal forest regions. The boreal barrens region is found in the Hudson Bay Lowland. This transition zone between the boreal forest and the tundra has no commercial value. Scattered patches of spruce and tamarack are surrounded by poorly drained land known as 'muskeg'. The commercial forest industry is located in the boreal forest region that extends from Québec to Manitoba. The geographic limits of the boreal forest region closely parallel those of the Canadian Shield. The main species are black and white spruce, tamarack (larch), balsam fir, Jack pine, white birch, and poplar. Some 50 communities depend on the forest industry based on the boreal forest region. By volume of wood cut, Ontario ranks just behind British Columbia and Québec. Ontario's mills produce pulp and paper, lumber, fence posts, and plywood. The pulp and paper industry in northern Ontario accounts for about 25 per cent of the national production, and Ontario is a leading exporter of newsprint and pulpwood to the US. In fact, most pulp and paper firms operating in northern Ontario are American-owned companies.

Road access is an extremely important factor in the forest industry. Trucks must bring the pulpwood to the mill and then the final product must be shipped to market. Pulp and paper mills are located in communities with access to transportation networks—Thunder Bay, Sault Ste Marie, Sturgeon Falls, Kenora, Fort Frances, Dryden, Marathon, Iroquois Falls, Kapuskasing, and Terrace Bay.

Technology in the forest industry has advanced over the years and greatly increased the productivity of workers in the mills and in the woods. When loggers first arrived in northern Ontario, they had only axes and saws. A logger with such equipment might cut five to 10 trees a day. By the middle of the twentieth century, the technology had advanced. Chainsaws then enabled a logger to cut 50 to 70 trees a day. In clear-cut operations today, mechanical harvesters, which can harvest hundreds of trees per day, have largely replaced the chainsaw, while in smaller operations, chainsaws remain.

Transportation of timber to mills has also changed. In the past, logging often took place in the winter months. Horses pulled the logs to frozen rivers where they were stored until the river melted in the spring. Tractors later took over the task of hauling the logs to the river. In the spring, the logs were floated to sawmills, which were often located at the mouths of the rivers. Today, most timber is used in the pulp and paper industry. These huge plants require a steady supply of pulpwood. This demand and the technological changes that have taken place have altered the nature of logging in four ways: (1) logging has become a year-round affair; (2) most logs are hauled to the mills by trucks; (3) trees are harvested before they reach maturity; and (4) mechanical tree harvesters are employed and usually clear-cut a wooded area.

Today, the forestry industry faces several challenges. From an economic perspective, local mills have been forced to close because of high electrical costs due to higher electricity rates. A second economic challenge facing the forest industry is the age of its pulp and paper

plants and a drop in demand for these products. Many mills in Ontario were built before World War II and continue to use old technology, which results in much higher discharges of toxic wastes into the environment. The consequences of such practices can be life-threatening. The worst cases of massive toxic discharges into streams and rivers occurred in the 1960s. At that time, a chemical plant that prepared the bleaching solution for the pulp plant at Dryden was releasing effluents containing mercury into the English–Wabigoon river system.[3] Since then, forestry mills have updated their operations to ensure a cleaner and safer natural environment. Another challenge is related to the US market for softwood lumber. The forest industry is in trouble because of the low prices for softwood lumber. In 2006, the sharp downturn in the US housing market produced a double whammy: fewer exports to the US and a tax on Canadian softwood lumber exports. The tax, which comes into effect when the price drops below US$355 per thousand board feet, is part of the new softwood lumber agreement. As US demand for lumber declines, so does the price of lumber. From May 2006 to May 2007, the price of softwood lumber had dropped from US$367 to US$224, thus triggering the maximum export tax of 15 per cent, which is collected by Ottawa (Anderson, 2006; Export Development Canada, 2007). By 2009, softwood lumber prices were low, falling below US$200, but within a year prices recovered so that by March 2010 the 52-week average reached US$241 (Natural Resources Canada, 2010a: Table 1).

From a sustainable forest perspective, the main challenge is to maintain a balance between logging and the regeneration of the forest. Since over 90 per cent of Ontario's forest lands are owned by the provincial government, private companies must obtain forest leases. The Ontario government, like most other provincial governments, has insisted that these private companies take on the responsibility for restoring trees to logged areas through forest management agreements. Efforts to hand-plant seedlings are helping to speed up the process of reforestation, but how successful these efforts have been can only be known 20 years from now. The practice of granting forest companies long-term timber leases is based on the assumption that it is in the companies' self-interest to manage the forests well. It remains to be seen whether this assumption and the corresponding leasing policy will ensure the protection and regeneration of Ontario's forests.

A second challenge is the changing nature of the boreal forest from a predominantly coniferous forest to a broadleaf one. Since the 1950s, the volume of timber harvesting has more than tripled. This level of logging has put so much pressure on the boreal forest that original species are disappearing. As a result, the boreal forest is shifting from coniferous to broadleaf species, and this might be expected to continue and increase over the longer term as species migrate north with climate change. Spruce, pine, and fir have been replaced by poplar and birch, which both are pioneer species, well adapted to regenerate in clear-cut portions of the boreal forest. The long-term consequences of this species shift are unknown, but such a shift clearly illustrates how forest companies have replaced forest fires as the main agent of change.

Photo 5.11

The boreal forest stretches across most of northern Ontario, providing the basis for a forest industry. Will the Canadian Boreal Forest Agreement (signed May 18th, 2010) between forest companies and environmental groups prove both environmentally sound and economically attractive?

These challenges have placed the forest economy of northern Ontario in a tailspin. Exports of lumber, pulp, and paper have reached record lows because of the depressed US market. In one year, from 1 April 2005 to 31 March 2006, nine mills closed (Natural Resources Canada, 2006). Kenora and Thunder Bay suffered two losses each—Kenora lost a sawmill and a newsprint plant while a fine paper mill and sawmill disappeared from Thunder Bay's industrial landscape. Only one new investment—a mere $14 million for a new boiler and kiln at Chapleau in northeastern Ontario—took place during this time frame. While mill closures have occurred in all regions of the country, the majority took place in Ontario and Québec. Conversely, forest companies have invested most heavily in Western Canada and British Columbia. The problem for northern Ontario has been many old plants with higher production costs than elsewhere, thus making them candidates for closure.

Mining Industry

Northern Ontario's mining industry is centred in the Canadian Shield, which provided ideal geological conditions for the formation of hard-rock minerals such as gold, nickel, and copper. In 2008, Ontario was the leading Canadian producer of gold and nickel, with the annual value of metallic minerals at nearly $6 billion (Natural Resources Canada, 2008). With the global recession that began at the end of 2008, however, Ontario's metallic mineral production for 2009 dropped to $3.8 billion; nonetheless, despite a nearly 34 per cent drop from the 2008 production value, for all minerals Ontario led the provinces and territories in production at $6.3 billion in 2009 (Natural Resources Canada, 2010b). Most production comes from Red Lake (gold), Hemlo (gold), Wawa (gold), Manitouwage (gold), Marathon (gold), Thunder Bay (gold, copper, zinc), and Sudbury (nickel, copper). Ontario's first diamond mine (the Victor Mine), located in the Hudson Bay Lowland some 500 kilometres north of Timmins, came into production in 2008. With a variety of minerals for several export markets, mining

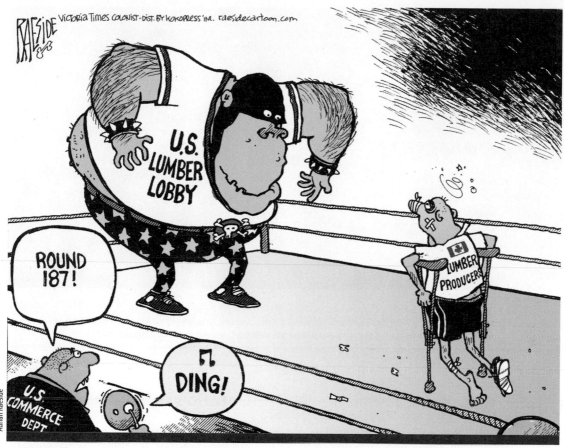

Adrian Raeside

Photo 5.12

The objective of the US lumber lobby is to curtail imports of Canadian softwood. The Canadian forest industry, over the years, has suffered from duties imposed by the US government. Just as the latest softwood lumber agreement was reached in 2006, US prices for softwood lumber dropped sharply. The combination of the US duties followed by falling prices has badly hurt the forest economy in northern Ontario, where plant closures and layoffs are far too common.

remains the most robust primary activity in northern Ontario.

Mineral production for the provinces and territories in 2009 is shown in Table 7.5, page 298.

Mineral production for the provinces and territories in 2009 is shown in Table 7.5, page 298.

The first important mineral discovery took place in 1883 when the copper-nickel ores of the Sudbury area were discovered during the building of the Canadian Pacific Railway. In 1903, a rich silver deposit near Cobalt was detected during the construction of the Temiskaming and Northern Ontario Railway. Rich gold deposits were uncovered near Timmins in 1909 and at Kirkland Lake in 1911. While new mineral finds kept the mining

economy of Ontario going, the overall pattern was one of boom and bust. During the 1960s, a large lead-zinc deposit was discovered in Timmins. A decade later, rich gold deposits were found at Marathon (the Hemlo gold mine) and at Pickle Lake. The Victor diamond mine, owned by the South African mining company De Beers, is now in production, and, in an agreement with Ontario, has committed 10 per cent of its production to the newly established Sudbury diamond mine factory. Ironically, while Sudbury's unemployment rate hovered around 10 per cent in 2009, no worker in Sudbury had the required skill set or was trained in diamond cutting. Instead, 27 highly skilled diamond cutters and polishers were flown in from Vietnam (Hoffman, 2009: B1).

Think About It

Is Ontario's ambition to become a key link in the global diamond trade a reasonable gamble or a pipe dream?

Photo 5.13

Mining forms a major part of northern Ontario's economy. The Williams gold mine near Marathon remains the largest gold-producing mine in Canada. In 2005, Ontario accounted for 60 per cent of Canada's gold production, followed by Québec at 20 per cent and British Columbia at 14 per cent (Chevralier, 2009).

Minerals are a non-renewable resource that is depleted over time. Therefore, mining communities can have a short lifespan. A mine closure can occur without warning, causing the sudden demise of a single-industry centre. Over the years, a number of ore bodies have been exhausted or production has ceased because new, lower-cost mines have been discovered. These two circumstances have forced some mines to close, resulting in a great outflow of miners and their families. For instance, Elliot Lake's uranium mine was closed, not because its ore was exhausted but because open-pit uranium mines in northern Saskatchewan could produce uranium oxide at a much lower price. Ontario Hydro, the principal buyer of Elliot Lake uranium, decided to purchase its supplies from the uranium mines in northern Saskatchewan. Other mines that have closed include an iron mine at Steep Rock and the silver mines at Cobalt. While former mining towns may remain, they undergo a drastic downsizing. With a stock of low-priced houses and an attractive wilderness setting, Elliot Lake made a modest recovery as a retirement centre. However, its population has declined steadily and is now only about 12,000 (see Table 5.8).

Ontario's Urban Geography

Ontario is the most highly urbanized province in Canada. Almost 9.5 million people—nearly 85 per cent of the province's total population—live in towns and cities. As well, 10 of Canada's 25 largest cities are in Ontario. With the exceptions of Sudbury and Thunder Bay, these census metropolitan areas (CMAs) are located in southern Ontario.

Since cities are the engines of the Canadian economy, Ontario's urban geography provides an enormous economic advantage. Cities are self-perpetuating: employers set up for business and that is where most employment is located. The pattern of urban growth from 2001 to 2006 varied considerably for these CMAs (Table 5.6). The fastest-growing cities in Ontario were Oshawa, Toronto, Guelph, and Ottawa–Gatineau. Sudbury and Thunder Bay, both located in the resource hinterland of the Canadian Shield, showed practically no growth from 2001 to 2006, well below the growth rate of all the CMAs in southern Ontario. In fact, in the previous five-year period, 1996–2001, these two cities lost population due to the stagnating economy in northern Ontario.

Table 5.6 Census Metropolitan Areas in Ontario, 2001–6

Census Metropolitan Area	Population 2001 (000s)	Population 2006 (000s)	Change (%)
Peterborough	110.9	116.6	5.1
Thunder Bay	122.0	122.9	0.8
Brantford	118.1	124.6	5.5
Guelph	117.3	127.0	8.2
Kingston	146.8	152.4	3.8
Sudbury	155.6	158.3	1.7
Oshawa	296.3	330.7	11.6
Windsor	307.9	323.3	5.0
St Catharines–Niagara*	377.9	390.3	3.5
Kitchener*	414.3	451.2	5.2
London	432.5	457.7	5.1
Hamilton	662.4	692.9	4.6
Ottawa–Gatineau*	1,067.8	1,130.8	5.9
Toronto	4,682.9	5,113.1	9.2
Total	9,012.7	9,691.8	7.2

*Statistics Canada has combined St Catharines and Niagara Falls as a single census metropolitan area although these cities still exist as separate political jurisdictions, as, of course, do Ottawa, Ontario, and Gatineau, Québec. Likewise, Statistics Canada includes Waterloo and Cambridge with Kitchener as a single CMA.

Source: Statistics Canada (2007c).

The eight largest cities in Ontario are located in southern Ontario (Figure 5.9), and of these the four largest (Toronto, Ottawa, Hamilton, and London) form the core of the three urban clusters in southern Ontario, namely, the Golden Horseshoe, southwestern Ontario, and the Ottawa Valley. Each of southern Ontario's three major urban clusters has a population of at least 1 million. As mentioned earlier, northern Ontario has fewer and smaller cities, each separated by long distances. It has two major cities (Sudbury and Thunder Bay) and no single urban cluster. Sudbury, the largest centre, had 158,258 residents in 2006, while Thunder Bay's population was 122,907.

The Golden Horseshoe

The Golden Horseshoe obtained its name because of its horseshoe-like shape around the western end of Lake Ontario and its outstanding economic performance over the years. This tiny area of the Great Lakes Lowland forms the most densely populated area of Canada. The Golden Horseshoe extends from the US border at Niagara Falls northward to Hamilton, Toronto, and then on to Oshawa. Nearly 7 million Canadians live, work, and play in Canada's largest population cluster, and many visitors come either as tourists or on business trips. Accounting for nearly one-quarter of Canada's population, the Golden Horseshoe contains numerous towns and cities, including Toronto, Hamilton, Oshawa, St Catharines, Niagara Falls, Burlington, Oakville, Pickering, Ajax, and Whitby. Toronto, the largest city in Canada, is its urban anchor, while Hamilton, with its steel plants, is the focus of heavy industry (Vignette 5.12), and Oshawa is Canada's leading automobile-manufacturing city.

Toronto

As the largest city in Canada, Toronto dominates the urban landscape of southern Ontario and plays a significant role in Canada's urban hierarchy. Toronto is the financial capital of Canada, housing the main offices of national and international banks and investment

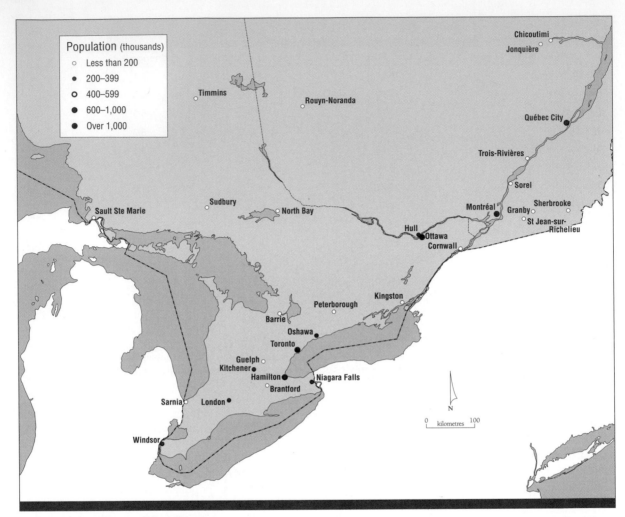

Figure 5.9   Major urban centres in Central Canada.
Most large urban centres in Central Canada are located in the Great Lakes–St Lawrence Lowlands, especially in southern Ontario.

Vignette 5.12   Hamilton: Steel City or Rust Town?

Hamilton, situated at the west end of Lake Ontario only 50 km from Toronto, is known as Steel City. Unfortunately, the North American steel industry has fallen on hard times and Hamilton is hurting. Canada's largest steel firms (Stelco and Dofasco) were purchased by US Steel and ArcelorMittal (a transnational from India) in 2007 and 2008, respectively. The companies were renamed US Steel Canada Inc. and ArcelorMittal Dofasco. Both global companies are struggling and new industrial development is not on Hamilton's horizon. As Trevor Cole (2009) wrote: 'As prosperity plumped nearby rivals such as Burlington, Oakville, Mississauga, Kitchener–Waterloo and—especially—Toronto, it skipped Hamilton completely, cruelly, until most of its big-name companies were gone, the stores along Barton Street deteriorated into dark and crumbling shells, downtown became a kind of forbidden zone, and even the Mafia couldn't make any money.' The future of Hamilton is unclear but McMaster University and close proximity to Toronto may provide the basis for a knowledge-based industry.

firms, as well as the Toronto Stock Exchange. Over the years, Toronto's population growth has outpaced that of most other cities, and by 2006 the population of the Greater Toronto Area (GTA) had reached 5 million. Over the last decade, the main factor driving Toronto's growth has been the arrival of immigrants from foreign countries. At the same time, people and businesses have spilled over Toronto's boundaries, causing a geographic expansion of the GTA. The spread of Toronto's population and businesses into neighbouring areas is due partly to lower land values and rents. Land values and rents are highest in downtown Toronto. Cities surrounding Toronto have significantly lower land values and can therefore attract businesses by offering lower office rents.

As the province's major cultural and entertainment centre, Toronto is a hub for the entertainment industry, which ranges from music and drama to professional sports, and plays an essential role in shaping Canadian culture. Tourists flock to Toronto to enjoy these world-class cultural and entertainment activities. Just how dependent the cultural and service industries are on tourists was revealed by the deadly outbreak of SARS in April 2003; as a result of a travel advisory issued by the World Health Organization, tourism receded and was estimated to have lost the city $1 billion in that month alone (Brean, 2003).

Like other major cities in North America, Toronto is trying to cope with a rapidly expanding population, much of which has moved into adjoining urban areas. One effort to deal with the administration of this urban area was to create a super-city known as the Greater Toronto Area (GTA). In 1998, the municipalities of the former Metropolitan Toronto (Toronto, North York, Etobicoke, Scarborough, East York, and York) officially formed a single city government with the hope that solutions to problems arising from rapid population growth and geographic expansion of the urban population could be found.[4] While many opposed this approach to mega-urban government, the objective is to reduce administrative overlap and thereby create a more efficient urban government. Whether or not this single administration approach is succeeding in serving such a large population

Dennis Flood, www.dennisflood.com

© Dennis Flood 2004

Photo 5.14

Toronto Eaton Centre is Canada's leading shopping centre.

remains to be seen. However, many problems remain, including traffic congestion. With so many people living outside of downtown Toronto but working in the city centre, massive numbers of them commute daily, creating traffic jams during the morning and afternoon rush hours. Traffic congestion leads to time lost travelling to and from work. A radical solution would be charging a toll on automobile traffic into and out of Toronto. London, England, imposed such a tax and it resulted in a reduction in traffic flow and a decrease in travel time.[5]

As the most popular destination for immigrants to Canada, Toronto has experienced remarkable ethnic diversification over the past 30 years, to the point where visible minorities are hardly in the minority. Visible minorities comprised 37 per cent of Toronto's population in 2001 compared to 26 per cent in 1991 and are projected to be in the majority by 2017.

Wave after wave of immigrants has come to Toronto, first from Europe, especially from Italy and the United Kingdom. More recent newcomers have come from Asian countries. According to the 2006 census, Toronto had 2.3 million immigrants who accounted for 46 per cent of the city's population; of these immigrants, nearly 14 per cent came from China and Hong Kong, 10 per cent from India, and 6 per cent from Italy and the Philippines (Table 5.7).

Toronto is known as a city of neighbourhoods, partly because immigrant groups have clustered in certain areas. Little Italy and Little Portugal are older, well-established ethnic neighbourhoods. More recent Asian, African, and Caribbean neighbourhoods have emerged. Recent immigration into Toronto has had an impact on its cityscape, including 'foreign' architecture and commercial activities designed to meet the demands of these new Canadians. Asian theme malls, which consist of a large number of small retail outlets, many of which are either Chinese restaurants or specialty stores, represent a dramatic change to Toronto's space. Unlike other suburban malls, no anchor store exists and, rather than a single developer who leases space, each retail unit is owned by the operator of that retail space. The failed development of an Asian theme mall in Richmond Hill indicates the resistance by local residents to a changing urban space with a new ethnic identity (Preston and Lo, 2000: 182–90). Toronto Island also represents a distinct neighbourhood that has survived confrontation with city planners who had decided to transform the area, accessible by a short ferry ride from downtown, into public open space. Its geography—essentially a sandbar extending into Lake Ontario that at one time was attached to the mainland—consists of a series of islands, most of which now is dedicated to parkland. However, residential communities exist on two islands (Ward's Island and Algonquin Island).

Ottawa Valley

Ottawa–Gatineau is not only a major population cluster in Canada, but is distinctive as an area where both official languages are used. As the nation's capital, federal government operations are found on both sides of the Ottawa River. In 2001, Ottawa had a population of just over 800,000 while Hull (now Gatineau) in Québec had nearly 260,000 citizens. Together, Ottawa–Gatineau represents the fourth largest metropolitan area in Canada, and in 2006 its population totalled 1,130,761. On the Ontario side, Greater

Table 5.7 Top 10 Countries by Birth of Immigrants* to Toronto, 2006		
Country of Birth	**Number of Immigrants**	**% of Immigrants**
India	221,930	9.6
China, People's Republic of	191,120	8.2
Italy	130,685	5.6
Philippines	130,315	5.6
United Kingdom	125,980	5.4
Hong Kong, Special Administrative Region	103,090	4.4
Jamaica	93,845	4.0
Pakistan	85,635	3.7
Sri Lanka	84,225	3.6
Portugal	76,304	3.3
Total of all immigrants to Toronto	2,320,165	100.0

*Immigrants are persons who are, or have ever been, landed immigrants in Canada. A landed immigrant is a person who has been granted the right to live in Canada permanently by immigration authorities. Some immigrants have resided in Canada for a number of years, while others are more recent arrivals.

Source: Statistics Canada (2009f).

Ottawa is the third-largest urban cluster in Ontario. Much of its growth has come from in-migration from other parts of Canada and from foreign countries. Many are attracted by the employment opportunities offered by the federal government and the business community. In 2001, 18 per cent of its population was foreign-born compared to 15 per cent in 1991. Visible minorities accounted for 14 per cent of population, with the three leading groups being blacks, Chinese, and Arabs, and West Asians.

The federal government, located in the national capital, is the major employer, followed by the high-technology sector. The federal government requires a wide variety of goods and services in its daily operations. This demand provides an opportunity for many small and medium-sized firms in the Ottawa Valley. By locating its departments and agencies in both Ottawa and Gatineau, the federal government has ensured that Ottawa's economic orbit extends to a number of small towns on both the Ontario and Québec sides of the Ottawa River. Most people live on the Ontario side, especially in Ottawa. Other urban centres on the Ontario side include Nepean, Gloucester, Kanata, and the municipalities of Rockcliffe and Vanier. Around Ottawa, in its periphery, are a number of smaller centres, including Pembroke, Cornwall, Brockville, and Kingston. However, with the exception of Pembroke, these cities are more closely tied to economic and social activities in Toronto and Montréal.

As the capital of Canada, Ottawa is the focus of national and international affairs and has subsequently developed into a national administrative centre with a large federal civil service. In its early days, the forest industry played a strong economic role, but it was eventually overshadowed by the activities of the federal government and then the high-tech industry. Still, the geographic location made this an ideal place for a pulp and paper mill: pulpwood was easily harvested from the interior forest lands and then transported along the Ottawa and Gatineau rivers to the Eddy Pulp and Paper Mill in Hull, Québec. An added advantage of this site is the availability of low-cost electrical power from the hydroelectric installation on the Gatineau River near the Chaudière Falls. Today, the E.B. Eddy plant produces fine and coated paper.

Described in the press as the 'Silicon Valley of the North', Ottawa has become an industrial leader in high technology. Ottawa firms specialize in telecommunications, computers, software and computer services, and electronics (Britton, 1996: 266). Northern Telecommunications (Nortel) was the leading high-tech company, specializing in fibre optics. In 2009, Nortel declared bankruptcy and its assets were sold to foreign buyers, led by Ericsson, headquartered in Sweden. The loss of Nortel has seriously hurt Canada's high-tech industry. It remains to be seen if Canada can produce another Nortel. Other leading figures in Ottawa's high-tech world failed and some, like Nortel, have been purchased by larger companies outside of Canada. As James Bagnall (2003) reported, 'With software maker Corel now in the hands of a California-based venture capital firm and fibre-optics star JDS Uniphase set to shift its headquarters next month to San Jose, Ottawa is swiftly reverting to its former role as a technology R&D centre, serving the whims of multinationals based elsewhere.'

Southwestern Ontario

Southwestern Ontario represents the third major urban cluster in southern Ontario. About 1 million people live in this area, which extends from Lake Erie to Lake Huron. Although this region does not have a major metropolitan city, London, with a population of 458,000 in 2006, is the largest city in southwestern Ontario. London is close to Toronto, Buffalo, and Detroit and several main urban centres within southwestern Ontario that form the Kitchener CMA (Cambridge, Kitchener, and Waterloo), as well as Guelph, midway between London and Toronto. Windsor and Sarnia are on the edge of the London/Kitchener urban cluster and fall into the orbit of Detroit.

As the leading city of southwestern Ontario, London provides administrative, commercial, and cultural services for the larger region. It is also the headquarters of several insurance companies, including London

Think About It

If a small country like Sweden can have a global player in the high-technology arena, why can't Canada have more than just RIM?

Life Insurance Company. Among its economic activities, London is noted for manufacturing, including the production of armoured personnel carriers and diesel locomotives by General Motors. London, therefore, has a sound and growing industrial foundation based on insurance, manufacturing, and high-tech industries. Such manufacturing firms pay relatively high wages to their employees, and consumer spending by these employees supports a strong retail sector.

Automobile assembly plants are located in Cambridge, Ingersoll, St Thomas, Alliston, and Windsor (Table 5.5 and Figure 5.8). In 2008, Toyota added an assembly plant at Woodstock. Auto-parts plants play an important role in the economy of southwestern Ontario while a number of high-tech firms, particularly in the Kitchener CMA, add to the region's economic diversity. This area comprising Kitchener, Waterloo, and Cambridge has been known for more than 20 years as Canada's 'Technology Triangle' where innovative technology is often developed in research institutes or universities. Such technology is then used to create commercial products.

Cities of Northern Ontario

In sharp contrast to cities in southern Ontario, those urban centres in northern Ontario find their economies stalled and their populations static or declining, so that many northern Ontarians are dissatisfied with their lot. Such grumbling is common in resource hinterlands. Claiming to be isolated and ignored by the provincial government, some of its citizens have called for a new province called Mantario (Matteo, 2006). Like most downward transitional regions, the resource base is losing its economic strength for three reasons: (1) as the best mineral and timber resources have been exploited, resource production is now more costly; (2) depressed prices for lumber have resulted in mill closures; and (3) as companies introduce more technology, fewer workers are required.

Timmins, located in the mineral-rich Canadian Shield, is in the centre of a gold belt, while Sudbury is in a world-famous nickel belt, where the mining and smelting of nickel and copper ores are core functions. Both cities began as single-industry towns, but over time they have become important regional centres with many service industries and more diversified economies, factors that have supported community stability. Even with these efforts, however, the two communities have been unable to maintain their populations over the past 10 years. For example, from 1996 to 2006, the populations of Sudbury and Timmins declined by 4.3 per cent and 3.0 per cent respectively (Statistics Canada, 2002, 2007c). For the past five years, Sudbury has enjoyed a slight revival in its population growth while Timmins continued to decline (Table 5.8).

Sault Ste Marie is a steel town located on the Great Lakes between Lake Superior and Lake Huron. Like most other centres in northern Ontario, its population trend over the last 10 years has been downward. For instance, Sault Ste Marie's population declined from 83,619 in 1996 to 80,098 in 2006 (Statistics Canada, 2002, 2007c). Along with Sudbury, North Bay, and Thunder Bay, Sault Ste Marie saw its population increase slightly from 2001 to 2006 (Table 5.8). The town's major firm is the Algoma steel mill, which has struggled to succeed in a highly competitive marketplace. Algoma is located at some distance from its major industrial customers in southern Ontario and the United States. With high transportation costs, the Algoma mill is at a geographic disadvantage. However, Algoma has made use of local railway and seaway connections to secure raw materials, such as iron ore, and to market its steel products. In 2001, Algoma Steel fell on hard times and filed for protection under the Companies' Creditors Arrangement Act. Algoma Steel was able to recover and is now a profitable firm. The measures were hard on workers, however, including a 9 per cent reduction in pay and a reduction of the workforce by 18 per cent (Bertin and Keenan, 2003). The smaller but more efficient Algoma operation focuses on direct strip production, which has positioned Algoma as one of the leaders in the North American hot-rolled sheet market. In June 2007 Algoma was acquired by Essar Steel Holdings Ltd., a division of the multinational

Table 5.8 Urban Centres in Northern Ontario, 2001–6

Centre	Population 2001	Population 2006	% Change 2001–6
Sudbury	155,601	158,258	1.7
Thunder Bay	121,986	122,907	0.8
Sault Ste Marie	78,908	80,098	1.5
North Bay	62,303	63,424	1.8
Timmins	43,686	42,997	−1.6
Kenora	15,838	15,177	−4.2
Temiskaming Shores*	12,904	12,927	−0.2
Elliot Lake	11,956	11,549	−3.4
Kapuskasing	9,238	8,509	−7.9
Kirkland Lake	8,616	8,248	−4.3
Total	521,036	524,094	0.6

*Temiskaming Shores is the newly restructured area comprising the three former municipalities of New Liskeard, Haileybury, and Dymond Township.
Source: Statistics Canada (2007c).

conglomerate, Essar Global. Algoma provides Essar with a strong platform for growth in the North American market.

Thunder Bay, at the western end of Lake Superior, is a major transshipment point and a key element in the east–west transportation system. Bulky products—particularly grain, iron, and coal—are shipped to Thunder Bay for loading onto lake vessels. In recent years, however, this function has diminished. In the past, for example, iron from mines at Atikokan passed through Thunder Bay on its way to steel mills located along the shores of the Great Lakes. This mining operation is now defunct. Grain and potash from the Canadian Prairies are the major products handled at Thunder Bay. The Canadian Wheat Board sends prairie grain by rail to Thunder Bay and then by grain carrier to eastern ports for export to foreign countries. Until 1996, Ottawa heavily subsidized grain shipments under the Crow Benefit. With the elimination of that subsidy, the advantage of sending grain by rail to Thunder Bay has diminished, thus weakening its grain transportation role. In addition, the rise of Asian markets for resources from Western Canada has cut into Thunder Bay's importance as a transshipment centre.

For more on the Crow Benefit, see Chapter 8, page 334.

Since 1996, Thunder Bay's population has declined. Its population dropped from 126,643 in 1996 to 122,907 in 2006 (Statistics Canada, 2002, 2007c), although, as shown in Table 5.8, the population of Thunder Bay increased slightly (by 0.8 per cent) from 2001 to 2006. Overall, thanks to marginal population increases in the larger centres—Sudbury, Thunder Bay, Sault Ste Marie, and North Bay—the cities in northern Ontario just held their own over the most recent census period (2001–6), but the smaller centres—Timmins, Kenora, Temiskaming Shores, Elliot Lake, Kapuskasing, and Kirkland Lake—all lost population.

Cities and towns in northern Ontario are struggling to diversify their economies. One option being pursued is tourism. Many small private firms—motels, summer camps, and fishing and hunting lodges—have sprung up (especially in smaller communities) to meet the demand from the growing number of tourists coming to experience the wilderness of northern Ontario. The urgency to diversify is driven by the continuing reduction in the size of the primary labour force. For example, the workforce at the nickel mines in Sudbury has fallen by half over the last 30 years. Sudbury has reacted by aggressively seeking to expand its service industry. The city has met with some success, partly because of its strategic location at the junction of three highways—the Trans-Canada Highway, Highway 69 (which leads to Toronto), and Highway 144 (which connects Sudbury with Timmins)—and partly because of its efforts to have federal and provincial agencies (the federal National Tax Data Centre and provincial Department of Mines) and institutions (Laurentian University and the Science North Museum) locate in Sudbury.

Many First Nations are found in northern Ontario. Some are situated in remote locations, such as the Kashechewan First Nation reserve. The residents of Kashechewan reserve, located near James Bay some 450 km north of Timmins, were evacuated three times in 2005–6 because of spring floods and polluted drinking water. The federal government was considering relocating the reserve to higher ground, but another option was to move the reserve to the Timmins area where access to drinking water, health, housing, schools, and employment would be greatly improved. Such an integrationist approach has some merit, but social engineering strategies also involve certain dangers. Reaction from Mayor Demeules of Smooth Stone, who was not consulted about the proposed relocation, was not enthusiastic about the proposed relocation to his community. With the closing in July 2006 of a timber mill, 230 employees lost their jobs. As the mayor stated: 'There's no work for my own people. You don't take a community in crisis and dump them on another community in crisis. This has to be a win-win situation. Our community has to be respected' (Campion-Smith, 2006). However, culture trumped economics. On 30 July 2007 Kashechewan Chief Solomon signed an agreement with Ottawa to redevelop the community at its present location.

Beyond the towns and cities of northern Ontario lies a sparsely populated area where fewer than 10,000 people live. Cree and Ojibwa Indians live on isolated reserves or in small Native settlements in this vast region of scrub forest and muskeg. Only a few gold mines, trapping cabins, and fishing lodges dot the landscape. For the Indian residents, fishing and hunting are an important source of food, while trapping and guiding provide cash income. Since this cash income usually does not meet their basic needs, transfer payments supplement their income. The development of the Victor diamond deposit near the Attawaspiskat First Nation reserve provides a potential economic boost for the band because De Beers agreed to hire local workers for both its construction and production operations. Two other proximate Aboriginal groups, Moose Cree First Nation and Taykwa Tagamou Nation, also signed agreements with De Beers for education, training, and compensation (Bone, 2009: 251). This open-pit mine opened in July 2008, under budget and ahead of schedule, and in 2009 was named 'Mine of the Year' by readers of the international trade publication *Mining Magazine*.

Vignette 5.13   Population Distribution in Northwestern Ontario

The large, sprawling Thunder Bay census division of northwestern Ontario stretches from Lake Superior to Hudson Bay and James Bay. This division has less than 150,000 people. Typically for northern Ontario, most people live in a few large urban centres while the remaining territory is virtually empty. Northwestern Ontario's population geography has three distinct zones:

- Zone 1: Most people (67 per cent) reside in the City of Thunder Bay. In 2006, this major CMA population cluster had 122,907 people.

- Zone 2: Most of the remainder (27 per cent) live south of the CN rail line where a number of small towns, villages, and Native reserves are located.

- Zone 3: North of the CN rail line lies the 'empty' population zone, which is beyond the reach of the national highway system and has only 2 per cent of the region's population.

SUMMARY

Ontario, located in Canada's most favoured physiographic region, faces serious economic challenges because of the global economic meltdown. Dangers lurk in the global economy for forestry and manufacturing. Both depend on US recovery, particularly in the housing market and in consumer demand for automobiles. Over the last five years, a declining US market for softwood lumber and paper products has devastated northern Ontario. Sawmills and pulp and paper plants have closed and those still operating confront a gloomy future with more shutdowns possible. Manufacturers also are looking at a bleak future. Fierce competition from foreign countries with low labour costs will continue to dog Ontario-based firms, forcing them to make hard decisions, such as to reduce their labour costs, search for niche production, and, if all else fails, either relocate to low-wage countries or close shop. Within North America's automobile manufacturing sector, the Big Three are reducing their production to better match their market share. Ontario has felt a few of these cuts, and the province also has made some gains, thanks to

its highly productive labour force, federal and provincial support, and Canada's public health-care program. Major gains came from Toyota and Honda. In 2008, Toyota opened a second plant at Woodstock, the midpoint between Cambridge and London, while Honda has a new engine plant near Alliston.

For Ontario to prosper and progress into the next decade as well as to retain its title as the economic engine of Canada, this dominant core region must see its forestry and manufacturing sectors recover, exports to the United States return to their previous levels, and trade with the rest of the world expanded. This is a tall order. Equally critical, Ontario must accelerate the development of its knowledge-based economy by investing heavily in innovation and research partnerships among universities, federal research institutions, and private industry. By succeeding in these arenas, Ontario will secure its place within the North American market by developing a stronger north–south economic axis while maintaining its traditional centrality as the core in the east–west axis.

CHALLENGE QUESTIONS

1. For much of its recent history, Ontario has fashioned its prosperity on trade with the United States. Is this 'all eggs in one basket' strategy still valid?
2. In 2003, the provincial government proposed closing its five coal burning thermoelectric generating stations and thus reduce smog in southern Ontario. By 2010, electricity is still produced at coal-burning plants. What prevented Ontario from acting on its 2003 proposal?
3. What is needed to pull the forest economy of northern Ontario out of its tailspin?
4. Is the Ontario government's 'green strategy' for electric production and for electric cars a move in the right direction, or is it going to price Ontario out of the global marketplace?
5. Hobson's choice: If the knowledge-based economy represents the future of Ontario, would you have 'saved' Nortel rather than General Motors and Chrysler?

FURTHER READING

Courchene, Thomas J., with Colin R. Telmer. 1998. *From Heartland to North American Region State: The Social, Fiscal and Federal Evolution of Ontario*. Monograph Series on Public Policy, Centre for Public Management. Toronto: Faculty of Management, University of Toronto.

From Heartland to North American Region State received the Donner Prize for the best book on public policy in 1998. While the book was written over a decade ago, the issue of Ontario's place within North America remains critical, and Ontario's automobile industry remains an integral though weakened part of the North American manufacturing system.

The premise of this book by two economists is simple: Ontario has become a regional economic

force within the North American economy. From a neo-liberal economic perspective, Ontario's interests now lie with the continental economy, not the national one. Put differently, Ontario benefited from Canada's national economy with its high tariffs but now the province's self-interest lies in the North American economy.

Since our world has changed so dramatically, consider two thoughts:

1. Do you think that these two economists might want to revise their emphasis on the North American market and place more attention to the fast-growing regions of the world?
2. Do you think that they would be so optimistic about Ontario's automobile industry in the continental economy? How about the global economy?

6

QUÉBEC

INTRODUCTION

'La belle province' retains its place as a cultural, economic, and demographic power within Canada. Of the six geographic regions, Québec has a francophone culture, thus distinguishing it from the rest of Canada, but also from the rest of North America. Equally important, Québec ranks second among the six regions in terms of economic output and population size, accounting for almost 20 per cent of the country's GDP and 24 per cent of its population, although this ranking is less secure as Western Canada and British Columbia are gaining ground.

Québec's culture adds a powerful dimension to its economic and demographic contribution to Canada. By virtue of its geography and history, Québec occupies a special place in Canada and North America. To begin with, the St Lawrence River opened North America to French settlement, exploration, and the fur trade. The lowlands along shores of this great river are the birthplace of the French presence in North America and the homeland of the Québécois. Even today, the St Lawrence River continues to figure prominently in Québec's economy and culture. The Port of Montréal owes much of its past economic prosperity to its role as a major transshipment point for goods moving to and from North America.

Since Confederation, Québec's geography has expanded northwards—first in 1898 and then in 1912. In 1927, however, with Britain recognizing a watershed boundary between Québec and Newfoundland, land was lost. Within this expanse are several Aboriginal homelands, including the semi-autonomous Inuit political territory of **Nunavik** and **Eeyou Istchee**, the lands of the James Bay Cree.

The *Key Topic* is Hydro-Québec, with special emphasis on the James Bay Hydroelectric Project. This project—which is only partially completed—has placed Québec in the forefront of hydroelectric production and has made Québec an essential source of power for New England. At the same time, the James Bay Project led to the first modern land-claim agreement in 1975, the James Bay and Northern Québec Agreement (JBNQA), which was followed 26 years later by the Paix des Braves. The JBNQA also laid the groundwork for the political evolution of the Inuit of Québec, who are close to achieving a new form of public self-government within a province.

CHAPTER OVERVIEW

This chapter will examine the following topics:

- Québec's physical geography and historical roots.
- The basic elements of Québec's population and economy in the context of the region's physiography.
- Québec's cultural and economic development within the context of the core/periphery model.
- Québec's position within Canada, its francophone culture, and its separatist movement.
- The effects of the James Bay and Northern Québec Agreement and the Paix des Braves on the Cree.
- Hydro-Québec's strategy of developing low-cost electrical power in northern Québec to stimulate industrial activities in southern Québec.
- Québec's changing role within Canada and North America.

View of Québec City at night from Lévis, Québec. Photo: Henry Georgi/All Canada Photos.

Québec's Cultural Place

History, with the early settlers who came from France to the St Lawrence Valley hundreds of years ago, has made this corner of North America the homeland of the French-speaking people. This alone represents a remarkable accomplishment. *La Fête nationale du Québec* (National Holiday of Québec) celebrates this accomplishment on 24 June each year. However, the road has not always been easy or pleasant, and the collective memory often recalls humiliations and resistance as well as hard-fought victories. Flowing through this collective memory is a fear of absorption into Anglo-American culture. Separatism is a response to that fear and to the desire to be independent. In spite of all that has passed, however, Québec remains part of Canada with its distinct culture and language recognized and respected by the rest of Canada more than ever before.

The French language and Québécois culture have generated a strong sense of belonging among its francophone citizens, forming the basis of ethnic pride, loyalty, and nationalism, which from time to time fuels the desire for an independent political state. Such feelings are particularly strong among the 'old stock'—the descendants of some 10,000 settlers who migrated from France in the seventeenth and eighteenth centuries. While their first loyalty is often to Québec, most Québécois also have a strong attachment to Canada. This sense of dual loyalty is unique in Canada and it accounts for Québec's interest in the concept of a political partnership within Confederation. In the past, federal politicians have avoided referring to Québec as a **nation** because they feared the acceptance of the word 'nation' would encourage the separatist movement as well as alienate the other provincial governments, which subscribe to the concept of 10 equal provinces. Times have changed and, in November 2006, the Canadian Parliament unanimously passed a motion by Prime Minister Harper that recognized the Québécois as a nation within Canada.

For a more complete discussion of the historic origins of the concept of a political partnership, see 'One Country, Two Visions', in Chapter 3, page 120.

Québec's culture is largely derived from the historical experience of **francophones** living in North America for over 400 years and of their being Canadians for more than 140 years. Events of the distant past, including massive immigration of British subjects to Québec in the early nineteenth century, the suppression of the rebellions in Lower Canada of 1837–8, and the 1885 execution of Louis Riel, and more recent ones, such as conscription crises during the two world wars, the War Measures Act of 1970, and the **patriation** of the Canadian Constitution in 1982 over the opposition of the Québec government, fuelled the desire for a separate state. Québec's role as the geographic heart of French language, customs, and heritage in North America is widely recognized in Canada and the US. Many French-speaking Canadians from the rest of Canada and Franco-Americans visit Québec to renew their sense of ethnicity. In sum, Québécois history and geography give Québec its raison d'être and its vision of Canada as a nation of two founding peoples (Paquin, 1999). Canada's cultural duality has led to tensions between French- and English-speaking Canadians that have often strained the bonds of Confederation, but the resolution of these strains has led to a 'compromising' type of federation. This complex concept of compromise is unique to Canada because our historic relations between the two founding European peoples have been tempered by the immigration of peoples from other European countries and then from around the world. As a dynamic concept, federalism continues to evolve.

For a broader historic discussion of two competing visions of Canada, see 'The French/English Faultline' in Chapter 3, page 114, and 'French/English Language Imbalance' in Chapter 4, page 162.

The vast majority of people living in Québec speak French. In 2006, Québec's population was 7.6 million, a sizable increase

Think About It

Are ideologies like railway tracks? If so, does the concept of two founding peoples—if accepted by Ottawa over the model of 10 equal provinces—inevitably lead to two independent states, Québec and the rest of Canada?

Think About It

Words have subtle meanings. Would it make a difference if Canada's Parliament had recognized Québec as a nation within Canada rather than the Québécois as a nation within Canada?

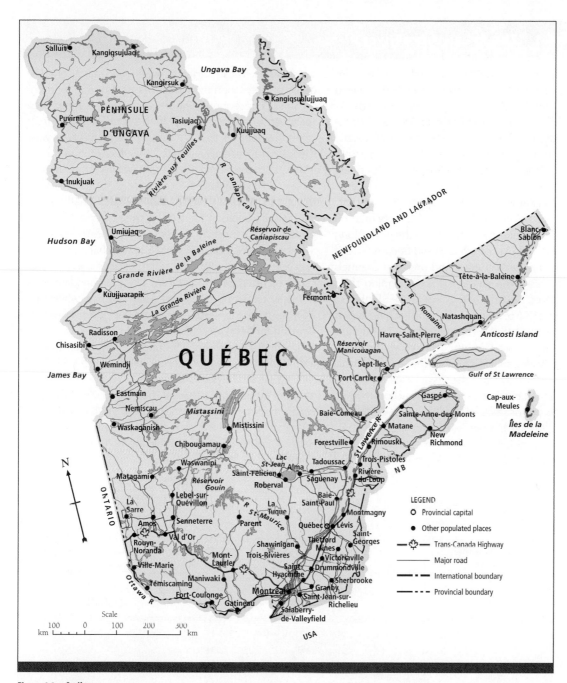

Figure 6.1   Québec.

Source: Atlas of Canada, 2006, 'Quebec', at: <atlas.nrcan.gc.ca/site/English/maps/reference/provincesterritories/quebec>.

Vignette 6.1   The St Lawrence River

The St Lawrence River provides a natural waterway into the interior of North America and, consequently, played a key role in the history of New France. Today, the St Lawrence is an essential part of North America's transportation system. Cities along its shores, particularly Québec City and Montréal, have benefited greatly from the waterway's role as a major trading route. Montréal's favourable location on the St Lawrence gives it an economic advantage and fuels the city's growth.

While waters from the Great Lakes empty into the St Lawrence River, ships could not reach the ports along the Great Lakes. As part of what would become the St Lawrence Seaway, canals, locks, and dredged channels were built to allow the passage of ocean-going ships into the heart of North America. For these ships, two major obstacles had to be overcome: the Lachine Rapids near Montréal, and Niagara Falls, which prevented ships from passing from Lake Ontario to Lake Erie. The building of the Lachine Canal (1825) and the Welland Canal (1829) allowed ships to reach the Great Lakes. Over time, these canals and locks were improved to accommodate the increasing size of ships and barges.

After World War II, there were vast improvements in ocean transportation. This was the era of the 'supertanker'. Even the average size of freighters had increased

substantially. These larger ships required much greater depth of water than their predecessors to avoid running aground. Consequently, more and more ocean-going ships were unable to sail past Montréal and enter the Great Lakes. The solution to this transportation bottleneck was the St Lawrence Seaway, which, when it was inaugurated in 1959, allowed ocean vessels to travel from the Atlantic Ocean to the Great Lakes—a distance of 3,700 km—from March to December. While Montréal lost its natural advantage as a transshipment point, the city remains an important port, especially for container traffic. The Seaway and the Port of Montréal (photo 6.1) face challenges from two sources. First, a long-term shift of grain and iron ore cargo occurred, with grain from Western Canada going primarily to Pacific ports and Asian markets while the downsizing of the North American steel industry saw the demand for iron ore shipments from northern Québec/Labrador decline sharply. Second, the short-term drop in traffic due to the global economic slump in trade has meant that, for the past decade, the Seaway has operated at about 60 per cent capacity. The magnitude of the decline in traffic, notably in grain and iron ore, is revealed by the drop in shipments from a high annual average tonnage of around 75 million in the late 1970s to 45 million tonnes in 2008 (Jenish, 2009: 41).

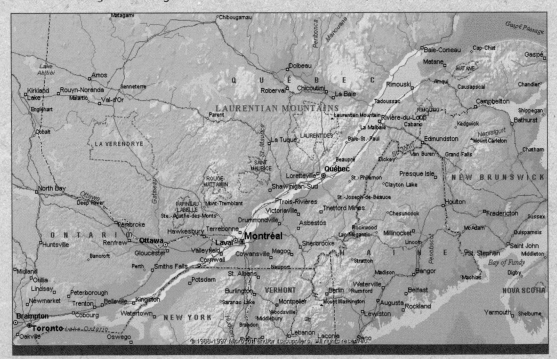

Figure 6.2   The St Lawrence River and Lake Ontario.

Source: <www.aquatic.uoguelph.ca/rivers/stlawmap.htm>.

from 7.2 million in 1996. Over 83 per cent of Québecers declared French as their mother tongue (Figure 6.3). These Québecers are known as the **Québécois**, although in recent years the term also has been applied to all Québec residents. The remaining Québecers include Aboriginal peoples, **anglophones** (those whose mother tongue is English), and **allophones** (those immigrants whose mother tongue is neither English nor French). Anglophones are concentrated in Montréal, Estrie (the Eastern Townships), and Outaouais (the Ottawa Valley). Allophones are concentrated mainly in Montréal. In northern Québec, Cree and Inuit form the majority. From time to time, social tensions surface between the French-speaking majority and its minority groups as **Québecers** continue to assert their desire to be '*maîtres chez nous*'. To a large degree, this goal of controlling their destiny within Canada has been achieved, and, as a result, Québecers are more comfortable with their position within Confederation, making relations with anglophones, allophones, and Aboriginal peoples more amenable. Within the business elites, the so-called **Québec Inc.**, francophones feel increasingly welcome at former Anglo bastions such as the Mount Royal Club and the University Club (Vignette 6.2)

Québec's Place within Canada

Québec's cultural, demographic, and economic position within Canada remains strong (Figure 6.3). Yet, Québec's economy and population have lost ground to the rest of Canada.

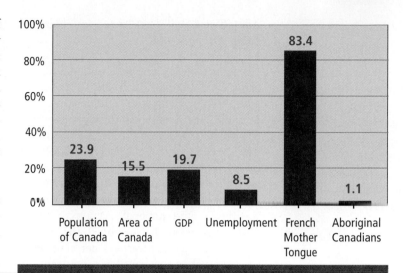

Figure 6.3   Québec vital statistics.
By population size and gross domestic product, Québec remains the second-ranking geographic region in Canada, though Western Canada and British Columbia are narrowing the gap.

Note: Unemployment percentage is for 2009; demographic data are based on the 2006 census.
Sources: Tables 1.1, 1.2, 5.1.

While still ranked second, Québec production dropped from 21.1 per cent of GDP in 2001 to 19.7 per cent in 2006. By 2008, Québec's GDP had fallen to 18.8 per cent (Statistics Canada, 2009b). Its population fared much better, only slipping from 24 per cent in 2001 to 23.9 per cent in 2006. This increase was due to a recent growth in the birth rate. For example, in 2003–4, it was 9.9 per 1,000 population compared to Canada's 10.6 birth rate; but by 2007–8 Québec's birth rate had reached the highest rate in many decades at 11.2, exceeding the national average of 11.1 (Statistics Canada, 2009f).

Vignette 6.2   Has the French/English Business Faultline Softened?

In its heyday, Club Saint-Denis was the place for cutting deals, trading gossip, and debating issues. An emblem of Québec's two solitudes at the time, it was once the place to be if you were a member of Québec Inc., the province's francophone business elite, while the Anglo business barons retreated to their own exclusive haunts. But changing times, society, and tastes claim their victims. And the iconic, now somewhat dowdy, private club founded 135 years ago, faced with an aging membership and onerous refurbishment costs, is preparing to shut its doors for good in mid-September 2010. A new rival in town—Club 357C—is plugged into the young francophone business crowd. Even Jacques Parizeau, one of the creators of Québec Inc., took note of the changes last year and declared it 'clinically dead'. Then, too, the breaking down of French–English business barriers over the past 20 years helped hasten its decline. Francophones have felt increasingly welcome at such former Anglo bastions as the Mount Royal Club and the University Club.

Annual national birth rates are found in Table 4.8, page 143.

Nonetheless, economic and population gains are taking place in British Columbia and Western Canada at a higher rate. For example, the combined 2006 populations of British Columbia and Alberta (7,686,215) now exceed that of Québec (7,546,100). The problem continues to be the rate of population increase. While Québec's population continues to increase, its rate, until 2007–8, was below the national average. Between 2001 and 2006, for example, the rate of population increase was 4.3 per cent while the national average was 5.4 per cent. In the previous five-year period (1996–2001), Québec's rate of population increase was only 1.4 per cent, and most of this was due to foreign migrants settling in the Montréal area

(Statistics Canada, 2009c). Three key questions are:

1. Will the long-term downward trend, which began in 1966 when Quebec's share of the total Canadian population was 28.9 per cent, continue?
2. If most population gain comes from immigrants, will this put more pressure on the issue of 'accommodation' of minorities?
3. Will the post-2006 'higher' Québec birth rate continue?

Concern over sluggish economic growth and low birth rates is not new to Québec. In 1995, the separatist leader, Lucien Bouchard, called on Québécoise women to have more children. Eleven years later, Bouchard, who had left politics for the business world, was worried about what he saw as an economic malaise in Québec. He led a group of business

Think About It

Has the decline in Québec's share of Canada's population eroded its influence in Ottawa and with the rest of the provinces?

Aerial Archives/Alamy

Photo 6.1

The Port of Montréal is a major container port and continues its more traditional role of exporting Prairie grain. Montréal has two geographical advantages over competing ports. First, it is close to major markets in Ontario and the US Midwest. Second, container traffic can continue to reach Chicago and other cities on the Great Lakes by lower-cost lake transportation. However, as container ships have increased in size, docking in Montréal's relatively shallow harbour has made this port less competitive with Atlantic ports, including Halifax.

leaders to prepare a manifesto, *Manifest—pour un Québec lucide* [Manifesto for a clear-eyed vision of Québec]. The manifesto sees the province slipping behind other provinces and states. It charts a different path for the province's economic future and calls for greater involvement of the private sector. Québecers, these business leaders believe, must work harder because other provinces and states have higher levels of worker productivity (Bouchard et al., 2006). Yet, this call for a different path may not resonate because Québec has always maintained a strong involvement in its economy as exemplified by Hydro-Québec.

Why has Québec's position within Canada weakened since Confederation? To begin with, Québec's geopolitical position within Canada has changed from one of four provinces in 1867 to one of 10 provinces and three territories in 2008. As the rest of Canada has grown at a more rapid rate, Québec's demographic position has shrunk from 32 per cent of Canada's population in 1871 to 23.9 per cent in 2006. Second, Québec's economy was hurt, at least initially, by the relocation of key financial business to Ontario in the 1970s, when separatism was on the march, culminating in the referendum of 1980. The fear of the anglophone business community was that an independent Québec would result in an unattractive investment climate and the loss of their economic/political power. The result was a number of corporate headquarters moved from Montréal to Toronto. On the upside, however, Québecers gained greater control over their economy and secured a stronger place for French in the workings of the business community.

Both within and outside Québec, that province's economic position has often been compared to that of Ontario, and, inevitably, Québec is found wanting. In past years the Québec economy has expanded at a slower rate than Ontario's economy, which accounts for out-migration and relatively high unemployment rates. Reflecting that economic situation, Québec's annual unemployment rate was higher than that of Ontario and much higher than those rates in western provinces, but by 2009 Ontario had a higher unemployment rate than Québec, indicating the more negative impact of the economic crash on Ontario than Québec (Table 5.1). According to the Royal Bank of Canada forecast for 2010, Québec should continue to exceed Canada's GDP and, even with Ontario's anticipated economic recovery in 2010, Québec's economic performance and unemployment rate will be better than those of Ontario for the second year in a row.

Québec's economy—as measured by its unemployment rate—has done relatively well within the six Canadian regions (Vignette 6.3). The liberalization of world trade has had many positive impacts. First, in 2009, the worldwide trade system hit Québec's manufacturing firms hard, and those that survived

Vignette 6.3   Has Québec's Economy Turned the Corner?

While Ontario and Québec are often lumped together and characterized, fairly, as Canada's manufacturing heartland, the structure of Québec's manufacturing sector has changed dramatically since the previous recession. 'Québec has gone through a transformation', said John Baldwin, director of the economic analysis division at Statistics Canada and one of Canada's top authorities on productivity (Scoffield, 2009).

Free trade with the United States encouraged all of North America to shift from the manufacturing of non-durable goods to durable goods, to take advantage of economies of scale and growing global markets, according to Baldwin. But Ontario's manufacturing and exports had always been concentrated in durable manufacturing—steel, cars, machinery, and equipment. Québec, on the other hand, with its textiles and shoes, was the centre of non-durable manufacturing for Canada. During the 1990s and especially in the past decade, Baldwin explains, Québec switched over, expanding into areas where Ontario was not as dominant—aerospace and pharmaceuticals. Québec had a painful adjustment, scaling back its textile sector and shutting down large parts of its pulp and paper industry in the past decade. But that restructuring is largely over, economists say.

became more efficient and more conscious of quality control. In turn, these firms, including the apparel industry, have regained market share in Canada and established a foothold in the American market. Second, large companies turned their attention to the global market in late 2009. Consequently, aerospace, metal refining, and pharmaceutical firms located in Québec—especially in the Montréal manufacturing core—have learned how to serve both North American and global customers. The construction and engineering giant, SNC-Lavalin provides such an example. Already 60 per cent of its business is international, especially from fast-growing countries (China, India, Brazil, and Russia) (Marowits, 2010).

Bombardier provides another example in its international sales of its high-speed trains. In 2006, Bombardier won $2.5 billion worth of train contracts in Canada, England, and South Africa. Similar sales continued in 2007, 2008, and 2009, culminating with a $4 billion sale of high-speed trains to China. Bombardier, now a well-established global company, plans to produce these trains in China using its Zefiro technology, which boasts speeds of 380 km/h. In June 2010, Bombardier's international presence was enhanced when the world tuned in to watch the World Cup soccer matches in South Africa where fans used the Bombardier Rapid Rail Link to reach the Orlando Stadium (SouthAfrica Info, 2008).

For more on Bombardier and SNC-Lavalin, see below, 'Québec's Economy', page 245, and Vignette 6.7.

The challenge for Québec is to continue on the path to create a more knowledge-based economy within its industrial core, to advance its global reach, and to ensure an atmosphere of harmony between its francophone majority and its minority groups. In the first years of the twenty-first century, Québec has moved forward on a number of fronts. The pace of economic activities is growing in some areas and declining in others. For the most part, transportation firms, especially Bombardier, have done well and pharmaceuticals remain strong. Hydro-Québec is increasing hydroelectric production with new dams in

northern Québec, the centrepiece being the $6.5 billion Romaine Complex. As well, this Crown corporation has negotiated long-term contracts with utilities in New England, and these agreements remain central to Hydro-Québec's strategy. Montréal remains Canada's east coast hub for container traffic between Europe and North America with traffic increasing from 2002 to 2008. While the global downturn led to a drop of 12 per cent in traffic from 2008 to 2009, an increase in traffic is expected in 2010 (Tower, 2010).

Petroleum refining continues as the second-largest manufacturing enterprise in Québec. The refining business is based on Middle East oil, which is shipped by tankers to Montréal and Québec City. The demand for gasoline, however, is declining as more drivers own smaller, more fuel-efficient cars. As a result, refineries are closing across North America with Montréal's Shell refinery having ceased operations in early 2010 (Vanderklippe, 2010).

Cultural development is moving forward largely because of Montréal's expanding role as a centre of the arts and theatre and Québec City's Winter Carnival, making it an international tourist destination. Major events are a key to the tourist industry and the cancellation of the re-enactment of the 1759 Battle of the Plains of Abraham at Québec City, while a concession to separatist groups, was a commercial setback. Still, such a reaction indicates the thin veneer covering the rapprochement since the 1995 referendum. Yet, tense relations with Aboriginal peoples, especially with the Mohawks and Cree, have softened. For the Cree, relations are moving towards a closer partnership with the provincial government as a result of hydroelectric development in the James Bay region, which offers benefits both for the province and for the Cree, for whom Subarctic Québec is their traditional homeland.

Québec's Physical Geography

Québec, the largest province in Canada, has a wide range of natural conditions and physiographic regions. Its climate varies from the mild continental climate in the St Lawrence Valley to the cold arctic climate found in Nunavik (Inuit lands lying north of the

Think About It

Distance provides a measure of Québec's enormous size. Using the scale in Figure 6.1, what is the distance between the two major cities in the St Lawrence Lowland? And what is the distance from Montréal to Kuujjuaq, the largest community in Arctic Québec?

fifty-fifth parallel). Four of Canada's physiographic regions extend over the province's territory—the Hudson Bay Lowland, the Canadian Shield, the Appalachian Uplands, and the Great Lakes–St Lawrence Lowlands—and each has a different resource base and settlement pattern. The Canadian Shield extends over most of Québec—close to 90 per cent—while the Appalachian Uplands and Great Lakes–St Lawrence Lowlands together form nearly 10 per cent. By far the smallest physiographic region is the Hudson Bay Lowland, which constitutes around 1 per cent of Québec's land mass.

The heartland of Québec is the St Lawrence Lowland. Formed from the Champlain Sea some 10,000 years ago, this physiographic region provides the best agricultural land in Québec. Settlers from France began farming along the edge of the St Lawrence River some 400 years ago. New France was established within this physiographic region and, by means of the St Lawrence, spread into the interior of North America. It remains the cultural core of Québec. The St Lawrence Lowland offers several natural advantages that give the region its central role in modern Québec. Arable land is one such advantage. By far the best agricultural land is in a small area between Montréal and Québec City. As well, most industrial plants are located in this same area.

The Appalachian Uplands physiographic region is an extension of the Appalachian Mountains, which stretch northward from the state of Georgia. While this ancient geological feature reaches into Québec and Atlantic Canada, its topography is much subdued and consists more of rugged hills and rolling plains. Most arable land in this region is in Estrie, where dairy farming prevails (photo 6.2). On the rugged Gaspé Peninsula, small communities dot the coastline, reflecting the inhabitants' orientation to the sea and the fishing industry. Because of the area's spectacular scenery, tourism has become an important source of income during the summer on the Gaspé Peninsula. Mining and forestry are other economic activities in this physiographic region. With the exception of the Lake Champlain gap in the Appalachian Uplands,

Vignette 6.4 Variations in Agriculture Resources in Québec's Physiographic Regions

In 2006, Québec had 3.5 million hectares of farmland with just over half in crops (Statistics Canada, 2008). The remaining land was in pasture. Within each physiographic region, climate and soils vary widely. Most agricultural land is found in the St Lawrence Lowland with a secondary core in the western section of the Appalachian Uplands (Estrie). Much farther north in the Clay Belt of the Canadian Shield, the shorter growing season limits crop agriculture to hay and other forage crops. Encouraged by climatic and economic factors, dairy farming dominates, making up more than half of the commercial farms in the province. However, dairy farming is more widespread outside of the St Lawrence Lowland. Over three-quarters of the farms in Estrie and close to 90 per cent in the Clay Belt are devoted to dairying. The predominance of dairy herds and the preference for cheese in the Québécois diet have resulted in a spinoff effect results in numerous local butter and cheese processing firms. Other specialty crops include apples, blueberries, other fruits, vegetables, tobacco, and sugar beets. In the marginal lands of St Lawrence Lowland and Estrie, stands of maple trees provide the sap for maple syrup.

Clara Parsons/Valan Photos

Photo 6.2

Estrie, formerly called the Eastern Townships, lies in the Appalachian physiographic region. Dairy farms are found in the rolling countryside, which is surrounded by wooded uplands. Hay is the principal crop and it is used as feed for dairy cows.

easy road access to the populous parts of New England is blocked by these rugged uplands. The Lake Champlain gap has therefore become a very important north–south transportation link between Montréal and points south in the US, especially New York City.

As the largest physiographic region in Québec, the Canadian Shield occupies nearly 90 per cent of the province's territory (Figure 5.4). The Canadian Shield is noted for its forest products and hydroelectric production—it has most of the hydroelectric sites in Canada because of a combination of heavy precipitation, large rivers, and significant changes in elevation. Near Montréal and Québec City, the Canadian Shield is a recreational playground where the rolling, rugged, forested upland with its numerous lakes has become a popular site for urbanites and tourists. Further north, single-industry towns based on mining and forestry dot the landscape. In the Lac Saint-Jean region of the Canadian Shield is a large flat deposit of clay—the Clay Belt. It is not rich or well drained, and the growing season is short, but in most years it can produce good crops of hay and other forage and sustain dairy and other livestock operations to supply the nearby mining and forest industries.

Beyond the **commercial forest** zone lie the lands of the Cree and Inuit, where the Cree must coexist with the massive La Grande hydroelectric project. The huge hydroelectric reservoirs initially resulted in a serious environmental problem for the Cree—the microbial breakdown of large amounts of submerged vegetation creates high levels of methylmercury contaminants. These contaminants enter the aquatic food chain and eventually find their way into fish. The Cree, who traditionally have consumed large quantities of fish, were confronted with the risk of mercury poisoning, but, as the microbial breakdown eased, the mercury problem has gradually diminished. In Arctic Québec, a number of Inuit settlements are found along the coasts of Hudson Bay and Hudson Strait. The Cree and Inuit communities are unable to provide sufficient employment opportunities, and consequently, these Native communities have high unemployment rates. The Québec government supports a hunting-and-trapping program, thus encouraging many Cree families to stay on the land. This popular program has important cultural and social values.

The Hudson Bay Lowland extends from Ontario along the southeast edge of James Bay near Rupert River. By far, this lowland is the smallest physiographic region in Québec. Few people live here.

With four climatic zones found in Québec, weather varies greatly. The four zones are Arctic, Subarctic, Atlantic, and Great Lakes–St Lawrence. The Subarctic climatic zone extends over 60 per cent of Québec followed by the Arctic, Great Lakes–St Lawrence, and Atlantic climatic zones.

The geographic extents of Québec's four climatic zones are shown in Figure 2.6, page 52, while the characteristics of each zone are discussed in the section 'Climatic Zones', Chapter 2, page 54.

For the most part, the weather across Québec falls within a predictable range for each climatic zone, but, like other regions of Canada, Québec has had its share of extreme weather events. While Québec commonly experiences heavy rainfall and freezing rain, two storms, in 1996 and 1998, were especially dramatic. In July 1996, the rainstorms in the Saguenay region caused extensive flooding and resulted in loss of life, extensive damage, and the evacuation of 15,000 people. Seven were killed in the flood, including two children when a mudslide buried their house. The loss of property from this storm was over $600 million. The damage was widespread with the communities of L'Anse Saint Jean, La Baie, Jonquière, Chicoutimi, and Hébertville hardest hit. Two years later, the worst ice storm in Canada's history struck southern Québec and eastern Ontario, causing great damage and disrupting electrical power. The physical damage was enormous, perhaps reaching $500 million. The ice storm began on 5 January 1998 and continued for several more days. Over 4 million people were without electricity, many for several weeks. Estrie was hard hit, including its largest city, Sherbrooke.

The section 'Extreme Weather Events' is found in Chapter 2, page 56.

Environmental Challenges

Québec, as an old industrial region, is confronted with a series of environmental problems, most of which stem from the discharge of agricultural, industrial, and mining wastes into the atmosphere and water bodies or from the construction of huge hydroelectric dams. Even its so-called pristine North has been affected as mineral exploration companies have failed to clean up their sites, leaving the landscape littered with abandoned machinery, chemicals, and oil barrels (Duhaime et al., 2005: 262). Then, too, Premier Charest's plan to expand the mining industry and thus create more high-paying jobs in this economically depressed area could cause further harm to the northern environment, especially the boreal forest (Moreault, 2009b).

Further south, in the most populated and industrial area near the St Lawrence River, marine biological life, including whales, are threatened by polluted waters. Once a pristine river, the St Lawrence became badly contaminated as great quantities of industrial toxic wastes were deposited directly into the river over the years. The result is a river still struggling with high levels of toxic chemicals—polychlorinated biphenyls (PCBs) and dichlorodiphenyltrichloroethane (DDT), as well as heavy metals such as lead, mercury, and cadmium and the introduction of the notorious **zebra mussel**, which has had a devastating impact on local aquatic ecosystems. As well, waste contaminants from the aluminum and magnesium smelters—known as polycyclic aromatic hydrocarbons (PAHs)—have had a serious impact on the ecological system of the St Lawrence. At one time, the population of the St Lawrence beluga whales totalled almost 10,000. Today, the population is under 1,000. Even more disturbing is the link to the health of local populations who obtain their drinking water from the St Lawrence River. While the St Lawrence River remains vulnerable to pollution, progress is taking place and the pollution level is declining (Moreault, 2009a). Unlike the declining whale population, the zebra mussel is increasing both in numbers and geographic extent. This mussel, native to the Caspian Sea, represents an environmental disaster. It has altered the structure and functioning of the St Lawrence and Great Lakes ecosystems; fouled and blocked conduits (water intakes, pipelines, tunnels); corroded ship hulls; and posed a threat to native species, including local fish populations. The cost of cleaning waterways and controlling its spread runs into millions each year (Jonich, 2009: 41, 46).

While greenhouse gas emissions represent an environmental challenge, Québec has embraced the need to reduce these gases. Unlike Ontario, which produces much of its electricity from coal-fired thermal plants, Québec's energy comes primarily from its hydroelectric stations and, to a much lesser extent, wind power. Even as the lowest per capita emitter of greenhouse gases among Canada's provinces, Québec leads in efforts to reduce its greenhouse emissions by charging motorists an extra 0.8 cent a litre for gasoline with the revenue from this tax used to reduce greenhouse gas emissions and to encourage public transportation (Dougherty, 2007). Only one other province, British Columbia, has also introduced a gasoline tax.

Québec's Historical Geography

Relative to some other parts of North America, the history of European settlement in Québec is long, rich, and complicated by the period of British rule and then the search for a place within Confederation (see Tables 6.1–6.3). Beneath the surface of this search lies the deeply fractured French/English faultline. Québec's historical geography can be divided into three periods: New France, British occupation, and Confederation.

New France, 1608–1760

The introduction of the French to North America began in 1534 when Jacques Cartier sailed into the Bay of Chaleur and set foot on the shores of the Gaspé Peninsula. Yet, the

Table 6.1   Timeline: Historical Milestones in New France

Year	Geographic Significance
1534	Jacques Cartier sails into the Bay of Chaleur and claims the land for France. The following year, Cartier discovers the mouth of the St Lawrence River, which provides access to the interior of North America.
1608	Samuel de Champlain, described as the 'Father of New France', founds a fur-trading post near the site of Québec City. Champlain was instrumental in the development of the fur economy, which provided the initial economic basis for New France.
1642	Paul de Chomedey de Maisonneuve establishes Ville-Marie on Île de Montréal, which is strategically situated at the confluence of the Ottawa and St Lawrence rivers. Later, Ville-Marie was renamed Montréal.
1759	The struggle between France and England over North America sees the British defeat the French army on the Plains of Abraham. The final battle between the French and English forces ends with the capture of Montréal by the British. In 1763, the formal surrender of New France to England takes place with the Treaty of Paris.

first permanent French settlement in Québec was established in 1608, when Samuel de Champlain founded a fur-trading post at the site of Québec City, thus establishing the French colony in North America. Through three decades, on foot and by ship and canoe, Champlain explored in the unknown heart of the continent through what are now six Canadian provinces and five American states, and by doing so, he created the territorial basis for the French Empire in North America. Although France eventually lost its North American colony, it left a cultural legacy in the form of the French language and

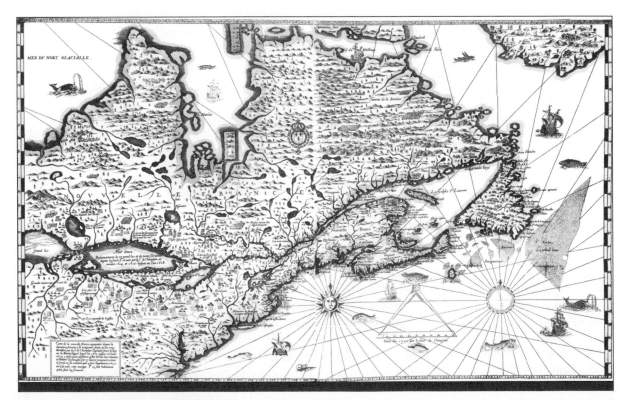

Figure 6.4   Champlain's map of La Nouvelle France, 1632.

Source: Bibliothèque nationale du Québec, <www.nsexplore.ca/maps1/novascotia1632champlain.jpg>.

the Catholic religion, and a French stamp on the landscape with its unique settlement pattern and arrangement of farms into long lots. But the vision of Champlain to strive for a French settlement in the New World that was founded on harmony and respect with the Huron Indians differed sharply with English and Spanish settlements. David Fischer (2008) argues that the 'humanist' values of Champlain, and the respect for the values and traditions of the 'other', have marked Canada

During the seventeenth and eighteenth centuries, France had control over vast areas of North America. Its core, however, was the St Lawrence Valley, from which New France developed a vast fur-trading empire. The wealth from the fur trade was enormous and was the reason for France's interest in the New World. Almost every male French settler wanted to participate in the fur trade, which left only a few to clear the forest and till the land. Indeed, several canoes full of beaver pelts could make a man extremely wealthy compared with the meagre returns obtained from the back-breaking toil of clearing land and breaking the soil. Frenchmen who were coureurs de bois (trappers) often lived with the Indians. A few were extremely successful and returned to France to enjoy their good fortune. Others remained in the fur trade or settled in New France.

Geography played a part in New France's success both as a fur-trading empire and as an agricultural colony. The St Lawrence River provided a route to the interior, which gave the French explorers and fur traders an advantage over their English rivals, who had to contend with crossing the Appalachian Mountains. With further exploration of the interior of North America, the French were able to expand their territory and secure more and better fur-trading routes. Thus, by portaging from the Great Lakes to the Ohio River and then to the Mississippi, the famous French explorer, René-Robert Cavelier, Sieur de La Salle, reached the mouth of the Mississippi River at New Orleans in 1682. Early in the eighteenth century, French fur traders, led by Pierre Gaultier de Varennes, Sieur de La Vérendrye, established a series of fur-trading posts in Manitoba. These trading posts, especially Fort Bourbon on the Saskatchewan

River at Cedar Lake (just east of the present-day community of The Pas), made it convenient for Indians to trade their furs with French traders rather than travel farther to the British fur-trading posts along the Hudson Bay coast.

New France also established a successful agricultural society. Once the land was cleared, the fertile soils in the St Lawrence Lowland provided a solid basis for essentially feudal agricultural settlement. Farming took hold in New France, particularly following the efforts of Jean Talon (1626–94), the greatest administrator of New France. By the early eighteenth century, the French had turned to farming, leaving the lure of the wilderness to a relatively small number of more daring souls. By then, New France had existed for over 100 years, and many of its people had been born and raised in the New World. They were no longer 'French' but *Canadiens*.

By the middle of the eighteenth century, almost all the lands in the St Lawrence Valley were under cultivation. Farmlands stretched in a continuous belt from Québec City to Montréal. As in the feudal agricultural system in France, peasant farmers (**habitants**) worked the land, paying their lords (**seigneurs**) both in kind and in labour. Through the efforts of the habitants and their seigneurs, New France became a successful agricultural colony, but one with a system of landownership and rural life that was quite different from that of the British colonies along the Atlantic seaboard (a difference that would eventually cause Britain to split Québec into Upper and Lower Canada in 1791; see Chapter 3).

The Seigneurial System
When the first Intendant of New France, Jean Talon, arrived in New France in 1665, he encountered a population of only 3,000 inhabitants, most of whom were men engaged in the fur trade. Talon had been instructed by Louis XIV to create a feudal agricultural society resembling that of rural France in the seventeenth century. Talon undertook three measures to achieve this goal. First, he recruited peasants from France. Second, he sent for young women—orphaned girls and daughters of poor families in France—to provide wives for the men of the colony. Third,

he imposed the French feudal system of land-ownership, known as the seigneurial system. In the seigneurial system, huge tracts of land were granted to those favoured by the king, namely, the nobility, religious institutions of the Roman Catholic Church, military officers, and high-ranking government officials. The seigneur was obliged to swear allegiance to the king and to have his tenants cultivate the lands on his estate. The tenants owed certain obligations to their seigneur: paying yearly dues (*cens et rentes*) to their seigneur, working the seigneur's land, especially in regard to road maintenance (*corvée*), and paying rent for using the seigneur's grinding mill and bake ovens (*droit de banalité*).

By 1760, there were approximately 200 seigneuries. Seigneuries, which were usually 1 by 3 leagues (5 by 15 km) in size, were generally divided into river lots (rangs). These long, rectangular lots were well adapted to the St Lawrence Valley for several reasons, the most important of which was that each habitant had access to a river, either a tributary of the St Lawrence or the river itself. At that time, most people and goods were transported along the river system in New France. For that reason, river access was vital for each habitant family.

After the British Conquest of New France, the seigneurial system was retained by the British more for political reasons—to gain the support of the principal source of power (the seigneurs and the Roman Catholic Church)—than for economic reasons. By the early nineteenth century, however, agriculture in Lower Canada had become a commercial venture, making the seigneurial system an anachronism. The seigneurial system was abolished in 1854 by the legislature of the Province of Canada.

British Colony, 1760–1867

Following the defeat of the French in 1760, the British ruled Québec for over 100 years. The British governor was installed at Québec City along with a regiment of British troops, while the fur trade continued to flourish and the agricultural economy went unchanged. Most French Canadians were peasant farmers. After the Conquest, their life on the land remained much the same. Their social and economic lives revolved around the parish church and a landholding system centred on the seigneuries. Life in the towns and cities, however, changed radically due to a massive influx of British immigrants, the powerful political position of English Canadians, and their control of the commercial and industrial sectors of urban places. By 1851, French Canadians accounted for only about half of the population of Montréal. From 1851 to 1951, enormous demographic changes took place. The saving grace for the Canadiens was their high fertility rate, which was known as the 'revenge of the cradles' (*revanche des berceaux*). Even though this rate began to slow, for 100 years Québec's birth rate remained higher than the birth rate in the rest of Canada. Thus, Québec's slower decline in fertility helped to compensate for the province's out-migration and for the growing English-speaking population.

Land hunger forced many **French Canadians** to migrate. By the middle of the nineteenth century, many had left the St Lawrence Lowland due to a land shortage.

Table 6.2	Timeline: Historical Milestones in the British Colony of Lower Canada
Year	**Geographic Significance**
1763	The Treaty of Paris awards New France to Great Britain.
1774	The British Parliament passes the Québec Act, which recognizes that Québec, as a British colony, has special rights, including use of the French language, the Catholic religion, and French civil law.
1791	The British Parliament approves the Constitutional Act that creates two colonies in British North America called Upper and Lower Canada
1841	Based on the Durham report, the British Parliament passes the Act of Union that reunites Upper and Lower Canada into a single colony and makes English the official language of the newly formed province of Canada.

Birth rates were so high in this rural society that there was not enough land left for the children of farm families. French Canadians migrated in three directions: to the Appalachian Uplands, where they either purchased farms from English-speaking farmers or found jobs in textile mills; to the Canadian Shield, where they tried to exist on extremely marginal agricultural land; or to New England's industrial towns, where most were employed in textile factories. By the early twentieth century large numbers of French Canadians, perhaps as many as one million, had left Québec for the United States, while only a small number settled in the Canadian West that was calling out for homesteaders. Through all of these economic and political changes, the vast majority of French Canadians maintained their language and Catholic religion. They turned to the Roman Catholic Church for both spiritual and political leadership. The Church, in turn, encouraged the people to stay on the land, far from the secularizing influences in towns and cities, where the English Protestants lived.

By the 1830s, political unrest was growing in both Upper and Lower Canada. Each colonial government was headed by a governor who was appointed in England and had absolute power. The governor administered the colony along with leading members of the community. This cozy arrangement not only concentrated power in the hands of a few but also led to blatant abuse by powerful elites. In Upper Canada, the political elite was known as the Family Compact, while in Lower Canada it was the Château Clique. In 1837, rebellions broke out in each colony (Vignette 6.5). Both rebellions, which called for political reforms and the curbing of the political power of the elites, were crushed by the British army.

Britain, however, was determined to remedy the political situation in its two colonies. In an attempt to identify the main sources of discontent in Upper and Lower Canada, the British government sent Lord Durham, a politician, diplomat, and colonial administrator, to North America. Durham recognized that the political solution lay in an elected government, where power was dispersed among elected representatives rather than concentrated within an

Lanbout/Dreamstime/GetStock.com

Photo 6.3

Roman Catholic churches are found in most communities across Québec, signifying the dominant role that the Roman Catholic Church once played in the social life and political affairs of the province. In fact, faith provided a key pillar for French-Canadian identity, the other being language. By the time of the Quiet Revolution, however, the Church's influence in Québec affairs had greatly diminished and many churches had lost many of their parishioners. Language remained the pillar of distinctiveness of Québec.

appointed elite. Durham also observed the French/English faultline, which he described as 'two nations warring in the bosom of a single state'. In Durham's report he recommended responsible (elected) government and the union of English-speaking people in Upper Canada with the French-speaking settlers of Lower Canada.[1] Lord Durham believed that assimilation of the French was desirable and possible, claiming that the French Canadians were 'a people with no literature and no history' (Mills, 1988: 637). He recommended that English be the sole language of the new Province of Canada, and that a massive immigration of English-speaking settlers be launched in order to create an English majority in Lower Canada. In response to Durham's recommendations, the Act of Union was passed by the British

Vignette 6.5   The Rebellions of 1837–8

In Lower Canada, Louis-Joseph Papineau, a lawyer, seigneur, and politician, was the leader of the French-speaking majority in the Assembly of Lower Canada. In 1834, Papineau issued a list of grievances known as 'The Ninety-Two Resolutions'. At this time, the economy was depressed and tensions between the French-Canadian majority and the British minority were growing. Papineau sought to shift political power from the British authorities to the elected Assembly of Lower Canada. He planned to use his majority in the Assembly to pass legislation, including tax bills. The British government rejected 'The Ninety-Two Resolutions'—it was just a matter of time before an armed uprising broke out. When it did, the British reacted with force. Even with the strong support of rural areas, Papineau and his Patriotes were soundly defeated. Nearly 300 rebels were killed in six battles. Papineau fled to the United States. A rebellion based on the same popular objections to elite rule also took place in Upper Canada at the same time and it, too, was crushed. Following a second uprising in Lower Canada in November 1838, the British captured hundreds of rebels and ultimately 12 men were sentenced and executed and another 58 were exiled to Australia (Bumsted, 2007: 169). The British government sought to remedy the unrest in both of its colonies. It began this process with a fact-finding mission headed by Lord Durham. The result and Britain's solution was the Act of Union in 1841.

Victor Last/Geographical Visual Aids

Photo 6.4

The Clay Belt is located within the Canadian Shield of northern Ontario and Québec. With the encouragement of the Roman Catholic Church, French-Canadian farmers began to settle these lands in the late nineteenth century. Life was hard and many farms were abandoned. Farm consolidation resulted in even fewer farmers by the end of the twentieth century.

Parliament in 1841, uniting the two colonies into the Province of Canada and thus creating a single elected assembly.

Lower Canada was now known as Canada East, and in spite of the political changes and the flood of immigrants from the British Isles, the French Canadians were not assimilated. Under the new form of British administration, the task of maintaining their French culture and language was not easy, but it was achieved because of several factors, the most important being their strong desire to remain Catholic and French-speaking. A second factor was the institutional support provided by the Roman Catholic Church. By providing spiritual guidance and schooling in French, the clergy played an essential role in cultural preservation. Geography and demography were other factors essential to the survival of French Canada in the nineteenth century. The overwhelming number and concentration of French-speaking people in Canada East provided a critical mass necessary for cultural survival. A high birth rate and high rate of natural increase for the Canadiens ensured an expanding population of French Catholics. Other factors were the rural nature of the French-speaking population, which isolated them from English-speaking residents of the major cities, and the emergence of a French-Canadian intellectual elite whose writings preserved the history and literature of French Canada. One of the most popular novels in early Québec was *Jean Rivard* (1874), published not long after Confederation. Written by Antoine Gérin-Lajoie, *Jean Rivard* is the story of a young French Canadian in Canada East who is advised by his *cure* on the advantages of becoming a farmer rather than a lawyer. The novel promotes the virtue of living a subsistence rural lifestyle in a remote area of Québec's Clay Belt far from Montréal with its many worldly temptations. Here, the French language, religion, and heritage could flourish.

Confederation, 1867–Present

Confederation, achieved in 1867, sought to unite two cultures—English and French—within a British parliamentary system. For Québec, Confederation provided a political framework offering three benefits: an economic union with Ontario, Nova Scotia, and New Brunswick; a political environment where Roman Catholicism and, to a lesser degree, the French language were guaranteed protection by Ottawa; and provincial control over education and language. George-Étienne Cartier, one of the Fathers of Confederation and a French-Canadian leader, viewed these provincial powers as a way for Québec to shape its own destiny within Confederation. Cartier may have identified a fourth benefit—since Québec and Ontario often had mutual economic interests, they could, by working together, influence federal policies and thereby shape the future of Canada.

Confederation also led to the expansion of the geographic size of Québec (Figures 3.4 and 3.6). Since Confederation, Québec's

Table 6.3	Timeline: Historical Milestones for Québec in Confederation
Year	**Geographic Significance**
1867	The Dominion of Canada is formed, the new state consisting of Québec, Ontario, New Brunswick, and Nova Scotia.
1898	Ottawa extends Québec's northern boundary to the Eastmain River, thus expanding Québec's territory well beyond its core area of the St Lawrence Lowland into the Cree lands of James Bay in the Canadian Shield.
1912	Ottawa adds the Territory of Ungava to Québec, thus extending Québec to Inuit lands of Nunavik. With the addition of these two northern lands in 1898 and 1912, Québec's territory more than doubles.
1927	In settling a dispute between Canada and Newfoundland, Britain rejects Canada's claim that the boundary should be placed just inland from the shore. Instead, Britain declares that the boundary is to follow the Hudson Bay and Atlantic Ocean watersheds. Québec does not recognize this boundary.

geographic size has increased greatly. It is now 1.5 million km². As Canada acquired more territory from the British government, Ottawa assigned to Québec parts of Rupert's Land lying north of the St Lawrence drainage basin. In 1898, the Québec government received the first block of Rupert's Land. The second was obtained in 1912.² Some of this land, however, was claimed by the British colony of Newfoundland. In its argument, Newfoundland demanded all of Québec's territory that drained into the Atlantic Ocean. Though the British Privy Council awarded this land, known as Labrador, to Newfoundland in 1927 (Figure 3.7), to this day the Québec government does not recognize the decision. As Premier Jean Charest said in September 2008, 'This is a traditional position that all governments have reiterated. There is a boundary line on which there is no agreement' (Robitaille, 2008).

Prior to World War II, Québec continued to project an image of a rural, inward-looking, Church-dominated society. In 1960, the Quiet Revolution unleashed the force of change that drove Québec into a modern industrial state. As with other social revolutions, the origin of changes began long before 1960. Yet, the election of the Liberal government of Jean Lesage marked a dramatic transformation in government, French-Canadian society, and the place of French within Québec. His government initiated major political innovations that accelerated the process of social and economic change. In effect, the provincial government replaced the Catholic Church as the leader and protector of French culture and language in Québec. The Quiet Revolution instilled a sense of pride and accomplishment among Québecers. The main reforms of the Lesage government were hinged on state intervention in the Québec economy through Crown corporations, and on the expansion of a French-speaking provincial civil service. The government's principal achievements were:

- nationalization of private electrical companies under Hydro-Québec;
- modernization and secularization of the education system, making it accessible to all;
- investment of Québec Pension Plan funds in Québec firms, thereby stimulating the francophone business sector;
- establishment of Maisons du Québec (quasi-embassies) in Paris, London, and New York, thus signalling to Ottawa that the Québec government wanted to represent Québec interests to the rest of the world.

For more on the Quiet Revolution, see Chapter 3, page 122.

With these accomplishments behind them, Québecers felt confident about their future. Lesage's 1963 campaign slogan, '*Maîtres chez nous*' (Masters in our own house), became a reality. For federalists in Québec, these achievements proved that a strong Québec could function within Canada, but for separatists they were not enough. The rise of separatism in Québec signalled that some Québécois felt only an independent Québec could adequately represent French-Canadian interests. For them the slogan became '*Le Québec aux Québécois*' (Québec for the Québécois). After two referendums, Québecers have, for the time being, turned away from pursuing political separation and are focusing more on economic and social concerns. Then, too, Québec nationalism, which is at the core of the separatist movement, has shifted somewhat from the goals and values of the 'old stock' francophones and has become more inclusive, i.e., embracing all French-speaking Québecers, including immigrants (allophones) and even bilingual anglophones. Even the federal government has referred to Québec as a 'nation' within Canada (Vignette 6.6).

Discussion of the French/English faultline in Chapter 3, page 114, includes more information on the historic twists and turns to Québec's place within British North America and then within Confederation.

Think About It

Why do Québec governments, regardless of political affiliation, advocate a vision of Canada as a 'partnership' between founding peoples?

Vignette 6.6   The 'N' Word

For years, federal politicians have avoided calling Québec a nation. With a language and territory different from the rest of Canada, Québec meets the basic requirements of 'cultural nation' within Canada. In November 2006, the House of Commons broke rank with the past by overwhelmingly passing a motion by Prime Minister Harper that recognized the Québécois as a nation within Canada.

Québec Today

Québec is a modern industrial society effectively operating within a francophone environment but dependent on exports to the rest of Canada, the United States, and global markets. Like Ontario, Québec is a major manufacturing centre, ranking just behind Ontario but well in front of the other four geographic regions. Unfortunately, Québec's economy did not benefit from the Auto Pact, which provided a huge boost to Ontario's manufacturing sector. On the other hand, Québec has many more natural sites for the production of hydroelectricity than does Ontario. Through a Crown corporation, Hydro-Québec, the province developed some sites, creating a surplus of electrical energy that it exports to the United States. Another key geographic advantage lies in Québec's position along the St Lawrence River, which provides low-cost access to the heart of North America. This location provides Québec with a hub position in the east–west and north–south transportation

Alphonse Tran/iStockphoto.com

Photo 6.5

Victoria Square in downtown Montréal is close to the city's financial institutions.

systems and accounts for the leading position of Montréal in container traffic from Europe to North America. Its fledgling knowledge-based economy is moving forward and is centred around three clusters of population, innovative firms, and technological institutions in Montréal, Québec City, and Sherbrooke. In spite of these achievements, Québec remains a 'have-not' province, receiving large equalization payments annually from the federal government. As well, its average personal income remains below the national average (Statistics Canada, 2009a).

One challenge for Québec in the twenty-first century is its relatively slow growth rate compared to Ontario and western regions. As a result, Québec's share of Canada's economy and population has slipped. If the current trend continues, both Western Canada and British Columbia will surpass Québec before mid-century. The political implications are considerable. Even though Québec is guaranteed 75 seats in the House of Commons, Québec's political clout will diminish as the total number of seats increases to accommodate population increases in other geographic regions.[3]

Another challenge, of course, is the maintenance of the province's French language and Québécois culture. Attracting immigrants ensures a positive rate of population growth, but some worry that immigrants could weaken the francophone culture and the linguistic balance between French- and English-speaking Québecers. Language laws that require immigrants to send their children to French schools, thereby ensuring the growth of the French-speaking population, appear to have solved the language question,[4] but the accommodation of newcomers remains a critical issue within Québec.

The last, and perhaps most important, challenge revolves around relations between Québec, Ottawa, and the rest of Canada. Language and culture remain 'hot spots' in the province's relationship with Ottawa and the other provinces. French is the issue around which most other concerns are framed—in other words, most political, social, and economic issues in Québec are seen from the perspective of language and culture. The French/English faultline is therefore a crucial component of Québec and must be well understood if it is to be adequately addressed.

Québec's Economy

Québec's economy is the second largest of the regions in Canada. Both Ontario and Québec are heavily dependent on the manufacturing sector. Fierce competition from low-wage countries has placed great pressure on this sector, especially on firms based on semi-skilled labour. From 2003 to 2005, manufacturing declined by 68,000 in Québec and 61,000 in Ontario, accounting for just under 90 per cent of the net loss in manufacturing jobs in Canada (Ferrao, 2006). This trend continued into 2008 (Cross, 2009) and then accelerated as the impact of the global recession hit.

In its report on the economic sector outlook for Québec, Service Canada (2006) confirmed that Québec manufacturing was in trouble:

> The manufacturing sector has faced multiple challenges since the beginning of the decade. They are reflected in its poor results in terms of output and especially employment. Between 2003 and 2005, employment across the sector declined at an average annual rate of 1.7 per cent. Never has there been such a prolonged decline, other than during periods of recession. Over the past three years, output, measured in real terms, has risen by a meagre 0.5 per cent. However—and this is an encouraging fact—manufacturing output has begun to recover in the last two years, albeit modestly, after declining for three consecutive years between 2001 and 2003.

Fortunately, other sectors of Québec's economy have done much better, and Québec continues to have a strong and diversified economy that depends heavily on foreign trade. Québec accounts for much of Canada's exports: 25 per cent of information technologies, 55 per cent of aerospace production, 30 per cent of pharmaceuticals, 40 per cent of biotechnology, and 45 per cent of high-tech exports. Access to the American market is crucial. Improved access after 1989 has had a positive impact on Québec's economy. No matter what the product, most is exported to the US, with a lesser amount to the rest of Canada and world markets. On the other hand, dairy products represent Québec's principal interprovincial export.

The spatial aspects of Québec's economy are similar to those of Ontario, and, like Ontario, Québec can be divided into two economic areas: an agricultural and manufacturing core in southern Québec and a resource-based periphery in the north. Québec's core is associated with the fertile St Lawrence Lowland and the hinterland with the Canadian Shield and the Appalachian Uplands (Figure 2.1). Consequently, Québec is similar to Ontario in its array of resources and access to US and global markets. Yet, there are differences. The climatic conditions in Ontario's part of the Great Lakes–St Lawrence Lowlands provide for a longer growing season, thus giving Ontario an advantage. On the other hand, Québec's part of the Canadian Shield contains much better natural conditions for the production of hydroelectricity. In addition, the close proximity of the Canadian Shield to Montréal and Québec City provides excellent recreation sites—lakes for summer activities and ski hills for winter—thus making the area one of North America's most popular recreational tourist areas.

The Canadian manufacturing belt extends from Windsor to Québec City, and the St Lawrence Seaway connects these two industrial cores (Vignette 6.1). Proximity to Canada's greatest trading partner, the United States, is extremely important for manufacturing. Windsor and other centres in southern Ontario have had the advantage of proximity to the American automobile-manufacturing centre of Detroit. Montréal, on the other hand, is farther from major American manufacturing cities. Nonetheless, many firms in Canada's manufacturing belt supply products to other manufacturing companies in this belt, indicating a high degree of economic integration. An example of such integration is the auto-parts industry. Over 20 firms in the Montréal area produce automobile parts and then ship them to assembly plants in Ontario and the United States.

Over the years, Québec-based firms have specialized in certain industrial sectors, including apparel and textiles, high-tech industries, metal refining, printing, and transport equipment. The apparel and textile industries formed the traditional heart of manufacturing in Québec, but since the FTA these industries have faced stiff competition from abroad because of the reduction in tariffs on apparel and textiles, which has allowed imports from low-wage countries to grab a major share of the North American market. In 2002, the final blow came with China's entry into the World Trade Organization (WTO). As a result, quotas on textiles and garments disappeared. With Canadian quotas on Chinese textiles removed in 2003, imports into Canada soared. In 1995, for example, China accounted for 6 per cent of textile imports, but by 2005 Chinese imports had reached 46 per cent (Wyman, 2006). The resultant decline in such manufacturing in Québec is a common phenomenon found in the rest of Canada and in other countries. By 2007, China was clearly the world's biggest and least expensive producer of quality garments, thus driving Québec and other Canadian garment firms from the marketplace.

High-tech industries have fared much better and provide hope for the future. In fact, many economic gains in Québec have come from this knowledge-based sector. These high-tech firms require a global reach, something that a select number of Québec companies have already achieved. Because Montréal has a critical mass of high-tech companies, several universities, and strong provincial support, it is the most important centre for the new economy in Canada. Leading components are found in aerospace, biotechnology, fibre optics, and computers (both hardware and software). Québec provides nearly half of Canada's information technologies, aerospace, and pharmaceutical exports as well as 40 per cent of its biotechnology exports. Québec also excels in engineering/construction on the international stage. SNC-Lavalin is such a firm (Vignette 6.7). Transportation firms, led by Bombardier and CAE Inc., have been particularly successful in recent years. Bombardier, while struggling in the airplane sector at the turn of the century, has countered this slowdown with major orders for its high-speed trains in both international and domestic markets. For instance, in June 2009, Bombardier secured a contract worth $1.23 billion from Toronto to build the next

Vignette 6.7   SNC-Lavalin: Building a New City and Transport System

SNC-Lavalin has a global reach, which is critical for Canadian export-dependent firms. The company has offices across Canada and in over 35 countries around the world and has projects in about 100 countries. In July 2009, SNC-Lavalin Group Inc., a world-renowned engineering/construction company, won a contract for $508 million for a massive urban-planning project in Algeria—to build a new city to be called Hassi Messaoud near Algeria's largest oil field. SNC-Lavalin will handle design and engineering, and manage construction for this future city that will house 80,000 people. Construction will take place over an eight-year period with a total estimated cost of about US$6 billion. More recently, SNC-Lavalin bought a 48 per cent stake in a Russian engineering company in order to improve its chances of involvement in Russian oil and gas operations. As a key player in the transportation system for the 2010 Winter Olympics, including the building of the rapid transit train from Vancouver's airport to the downtown, SNC-Lavalin had the experience and Russian contacts to win a similar contract for the 2014 Winter Olympic Games in Sochi.

generation of streetcars by 2012 (Bombardier, 2010). These streetcars will feature air conditioning and accessibility to the disabled. Eight months later, in February 2010, Bombardier won a huge contract worth $11.4 billion from France's national railway. In the meantime, orders for its aircrafts have finally picked up as the global economic recovery appears in sight: in February 2010, US Republic purchased a fleet of Bombardier's C series aircraft (Bombardier, 2010).

Founded in 1947, CAE is another anchor in Montréal's transportation sector. The company is a world leader in providing simulation and modelling technologies and integrated training solutions for the civil aviation industry and defence forces around the globe. The company employs about 4,000 highly trained workers in the Montréal area and another 2,000 in other parts of the world. Approximately one-third are engaged in research and engineering, designing, and testing new products, such as a robot that strips paint from aircraft bodies. Its main product, however, is a flight simulator manufactured at its plant in Saint-Laurent near Montréal. Unlike most manufacturing firms, most of CAE's sales do not go to the United States but to customers in Europe, the Middle East, and Asia.

As other regions in Canada, Québec is moving quickly towards a knowledge-based economy with global connections. Each region has a different emphasis, with Ontario focusing on automobile research, including the electric car. Québec has already established itself in several key areas—large-scale engineering projects, cutting-edge transportation production, pharmaceutical research, and biotechnology. Most are located in Montréal. However, the key to international exports requires branch locations in other countries. With research costs high, governments must invest heavily in universities and public research institutes as well as provide tax incentives for private research firms. Most funds come from the federal government with minor support from provinces. Of the provincial governments, Québec financial support is by far the greatest. With the cost of such investments high, the question governments must answer is, 'what research areas should be emphasized?'

Federal and provincial tax credits for scientific research are discussed in Vignette 4.8, 'Public Support for Scientific Research and Knowledge-based Nodes', page 168.

Industrial Structure

As a core region of Canada, Québec has the second-largest number of workers—23 per cent of the country's workers. Its industrial structure is very similar to that of Canada's other core region, Ontario. First, from 2005 to 2008, both regions saw their workforces increase in size. From 2005 to 2008, Québec's workforce increased from 3.7 million to 3.9 million. Second, the division of Québec's and Ontario's industrial labour forces into the three principal sectors (primary, secondary, and tertiary) indicates that less than 3 per cent of workers in each province are engaged in the primary sector (most in agriculture), just under 22 per cent are in the secondary sector (most in manufacturing), and approximately 77 per cent are in tertiary industries. Third, the tertiary sectors of Québec and Ontario increased their share of the total labour force from 2005 to 2008 at the expense of their primary (forestry) and secondary (manufacturing) sectors.

But what aspect of the Québec and Ontario industrial structures distinguishes them from those of the other geographic regions? The answer lies in their secondary sectors. As core regions, both Québec and Ontario have strong secondary industrial sectors dominated by manufacturing. In 2008, Québec's secondary sector accounted for 20 per cent of its labour force compared to 21 per cent in Ontario (Tables 6.4 and 6.5).

Table 6.4 Employment by Industrial Sector in Québec, 2008

Industrial Sector	Workers (000s)	Workers (%)	Percentage Change from 2005
Primary	94.2	2.5	−0.2
Secondary	792.3	20.4	−1.8
Tertiary	2,995.2	77.1	+2.0
Total	3,881.7	100.0	

Source: Statistics Canada (2009d).

Table 6.5	Shift of Ontario and Québec Industrial Structures, 2005–8 (%)			
Sector	Ontario 2008	Ontario 2005	Québec 2008	Québec 2005
Primary	1.8	2.0	2.5	2.7
Secondary	21.0	23.6	20.4	22.2
Tertiary	77.2	74.4	77.1	75.1
Total	100.0	100.0	100.0	100.0
Workers (000s)	6,687.4	6,398	3,881.7	3,717

Source: Statistics Canada (2006a, 2009d).

In comparison, the four other geographic regions, the so-called hinterland, had smaller secondary sectors. In 2008, both British Columbia and Western Canada had 18 per cent of their labour force in the secondary sector, while Atlantic Canada had 16 per cent and the Territorial North approximately 2 per cent. Yet, this differential declined from 2005 to 2008. In 2005, Québec and Ontario had higher percentages of workers in the secondary sector, at 22.2 per cent and 23.6 per cent, respectively. Both core manufacturing regions suffered from the economic downturn. From 2005 to 2008, Québec and Ontario recorded losses in the number of secondary sector workers. For Québec, the figure dropped by 5 per cent, from 827,000 to 792,000; the Ontario secondary workforce declined by nearly 7 per cent, from 1.5 million to 1.4 million.

Key Topic: Hydro-Québec

Since the Quiet Revolution, successive Québec governments have played an active role in shaping the province's industrial economy and its energy export strategy. Hydro-Québec has been central to this strategy by harnessing the province's vast water resources, especially in the Canadian Shield, and arranging long-term sales to utilities in New England (Vignette 6.8). This approach to economic development, initiated by the Lesage government and continued by successive governments, has pursued two goals— one economic, the other political. Its economic objective has been to stimulate economic growth through state intervention in the marketplace. Hydro-Québec has undertaken construction of huge hydroelectric projects, developed high-voltage transmission systems, and offered low electric rates to industrial firms. Its political goal was to increase Québec's public and private ownership of its economy within the francophone business community. This strategy has been highly successful. Hydro-Québec is Canada's largest electric utility, and the expertise gained from huge hydroelectric construction projects has allowed its contractors, such as SNC-Lavalin, to undertake similar projects around the world. Another spinoff was the breakthrough **transmission technology** that enabled Hydro-Québec to transmit electrical power from Churchill Falls[5] and James Bay to markets in southern Québec and New England with acceptable levels of electric leakage. In 2009, Hydro-Québec took a bold step by seeking to purchase New Brunswick

Vignette 6.8   Early Years of Hydro-Québec

Created in 1944, Hydro-Québec was a minor force in the Québec economy until the Lesage government came to power in 1960. At that time, the Québec government announced its intention to purchase the private electricity companies in Québec and place them under the umbrella of Hydro-Québec. The Crown corporation soon extended its activities to cover the whole province by purchasing the shares of nearly all remaining privately owned electrical utilities and taking over their debts. With a virtual monopoly to generate and distribute electricity in the province, Hydro-Québec undertook the task of expanding Québec's hydroelectric capacity.

Power, a deal that fell through in March 2010, and began negotiations with Nova Scotia and Prince Edward Island to acquire their electricity utilities. The strategy is to gain control over the Maritimes power grid and thus to achieve unfettered access to the New England market.

For discussion of Hydro-Québec's failed attempt to purchase New Brunswick Power, see Vignette 9.6, 'Can Hydro-Québec Mend Atlantic Canada's Fractured Geography?', page 377.

While the completion of the James Bay Project remains central to Hydro-Québec's plans, its next construction project calls for the harnessing of the Romaine River (photo 6.6), which flows from the Canadian Shield to the St Lawrence River on the North Shore near Havre-Saint-Pierre, some 200 km east of Sept-Îles (Hydro-Québec, 2009c). The cost is estimated at $6.5 billion, and construction will extend over a 10-year period. The Romaine complex will enable Hydro-Québec to secure Québec's energy future and to increase its exports to markets outside Québec. The commissioning of the first hydroelectric generating station is scheduled for 2014 and the fourth for 2020. By then, the Romaine hydroelectric system will have an installed capacity of 1,550 MW (Canadian Hydropower Association, 2009b). For this project, Hydro-Québec will also erect 500 kilometres of transmission lines and carve out more than 150 kilometres of roads (Yakabuski, 2008). Because the river is sharply incised, the total area of the four reservoirs will be only 280 square kilometres.

Meanwhile, construction of the Eastmain Diversion Project continues with a completion date of 2012 (Hydro-Québec, 2010). This project entails diverting a portion of the water in the Rupert River watershed into the Eastmain River watershed and then into La Grande Basin. The project consists of a 768-MW generating station—Eastmain-1-A powerhouse—near the existing Eastmain-1 powerhouse, and diverting part of the flow of the Rupert River into these two facilities,

Hydro-Québec

Photo 6.6

The waters of the Romaine River flow into the St Lawrence River near Havre Saint-Pierre.

then through the future Sarcelle powerhouse, on to existing reservoirs, and then to existing powerhouses Robert Bourassa; LG-2A, and LG-1, before flowing into James Bay.

Hydro-Québec's Industrial Strategy

With hydroelectric power developments in Québec's Canadian Shield (Figure 6.5), an industrial strategy was born. The vast electrical power generated by the first phase of the James Bay Project provided the provincial government with an opportunity to attract energy-hungry industries into southern Québec by offering them special, low electricity rates, and to export surplus power to energy-hungry utilities in New England. Such an industrial strategy takes on the spatial form of the core/periphery model where the hinterland supplies the energy for industrial users in the core. An early version of this strategy took place in 1957 when Reynolds Aluminum built a smelter at

Vignette 6.9   Natural Advantages for Hydroelectric Developments in Québec

The natural setting of the Canadian Shield in Québec provides numerous technical potential hydroelectric sites, giving the region a clear lead over other provinces (Table 6.6). Hydroelectric developments depend on three factors: precipitation, topography, and access to market. The Canadian Shield provides two of the factors. First, annual precipitation often exceeds 800 mm. thus providing a regular source of water for the lakes and rivers. Second, favourable topography with entrenched river valleys that begin at elevations of 1,000 metres; deep lakes and reservoirs store water and thereby ensure a steady flow for the power plants; and steep rock walls blasted out of the Canadian Shield facilitate dam and diversion projects (photo 6.7). Distance to market was a problem but the innovation of high-voltage transmission lines in the mid-1960s provided a solution, making it possible to reach the large but distant markets, such as southern Québec and the surrounding provinces and American states. The main advantages of hydroelectric developments are: the generation of clean, renewable, low-cost power; the long life of the facilities; low operating costs; job creation during the construction phase; and zero air pollution or GHG. However, there are drawbacks: the initial high capital investment; the long construction period; and the extensive and time-consuming environmental studies.

Table 6.6 Technical Potential Hydro Power by Province and Territories, 2006	
Province/Territory	Technical Potential in Megawatts
Québec	44,100
British Columbia	33,137
Yukon	17,664
Alberta	11,775
Northwest Territories	11,524
Ontario	10,270
Manitoba	8,785
Newfoundland and Labrador	8,540
Nova Scotia	8,499
Nunavut	4,307
Saskatchewan	3,955
New Brunswick	614
Prince Edward Island	3
Canada	163,173

Source: Canadian Hydropower Association (2009a).

Hydro-Québec

Photo 6.7

Construction work for penstocks through which the Eastmain River will flow to generate electricity for power stations located on La Grande Rivière. Note the steep rock walls carved out of the Canadian Shield by construction workers.

Baie-Comeau because Hydro-Québec provided it with low-cost power. In return, the company added to the value of production in the province and, more importantly, it provided jobs in an economically depressed area. Today, Reynolds Aluminum, known locally as Société canadienne des métaux Reynolds, employs approximately 2,500 workers.

A more recent version of this strategy is the decision by the Norwegian solar company Renewable Energy Corp. to take advantage of the province's cheap and abundant electricity. REC, one of the world's biggest manufacturers of silicon and wafers for solar power applications, plans to build its plant in Bécancour, just across the St Lawrence River from Trois-Rivières, with construction starting in 2010. REC has signed a long-term power contract with Hydro-Québec at what the Norwegian firm calls 'a competitive industrial rate' (Blackwell, 2008: B1).

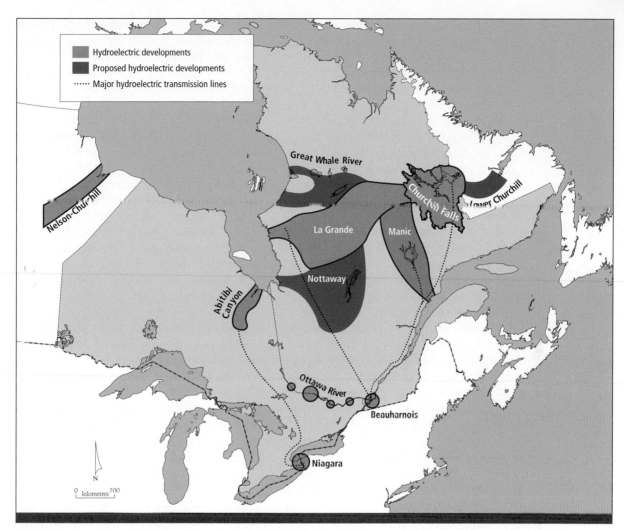

Figure 6.5   Hydroelectric power in Central Canada.
Québec's dominant role in the production of hydroelectric power in Canada is due to three natural factors found in Québec's Canadian Shield: (1) abundant precipitation, (2) natural reservoirs, and (3) high elevations. Geography dictates that the Churchill Falls hydroelectric facility, while located in Labrador, sells virtually all its electric power to Hydro-Québec.

This 'interventionist' industrial strategy of the provincial government took hold during the Quiet Revolution of the 1960s. The provincial government took a more active role in Québec's economy at a time when Hydro-Québec, after its takeover of the private power companies in the province and the construction of the large hydroelectric project at Churchill Falls in Labrador and the Manic-Outardes hydroelectric complex on the Manicouagan River, had a large surplus of power. With that power and the capacity to offer low rates, the government lured industrial companies into southern Québec. With the completion of the first—La Grande—phase of the James Bay Project in 1985, the provincial government took a more aggressive stance by offering risk-sharing contracts and power rebates to industrial firms requiring large amounts of power in their processing operations, such as metallurgical companies. As a result, industrial firms either expanded their operations or new firms located along the St Lawrence River.

Hydro-Québec is able to provide industrial firms with low-cost energy for three reasons:

- Northern Québec can produce vast quantities of low-cost electrical power.
- Hydro-Québec has a long-term contract to buy power from Churchill Falls in Labrador at 1969 prices.

- Hydro-Québec has control over its price structure and can set extremely low power rates for its industrial customers. In fact, these rates were so low that, in 1992, American magnesium firms in competition with Norsk Hydro's magnesium plant in Québec filed a 'dumping lawsuit', arguing unfair market advantage over US producers (Koplow, 1994).

Québec's industrial strategy attracted metallurgical firms but even so, global competition is a threat. Norsk Hydro provides such an example.[6] Norsk Hydro is involved in a large magnesium-smelting operation near Trois-Rivières and such smelting operations require huge amounts of electrical power. In 1988, Norsk Hydro signed a 25-year contract with Hydro-Québec to purchase electrical power at very low rates. By the mid-1990s, the demand for magnesium die castings for automobile engines was growing at an annual rate of 15 per cent and the price of magnesium rose to over $1.30 a pound, causing Norsk Hydro to expand its smelter on the south shore of the St Lawrence River. By 2001, automobile production began to slow and magnesium prices, which had reached $1.50 a pound, began to fall. At the same time, magnesium imports from China began to undercut North American producers. As China grabbed a larger and larger share of the North American market, the magnesium refining industry in Québec was threatened. In 2003, Norsk Hydro's plant at Bécancour and the Noranda state-of-the-art magnesium plant at Danville lost their profit margin because the North American price for magnesium dropped below $1 a pound. In 2003, Noranda announced that it would temporarily close its plant and not reopen it until the price for magnesium reaches $1.30 a pound. Noranda may have a long wait because China can supply the North American magnesium market at prices well below $1 a pound.

Hydro-Québec's James Bay Project

The massive James Bay Project calls for the production of hydroelectricity from all the rivers that flow into James Bay from Québec territory. This hydroelectric project was announced in 1971 by Premier Robert Bourassa. The James Bay Project is divided into three separate river basins (La Grande, Great Whale, and the Nottaway-Broadback-Eastmain-Rupert basins). The project involves about 20 rivers and affects an area one-fifth the size of Québec. Construction of the first phase of the James Bay Project, La Grande, began in 1972 and was completed 10 years later at a cost of $15 billion (Bone, 2009: 157). The La Grande project involved diverting waters from three other rivers (Eastmain, Opinaca, and Caniapiscau) into La Grande Rivière. Electrical energy generated from the three power stations in La Grande Basin is more than 10,000 MW each year. The power is sent to southern markets via transmission lines suspended from large steel towers at high voltage of 735 kV. At the market, the voltage is reduced to levels suitable for local distribution lines.

The first phase of the James Bay Project, however, raised considerable controversy. It evoked an unprecedented response from Aboriginal peoples and environmental organizations. For them, the James Bay Project unleashed social and environmental problems that remain unresolved. For example, the project resulted in an unexpected high level of mercury in the reservoirs, which has had serious implications for the Cree, who consume fish on a regular basis. Hydro-Québec maintains that the environmental impacts have been mitigated to an acceptable level through modifications to the design of the project. Remaining environmental impacts, such as high mercury content in the waters of the reservoirs, the Crown corporation contends, will diminish over time.

In 1985, the second phase of the James Bay Project was announced. The Great Whale River project, located just north of La Grande Basin, was to consist of three powerhouses, four reservoirs, and the diversion of two rivers. In addition, another generating station (LG-1) was to be located at the mouth of La Grande Rivière. Opposition from the Cree and the Sierra Club, who mounted a joint public relations campaign, had an effect on public opinion in New England and New York—the

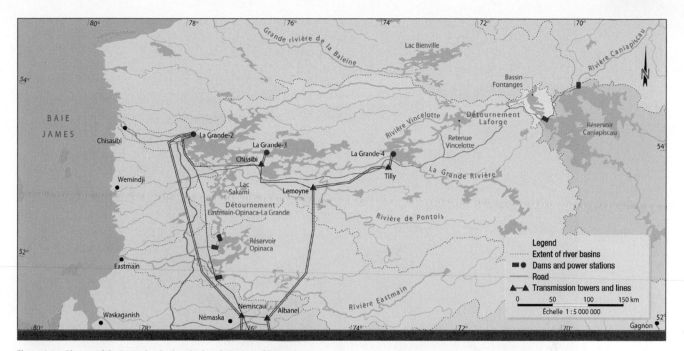

Figure 6.6   Phase 1 of the James Bay Project: La Grande Basin and Eastmain-1.

La Grande's flooded lands (blue-coloured reservoirs) lie east of the LG-2 dam. Downstream is the newly constructed community of Chissibi. In 2007, waters from the Eastmain River were diverted into La Grande Basin.

Source: Hydro-Québec (2006a).

principal export markets for Québec electric power—and when a new natural gas pipeline from Alberta to New England came on stream to provide a low-cost alternative the government of Quebec announced in 1994 that the project would not proceed until the demand (price) for electricity in New England improved. From the very beginning, the Cree opposed the James Bay Project because of its effect on their hunting grounds. The Cree, joined by the Inuit of Arctic Québec, forced a land-claim settlement known as the James Bay and Northern Québec Agreement.

In 2007, the third phase saw the addition of the $2 billion Eastmain-1 project to the huge James Bay Project. Located south of the original development on La Grande Rivière, the three generating units of Eastmain-1 have a total installed capacity of 507 megawatts. Construction began in 2002 following the 2001 Paix des Braves agreement with the Québec Cree. The main components of the Eastmain-1A-Sarcelle-Rupert project are the diversion of some water from the Rupert watershed, two powerhouses, the main dam across the Eastmain River, the

spillway, and the dikes for reservoir closure (photos 6.7, 6.8, and 6.9). A 315-kV transmission line links the powerhouse to Nemiscau substation. With the completion of the dam in 2006, water from the Eastmain River formed a reservoir 35-km long with a total area of 603 km². After passing through the Eastmain powerhouse, water flows to Opinaca Reservoir and then to Robert Bourassa Reservoir. The same water is utilized four times at Eastmain-1 or Eastmain1A, Sarcelle, Robert Bourassa or La Grande-2-A, and La Grande-1 power stations, before flowing into James Bay.

Hydro-Québec's Exports to New England

Geography favours the New England market for two reasons—distance and price. In terms of distance, the length of the transmission line is significant. The major market of Boston is only 400 km from Montréal and New York City is 500 km. In comparison, Toronto is also about 500 km from Montréal. Price differential is significant, with the energy-short

Hydro-Québec

Photo 6.8

Diversion of waters from the Eastmain and Rupert rivers to La Grande Basin enabled the increase of electric power from existing generating stations. To facilitate the flow of water and navigation, the land has been cleared of trees. About 45,000 cubic metres of wood having a commercial value are salvaged by the Crees and used in their sawmills.

Think About It

If you were the Premier of Ontario, would you seek an agreement for energy from Québec to solve your energy shortfall or would you build more power facilities in Ontario? Conversely, if you were the Premier of Québec, why would you prefer New England agreements to one with Ontario?

New England region having approximately double the electrical rate found in Ontario (the Boston rate is approximately 12 cents/kilowatt hour while Toronto is closer to 6 cents). With Québec's surplus of hydroelectric power, exports provide a key economic factor for building megaprojects in northern Québec. As well, long-term agreements to purchase Québec electricity help pay for the construction costs of these massive hydroelectric projects. In 2009, Hydro-Québec gained approval from the US Federal Energy Regulatory Commission to build a major transmission line from Québec to New England and, in the same year, sought to purchase New Brunswick Power, a deal that later fell through. The last step would be to reach a long-term agreement with US utility companies to purchase Québec hydroelectricity. If long-term agreements are reached, the prospects of Ontario obtaining electricity from Québec become very problematic.

Hydro-Québec and the James Bay and Northern Québec Agreement

In 1971, 6,000 Cree in northern Québec lived as eight bands scattered across 375,000 km² of rivers and forest. They were under the administration of the federal Department of Indian Affairs and Northern Development. The James Bay Project threatened to flood their lands. This threat united the eight bands. When construction began in 1972, the Cree asked the Inuit to join them in taking legal action to halt the construction until the Cree and Inuit land claims were addressed. This action forced the Québec government and the Aboriginal claimants to the bargaining table. The result was the **James Bay and Northern Québec Agreement** (JBNQA). Under this agreement, both the federal and Québec governments became responsible for providing the 'treaty' benefits. As the first modern land-claim agreement in Canada, this

1975 agreement provided land, cash, and the power to administer cultural matters (education, health, and social services) to Aboriginal peoples. In exchange, the Cree and Inuit surrendered their Aboriginal claims to northern Québec and agreed to allow construction of La Grande project to proceed.

In combination, these events—the negotiations, the agreement, and the construction project—have forever altered the lives of the Cree and Inuit. Both groups, now living in settlements, are more involved in the modern industrial society than ever before. Many are employed in businesses run by Cree and Inuit organizations, while others work in construction activities in the growing Cree and Inuit settlements and for Hydro-Québec. Still others are involved in the administration of their cultural affairs through the Cree Regional Authority and the **Kativik Regional Government**. In comparison with other Aboriginal peoples in Québec, the economic situation of the Cree and Inuit is much improved and certainly much better than that

Photo 6.9

The James Bay Project generates huge amounts of electrical energy. At the same time, the project has altered the natural environment by diverting rivers, changing their seasonal flow, and creating enormous reservoirs in Québec's Canadian Shield.

of those without such an agreement (Simard et al., 1996). Still, the Cree felt that both Ottawa and Québec had failed to honour their responsibilities under the JBNQA. When Québec announced plans for the second phase of the James Bay Project in 1985, relations between the Cree and the Québec government were so confrontational that the Cree actively opposed the Great Whale River project and took the Québec government to court over a number of issues related to the earlier agreement. Without a doubt, tensions between the Cree and the Québec government reached a low point in the latter part of the twentieth century, marking a particularly strained Québec version of the Aboriginal/non-Aboriginal faultline.

Turnaround

In 2001, the Cree reached an agreement with the Québec government for the economic development of the resources of northern Québec. This agreement, known as the Paix des Braves, opens the door to a major diversion of the Rupert and Eastmain rivers into La Grande Basin, thus adding more water for its hydroelectric plants. The acceptance of such an agreement is an astonishing reversal for the Québec Cree, who had bitterly opposed the project and who had mounted numerous national and international protests against further hydroelectric developments in their traditional lands. Some argue that the Cree now recognize that their participation in northern economic development is their only option. But feelings run high because efforts to protect the land for a hunting and trapping lifestyle have faltered. Paul Dixon, the Cree trapper representative, put it this way: 'They [Québec] promised the traditional way of life would continue undisturbed. Today, the whole territory has been slated for development' (Roslin, 2001: FP7).

Some, especially younger Cree, have chosen an urban lifestyle. For them, living on the land is no longer a viable option. Some say that the Cree leaders had to make a deal. Faced with a rapidly growing population, high unemployment rates, a critical shortage of public housing, and a desperate need for sewer and water systems, the Cree leaders had to seek an agreement with the Québec provincial government. Québec wanted to develop the northern resources and the Cree needed revenue to operate their communities and to find work for their people. Under the terms of the agreement, the Cree receive $3.6 billion over 50 years (roughly $70 million a year), but these funds release Québec from its obligations for economic and community development associated with the James Bay and Northern Québec Agreement. A newly created Cree Economic Development Agency will administer these funds with a mandate to foster

Figure 6.7   Cree communities of Québec.

Source: Based on <www.ottertooth.com/Native_K/jbcree.htm>. Copyright © Brian Back. All rights reserved.

growth of Cree businesses. The agreement also required the Cree to drop their lawsuits against the Québec government for failure to meet its obligations under the James Bay and Northern Québec Agreement. Whether or not the Cree will benefit from this model of economic development remains to be seen. What is clear, however, is that similar agreements are taking place across the country and all are designed to allow Aboriginal people to participate in economic development taking place in their traditional lands.

Tourism

Tourism is extremely important to Québec's economy In fact, Québec has combined its natural beauty, historic past, and francophone culture to draw more tourists each year for over a decade, and tourist dollars spent in the province also have increased. In 2004, 7.2 million tourists visited different areas in Québec. Approximately half originated in the province. Of those coming from outside of the province, nearly 40 per cent came from the United States, 33 per cent from other provinces, and 26 per cent from other foreign countries, especially France, the UK, Germany, and Japan (Québec, 2006b). Since Québec is ideally located to cater to tourists from Ontario, New England, and Western Europe, its future as a world-class tourist destination seems secure, although a Canadian dollar at parity with the US dollar and global economic troubles can affect the tourist industry.

Montréal and Québec City are major attractions for tourists seeking an urban vacation with a francophone atmosphere, while the Laurentides, a low range of mountains bounded by the Saguenay, St Lawrence, and Ottawa rivers, attract visitors looking for a summer or winter playground in the forests and lakes of the Canadian Shield. In addition to the natural beauty, the Laurentides are but a short distance from Montréal and relatively close to New England and to major American cities such as Boston and New York. The climate in this area is ideal for the tourist business—the summers are hot, and heavy snowfall provides ideal conditions for winter sports. Mont Tremblant Resort, for example, is a world-class tourist destination for both winter and summer activities. Québec City has cashed in on its cold winter climate to host a wild ice and dangerous skating race called the Red Bull Crashed Ice, which attracts skaters and visitors from around the world. The skaters must navigate a narrow winding course some 430 metres long with a vertical drop of over 60 metres. While Québec has the setting for summer and winter recreation activities, the tourist industry is vulnerable to external events. With the terrorist attacks on New York and Washington and the resulting 'War on Terrorism' announced by Washington, the tourist industry lost some American tourists.

Southern Québec

Southern Québec is the economic, social, and political core of Québec, while northern Québec is a sparsely populated resource hinterland. The relationship between the two regions is a provincial version of the core/periphery model and overlaying that model are the francophone presence in southern Québec and an Aboriginal majority in northern Québec.

The more physically favoured lands are found in southern Québec, especially in the valley of the St Lawrence River. Southern Québec contains two physiographic regions: the Appalachian Uplands and the St Lawrence Lowland. Bordered on the north by the Canadian Shield and on the south by the United States, southern Québec is only a small part of the territory of Québec, but it contains over 90 per cent of Québec's population and agricultural lands and is the industrial heartland of the province. The human and physical characteristics of each physiographic region in southern Québec are presented below.

Appalachian Uplands

The northern edge of the Appalachian Uplands region faces the St Lawrence Lowland. Because of this geography, the Appalachian Uplands region was settled at different times and in different ways. First settlements took place along the Gaspé coast. In the sixteenth century the waters off the Gaspé Peninsula attracted fishers from

Spain, Portugal, England, and France. Even today, fishing in this part of Québec provides a way of life, though many supplement their income with farming and wage employment in the small coastal settlements. The main settlements are scattered along the Gaspé coast and the south shore of the lower St Lawrence River, leaving the rugged interior with few settlements. The largest urban centres are along the south shore. Rimouski, with a population of nearly 50,000 in 2006, is by far the largest city in the region. Rivière-du-Loup and Matane are medium-sized towns with populations roughly half the size of Rimouski. Towns are much smaller along the Gaspé coast, as the area has a very limited resource base. Percé, now an important tourist town, is the largest of the small centres along the Gaspé coast with a population of about 4,000. This French-speaking area has place names like New Richmond, New Carlisle, and Chandler, the legacy of early English settlers, some of whom had relocated along the Gaspé coast after the American Revolution. Over the last half-century, the lack of jobs caused many to migrate out of this area, which seriously reduced the size of English-speaking communities. In 2006, English-speaking residents constituted less than 5 per cent of the region's population.

Estrie (the Eastern Townships) is a pocket of communities in the rolling land of the Appalachian Uplands located east of Montréal. Estrie was settled after the American Revolution by British Loyalists. The British organized the surveyed land into rectangular townships rather than the French long lots found in the St Lawrence Valley. Much of the land was ill-suited for cultivation. Within several generations, the more marginal lands were abandoned when English-speaking owners left to look for jobs in Montréal and Boston or to try homesteading on the American frontier. Because of a land shortage in the St Lawrence Valley, French Canadians began to move into the Eastern Townships. French-Canadian migrants, often sons of farmers living in the St Lawrence Lowland, either bought or took over abandoned farms from the original English-speaking settlers.

The physical geography of Estrie is much more conducive to economic development and agricultural settlement than are the Gaspé coast and south shore of the St Lawrence. From the 1870s to the 1970s, mining at Thetford Mines and Asbestos was a key sector of the regional economy. Since the 1970s, asbestos mining has fallen on hard times because this mineral, formerly used as insulation in the construction trade, has proven hazardous to human life. Still, mining of this deposit continues, though sales are mainly to developing countries. Agriculture and forestry have continued to be durable in this area. Dairy farming is pursued in the broad valleys, and logging in the forested uplands. Overall, Estrie is the most prosperous area of the Appalachian Uplands. Sherbrooke, which has grown over the years and is the largest urban centre in the Appalachian Uplands, exemplifies the relative well-being of the region. With a population of 186,952 in 2006, Sherbrooke has become an important regional centre (Table 6.9). From 2001 to 2006, its population increased by 6.3 per cent, making it the second-fastest-growing city in Québec after Gatineau (6.8 per cent). Montréal was third at 5.3 per cent. This demographic increase over the years is reflected in its economic growth. Proximity to Montréal has often worked in Sherbrooke's favour, allowing it to engage in the textile industry in the nineteenth century and now in the high-technology industries.

St Lawrence Lowland

Most of Québec's agricultural and industrial production is in the St Lawrence Lowland. In fact, this area functions as the province's core. It contains Québec's largest market and is close to transportation networks and the St Lawrence River, which facilitate trade with foreign countries.

The warm to hot summer climate coupled with abundant rainfall makes the St Lawrence Lowland the most favoured region in Québec for agricultural activities. Livestock farms specializing in cattle, hogs, or sheep are common, while some farmers concentrate on dairy, poultry, and egg production. Livestock farmers grow forage crops for winter feed. During the summer, the cattle graze on pastures. Specialized crops, particularly vegetables and fruit, are also popular and are sold mostly in the major urban centres of the province.

Though farmers in this region engage in a variety of agricultural activities, dairy and vegetable farming predominate. With 37 per cent of Canada's one million milk cows, Québec leads in the production of dairy products. While each province has its own milk marketing board, the National Milk Marketing Plan calculates the allocation of milk quota for each province, with Québec receiving nearly 40 per cent of the Canadian market. Moreover, the processing of agricultural products provides an added benefit for the Québec economy. Large butter and cheese firms rely on the dairy farms. Often the processed food products are designed for the provincial market. Because of a ready market for unpasteurized cheese products in Québec, a few cheese firms began producing *fromage au lait cru*, cheese made with unpasteurized milk. These cheeses compete favourably with European imports, including popular brands from France.

Dairy farmers have fared relatively well under Québec's fluid milk marketing board and the national marketing system. In 2002, cash sales from milk and cream amounted to $1.5 billion in Québec. Under NAFTA and WTO rules, however, Canada is under pressure to dismantle its marketing boards. Without a regulated dairy industry, dairy farmers in Québec would face stiff competition from American dairy imports. In 2002, the World Trade Organization ruled that Canadian dairy exports are illegally subsidized because the national marketing system's price for Canadian milk production is above market price. While 95 per cent of Canada's total milk production is consumed domestically, the value of Canadian dairy exports amounts to about $250 million. According to the WTO ruling, Canada will have to reduce its exports by around half of the current figure of $250 million.

Barrett & MacKay/All Canada Photos

Photo 6.10

The rolling terrain of the subdued Appalachian Uplands in Estrie is well suited for dairy operations.

Manufacturing is concentrated in the Montréal area but extends eastward to Estrie and northeastward to Québec City. As part of the Canadian manufacturing axis, this industrial zone serves as the engine driving both the provincial and the national economies. The manufacturing sector is very sensitive to global trade. While information technology, aerospace, pharmaceutical, and biotechnology firms have benefited from the liberalization of world trade and NAFTA, they are now tied to world demand. To stay ahead of foreign competitors, these firms have spent heavily in research, making Montréal the leading research centre in Canada (Canada's Innovation Leaders, 2006).

Since 1989, these firms have expanded, although in the early years of the twenty-first century some businesses, such as the aerospace industry, have contracted due to a reduction in worldwide demand. These high-technology industries employ highly skilled and well-paid workers. Most of Québec's manufacturing sector is in Montréal, where high-tech companies are now the leading edge in manufacturing. These companies were responsible not only for Montréal's impressive economic recovery in the late 1990s but also for transforming the city into one of the leading high-tech centres in North America. This shift from traditional to high-tech manufacturing has two consequences for Montréal's labour force. First, the demand for highly skilled workers is increasing, creating labour shortages and a search for skilled immigrants. Second, the demand for semi-skilled workers is decreasing, thereby contributing to the relatively high unemployment rate in the region and province.

The traditional sector, composed of textile, knitting, leather, and clothing firms, was once the mainstay of manufacturing, but this sector has been declining for some time and its future is uncertain. As a labour-intensive industry where education and a good command of English or French is not necessary, many low-income, immigrant families have members, especially females, working in this industry. Now struggling to compete with foreign imports, the traditional manufacturing firms have sought to lower their costs of production by reducing the size of their labour force, introducing more efficient machines, and seeking niche markets in North America. While Montréal remains the focus of Canada's clothing and textile industry, its grip on market share is slipping for two reasons. First, established firms are either shifting their production from Canada to countries with low wages, or these firms are subcontracting various elements of production in these low-wage countries. Second, in accordance with WTO rulings, Ottawa has continued to reduce tariffs on textiles, clothing, and related products from developing countries. While these measures may stimulate economic development in foreign countries, including China, the negative impact on Québec firms is contributing to the collapse of the textile and garment industry in Montréal and other Québec centres.

With the automobile and aerospace industries facing a slump in demand, both industries have had to reduce production capacity. In 2002, Québec lost its only automobile-assembly plant when GM closed its plant at Sainte-Thérèse, laying off more than 1,000 workers. Québec's aerospace industry has been a major player on the world scene. Assisted by generous funding from Ottawa and the Québec government, Bombardier is Canada's leading aerospace firm. The firm accounts for around one-quarter of business jet sales in the world and employs nearly 20,000 workers. Bombardier produces passenger jets (the Challenger and the Regional Jet) and smaller business jets (Learjets). The Regional Jet, a stretched version of the Challenger, is the leading aircraft for short-haul markets in North America and Europe. The success of this firm is due largely to its ability to sell its product in the global market. Two international factors have hurt Bombardier's aerospace division. First, the 9/11 terrorist attacks caused a drop in air travel. Second, the 2008 global economic downturn has affected Bombardier and other aerospace firms. They face a difficult future. Bombardier, however, has achieved impressive gains in train technology and production, as noted earlier.

By the end of 2009, manufacturing in Montréal and the rest of the St Lawrence Lowland had just endured a challenging period of economic adjustment. In the greater

Montréal region, both traditional and high-technology firms reduced their operations and some plants closed. During this painful period of adjustment, economic growth in the eastern segment of Canada's manufacturing core slowed. Fortunately, the tertiary sector expanded, thus allowing the total labour force to expand. By early 2010, the economic climate had improved and manufacturing was on the upswing.

Northern Québec

Northern Québec lies beyond the ecumene of the St Lawrence Valley and in the terrain of two physiographic regions—the Canadian Shield and the Hudson Bay Lowland. Here are the traditional homelands of the Cree, Naskapi (Innu), and Inuit. Its resource-based economy—mining and forestry—depends on foreign markets for much of its production. With the housing crisis in the US, exports have dropped for lumber while pulp and paper products face an ever-diminishing demand. As an old resource hinterland, northern Québec is troubled by a depressed forest economy, a declining population base, and high unemployment rates. On the other hand, northern Québec is the site of exciting political developments among the Inuit and Cree. One such development is taking place in Nunavik, the political territory of the Inuit. This territory soon will have a public style of regional government, like that of Nunavut, rather than an ethnic one (such as First Nations governance structures), which means that all residents are treated equally. The political significance is twofold. First, public government offers more opportunities for political autonomy within Québec. Second, it allows for closer integration with Québec and therefore better access to its programs. Since the signing of the James Bay and Northern Québec Agreement, **Makivik Corporation** has been responsible for managing the political and economic interests of the Inuit of Nunavik (Vignette 6.11), who are known as Nunavimmiut. Makivik has led negotiations for the Inuit with the governments of Québec and Canada regarding the creation of a Nunavik government (Vignette 6.10).

With a faltering economy, northern Québec's population has taken on four demographic characteristics that are strikingly different from southern Québec:

- an aging population;
- a net out-migration, especially of younger members of its population;
- few immigrants;
- Rapidly expanding Inuit and Cree populations.

As a resource hinterland lying in the Canadian Shield, the main economic activities are associated with forestry, mining, and hydroelectric generation. Today, such economic activities require relatively little labour. Consequently, the region is sparsely populated, with a few large mining towns such as Chibougamau. Further north in the James Bay region, most people are Aboriginal Québecers. While tourism flourishes in the Laurentides, most of northern Québec is too remote to attract tourists. Attempts at agricultural settlement have had marginal success in the Lac Saint-Jean area, but the short growing period and thin soils associated with the Canadian Shield prohibit commercial agriculture. Northern Québec's geography is best suited as a resource frontier. Low-cost electricity makes the smelting of bauxite (alumina ore) a major processing industry in the Saguenay region and along the north shore of the St Lawrence. Alcan, the world's second-largest aluminum producer, continues to expand its production capacity in Québec. Low-cost electrical power and an ocean shipping route to the Atlantic Ocean provide key location factors.

Agriculture

The settlement of the Clay Belt demonstrates the difficulty of farming in the northern area of Québec. The Clay Belt occupies an enormous area of northwestern Québec and northeastern Ontario. The Canadian Shield acquired this relatively thick layer of sand, silt, and clay sediments near the end of the last ice age about 10,000 years ago. The rivers flowing to Hudson and James bays were blocked by the remnants of the Laurentide ice sheet. Gradually, a huge glacial lake called Lake Barlow-Ojibway was formed and lasted

Vignette 6.10   Mapping the Road to Nunavik

The land now known as Nunavik became part of Canada in 1870 when Rupert's Land was transferred from Great Britain to Canada. In 1912, the Boundaries Extension Act assigned Nunavik to Québec on condition that outstanding Aboriginal rights be settled.

In 1971, the announcement of the James Bay Project triggered a court case over Aboriginal rights. In 1973, the court challenge led to Québec agreeing to fulfill its obligation and resulted in the signing of the James Bay and Northern Québec Agreement on 11 November 1975. One result was the creation of two corporations for the Inuit—Makivik Corporation and the Kativik Regional Government. Makivik administers the Inuit compensation funds from the JBNQA and represents the Inuit in political and economic matters, such as negotiating the Nunavik Inuit Offshore Land Claim (resolved in 2008). On the other hand, Kativik provides the various public services to residents of Nunavik, including education and security. These two organizations have allowed the Inuit to manage their own affairs and thus gain administrative experience for over 30 years. In the process, the dream of a regional government emerged. The breakthrough for this radical political goal came in 1983 when Premier René Lévesque stated unequivocally that an Inuit regional government within Québec was possible.

Like other Aboriginal peoples, the Québec Inuit are seeking a form of political autonomy within the existing structure that would respond to their needs, desires, and aspirations. To achieve that goal, three formidable challenges had to be resolved. First, how can a regional government function within a province? This begs the question of the division of powers between the province of Québec and the soon-to-be-formed government of Nunavik. Second, how can Nunavik (a non-ethnic government) treat all its residents equally and still promote the Inuit culture? Lastly, how can 11,000 people living in 14 communities scattered over 500,000 km² govern themselves and also generate sufficient revenue to pay for their government?

The road to Nunavik is difficult but not impossible. Somehow the structure, operations, powers, and design of this new form of government within a Canadian province can be achieved. Since 2001, negotiations continued between Ottawa, Québec City, and Makivik Corporation. In 2003, a framework agreement was struck with the purpose of creating a new form of government in Nunavik. By 2007, an Agreement-in-Principle was signed by federal and Québec officials with the final agreement scheduled for approval in 2011. After a two-year transition period, elections for the Nunavik Regional Government will take place and the Kativik Regional Government will pass into history (Bone, 2009; Kativik Regional Government, 2009).

Vignette 6.11   Nunavimmiut Benefit from Resource Profit-Sharing Agreement

The Raglan Agreement, signed in 1995 by Falconbridge Ltd, Makivik, and the communities of Salluit and Kangiqsujjuaq, has generated $65.4 million for Nunavik residents. Most benefits go to residents of Salluit and Kangiqsujjuaq, the two Inuit communities closest to the nickel mine at Raglan. Salluit receives the greatest share because it is closest to the mine and the port at Deception Bay. Last year, after $14 million was shared among Salluit's 1,100 beneficiaries, each adult got $15,000 and every child received $3,500. In Kangiqsujjuaq, beneficiaries, about 520 in all, received cheques for $4,700, with $4,000 coming from the Raglan profit-sharing agreement and the rest from other enterprises of Makivik, which are shared by all beneficiaries in Nunavik.

for several thousand years over an area of about 200,000 km². Lake sediments, especially minute particles of clay, sand, and silt, were deposited. When the glacial lake drained, much of the land was covered by peat, making it unsuitable for agriculture. In fact, less than 5 per cent of the Clay Belt contains arable land, only a portion of which has been cultivated. As a result, farmland in the Clay Belt is in scattered pockets, separated by large areas of forested country.

Forest Industry

Québec has 22 per cent of Canada's productive forest lands. In terms of productive forest, Québec ranks first among Canada's geographic regions, but it ranks second behind British Columbia in total volume of wood cut. However, Québec leads British Columbia in pulp and paper production and the output of newsprint. Québec's advantage is its close proximity to major US cities, especially New York, where, historically, the demand for paper and newsprint has been extremely high. For US buyers, Québec softwood is preferred over US softwood lumber because the cold climate in northern Québec results in slower growth, which increases the strength of the wood.

In total, Québec accounts for nearly 760,000 km² of forest lands. These lands are found in each of the four physiographic regions of Québec, but the vast majority of commercial forest is in the Canadian Shield south of 53° N. The boreal forest is concentrated in the southern half of the Canadian Shield, while mixed and hardwood forests are found in the St Lawrence Lowland and the Appalachian Uplands. The Hudson Bay Lowland has few commercial stands. While timber is cut in the northern hinterland, most logs are processed at mills located at the mouths of tributaries flowing into the St Lawrence River. For instance, logs are floated down the St Maurice River to the pulp and paper mill at Trois-Rivières. The main exceptions are the Lac Saint-Jean and Saguenay regions, where there are many sawmills and eight large pulp and paper mills. Québec has a total of 64 pulp and paper mills.

While the forest industry contributes significantly to the province's economy in both value of production and employment, it has fallen on hard times due at first to the US duty on softwood lumber imposed in 2001 and later to the collapse of the US housing market in 2006. Employment is one measure of its importance, especially to small towns in the boreal forest zone where logging and sawmills dominate the economy. While employment has fallen, in 2008 Québec's forest industry directly employed 85,100 people (Natural Resources Canada, 2009b). The value of forest production is another measure. Falling prices for paper and high duties on softwood exports to the US have resulted in a contracting industry. From 2001 to 2005,

the number of sawmills declined from 291 to 262. More bad news arrived in 2006 when Domtar closed three sawmills in the Abitibi area until softwood lumber prices increase. A measure of the decline is revealed in the value of forest production, which was $18 billion in 2001 but had dropped to $13.5 billion by 2003; by 2007, the figure had climbed to $19.9 billion, a result of the removal of the US duty (Québec, 2003, 2006a; Natural Resources Canada, 2009b). With over 80 per cent of Québec's forest products exported to the United States, dependence on that single market is crucial. With the United States imposing a 27 per cent duty on Canadian softwood lumber from 2001 to 2006, the Québec lumber industry suffered. In late 2006, the US duty was removed, but the US housing industry had collapsed, causing the price for softwood lumber to drop precipitously.

Mining Industry

Mining has always been important in northern Québec. In 2010, the value of its metallic mineral production was $4.6 billion, down from $5.2 billion in 2008 as a result of the recession, but this placed Québec first among Canada's six regions. Québec accounted for 19.3 per cent or $6.2 billion of Canada's $32.2 billion total (metallic and non-metallic) mineral production in 2009 (Natural Resources Canada, 2010, 2009a). Among provinces and territories, this was second, just behind Ontario's 19.7 per cent.

Table 7.5, page 298, shows 2009 mineral production for the provinces and territories.

One reason for the importance of mining is the presence of the Canadian Shield, which contains many mineral deposits. In particular, its wealth is concentrated in the Labrador Trough where many rich ores, especially iron deposits, are found. The Labrador Trough extends for about 1,100 km southeast from Ungava Bay through both Québec and Labrador. Further south, it turns southwest past the Wabush and Mount Wright areas to within 300 km of the St Lawrence River.

Several northern Québec communities are single-industry towns and rely on mining for their existence. Extraordinary measures are taken to survive. Residents of Malartic, for example, are relocating outside of town because one of the largest gold deposits in North America exists under the town. As in other resource towns, times are tough. For example, half of Malartic's labour force are either unemployed or are on welfare. But this $1 billion project will breathe new life into the community by creating nearly 500 permanent jobs and some 800 construction jobs (Séguin, 2009).

Mine (and town) closures are not unusual events. The cyclical nature of the mining industry, due to its dependence on world markets, poses a problem for resource communities. Low demand means layoffs and even the closure of mines and ore-processing mills. The iron mine at Gagnon, for example, was closed in 1985, spelling the end to this single-industry town. This boom-bust cycle driven by fluctuations in world prices is particularly hurtful to the narrowly based economies of resource hinterlands. The world demand for mineral products reached a peak in the 1960s, declined sharply in the 1980s, recovered somewhat by the late 1990s, and then began to slide downward in 2002. Since then, demand (and prices) for minerals has climbed significantly. Such price fluctuations have had a profound impact on Québec's iron-mining industry. Québec iron mines account for much of the iron ore produced in North America. Québec's low-grade iron ore is converted into high-grade iron pellets and these pellets are the main raw material for North American and European blast furnaces and electrical mini-mills. The iron pellets also allow the mining companies to reduce their transportation costs from the mine site to the nearest ocean shipping point (Port Cartier or Sept-Îles). In 1982, the slowdown in the world economy caused the Iron Ore Company of Canada to close its pellet plant at Sept-Îles, thus concentrating its pellet-making operations in Labrador City. Sept-Îles remains the key transshipment point for iron ore (from rail to ship).

Commercial mineral deposits in the hard rock of northern Québec often consist of gold, iron, and copper. There are two main mining areas within this portion of the Canadian Shield. In the west, gold and copper are mined at Noranda and Val-d'Or, the centres for much of this production. In the northeast, iron is mined. Deposits of iron were first reported in 1895 by A.P. Low of the Geological Survey of Canada, the first geologist to investigate the region's mineral potential. At that time, however, these deposits had no commercial value because more accessible mines could supply the needs of the iron and steel companies.

All that changed in the 1940s. American steel producers could no longer count on domestic supplies of iron ore, so they sought more reliable sources, including those in northern Québec and Labrador. Through a process of market integration initiated by US steel companies, Québec's resource hinterland became dependent on a particular group of steel companies in the industrial heartland of the United States for its economic well-being. Two mining companies, Québec Cartier Mining Company and the Iron Ore Company of Canada, developed iron mines in isolated areas of northern Québec and Labrador. By 1947, plans were laid for an open-pit mine in northern Québec near the border with Labrador. The Iron Ore Company built a town (Schefferville) for miners and their families; transmitted power from Churchill Falls to operate the mine and the town; and built a railway (the Québec North Shore and Labrador Railway) to deliver the iron ore to the port at Sept-Îles, from where the ore was eventually transported to supply US steel mills in Ohio and Pennsylvania. The demand for iron ore rose in the 1960s, resulting in the establishment of three more mining towns—Wabush and Labrador City in Labrador and Fermont, Québec. At the same time, Québec Cartier Mining Company built a similar iron-mining operation by constructing the Cartier Railway from Port Cartier on the St Lawrence River to the resource town of Gagnon. But by the 1980s world steel production had surpassed the demand, causing a severe slump in the demand for iron ore. To add to this economic problem, US steel plants were now less efficient than the new steel mills in Brazil, Canada, Korea, and Japan, and lower-cost iron mines had opened in Australia and Brazil. As lower-priced steel from these countries undercut the price of US steel, American steel companies had to reduce their output, close plants, and sell their shares in the two mining companies. The repercussions for the mine workers in Québec and Labrador were severe. Production from these northern mines fell by half in the early 1980s and hundreds of workers were laid off. In 1983 and 1985, the mines at Schefferville and Gagnon were closed. During the 1990s, both companies restructured their operations to reduce costs. As demand from American iron and steel plants increased in the 1990s, these two companies were ready to supply this North American market. In 2010, the Iron Ore Company of Canada operates a pellet plant and mine at Labrador City, Wabush Mines has a mine and pellet plant at Wabush, and ArcelorMittal Mines Canada, which purchased the Quebec Cartier Mining Company in 2008, has two mines, Mont-Wright and Fire Lake, plus a pellet plant at Port Cartier.

Québec's Urban Geography

Over 80 per cent of Québec's population live in urban centres. Two of the largest metropolitan cities in Canada, Montréal and Québec City, are located in the province. Other major population centres are Gatineau, Sherbrooke, Saguenay, and Trois-Rivières. In total, these six urban clusters have a population of 5.1 million (Table 6.7).

As in Ontario and most other regions of Canada, migration has played a major role in Québec's urbanization. Push and pull factors have attracted rural Québecers to cities. The key push factors in rural Québec have been limited job opportunities, a shrinking labour force in the primary sector, and an increasing number of young people entering the workforce.

From 2001 to 2006, the population of the major cities, with the exception of Saguenay, grew, led by Gatineau (formerly Hull), Sherbrooke, and Montréal (Table 6.7). The population growth in Gatineau, across the Ottawa River from the nation's capital, was largely due to the employment and business opportunities generated by the federal government. Saguenay (formerly Chicoutimi–Jonquière) saw its population drop by 2.1 per cent. Located in the resource hinterland of the Lac Saint-Jean region, this city has been losing population for several decades. The downward trend is related to its two main economic activities, the aluminum plants and the forest mills. As in other resource hinterlands, employment prospects in these two industries have been declining, especially in the forest industry.

Montréal

Montréal, the metropolis of the province, is the industrial, commercial, and cultural focus of Québec. Montréal is the largest census metropolitan area in the province with a population in 2006 of 3.6 million. Nearly half

Table 6.7 Major Cities in the St Lawrence Lowland

City	Population 2001	Population 2006	% Change 2001–6
Montréal	3,451,027	3,635,571	5.3
Québec City	686,569	715,515	4.2
Gatineau	261,704	283,959	8.5
Trois-Rivières	137,507	141,529	2.9
Saint-Jean-sur-Richelieu	79,600	87,492	9.9
Shawinigan	56,412	56,434	0.0
Saint-Hyacinthe	54,275	55,823	2.9
Sorel-Tracy	47,802	48,295	1.0
Salaberry-de-Valleyfield	39,028	39,672	1.7
Total	4,813,924	5,064,290	5.0

Source: Statistics Canada (2007).

of Québec's population lives in the Montréal CMA, which includes the cities of Laval (population 368,709) and Longueuil (229,330) as well as the municipalities of Beaconsfield, Baie-D'Urfé, Côte-Saint-Luc, Dollard-Des-Ormeaux, Dorval, Hampstead, Kirkland, L'Île-Dorval, Montréal-Est, Montréal-Ouest, Mont-Royal, Pointe-Claire, Sainte-Anne-de-Bellevue, Senneville, and Westmount.

In recent years, the rapid rate of increase of urban sprawl and commuter satellite towns has resulted in a greater rate of population increase in most satellite towns than in Montréal. This pattern of urban growth presents many challenges for Montréal, especially in the areas of transportation, public services, and the environment. One challenge is providing better transportation service, which often means building highways. For example, the Laurentian Autoroute (Highway 15) contributed to the development of Sainte-Thérèse, Blainville, Mirabel, and even Saint-Jérome, which is situated in the resort country of the Laurentides. In most cases, urbanites have fled to the suburbs, attracted by lower-cost housing and the amenities of suburban life. These suburbanites often work in Montréal. Each weekday morning and late afternoon, commuters generate huge volumes of slow traffic and crowded buses in and around Montréal.

Given its strategic location and economic size, Montréal serves as the transportation hub of Québec, making it the **regional core** of the province and the rest of Québec the periphery. At the national scale, Montréal is part of the national core because it is part of Canada's manufacturing belt.

Following the Free Trade Agreement, Montréal's manufacturing sector had to respond to strong foreign competition. Labour-intensive manufacturing firms experienced great difficulty in competing with foreign firms that had substantially lower labour costs. Montréal's manufacturing firms made two major changes: labour-intensive plants substituted machinery for workers to increase productivity, and high-technology firms expanded. By the late 1990s, Montréal's economy had become much more specialized in aerospace, computers, fibre optics, multilingual software, telecommunications, and other areas of industrial

research and development. The provincial government has taken a leading role by providing subsidies for high-tech firms that relocate to Montréal and other Québec cities. In 2006, Montréal and Québec City were ranked first and fifth among Canada's top research communities (Canada's Innovation Leaders, 2006).

Montréal and Toronto

Beginning in the 1970s, Toronto quickly replaced Montréal as the premier city in Canada. Montréal was no longer the largest city and the financial capital of the country. The principal reason for this shift in metropolitan power was the strong economic growth in Toronto and in southern Ontario powered by the Auto Pact and the resulting expansion of the automobile industry. A secondary factor was the economic and demographic fallout from the political unrest in Québec and the very real possibility of Québec separating from Canada. The unsettled political environment leading up to the 1980 Québec referendum worried the anglophone community. Many anglophones and some corporations moved to Toronto. While the francophone business community and the provincial government kept the Montréal economy growing, by 1981 Toronto had a population of 3.0 million compared to Montréal's 2.8 million (Table 6.8). Three other economic factors explain Montréal's slow growth and Toronto's much faster growth:

1. Montréal's economy was much more dependent on labour-intensive manufacturing (which was in decline) such as that in the textile industry.
2. Montréal's industrialization had begun much earlier than Toronto's and, for that reason, its manufacturing firms tended to be older and less efficient than those in Toronto.
3. Toronto has a national hinterland while Montréal has a provincial one and, for that reason, Toronto has a much larger 'external' demand for its wholesale services industry.

For those reasons, Toronto's growth rate exceeded that of Montréal. Even in the 10-year period from 1951 to 1961, when both cities grew at phenomenal rates—Montréal at 4.3 per cent per year and Toronto at 5.1 per cent—Toronto's rate was higher. From

Think About It

Does expanding the highway system improve automobile flow to suburbs and satellite commuter communities, or does it facilitate more urban sprawl?

1971 to 1981, the rate of population increase slowed in both cities, with that of Montréal close to zero. During that time, Montréal's annual growth rate was only 0.3 per cent, while Toronto's was 1.9 per cent. This gap continued in the early 1990s with the annual rates for Toronto and Montréal at 1.9 per cent and 1.3 per cent, respectively. By 2006, Toronto's population had reached 5.1 million while Montréal's was 3.6 million (Table 6.8).

Québec City

Québec City has the next largest urban concentration in the province, totalling nearly 700,000 people. Close to the Laurentides, Québec City has a magnificent physical setting on high banks just above the St Lawrence River. It is the only walled city in North America and features buildings over 300 years old. In 1985, Québec City was selected as a World Heritage Site by UNESCO.

The economic base of Québec City revolves around three functions. First, it is a government and university town. As the seat of government for the province, Québec City employs a large number of civil servants. Second, it has become a world-class tourist centre. The Old World charm of Québec City draws tourists from around the world, while special events such as its Winter Carnival are very popular. Third, Québec City is only minutes away from excellent seasonal recreation areas—from skiing in the winter to water sports in the summer. The economic base of Québec City remains heavily dependent on its political and cultural roles, but it also has a number of other economic functions: it is a port and rail centre, as well as a centre for resource processing, metal fabricating, and manufacturing. High-tech industries now play an important role in Québec City's economy, including the highly specialized photonics and optics sector.

Urban Development in the Western Appalachian Upland

To the east of Montréal lies the subdued landscape of the Appalachian Uplands. The Western Appalachian Upland consists of a

Table 6.8   Population Change: Montréal and Toronto, 1951–2006 (000s)

Year	Montréal	Toronto	Difference
1951	1,539	1,262	277
1961	2,216	1,919	297
1971	2,743	2,628	115
1981	2,828	2,999	−171
1991	3,127	3,893	−766
2001	3,426	4,683	−1,257
2006	3,636	5,113	−1,477

Sources: Statistics Canada (2002, 2007). Adapted from Statistics Canada publication *Population and Dwelling Counts, 2001 Census,* Catalogue 93F0050XCB2001013, Released 16 July 2002, http://www.statcan.ca/bsolc/english/bsolc?catno=93F0050X2001013.

number of small cities that are being drawn, more and more, into the orbit of Montréal. The rate of population increase for these cities from 2001 to 2006 outstripped those of the St Lawrence Lowland (Tables 6.9, 6.10).

First settled by Loyalists following the American Revolutionary War, this physiographic region was named the **Eastern Townships**. These hilly lands lie between the St Lawrence Lowland and the border with the United States. The area is now known as **Estrie** (meaning east of Montréal). Because of the Loyalists, the land survey and naming

Photo 6.11

Founded in 1608, Québec City is one of the oldest cities in North America. With the ancient wall enclosing Old Québec in the foreground, the Château Frontenac dominates the background. On 3 July 2008, Québec City celebrated its 400th anniversary since its founding by French explorer Samuel de Champlain.

Luc-Antoine Couturier Photographe

Table 6.9 Major Cities in the Western Appalachian Upland, 2001–6

Cities	Population 2001	Population 2006	% Change 2001–6
Sherbrooke	175,950	186,952	6.3
Drummondville	72,778	78,108	7.3
Granby	63,069	68,352	8.4
Victoriaville	46,908	48,893	4.2
Thetford Mines	26,721	26,107	–2.3
Total	385,426	408,412	6.0

Source: Statistics Canada (2007). Adapted from Statistics Canada publications *Population and Dwelling Counts, 2001 Census,* Catalogue 93F0050XCB2001013, http://www.statcan.ca/bsolc/english/bsolc?catno=93F0050X2001013 and *Population and Dwelling Counts, 2006 Census,* Catalogue 97-550-XWE2006002 http://www.statcan.ca/bsolc/english/bsolc?catno=97-550-XWE2006002.

of the towns took on a British flavour. Unlike the land grants in the St Lawrence Valley, land was surveyed into townships similar to the method used in New England. However, farming in the Appalachian Uplands was difficult and by the middle of the nineteenth century English-speaking residents began to leave for Montréal and other cities in Canada. At the same time, land shortages in the St Lawrence Lowland caused French Canadians to move into the Eastern Townships.

Beyond the Urban Cores

Québec's urban geography takes on a distinct regional character. Southern Québec has experienced modest growth with the highest rate of increase from 2001 to 2006 taking place in the Western Appalachian area (Table 6.9). On the other hand, beyond this area and the St Lawrence Lowland, most urban centres saw their populations decline (Table 6.10). Most of these cities and towns suffered population losses caused by out-migration, just as those in Estrie and along the St Lawrence gained population by in-migration. This demographic pattern is associated with weak regional economies found in the hinterlands, especially the depressed state of the forest industry. From 2001 to 2006, population declines took place in the urban centres of northern Québec, the North Shore, and the Gaspé Peninsula. Two forces were at play.

First, the resource-based economies in these outlying areas of Québec are contracting. Within the new North American marketplace, resource companies that export most of their products are under great pressure to reduce their costs, which translates into reducing the size of the labour force or, in the worst-case scenario, closing their operations. This global-induced trend was reinforced by the 2008 global crisis that saw demand for both forest products and minerals decline sharply. Urban centres are affected by the negative spinoff from industrial restructuring, resulting in a smaller population, fewer well-paying jobs, and, therefore, fewer customers. Such a

Table 6.10 Population of Cities of Northern Québec, the North Shore, and Gaspé Peninsula, 2001–6

City	Population 2001	Population 2006	% Change 2001–6
Northern Québec			
Saguenay	154,938	151,643	–2.1
Rouyn–Noranda	39,621	39,924	0.8
Alma	32,930	32,603	–1.0
Val-d'Or	32,423	32,288	–0.4
North Shore			
Baie-Comeau	30,401	29,808	–2.0
Sept-Îles	27,623	27,827	0.7
Gaspé Peninsula			
Gaspé	14,932	14,819	–0.8
Percé	3,614	3,419	–9.5
Total	336,482	332,331	–0.1

Source: Statistics Canada (2007). Adapted from Statistics Canada publication *Population and Dwelling Counts, 2006 Census,* Catalogue 97-550-XWE2006002 http://www.statcan.ca/bsolc/english/bsolc?catno=97-550-XWE2006002.

downward spiral reduces the size of city markets and forces some local businesses to close.

Second, the demographic outcome of a declining regional economy where few job opportunities exist leads to out-migration, especially by younger and more educated people. Such migration causes a contraction of the local population and a loss of potential community leaders, and signals a general economic malaise that results in a relentless downward spiral of both the regional economies and urban populations.

The French/English Faultline in Québec

After the Quiet Revolution, the Québec government took legislative measures to ensure that the French language and the Québécois culture prospered within the province. From the Québec government's perspective, such action was necessary because Québec was surrounded by a sea of English-speaking North Americans. Accordingly, the government passed a series of language laws that obliged businesses to use French and required allophone parents to send their children to French schools. For a time, these language laws were a major sore point between French- and English-speaking Québecers. Now the focus has shifted to relatively minor issues, such as English-language street signs in Montréal (Murphy, 2007). After the 1995 referendum, the political intensity of the sovereignty discourse gradually abated (and language issues are less of a sore point within Québec because of a general acceptance of the primacy of the French language and the growing number of anglophones, allophones, and francophones who are bilingual). Evidence of the primacy of the French language is revealed in a remarkable shift over the last 50 years—from 2001 to 2006, 75 per cent of newcomers who do not speak either French or English chose French; before 1961 the reverse was true (Statistics Canada, 2009e). Another sign of the well-being of the French language in Québec is that Lucien Bouchard and the other authors of the manifesto, *Pour un Québec lucide*, called on the government to encourage the learning of English: 'The Government must also make far greater effort to ensure that all Québecers speak and write English, as well as a third language' (Bouchard et al., 2006: 8). With language now serving as the key pillar of Québécois identity, this faultline remains relatively dormant for now, but it is ready to ignite at any sign of a threat.

SUMMARY

Culture remains the fundamental distinguishing feature of this region of Canada. Québec's francophone culture and French language remain strong. As the heartland of francophones in Canada and North America, Québec has a special role to play. Cultural events like *La Fête nationale du Québec* evoke a sense of ethnic nationalism and a love of the land and its people, which is popularly expressed as 'j'ai le goût du Québec'. Within this cultural context, the French language serves as a linchpin.

Economics and demography are doing well but not as well as in most other regions. As one of the original British colonies to form Canada in 1867, Québec's position within Confederation is weakening as its share of Canada's population and economic output declines. The reason is simple: while Québec's economy and population have expanded, other regions of Canada—Ontario, British Columbia, and Western Canada—have expanded at a more rapid rate. If this trend continues, then Québec's place will be seriously eroded. Yet two bright lights suggest a turnaround. First, Québec has recognized the importance of the knowledge-based economy and is devoting more public funds for scientific research initiatives than any of the other five regions. Already, Québec has a few internationally established high-tech companies that can compete on the world stage; the challenge is to expand in this vital economic sector. Second, Hydro-Québec is flexing its muscle by building more hydro dams, producing more electricity, and seeking to control the access route to the lucrative New England energy market. Its long-term strategy may be to become owners of the power systems in the Maritimes.

CHALLENGE QUESTIONS

1. Why was the Premier of Newfoundland and Labrador so upset about Hydro-Québec seeking to gain control over New Brunswick's power grid?
2. What are the political implications for Québec if its share of Canadian population and GDP continue to decline?
3. Why was the Paix des Braves so fundamental to continued development of the James Bay Project and why did the Cree support this agreement?
4. When Nunavik gains its desired form of self-government, will this political development result in a semi-autonomous territory within Québec?
5. How can Québec be both a core region and a 'have-not' province?

FURTHER READING

Bourassa, Robert. 1985. *Power from the North.* Scarborough, Ont.: Prentice-Hall.

Power from the North not only tells the story of the James Bay Project from the pro-development perspective of the late Liberal Premier, Robert Bourassa, but the book also provides contextual insights into Québec society (and, for that matter, Canadian society) and its view of Aboriginal peoples and their place (or lack of place) within Canada at that time. Five themes resonate in *Power from the North*. The first three themes reflect the pro-development perspective of Bourassa while the last two are 'unexpected' consequences that still resonate within Québec:

- a strong pro-development sentiment;
- a continental energy policy benefiting both Québec and New England;
- a provincial development strategy based on supplying low-cost electrical energy to industrial firms willing to relocate to southern Québec;
- a triggering of an angry Aboriginal response to this project that violated their traditional hunting grounds;
- a legal battle that resulted in the first modern land-claim agreement, the James Bay and Northern Québec Agreement.

Students should recognize that *Power from the North* describes a classic example of old-style development schemes. While huge quantities of electrical power were generated from this project and then shipped to southern markets, the James Bay Project was very controversial because of its impact on the traditional hunting economy of the Cree and on the environment (McCutcheon, 1991; Richardson, 1975; Salisbury, 1986; Turgeon, 1992). With the reconfiguration of the landscape by huge dams, river diversions, and reservoirs, the environment has changed dramatically and the Cree have been drawn into a new way of life where the traditional hunting economy is less prominent, especially among young Cree. But several decades ago, Cree and environmental groups banded together to fight this project tooth and nail to save the traditional hunting economy of the Cree and, by doing so, save the environment. In this sense, the James Bay Project took on a larger dimension, that is, a bitter choice between development and the well-being of the environment and Aboriginal peoples. In 1999, James Hornig edited a collection of papers on James Bay some 30 years after the project was announced. With the benefit of hindsight, these articles present a more balanced and less inflammatory assessment of the social and environmental impacts of the James Bay Project.

Students should also recognize that the project has taken new directions. The 2001 Paix des Braves marked a U-turn in relations between the Cree and the Québec government. No longer enemies but now partners, the two parties plan to develop northern resources together. The first joint endeavour was the $2 billion Eastmain-1 Project, which began in 2002 and was competed in 2007. As a consequence of this new partnership, provincial dollars are flowing to the Cree communities to help pay for new houses, roads, and other urban amenities, and Cree workers found jobs in Eastmain construction work and business contracts. While not all Cree are happy with this arrangement, the majority voted to support the Paix de Braves.

7

BRITISH COLUMBIA

INTRODUCTION

British Columbia remains a powerful regional force within Canada. Its population continues to increase well above the national rate while its export-oriented economy expands and contracts with global trade, especially trade with Pacific Rim countries.

British Columbia's location on the Pacific coast binds BC to the rest of Canada through global trade. BC benefits from trade in two ways—from the booming export-based economy of Western Canada and from the expanding economies of China and other Pacific Rim countries. While BC's economy is more and more dependent on its vital transshipment role between Asia and North America, its traditional economic base, the forest industry, has lost some of its lustre. Yet, its spectacular scenic beauty, especially its interface between sea and mountains, supports a vital tourist industry; its natural resources remain critical to this region's economic development and diversification. However, BC's expanding knowledge-based industries hold the key to its future and are pushing other economic sectors into the information age. In the heart of these activities, Vancouver serves as a transportation hub between Canada, the United States, and the Pacific Rim countries.

British Columbia's place in Canada represents a paradox. On the one hand, its natural orientation is southward, following the grain of its physical geography and its ready access to the Pacific Ocean. On the other hand, its political orientation is eastward. In this context, geography tends to either dull or inflame relations with Ottawa and Central Canada. In the minds of its residents, Ottawa only has time for Central Canada. This paradox shows its face in the centralist/decentralist faultline.

CHAPTER OVERVIEW

The key topic is the forest industry. Other topics in this chapter focus on:

- British Columbia's physical geography and history.
- BC's economy and population, especially as these factors relate to the region's physical setting and the core/periphery model.
- The BC version of the centralist/decentralist faultline.
- Resources and their changing role in the BC economy.
- The importance of international trade, the Asia-Pacific Gateway, and Prince Rupert's new role as a major port for container cargo.
- Whether or not BC remains an upward transitional region or has become an industrial core.

Vancouver Rowing Club Marina with North Shore Mountains in the background, Vancouver, British Columbia. Photo: Ian Cook/All Canada Photos.

British Columbia within Canada

British Columbia lies at the western edge of Canada's land mass. Its geographic size consists of nearly 10 per cent of Canada's land mass. One physiographic region, the Cordillera, dominates the landscape while the Interior Plains occupies a small portion of its northeast corner. The southern half of British Columbia has access to the Pacific Ocean while the **Alaska Panhandle** blocks sea access for the northern half.

British Columbia is an emerging giant within Canada's economic system. This west coast region's economy is heavily based on its natural resources and the export of those resources and those produced in Western Canada, namely coal, grain, and potash. British Columbia's scenic beauty supports a vibrant tourist industry; and its expanding knowledge-based industries, including the film and high-technology industries, help drive economic growth in new directions. Trade is crucial. Most exports go to the United

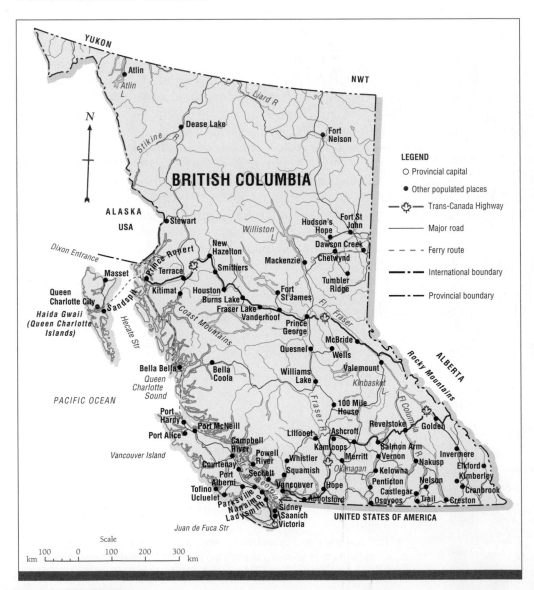

Figure 7.1   British Columbia.

Source: *Atlas of Canada,* at: <atlas.nrcan.gc.ca/site/English/maps/reference/provinceterritories/british_columbia>.

Photo 7.1

The new seven-lane Pitt River Bridge just east of Vancouver was completed in late 2009. As part of the Asia-Pacific Gateway and Corridor Initiative, this bridge will serve to improve access to Vancouver along the north shore of the Fraser River, with easier and quicker flow by trucks of exports to ports in the Lower Mainland and then to Pacific Rim countries. Supported by both the provincial and federal governments, the Pitt River Bridge represents one phase of joint government efforts to create a superhighway corridor from Calgary to Vancouver.

States, though exports to Pacific Rim countries continue to climb. Lumber, pulp, natural gas, and coal are the province's four main exports. Imports, especially from China, Japan, and South Korea, flow through Vancouver to markets across Canada. The expansion of CN and CP rail lines and twinning of sections of the Trans-Canada Highway are facilitating access to the Port of Vancouver and exports to Asian countries. While Prince Rupert remains in the shadow of Vancouver, new facilities at the Port of Prince Rupert signal its arrival as a potential major player in international shipping. The **British Columbia–Alberta–Saskatchewan Trade, Investment, and Labour Mobility Agreement** (TILMA) is another sign of BC's interest in promoting trade, in this case, with provinces in Western Canada.

BC's strong place in Canada is revealed by its share of Canada's GDP and population (Figure 7.2). Since 2001, British Columbia's rate of economic and population growth has outperformed the national rates even with

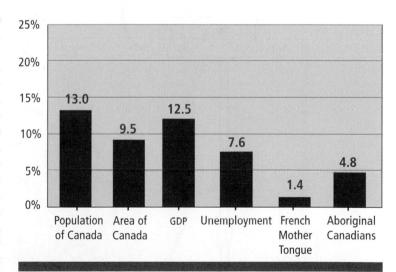

Figure 7.2 British Columbia vital statistics.

BC's strong place in Canada is revealed by its share of Canada's GDP and population. The Aboriginal population and French by mother tongue show the relative weak position of Aboriginal peoples and French-speaking Canadians in the province. The small percentage of Aboriginal peoples, however, does not reflect the extent of First Nations' potential ownership of land in BC through ongoing land-claim negotiations.

Note: Unemployment percentage is for 2009; demographic data are based on the 2006 census.

Sources: Tables 1.1, 1.2, 5.1.

a depressed forest industry, though a divide exists between urban-core BC and rural-hinterland BC where the forest industry dominates the economy. Like other regions, BC was hit by the global economic crisis and its unemployment rate jumped from 4.2 per cent in 2007 to 7.6 per cent in 2009 (Table 5.1). With ever-increasing immigration, especially from China and Hong Kong, BC's population has become more diverse. The City of Richmond, located just 25 minutes from both downtown Vancouver and the US border, is a popular destination for Chinese immigrants.

See Chapter 4, 'Population Trends and Demographic Faultlines', page 155, for background discussion pertinent to British Columbia.

Statistics Canada reports that 4.8 per cent of BC's population was Aboriginal in 2006 while French-speaking residents of BC formed only 1.4 per cent of the population (Table 1.2). The 2011 census results are anticipated to show an even larger percentage gap between the Aboriginal and French-speaking residents of BC, chiefly the result of the rapidly increasing Aboriginal population.

Population is a measure of political power, i.e., the larger the population, the greater the political clout in Ottawa. Yet, political friction between BC and the federal government prevents this simple relationship from taking form. The struggle for political power and respect is ongoing and underscores the centralist/decentralist faultline discussed

Photo 7.2

British Columbia received a much needed economic boost when Vancouver was selected to host the 2010 Olympic and Paralympic Winter Games. At a cost of $2 billion, construction of the many winter facilities and associated transportation systems kept the Vancouver economy humming. In the aftermath of the Winter Olympics, the completed winter sports facilities provide the Greater Vancouver area with another facet to its tourist industry. The ski resort of Whistler, already an international destination for skiers, will see its reputation enhanced.

in this chapter. From time to time, various signs of disenchantment with Ottawa emerge. To those living in British Columbia, the province fits comfortably into the Pacific Northwest. In a sense, the Rocky Mountains are more than a physical divide. One expression of this regionalism that transcends the forty-ninth parallel is the concept of **Cascadia**—Oregon, Washington, and British Columbia. In part, Cascadia is a reaction to the negative feelings towards Ottawa. Within BC there is another faultline—the Aboriginal/non-Aboriginal faultline—whereby Aboriginal peoples are demanding more power through land-claim agreements.

British Columbia's Physical Geography

The spectacular physical geography of British Columbia is perhaps the region's greatest natural asset. The variety of its physiographic features is unprecedented. Then, too, British Columbia is famous for its mild west coast climate. The combination of two contrasting climates (west coast and interior climates) with mountainous terrain has resulted in a wide variety of natural environments or ecosystems. Three examples of natural diversity are rain forests along the coast, desert-like conditions in the Interior Plateau, and alpine tundra found at high elevations in many BC mountains.

The physical contrast between the wet BC coast and its dry interior is largely due to the effect of the Coast Mountains on precipitation. Easterly flowing air masses laden with moisture from the Pacific are forced to rise sharply over this high mountain chain, and consequently most moisture falls as orographic precipitation on the western slopes while little precipitation reaches the eastern slopes (Vignette 7.1).

The climate of the west coast is unique in Canada. Winters are extremely mild and freezing temperatures are uncommon. Summer temperatures, while warm, are rarely as high as temperatures common in the more continental and dry climate of the Interior Plateau of British Columbia. Moderate temperatures,

high rainfall, and mild but cloudy winters make the west coast of British Columbia an ideal place to live and a popular retirement centre for those Canadians wanting to escape long cold winters.

The Pacific Ocean has a powerful impact on BC's climate, resource base, and transportation system. Unlike in Atlantic Canada, the continental shelf in BC extends only a short distance from the coast. Within this narrow zone there are many islands, the largest being Vancouver Island followed by Haida Gwaii (formerly known as the Queen Charlotte Islands).[1] The riches of the sea include salmon, which return to the rivers, such as the Fraser and the Skeena, to complete their life cycle. Most of BC's natural wealth, however, is not in the sea but in the province's diversified physical geography, which provides valuable resources, particularly forests, minerals, and rivers.

Vignette 7.1   BC's Precipitation: Too Much or Too Little?

British Columbia receives the greatest amount of precipitation along its Pacific coast. Inland, the annual precipitation decreases sharply. In simple terms, there are two precipitation areas in British Columbia—one in the Pacific climatic zone, which receives from 800 to 2,000 millimetres of precipitation per year, the other in the Cordillera climatic zone, where less than 800 millimetres fall each year. These figures are average amounts of rain and snow. Fluctuations do occur. In August 2006, for instance, the west coast received very little precipitation. The small fishing/tourist town of Tofino, which is situated on the windward outer coast of Vancouver Island and is noted for its heavy rainfall, ran out of drinking water for the first time in its history. Two months later, in November, warm, moist subtropical air from Hawaii brought high temperatures and record amounts of precipitation to the BC coast. Known as the Pineapple Express, heavy rainfall from this subtropical air mass flooded the low-lying Chilliwack area in the Fraser Valley, forcing an evacuation order for the residents of 200 houses. Elevation plays a role, too, as the skiers and snowboarders at the 2010 Winter Olympics discovered. While the higher mountains at the Whistler alpine ski events (around 2,000 metres) experienced some postponements because of rain and fog, conditions were much warmer and sloppier for snowboard and non-alpine skiing events at Cypress Mountain (just over 900 metres) in the West Vancouver area.

BC's Physiographic Regions

Think About It

From studying Figure 7.3, would you conclude that the mountain chain found on Vancouver Island and on Haida Gwaii is geologically related and that a submerged mountain chain exists under the waters separating these two islands?

Most of British Columbia lies in the physiographic region known as the Cordillera—a combination of mountains, plateaus, and valleys. British Columbia's narrow and sometimes deep continental shelf is, in fact, a submerged mountainous ocean bottom. Vancouver Island and Haida Gwaii, for example, form the Pacific edge of the Cordillera and the boundary between the Pacific and North American tectonic plates extends underwater along the west coast of these islands. Not surprisingly, then, Canada's largest earthquake (magnitude 8.1) took place in 1949 off the coast of the Haida Gwaii archipelago. A small portion of northeastern British Columbia, known as the Peace River country, extends into the Interior Plains (Figure 2.1). Here, the geological structure is part of the Western Sedimentary Basin.

The Cordillera is a complex physiographic region. Extending from southern British Columbia to Yukon, this region encompasses over 16 per cent of Canada's territory. As explained in Chapter 2, the Cordillera was formed by severe folding and faulting of sedimentary rocks. This coastal zone is subject to earthquakes because of tectonic movement. The Cordillera has at least 10 mountain ranges, the most prominent of which are the Coast Mountains, which extend northward from Vancouver to the Alaskan panhandle, and the Rocky Mountains, which stretch from the US–Mexico border almost to Yukon (Figure 7.3). Other north–south mountain ranges include the Insular Mountain Range, which rises above the sea to form Haida Gwaii and Vancouver Island. The zone between the Insular Range and the Coast Mountains forms the Inland Passage. Sheltered by Vancouver Island and Haida Gwaii, the Inland Passage is the only body of water along the shores of British Columbia, Washington, Oregon, and California protected from the direct impact of the Pacific Ocean. Because of this protection, cruise ships, ferries, and private vessels travel along the Inland Passage in the waters of Georgia and Hecate straits.

Two more mountain ranges, which lie in the southern interior of BC, are the Cascade Mountains, in south-central BC along the US border, and the Columbia Mountains. The Columbia Mountains consist of three parallel, north–south mountain ranges—the Purcell, Selkirk, and Monashee—and a fourth range, the Cariboo Mountains, forms the Columbia's northern extension. Further north are four mountain ranges—the Hazelton, Skeena, Ominica, and Cassiar mountains.

The Interior Plateau separates the Coast Mountains from the mountains of the interior (Figure 7.3). North of the Interior Plateau is the Stikine Plateau. The topography of these plateaus consists of gently undulating land with occasional deeply trenched river valleys. The Fraser Canyon, one of BC's best-known landforms, drops about 300 to 600 m below the Interior Plateau and extends from just south of Quesnel to Hope. Just south of Lytton, the canyon walls rise about 1,000 m above the river. The Fraser Canyon is a major fault zone that separates the Cascades from the Coast Mountains. The Fraser River breached the Coast Mountains at the gorge known as Hell's Gate (photo 7.3).

Because of the rugged nature of the Cordillera, there is little arable land. Only about 2 per cent of the province's land is classified as arable. British Columbia's largest area of cropland lies outside of the Cordillera physiographic region in the Peace River country. Within the Cordillera, most arable land is in the Fraser Valley, while a smaller amount can be found in the interior, especially the Okanagan and Thompson valleys. This shortage of arable land poses a serious problem for British Columbia. With urban developments spreading onto agricultural land, British Columbia lost some of its most productive farmland. From the end of World War II to the 1970s nearly 6,000 hectares of prime agricultural land were lost each year to urban and other uses. With only 5 per cent of BC's land mass classified as cropland, the provincial government responded to the serious erosion of its agricultural land base by introducing BC's Land Commission Act in 1973. This Act formed the **Provincial Agricultural Land Commission**, which is charged with preserving agricultural land from urban encroachment and encouraging farm businesses (Provincial Agricultural Land Commission,

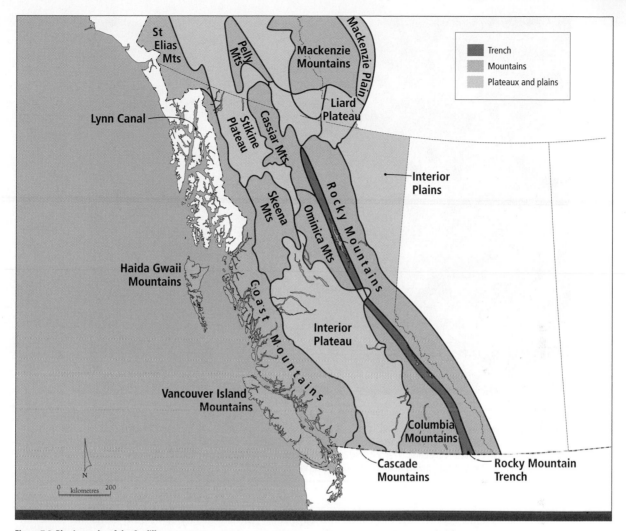

Figure 7.3 Physiography of the Cordillera.
British Columbia's complex physiography is evident in the physiographic sub-regions of the Cordillera. The difficulty of constructing east–west transportation routes can be appreciated if one considers the number of north–south mountain ranges that must be traversed. For that reason, the importance of the Pacific Ocean for transportation prior to the completion of the CPR in 1885 becomes clear. As well, the grain of the land makes north–south land transportation construction into the Pacific Northwest of the United States relatively simple. Herein rest the natural factors behind the political concept of Cascadia and the feeling that BC is California North.

2009). With continuing pressure from urban land developers, this legislation remains under fire from market economy-oriented groups such as the Fraser Institute, the central argument being that more 'value' can be derived from non-agricultural use of the land (Katz, 2010).

Climatic Zones

British Columbia has two climatic zones, the Pacific and the Cordillera. Because of the extremely high elevations in the Coast Mountains, few moist Pacific air masses reach the Interior Plateau. The spatial variation in precipitation is remarkable. Heavy orographic precipitation occurs along the western slopes of the Insular and Coast mountains where 2,000 mm of precipitation fall annually (Vignette 7.1). In sharp contrast, the Interior Plateau receives less than 400 mm per year. In the Thompson Valley and Okanagan Valley of the Interior Plateau, hot, dry conditions result in an arid climate with sagebrush in the valleys and ponderosa pines on the valley slopes. Most rain falls in the winter. For those along

Think About It

Should the BC government allow development of agricultural land for greater individual profit, or does preservation of scarce farmland achieve more important social goals?

Photo 7.3

At Hell's Gate, the canyon walls of the Coast Mountains rise 1,000 metres above the rapids. Located near Boston Bar, the Fraser River found a way through the Coast Mountains to the Pacific Ocean. On the right, a fishway enables salmon to head upstream to their spawning grounds while an observation station is found on the upper left. In 1913, the blasting of a route for the Canadian Pacific Railway caused a rockslide that blocked the migration of salmon to their spawning grounds. As the principal transportation route, the Canadian Pacific Railway occupies the lower reaches of the canyon while the Trans-Canada Highway is located in the upper reaches.

the Pacific coast, the so-called **Pineapple Express**, which originates over the warm waters around Hawaii, brings torrential rains but also relatively warm weather in the winter months, while those inland receive heavy snowfalls.

The three types of precipitation are described in Vignette 2.8, page 52.

The Pacific coast of British Columbia has the most temperate climate in Canada, dominated by the constant flow of moist Pacific air masses. The result is mild and often wet, cloudy weather. Because eastward-moving Pacific air masses must rise above the Insular Mountains and the Coast Mountains, orographic precipitation frequently occurs. This, combined with frontal precipitation, gives the west coast a high level of annual precipitation.

Known locally as 'liquid sunshine', the annual precipitation along the west coast of Vancouver Island can exceed 3,000 mm in some locations. Further east, the annual precipitation declines. At Vancouver, located near the Coast Mountains, the annual precipitation declines to about 1,000 mm. Victoria, however, lies in the partial rain shadow of the Insular Mountains of Vancouver Island and thus avoids the heavy orographic rainfall that affects Vancouver. Consequently, the capital city receives nearly 40 per cent less rainfall than Vancouver.

The Cordillera climatic zone consists of a number of microclimates. Changes in elevation (height above sea level), latitude (from 49° N to 60° N), variation in topography from mountain ranges to plateaus, and distance from the Pacific Ocean to the Rocky Mountains (600 km) control each microclimate, and in turn affect vegetation and soil conditions within these microclimates. In general, these areas become drier as distance from the Pacific Ocean increases and cooler as either elevation or latitude increases. The range of natural vegetation and soil types is enormous. In the Thompson Valley near Kamloops, an arid microclimate results in a desert-like grassland vegetation with chernozemic soils. In sharp contrast, huge trees in the coastal rain forest grow in podzolic soils. Different still are the higher elevations of the Rocky Mountains extending beyond the treeline, where grasses, mosses, and lichens grow on cryosolic soils. For much of northern British Columbia, the northern coniferous forest (evergreen, needle-leaf trees) is the most common natural vegetation.

Because the Cordillera stretches from southern British Columbia to Yukon, the climate turns into a Subarctic mountain climate north of 58° N. Higher elevations in these latitudes have Arctic climatic conditions. South of 58° N, however, summers are longer and warmer, partly because the eastward-moving Pacific air masses are more dominant and partly because of greater annual solar energy.

Natural Vegetation Zones

The natural vegetation of British Columbia is far from homogeneous (Figure 2.7). Along the west coast, the mild, wet climate encourages a rain forest. The Insular Mountains of Vancouver Island and the Coast Mountains along the mainland both rise abruptly from the Pacific Ocean and therefore receive much rainfall throughout the year. The land responds with lush evergreen and deciduous trees. Characteristic tree species are western hemlock, Douglas fir, western red cedar, and Sitka spruce. Under this vegetation cover, podzolic soils are common. These soils are strongly acidic and low in plant nutrients, making it necessary for farmers to use fertilizers. Fertilizers are used because nutrients are washed away by heavy rainfall, and the needles of the coniferous trees, when dropped, add to the ground cover and produce an acidic soil. This coastal region contrasts with the sagebrush and yellow grasses in the lower elevations of the southern

Interior Plateau, where higher temperatures occur. Ponderosa pines are the predominant species in the southern Interior Plateau; at higher elevations and in more northerly areas, lodgepole pine forest replaces ponderosa pine along with other commercially valuable trees, including white and Engelmann spruce, Douglas fir, western hemlock, western red cedar, and two true firs—amabilis fir and subalpine fir. The series of forested mountain ranges, aligned north to south, include the Ominica Mountains located north of Prince George and the Columbia Mountains centred on Revelstoke. Both mountain ranges are situated west of the Rocky Mountain Trench while the Rocky Mountains lie to its east (see Figure 7.3).

Environmental Challenges

British Columbia faces several environmental challenges. These challenges can be divided into those caused by human activities and those caused by nature. Human activities have subjected the seemingly limitless natural riches to mismanagement and wasteful practices that have led to resource loss, environmental degradation, and land-use conflicts. Without **sustainable resource use**, the future of BC's natural resources is not bright. One challenge comes from clear-cutting of the forest. Clear-cutting, the harvesting method most widely used in BC whereby every tree within a large area is cut down, remains a controversial practice. In the past, huge areas were stripped of their trees. Now provincial regulation restricts such logging to 40–60 hectares. The alternative to clear-cutting is selective cutting. However, logging companies claim that selective logging is too expensive and would not allow them to compete in world markets. In the past, clear-cutting extended right to the banks of streams and rivers, leaving surrounding land vulnerable to rapid soil erosion and stream sedimentation. Under such conditions, fish habitat is damaged and spawning grounds may be destroyed. The Nisga'a claim (Nisga'a, n.d.) that salmon spawning grounds along the Nass River have been damaged by such logging practices, thus significantly reducing the salmon run.

Think About It

Why do residents of White Rock near the US border receive much less annual precipitation than do residents of North Vancouver? Annual average precipitation for White Rock is 1,064 mm and that for North Vancouver is 1,770 mm. The locations of these communities are found in Figure 7.7.

Changing weather patterns pose the chief natural challenge. Exceptionally dry summers over the last seven years, for example, have resulted in vast forest fires. Forests in the interior become tinder dry, and the Okanagan Valley has three characteristics that can lead to fires spreading rapidly—strong winds, steep slopes, and a combination of dry underbrush, grass, and lodgepole pine trees. On 23 August 2003, such a fire reached Kelowna. As a forest fire raged out of control and swept into its suburbs, many thousands of residents ultimately had to abandon their homes and more than 250 houses were destroyed (Boei and O'Brian, 2003). Six years later, in July 2009, a forest fire forced the evacuation of more than 10,000 Kelowna residents and destroyed a number of homes (CTV News, 2009).

Another natural impact on the environment has been the destruction caused by the pine beetle on the lodgepole and ponderosa pine trees in the Interior Forest. Vast areas have been affected and there is no way to stop the spread of the pine beetle. The lifespan of an individual mountain pine beetle is about one year. Pine beetle larvae spend the winter under bark, feeding on the tree, which is often a mature lodgepole pine. The adult pine beetle emerges from an infested tree and seeks another host. Logging activities are accelerating to salvage these trees because the beetles do not affect the wood's strength, its gluing characteristics, or its ability to be finished. While the accelerated logging means more timber for the sawmills in the short run, the loss of such a huge section of the Interior Forest means that mature timber will be in short supply in the future. For instance, a lodgepole pine matures in approximately 80 years. Why is the pine beetle so active? In the past, cold winters kept this pest under control. In recent years, however, pine beetles have spread and multiplied as a result of milder winters. Temperatures of at least −38°C for four days or longer are required to kill off pine beetle infestation. If cold winters return, then the pine beetles will be controlled. On the other hand, if milder winters are a feature of climate change, then there will be no stopping the pine beetle from spreading across Canada's boreal forest.

Already, the pine beetle has crossed the BC border into Alberta.

While the BC government has allowed massive clear-cutting of beetle-infested stands, the problem for the lumber companies is that the wood is deteriorating very quickly. An alternative solution is to turn the beetle-killed trees into electricity. BC plans to convert the dead trees into electricity by using the wood to fuel local electric generators situated at sawmills and other strategic locations (VanderKlippe, 2007a). Like other bioenergy plans, the BC plan will require subsidies, and BC Hydro's Bioenergy Call for Power program is providing such support (BC Hydro, 2009). In this case, the cost of electricity produced from hydroelectric stations is much less. For example, the cost at a sawmill where the wood is found would be around $82 a megawatt. When transportation of the raw material is factored in, the cost rises dramatically—to about $120 a megawatt (ibid.).

British Columbia's Historical Geography

Indians lived along the Pacific coast of British Columbia for over 10,000 years before European explorers reached the northern Pacific coast in the mid-eighteenth century. The Spanish had already sailed northward from Mexico to California, but the Russians were the first to reach Alaska and establish fur-trading posts along its coast. In 1778, Captain James Cook established Britain's interest in this region by sailing into Vancouver Island's Nootka Sound, where he and his sailors found the Nootka village of Yuquot. The Nootka, now known as Nuu-chah-nulth, fished for salmon and hunted the sea otter. Upon landing, Cook engaged in trade for sea otter pelts, which opened up a profitable trade with China, although Cook, among the greatest of nautical explorers, did not live to see this trade flourish—he was killed in a skirmish with natives in the Hawaiian Islands on the return voyage. After the Royal Navy published Cook's record of his voyage, British and American traders came to the Pacific Northwest to seek the highly valued sea

Think About It

While forests are a renewable resource, do you think that clear-cut logging—even with tree planting—makes the sustainability of this resource problematic?

Library and Archives Canada C11201

Photo 7.4

Captain James Cook's ships moored in Nootka Sound in 1778, as depicted in a watercolour by M.B. Messer. Four years earlier, the Spanish explorer Juan Hernandez sailed along the BC coast. However, there is the possibility that Francis Drake, while on a secret mission to find a western entrance to the Northwest Passage, reached these waters in 1579 (Hume, 2000: B1).

otters. Russian fur traders, based in Alaska, also harvested sea otters. Spain, which considered the lands Spanish territory, was disturbed by these interlopers and sent a fleet northward from Mexico in 1789. At Nootka Sound on the west coast of Vancouver Island, the Spanish seized several ships and built a fort to defend their claim. In 1792, Captain George Vancouver of Britain's Royal Navy sailed around Vancouver Island. In the following year, Alexander Mackenzie of the North West Company travelled overland from Fort Chipewyan to just south of Prince George and then to the Pacific coast near Bella Coola, which is just over 400 km north of Vancouver. Under the Nootka Convention (1794), the Spanish surrendered their claim to the Pacific coast north of 42° N, leaving the British and Russians in control.

In the early nineteenth century, the North West Company established a series of fur-trading posts along the Columbia River. From 1805 to 1808, Simon Fraser, a fur trader and explorer, explored the interior of British Columbia on behalf of the North West Company. He travelled by canoe from the Peace River to the mouth of the Fraser River. As elsewhere, the strategy of the North West Company was to develop a working relationship with local Indian tribes based on bartering manufactured goods for furs. After 1821, when the North West Company merged with its rival, the Hudson's Bay Company, the HBC took charge of the Oregon Territory, which extended from the mouth of the Columbia River to Russia's Alaska. In the late nineteenth century, a bitter dispute arose over the boundary of the Alaska Panhandle and access to Dawson City and the Klondike gold rush. The boundary dispute was settled in 1903 by a six-man tribunal, composed of American, Canadian, and British representatives. The settlement totally favoured the United States, suggesting that Canada's interests may have been sacrificed by Great Britain, which sought better relations with the United States.

In 1843, American settlers began to arrive on the coast from the eastern part of the United States. In the same year, the HBC relocated its main trading post from Fort Vancouver at the mouth of the Columbia River to Fort Victoria at the southern tip of Vancouver Island. The increasing number of American settlers who came west along the Oregon Trail represented a challenge to the authority of the Hudson's Bay Company. A few years later, the United States claimed the Pacific coast northward to Alaska, where Russian fur-trading posts existed. In 1846, Britain and the United States agreed to place the boundary between

Think About It

If Canada had obtained the power to make international treaties in 1867, do you think its goal of access to Yukon from the Pacific Ocean along Lynn Canal (Figure 7.3) would have been more likely?

Ian Cook/All Canada Photos

Photo 7.5

The tall trees of the coastal forest provided the perfect raw material for Northwest Coast First Nations to express their culture in the form of totem poles.

the two nations at 49° N and then to follow the channel that separates Vancouver Island from the mainland of the United States. While the loss of the Oregon Territory in present-day Washington and Oregon was substantial, Britain was fortunate to hold onto the remaining lands administered by the Hudson's Bay Company. Britain recognized that its hold on these lands through the HBC was tenuous and could not withstand the political weight of the growing number of American settlers. Indeed, without the presence of the HBC and Britain's negotiating skills, Canada might well have lost its entire Pacific coastline.

The gold rush of 1858 brought about 25,000 prospectors from California to the Fraser River. Prospectors walked upstream along the sandbanks and sandbars of the Fraser and its tributaries, panning for placer gold (small particles of gold in sand and gravel deposits). The major finds were made in the BC interior, where the town of Barkerville was built near the town of Quesnel. By 1863,

Barkerville had a population of about 10,000, making it the largest town in British Columbia. To ensure British sovereignty over territory north of 49° N, the British government established the mainland colony of British Columbia in 1858 under the authority of Sir James Douglas, who was also governor of Vancouver Island. In 1866, the two colonies were united. During his time as governor, land claims were settled on Vancouver Island but not on the mainland (Vignette 7.2).

For further discussion of Aboriginal rights and the modern treaty process, see Chapter 3, 'From Hunting Rights to Modern Treaties', page 101, and Vignette 3.9, 'From a Colonial Straitjacket to Aboriginal Power', page 106.

Confederation

By the 1860s, the British government was actively encouraging its colonies in North America to unite into one country. Once the

Vignette 7.2 Who Owns BC?

How much of BC do First Nations peoples own? The question still is only partially resolved, but it has terrific implications for BC and its 200,000 Aboriginal people who form 198 First Nations. BC does not know who has rights to what lands—an uncertainty that stifles investment, cripples community development, and plagues the province with lawsuits. Other than the 14 pre-Confederation Douglas treaties signed between 1850 and 1854 with the Coast Salish on Vancouver Island and Treaty No. 8 of 1899, which extended into northeast British Columbia, the BC government for many years denied Aboriginal land claims.

In 1993, the provincial government accepted the concept of Aboriginal title and the process of negotiating treaty began. For BC, the challenge was to whittle down the total size of the land claims, which exceeded the size of BC, down to around 10 per cent of the claimed 'traditional' territory. In exchange for land surrender, cash and other benefits and rights are acquired by the First Nations. The stakes are very high, and the complex negotiations mean that progress is slow. By 2010, 116 of the 198 BC First Nations who occupy 1,700 reserves remained in the process of negotiating treaty. Two agreements have been reached. The first modern treaty was signed in 2000 with the Nisga'a, who occupy the Nass Valley north of Prince Rupert (Vignette 7.3). In the highly urban area of the Lower Mainland, the Tsawwassen First Nation signed the second one in 2006. Near Prince George, negotiation with the Lheidli T'enneh First Nation reached the final stage but the agreement did not receive a majority of the votes of the band members.

first four colonies were united in 1867, Ottawa adopted the British strategy to create a transcontinental nation. An important part of that strategy was to lure British Columbia into the 'national fabric'. The Canadian Pacific Railway was the first expression of this national policy.[2] Ottawa promised to build a railway to the Pacific Ocean within 10 years after British Columbia joined Confederation. In Fort Victoria, however, some wanted to join the United States. By the middle of the nineteenth century, British Columbia had developed significant commercial ties with Americans along the Pacific coast. San Francisco

Vignette 7.3 Facts about the Nisga'a Agreement

Significance:
The political significance of the Nisga'a Agreement goes back to 1973 when the Calder decision was made by the Supreme Court of Canada. This split decision opened the door for land claims. According to Frank Cassidy (1992: 11) the Supreme Court decision represents 'a centrepiece in the historical development of the province of British Columbia'.

Location:
The Nass Valley extends from the Pacific Ocean across and beyond the Coast Mountains approximately 150 km inland. This valley is the homeland of the Nisga'a. Prince Rupert is the closest large city and it lies some 100 km southwest of the headwaters of the Nass Valley.

Population:
The total population of the Nisga'a is almost 10,000, with approximately 6,000 living in British Columbia and the rest in other parts of Canada and the United States. Some 2,400 Nisga'a reside in the Nass Valley in the villages of New Alyansh (Gitlakdamiks), Canyon City (Gitwinksihikw), and Kincolith (Gingolx).

Agreement:
The Nisga'a and the federal and BC governments approved and signed the agreement in 1999. In exchange for surrendering their traditional lands, the Nisga'a received title to almost 2,000 km^2 in the Nass Valley of British Columbia; access to forest and fishery resources; self-government powers, including a Native judicial system and policing; and nearly $500 million in cash, grants, and program funds from Victoria and Ottawa. British Columbia supplies the land and 30 per cent of the cash settlement, while the federal government pays 70 per cent of the cash settlement. Under this agreement, the Nisga'a are no longer under the Indian Act and therefore cease to receive benefits as status Indians.

was the closest metropolis and, with its railway to New York, offered the simplest and quickest route to London. In 1859, Oregon became a state and Washington was soon to follow. Commercial links with the United States were growing stronger. But the majority of people in Fort Victoria wanted to remain British. In 1871, British Columbia chose to become a province of Canada (Figure 3.5). British Columbians, however, had to wait 14 years for the Canadian Pacific Railway to reach the Pacific coast at Port Moody on Burrard Inlet, near Vancouver. Later, the railway was extended 20 km westward to the small sawmill town of Vancouver, where there was a better harbour and terminal site for the railway (Figure 7.4).

When British Columbia joined Confederation in 1871, its official population was 36,247 (McVey and Kalbach, 1995: 35). This figure likely underestimated the number of Indians and prospectors who were living in the more remote areas of the province.[3] At the time of the first comprehensive census of British Columbia in 1881, there were approximately 4,200 Chinese, 19,000 white settlers (American, Canadian, and British), and about 30,000 Indians. Most British settlers lived around Fort Victoria. Beyond Fort Victoria, the vast majority of the inhabitants were Aboriginal peoples. The BC government denied that First Nations had a claim to land (Vignette 7.3). In the previous decade, about 25,000 prospectors, mainly Americans, had been scattered in small camps along the Fraser River or at Barkerville. Most Americans had left at the end of the gold rush.

Post-Confederation Growth

At first, Confederation had little effect on British Columbia. The province was isolated from the rest of Canada, Canada's fledgling factories, and even the halls of power in Ottawa. Goods still had to come by ship from San Francisco or London. When the Canadian Pacific Railway was completed in 1885, British Columbia truly became part of the Dominion, and BC's role as a gateway to the world began.

The main line of the Canadian Pacific Railway and its many branch lines were responsible for the formation of many of the province's towns and cities and for providing access to its forest and mineral wealth. The Esquimalt and Nanaimo Railway, the Canadian National Railways, and British Columbia Rail (BC Rail) added to the rail network in British Columbia. In 1886, the Esquimalt and Nanaimo Railway was built, connecting the coalfields of Nanaimo with the capital city of Victoria, and stimulated logging and sawmilling along its route. With the closure of the coal mines in the early 1950s, the railway lost its main function and now plays a minor role in the transportation system of Vancouver Island. By 1914, the Grand Trunk Pacific Railway (later the Canadian National Railways; today, CN) provided a trans-Canada rail route to Prince Rupert and an alternative rail service to Vancouver. The Pacific Great Eastern Railway (later BC Rail) was incorporated in 1912 but laid few rails until the early 1950s, when the provincial government made a commitment to complete the railway in order to facilitate resource development in the interior of British Columbia. By 1956, this rail line extended from North Vancouver to Prince George. Two years later, BC Rail extended its rail system to Dawson Creek and Fort St John in the Peace River country, and by 1971 it reached Fort Nelson in BC's forested northeast. CN purchased BC Rail in 2004 and proceeded to expand its facilities at Prince Rupert (Figure 7.4).

After the completion in 1885 of the CPR line, Vancouver grew quickly and soon became the major centre on the west coast. By 1901, Vancouver had a population of 27,000 compared to Victoria's 24,000. As the terminus of the transcontinental railway, Vancouver became the transshipment point for goods produced in the interior of BC and Western Canada. As coal, lumber, and grain were transported by rail from the interior of British Columbia and the Canadian Prairies, the Port of Vancouver spearheaded economic growth in the southwest part of the province. It was then possible to tap the vast natural resources of BC and ship them to world markets. By the twentieth century, Vancouver

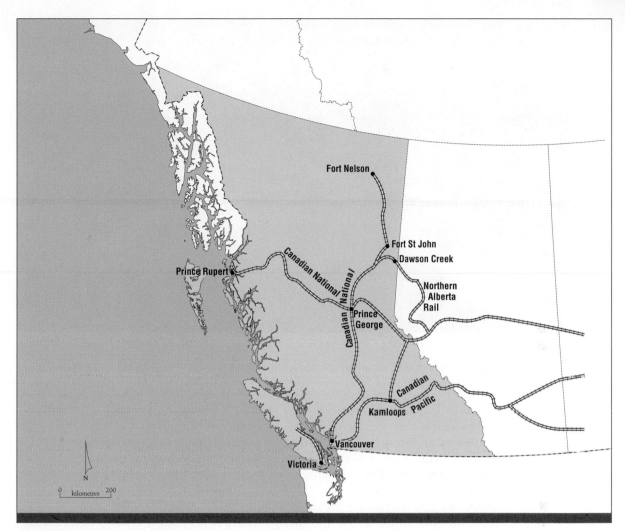

Figure 7.4 Railways in British Columbia.
Railways have played a key role in BC's economic development. The first railway to cross the mountains of the Cordillera was the Canadian Pacific Railway in 1885. To open the northern interior of the province, Victoria built the Pacific Great Eastern Railway (now known as British Columbia Rail). After several extensions, BC Rail joined North Vancouver to Fort Nelson in 1971. An eastern link joins Dawson Creek to the Northern Alberta Railway. BC Rail was owned and operated by the province from 1918 to 2004, when it was sold to Canadian National. By 2008, Prince Rupert, the western terminus of the CN line, had the capacity to handle container traffic, thus relieving the congestion at the Port of Vancouver and providing another transportation route into the interior of North America.

had become one of Canada's major ports. Located on Burrard Inlet, Vancouver has an excellent harbour. Unlike Montréal, it is an ice-free harbour, thanks to the warm Pacific Ocean. As Canada's major Pacific port, Vancouver became the natural transportation link to Pacific nations. With the opening of the Panama Canal in 1914, British Columbia's resources were more accessible to the markets of the United Kingdom and Western Europe.

Between 1885 and the end of World War I in 1918, British Columbia underwent a demographic explosion. By 1921, BC had over half a million inhabitants. The province had been gradually transformed from a fragile political entity in 1871 into a self-confident political and social region within Canada. During this time, the populations of both European and Asian origin increased about tenfold. The European population had reached nearly half a million, while the Asian population was about 40,000. At the same time, a combination of disease and social dislocation caused the number of Aboriginal people to decline sharply, perhaps from 40,000 to 20,000.

Al Harvey/ Slide Farm

Photo 7.6

Vancouver has a magnificent harbour that has facilitated trade with Pacific Rim countries. With Vancouver in the background, a cruise ship sails westward through Burrard Inlet and passes under Lions Gate Bridge. Stanley Park is located on the south side of Lions Gate Bridge (on the right) while West Vancouver is on the north side (left foreground).

This magnitude of demographic decline was matched in other regions of Canada. While a number of factors were at work, smallpox, tuberculosis, and other communicable diseases caused the greatest losses.

While BC's economy and population continued to grow in the 1920s, the Great Depression of the 1930s caused the province's economy to stall and unemployment to rise sharply. Economic disaster struck British Columbia in 1929. Exports of Canadian products from BC, so necessary for its economic well-being, slowed and prices dropped. In the Prairies, the collapse in agricultural prices was accompanied by prolonged drought, turning the land into a dust bowl. Many Prairie farmers abandoned their farms and fled to British Columbia, adding to the burden of unemployment in that province.

World War II and the Post-War Economic Boom

World War II called for full production in Canada, thereby pulling British Columbia's depressed resource economy out of the doldrums. Military production, including aircraft manufacturing, greatly expanded BC's industrial output. As well, resource industries based on forestry and mining (especially coal and copper) were producing at full capacity.

When the war ended in 1945, BC's resource boom continued. With world demand for forest and mineral products remaining high, the provincial government focused its efforts on developing the resources of its hinterland, the central interior of British Columbia. The first step was to create a transportation system from Vancouver to Prince George, the major city in the central interior. The highway system was improved and extended from Prince George to Dawson Creek in 1952. But the completion of the Pacific Great Eastern Railway to Prince George in 1956 and then to Dawson Creek in 1958 opened the country, allowed for exports to foreign markets, and triggered economic growth, especially in the forest industry. With rail access to Vancouver, forestry, as well as other resource industries, expanded rapidly, thereby leading to the integration of this hinterland into the BC and global industrial core.

Over the past two decades, BC's increasing economic strength, partly driven by the Asian economy, has outpaced that of all other regions in Canada. Trade is a dynamic force propelling British Columbia's economy, and as China and other Pacific Rim countries have led the world in economic growth, this growth has created new markets for Canadian products shipped through Vancouver. Trade opportunities are almost endless between British Columbia and the population of 2.5 billion people in the Pacific nations. Vancouver and, to a lesser degree, Prince Rupert and other Pacific ports serve as trade outlets for coal, lumber, potash, and grain from the interior of British Columbia and the Prairie provinces to reach world markets. In BC's export-oriented economy, it is usually much less expensive to ship the raw material than the finished product. This has led to trade focused on resources. High labour costs, a relatively small local market, and distance from world markets have inhibited the development of manufacturing in BC.

British Columbia Today

British Columbia has evolved from a resource-dependent economy into a much more diversified one. Still, BC's natural wealth continues to provide a base for its new economy. While the role of its resource industries has diminished, forestry and mining alone produce much wealth and products for further processing. But without a doubt, the emerging economy has an 'urban/global' look and thrust to it. This economy is more focused on global trade, tourism, high technology, manufacturing, and even filmmaking. In the years leading to the 2010 Winter Olympics, the construction industry added another dimension to BC's economy. Innovative industries, such as fuel cell development by the federally funded National Research Institution and Ballard Power Systems, are at the cutting edge of new technology. International trade, as noted above, also is a key element in BC's economic structure. Symptomatic of the global slowdown, trade through BC ports dropped an estimated 12 per cent in 2009 (Ebner, 2009: B3). Yet, employment numbers continued to rise, with job growth greatest in the construction industry and least in manufacturing. While forestry (logging) was the only industry to suffer job loss, wood manufacturing continued to expand (White et al., 2006; BC Stats, 2009a).

British Columbia's physical geography, especially the sea/mountain interface, provides a solid basis for a tourist industry and explains why BC is recognized as a world-class tourist destination. One tourist attraction involves the **Inland Passage**, a popular route for cruise ships heading for Alaska. The Winter Olympics of 2010 in Vancouver and the ski resort of Whistler sparked tourism, with the economic fallout spilling over into other sectors of BC's economy. Filmmaking, too, takes advantage of BC's physical geography. Vancouver is known as 'Hollywood North', and the Vancouver area has become a major film production centre. Filmmaking represents one aspect of the emerging highly diversified urban 'knowledge economy' that has taken hold in Toronto and Montréal as well. The Vancouver film industry generates over $1 billion annually and directly employs about 35,000 people (BC Facts, 2009). High-technology industries—manufacturing and service—have made the greatest impact on BC's urban economy. Manufacturing involves those processes that require a substantial amount of technology. Three examples are telecommunications, pharmaceuticals, and scientific instruments. The service sector includes the application of high technology to customers through computer, engineering, and medical services.

BC's population increase has caused a shift of the population centre of Canada towards the west and to BC in particular. In turn, BC has gained more seats in the House of Commons. This has changed relationships between central and eastern Canada and British Columbia. In short, economic, demographic, and political power has shifted to the west.

Economic Structure

Is British Columbia a core or a periphery? No doubt this Pacific province is an emerging powerhouse within Canada, but its manufacturing sector is tiny compared to that found in Ontario and Québec. While all regions are witnessing a shift from primary and secondary industries to tertiary ones, Ontario and Québec have relatively large secondary sectors based on manufacturing. In that sense, BC is not a core region. Yet, postmodern economies across the globe are, like Canada's six regions, retreating from primary/secondary activities and expanding into tertiary ones. Economic advancement is now largely driven by knowledge-based activities that flourish mainly in the tertiary sector. In the post-industrial information society, these advances differ from past ones because they are more heavily based on scientific breakthrough and take the form of technical advances and digitized information. Some of BC's knowledge-based companies are local firms while others are associated with multinational corporations. Microsoft, for instance, with its home base in Seattle, opened its first Canadian Development Centre in Richmond, BC, employing 300 researchers from around the world. However, the knowledge-based economy is not limited to the computer world; it permeates all avenues of the economy. In the primary sector, for example, technical advances in oil and gas operations permit **horizontal drilling**, which, when combined with **hydraulic fracturing**, has unlocked natural gas deposits from the Horn River shale formation in the Interior Plains of northeast BC.

For more on economic change in the twenty-first century, see Chapter 4, 'Economic Structure', page 166.

While employment statistics presented in Table 7.1 indicate a jump in the percentage of workers in the secondary sector (which includes both construction and manufacturing activities) and thus seem to contradict the long-term shift of workers to the tertiary sector, the explanation lies in the construction boom associated with the 2010 Winter Olympics as well as the building of the **Sea-to-Sky Highway** and the rapid-transit facility known as the **Canada Line**. With the completion of these transportation and sports facilities by 2009, construction declined sharply. During the same period, employment in manufacturing declined from 198,000 in 2005 to 187,000 in 2008 (Statistics Canada, 2006, 2009).

While BC is no longer seen as a resource-based economy, its natural resources, especially natural gas and forests, provide much of the wealth generated by the economy. BC, like Western Canada, is somewhere between a resource-based economy and an industrial one. In Friedmann's classification, BC is an upward transitional region (Table 1.4). Yet, the shift towards a more balanced economy is revealed in a comparison of the 1991 and 2008 economies. In 1991, 17 per cent of the province's GDP came from the extraction and processing of natural resources, but by 2007 this figure had fallen to 14 per cent (BC Stats, 2009a: 6). In 2007, BC's service sector accounted for 75 per cent of provincial GDP and 78 per cent of total employment (ibid., 4, 6). On the other hand, BC's exports exhibit one of the classic characteristics of a resource hinterland: close to 90 per cent of its resource exports are either unprocessed or semi-processed goods. Coal, lumber, and natural gas are the three leading exports and they fall into that category. Prices for coal and natural gas were at all-time highs in 2007 while lumber prices were near the bottom of their price cycle due to the lack of demand from US customers. By 2009, prices for all commodities had hit bottom due to the global downturn, but then prices, including for softwood lumber, began to surge with the start of global recovery.

The leading edge of economic change is found in the Vancouver and Victoria urban clusters, where high-technology and tourism services are concentrated. Beyond those clusters, the resource economy remains strong. The trend, as observed in the economic structure, reveals a shift of workers from primary/secondary sectors to the tertiary sector. Yet, as Table 7.1 shows, the secondary sector increased its share of the total workforce from 17.7 per cent in 2005 to 18.3 per cent in 2008. With the construction boom associated with the Winter Olympics at an end, the size of the secondary sector is expected to return

Table 7.1   Employment by Economic Sector in British Columbia, 2005 and 2008

Economic Sector	Workers, 2005 (000s)	Workers, 2005 (%)	Workers, 2008 (000s)	Workers, 2008 (%)
Primary	76.2	3.6	79.1	3.4
Secondary	376.5	17.7	422.4	18.3
Tertiary	1,677.8	78.7	1,812.8	78.3
Total	2,130.5	100.0	2,314.3	100.0

Sources: Statistics Canada (2006, 2009).

to its smaller 2005 figure and the long-term trend of a growing tertiary sector is expected to reassert itself.

Within the secondary sector, the number of workers in manufacturing was less than those employed in construction industry in 2008. Manufacturing in British Columbia remains based on the processing of forest and other primary resources. The value of manufacturing has increased but very slowly. In 1999, the value of manufacturing was $37 billion; in 2007 it had reached $43 billion worth of goods. With such an abundance of natural resources, why does manufacturing play such a relatively weak role in the BC economy? One reason is that most resources are exported in a raw or semi-processed form. A second reason is the small size of BC's population, which translates into a small market. Exports to foreign markets are difficult because of the relatively high cost of manufactured products. High manufacturing costs are associated with two factors: (1) the absence of economies of scale, which would reduce per-unit costs of the manufactured product, and (2) the relatively high cost of labour. Technological breakthroughs, however, can create global demand for superior products. Such breakthroughs generally require long-term investment in research and development from the public sector.

In 2008, the tertiary sector accounted for 78.3 per cent of employment in BC (Table 7.1). Employment in transport, trade, finance, and other services indicates the relative strength of the service industry. The service sector owes its strength to several factors, including British Columbia's growing domestic market, strong participation in world trade, a vibrant tourist industry, and its active **producer services** firms.

The evidence is clear—British Columbia's economy is changing, but growth in the manufacturing sector remained elusive until recently. In the past two decades, however, BC has experienced growth in high technology. As the front edge of the structural shift taking place in the manufacturing industry across Canada, high-tech firms are playing a greater role in BC's economy and are providing jobs for highly skilled workers. High technology involves cutting-edge research in the manufacture of electronics, telecommunications equipment, and pharmaceuticals. For example, Ballard Power Systems Inc., based in Burnaby, is a world leader in the development of fuel cells that produce electrical power with virtually no pollution. BC's high-technology industry ranked third in 2005 behind forestry and construction.[4]

No doubt, manufacturing is a fundamental element in Friedmann's theoretical version of a core region (see Chapter 1, especially Table 1.4). But times are changing and the traditional manufacturing activities dependent on unskilled and semi-skilled workers are in trouble because such activities as textile manufacturing have shifted offshore to countries where wages are much lower. High-tech manufacturing employing highly skilled workers is the wave of the future for industrial countries like Canada. As discussed in Chapters 5 and 6, these core regions have lost many 'low-end' firms and jobs. The good news for Ontario and Québec is that most high-technology firms are located in these provinces; the bad news is the mismatch between the skill level of those losing their jobs in low-end manufacturing and the labour needs of high-tech firms.

Think About It

With large high-quality timber reserves in BC, why is it so difficult for value-added forest manufacturing firms to increase their exports around the world? Should BC export logs to China rather than process them at home?

British Columbia's Wealth

British Columbia's economic prosperity rests on two economic pillars: its geographic location as Canada's trade window to the Pacific and its natural resources. Key exports are commodities—forest and energy products from BC; coal, grain, and potash from Western Canada. Imports consist of a variety of consumer goods, including automobiles and textiles, and they flow to all parts of Canada. In 2007, exports worth $32 billion and imports worth $39 billion passed through BC ports. Four countries, the US, China, Japan, and South Korea, account for over 80 per cent of exports and imports moving through BC ports (Table 7.2).

Pacific Rim Trade

Trade with Pacific Rim countries is growing rapidly. Much of BC's future economic well-being—as well as that of Western Canada—depends on access to these Asian markets, which provide two advantages. First, they represent new markets for BC and Western Canada commodities. Second, exports to Asia lessen dependency on the US market. While the US remains the most important trading partner, China, Japan, and South Korea occupy the next most significant positions. In terms of exports in 2007, the United States dominated trade by accounting for just over 60 per cent of exports through BC ports while the three Asian countries received nearly 23 per cent. Imports tell a different story, with the trade figures for the United States and the three Asian countries at 41 per cent and 33 per cent, respectively (BC Stats, 2009a).

Looking to the future, while trade with Japan and South Korea are likely to increase slowly, trade with China is anticipated to accelerate rapidly due to China's economic growth, which will soon make it the world's second-largest economy. Accordingly, BC exports to China are expected to increase sharply. From 1998 to 2008, BC exports to China increased from 1.7 per cent to 6 per cent (BC Stats, 2009b). By 2020, China's import of Canadian products through BC ports may well double, perhaps reaching 15 per cent (Industry Advisory Group, 2006: 1); this could easily reach 30 per cent of total exports when Japan, South Korea, and Taiwan are included. The concept of a transportation corridor across Western Canada to Pacific ports is fundamental to Canada's and BC's future economic well-being. To take advantage of this potential expansion of trade, the BC and federal governments, in conjunction with governments in Western Canada and the two railway companies, CN and CP, are working together through the **Asia-Pacific Gateway and Corridor**: the highway and rail corridors from Western Canada to ports on BC's coast. Over the four-year period, 2006–7 to 2009–10, BC and Ottawa have committed a total of $6 billion (BC Ministry of Transportation, 2007a). One project that eliminated a critical bottleneck is the recently completed Pitt River Bridge (photo 7.1). CN and CP are investing additional funds in rail line upgrading and port expansions. Cost-sharing projects between the province and the federal government include expansion of container facilities at major ports, notably Prince Rupert, and highway improvements at critical bottlenecks, such as the Pitt River Bridge and South Fraser Perimeter Road.

Think About It

Does the federal commitment of $3 billion for the $6 billion Asia-Pacific Gateway and Corridor project counter the perception that Ottawa does little for BC?

Table 7.2	Exports through British Columbia, 2007 and 2009			
Country	2007 ($ millions)	2007 (%)	2009 ($ millions)	2009 (%)
US	19,009	60.4	12,915	51.3
Japan	4,102	13.0	3,460	13.7
China	1,744	5.5	2,574	10.2
South Korea	1,308	4.2	1,662	6.6
Other countries	5,296	16.9	4,578	18.2
Total	31,459	100.0	25,189	100.0

Source: Adapted from Schrier (2010, 2).

Container traffic between Asia and North America is growing rapidly. Over two decades ago, Peter Nemetz (1990) foresaw that BC and the rest of Canada could reap many economic benefits through trade with the growing industrial giants in Asia. Within the geopolitical world, BC has a decided advantage because its ports, especially Prince Rupert, provide the shortest North American sea links to Asia. However, since increasing amounts of this cargo come in the form of containers, port infrastructures to handle **container** shipments are critical. According to the BC government, container traffic to all west coast ports is forecast to rise a staggering 300 per cent by 2020 (BC Ministry of Transportation, 2007b). In 2004, the Port of Vancouver had the capacity to handle 1.7 million TEU of container cargo (TEU refers to containers measuring 20 feet), Montréal, 1.2 million TEU, Halifax 0.5 million, and the Fraser River 0.25 million (Padova, 2004). The Asia-Pacific Gateway and Corridor strategy, a largely federal program, aims at expanding BC's transportation system, including its container port facilities. One goal of this strategy has been to increase container traffic imports from the Pacific Rim countries from 2 million TEU in 2005 to 9 million TEU in 2020 by expanding port facilities at Vancouver and by constructing a container port at Prince Rupert, which opened in 2007. As a result, Prince Rupert in 2009 recorded its highest volume throughput since 1997, despite declining container traffic through other North American west coast ports (BC Ministry of Transportation, 2007b; Prince Rupert Port Authority, 2007, 2010).

Natural Resources

Natural resources represent the basis of BC's wealth. Resource development, led by forest products and the export of forest and other resources, has provided the economic thrust for the province's growth. These resources include forests, fish, minerals, petroleum, water power, and physical beauty, and have driven growth in both the primary and secondary sectors. Moreover, a portion of the tertiary sector, the tourism industry, relies largely on the province's

---

Vignette 7.4   Is the Manufacture of BC Natural Resources Important?

For BC, the key to manufacturing is related to making products from its natural resources. As David Emerson, the former federal Minister of Trade and Industry, observed, 'There's almost unlimited scope to spin technology off our resource base' (Ebner, 2009: B2). Emerson suggests that BC should follow Finland's example with forestry, where massive public funding is directed to research and development within the Finnish Forest Research Institute of the Ministry of Agriculture and Forestry. To make such a commitment, the BC government has to recognize that a critical link exists between research and the region's most prominent industry. Put differently, does a long-term commitment of public funds for knowledge-based firms and universities to apply their research and development skills and techniques to the forest industry constitute a critical element in BC's future?

---

natural beauty and resources through ecotourism, hiking, skiing, sports fishing and hunting, and sailing, kayaking, and canoeing along BC's island-dotted coast of fjords and inlets. BC natural resources are found in its hinterland, far beyond BC's core. As in other hinterlands, the economic position of primary industries is declining and forms a small percentage of the total economy (Table 7.1).

Fishing Industry

British Columbia, like Atlantic Canada, has an important fishery. The BC fishery, as a renewable resource, ranks fourth in value of production among resource industries behind mining (including natural gas), forestry, and agriculture. Fish processing plants employ 25,000 full-time and part-time employees. More than 80 species of fish and marine animals are harvested from the Pacific Ocean, freshwater bodies, and aquaculture areas. Salmon is the most valuable species, followed by herring, shellfish, groundfish, and halibut. In 2006, the landed value from the sea reached $789 million, while the processed value was $1.3 billion (BC Stats, 2009a: 9). Most salmon today—ironically, Atlantic salmon—comes from fish farms. In 2006, the wholesale value

---

Think About It

As the CEO of one of the potash mines near Saskatoon, would you select Prince Rupert over Vancouver to ship your produce to China?

Table 7.3 Fishery Statistics by Species, 2007		
Species	Wholesale Value ($ millions)	Percentage
Farm salmon	479	37.0
Shellfish	198	15.3
Wild salmon	190	14.6
Groundfish	170	13.1
Halibut	169	13.0
Herring	52	4.0
Tuna	26	2.0
Other species	12	1.0
Total	1,296	100

Source: Adapted from BC Stats (2009a: 'Fisheries', 9). At: <www.bcstats.gov.bc.ca/data/qf.pdf>.

of farmed salmon was $443 million, compared to $228 million for wild salmon (Table 7.3). In recent years, excessive harvesting has reduced fish stocks and, in turn, the number of landings (fish caught). At the same time, production from fish farms has made fresh salmon available to the consumer throughout the year, thus popularizing the product. Environmentalists, however, have long expressed concern about fish farming—estuaries and inlets are subject to pollution from waste; fish meal used as feed in aquaculture results in the disruption of food chains and depletion of species in other parts of the globe; and predators break through the netting where farmed fish are kept, thus permitting their escape to the wild and the unknown effects of breeding with wild species of salmon. In January 2004, fears surfaced in the media that salmon from fish farms contain high levels of PCBs and other pollutants that could increase the risk of cancer, and such fears can affect sales of farmed salmon (Mittelstaedt, 2004; David Suzuki Foundation, 2009).

Unfortunately, BC shares another similarity with Atlantic Canada—overexploitation of fish stocks, particularly the valuable salmon stocks. Pressure on the fish stocks, especially the salmon stocks, comes from four sources—Canadian commercial fishers, American

© Natalie Forbes/Corbis

Photo 7.7

West coast fish canneries, like this one at Sechelt on BC's **Sunshine Coast** some 20 km north of Gibson, have long played a key role in British Columbia's economy. However, given the presence of American and Canadian sports fishers, as well as Aboriginal fishers who have special rights, the salmon fishery could suffer the same fate as the Atlantic cod fishery.

commercial fishers, the Aboriginal fishery, and the sports fishery. All want larger catches.

Management of the salmon fishery falls under the Pacific Salmon Treaty, which was signed by Canada and the United States in 1985 and sets long-term goals for the sustainability of this resource. The Pacific Salmon Commission, formed by the governments of Canada and the US to implement the Pacific Salmon Treaty, does not regulate the salmon fisheries but provides regulatory advice and recommendations to the two countries. In Canada, the Department of Fisheries and Oceans (DFO, now called Fisheries and Oceans Canada), in consultation with the Pacific Salmon Commission (www.psc.org/), has set annual quotas for salmon in Canadian waters.

Salmon are migratory fish, so regulating salmon fishing is particularly challenging. Like other fish, they are common property until caught. This principle is based on the 'rule of capture'. Fishers therefore try to maximize their share of a harvest so no one else will take 'their' fish. The problem is complicated further because the Canadian government cannot regulate the 'Canadian' salmon stocks, i.e., those that spawn in Canadian rivers, because they migrate to American waters, where the American fishing fleet harvests them.

The net result is that the salmon stocks are threatened. This problem is commonly referred to as the **tragedy of the commons** and is compounded by the economic cycle, whereby resource extraction accelerates during periods of high global demand for natural resources.

See Chapter 9, 'Overfishing and the Cod Stocks', p. 385, for an account of the over-exploitation of the northern cod in the waters of the Grand Banks.

In 2000, Ottawa tried to alleviate the pressure on salmon stocks by reducing the fleet of 4,500 fishing vessels by about one-third. However, this announcement sparked a strong reaction and little was accomplished. At the same time, Ottawa allowed Indian fishers, who have treaty rights to harvest fish for subsistence purposes, a share of the commercial stock. In 1999 Ottawa successfully negotiated a new Pacific Salmon Treaty with the United States, which extends to 2018 (Fisheries and Oceans Canada, 2009). Ottawa is able to exert some management of fish stocks in the Pacific Ocean because of its 200-mile fishing zone and because of its role in the Pacific Salmon Commission. Salmon fishing on the Pacific coast is regulated, based on the Pacific Salmon Treaty, which determines the size of the catch taken by each nation.

Salmon catches by fishing fleets fluctuated during the 1990s and early 2000s, not because of conservation measures that restrict the allowable amount of catches but simply because the number of salmon varied from year to year. Salmon (chinook, sockeye, coho, pink, and chum) spend several years in the Pacific Ocean before returning to spawn in the Fraser and other rivers. The principal BC salmon-spawning rivers are the Fraser and the Skeena. One of the largest harvests took place in 1997 when 487,000 tonnes of salmon were caught. Two years later, the figure was 17,000 tonnes! In 2009, the Fraser River experienced one of the biggest salmon disasters in recent history. Between 10.6 million and 13 million sockeye were expected to return to the Fraser but less than 2 million arrived (Hume, 2009).

Ottawa, which is responsible for managing salmon stocks within Canada and for negotiating international fishing agreements, must deal with four perplexing and inter-related issues:

1. the natural cycle (up to five years) of salmon to spawn in rivers, migrate to the sea, and then return to the rivers to spawn again;
2. the harmful effects of the forestry and hydroelectric industries on salmon spawning grounds and sea lice from salmon farms on juvenile salmon;
3. the division of the salmon catch among commercial, Native, and sports fishers;
4. the harvesting of 'Canadian' salmon by American fishers in international waters.

Management of fish resources is based on the estimated size of the fish stock. The problem of determining the size of the stock, and therefore the quota to be set for each year's catch, is complicated by the natural population cycle of the fish. For salmon stocks, this problem is complicated by their migratory life cycle, from spawn in Canadian rivers, such as the Fraser and Skeena, then to most of their adult life in the Pacific Ocean, and finally a return to their original spawning ground to spawn and thereby begin the cycle for another generation. Population fluctuation is illustrated by pink salmon catches in the past. Over one 10-year period, the size of catch varied from a low in 1975 of 38,000 tonnes to a high of 108,000 tonnes in 1985. In 1994, an estimated 2.3 million fish did not return to BC rivers to spawn. What caused this decline is unknown, but a similar drop in 1995 in the number of salmon returning to the Fraser River forced Ottawa to close fishing on the Fraser. Several factors account for decline in the salmon stocks:

- *Pollution of fish habitat*. Forestry companies have affected salmon habitat through their logging practices, which have blocked streams and thereby interfered with migration routes to spawning areas. Pulp and paper plants also discharged toxic wastes into rivers and the ocean. Fish farms, too, play a role because parasitic sea lice from these farms may be killing the juvenile salmon before they mature and return to the Fraser River to spawn (Morton et al., 2004).
- *Warming ocean temperatures*. With global temperatures increasing, the North Pacific Ocean temperatures may have risen above levels suitable for salmon. Also, the El Niño effect provides a natural explanation for the declining salmon stocks: salmon tend to survive poorly in a warmer ocean and better in a cooler environment due to changes in ocean plankton and fish community composition (McKinnell et al., 2009).
- *Overfishing*. The most likely explanation for the variation in salmon stocks is overfishing. With a combination of a larger fishing fleet and the application of new technology (radar and sonar equipment), the capacity to track and catch salmon and other fish has improved greatly. Since larger fish are normally caught in the fish nets, too few adult fish are left to reproduce and thus replenish the fish stocks.
- *High fish quotas*. Within Canadian waters, DFO is responsible for estimating the size of the fish stocks and, based on this information, sets fish harvest quotas. Since estimating the size of the fish stocks is a chancy business, being based on historic records and insufficient sampling of fish stocks, DFO estimates tend to be on the high side. Quotas are set from these estimates, though pressure from fishers and the Americans has resulted in some past quotas set too high.
- *The Aboriginal fishery*. First Nations peoples along the BC coast have always harvested fish for subsistence and cultural purposes. More recently, Indian fishers have sold some fish, mainly salmon, as a means of generating cash income. They are seeking a share of the fish quotas in order to participate more openly in the market economy. But how large a share should they receive? This is a difficult question, yet an Aboriginal commercial fishery was created in 1992 (see Vignette 7.5). As modern treaties are concluded with BC's First Nations, it is hoped that the Aboriginal fishery as it now exists will disappear and be replaced by individual harvesting agreements with each First Nation. As part of the Nisga'a agreement, for example, their share of the fishery harvest for commercial and subsistence purposes was defined.

Vignette 9.4, 'Mi'kmaq Become Lobster Fishers', page 370, discusses the outcome of the Supreme Court's Marshall decision.

Vignette 7.5   Aboriginal Fisheries Strategy: A Temporary Arrangement?

First Nations people along the Pacific coast based their economy on the sea. After BC joined Confederation, fishing for subsistence was generally accepted, but commercial fishing was controlled by Ottawa and Aboriginal fishers could not sell their catch. The federal government's Aboriginal Fisheries Strategy (AFS) was established in 1992 following the 1990 Supreme Court of Canada decision in *Sparrow*, which recognized the 'existing aboriginal right' to fish for food and ceremonial purposes. This new strategy rankled other commercial fishers, who argued that there should be but one commercial fishery. When DFO assigned commercial fishing licences and allocated fixed quotas of salmon to several First Nations along the Fraser River and then opened the annual salmon harvest to First Nations before it permitted fishing by the general commercial fleet, the reaction was swift and soon became a political issue. The commercial fishers argued that the 'advanced' fishing date for First Nations was a violation of the equality guarantees in the Charter of Rights and Freedoms. Only with the Supreme Court's 1999 decision in *Marshall*, a case originating in Nova Scotia, did

it become clear in law that Aboriginal peoples have a right to fish for commercial profit.

A final agreement with the Lheidli T'enneh in and near Prince George, which included a sub-agreement involving allowable salmon catches for commercial and personal/ceremonial purposes, might have provided a template for other BC First Nations who have made a claim for a commercial fishery, so that eventually the AFS would be replaced by individual fishing agreements with each First Nation. As it turned out, however, on 30 March 2007 the Lheidli T'enneh, by a 123–111 vote, refused to ratify the agreement. Reasons for the failure of ratification included the ramifications of extinguishment of Indian title, taxation issues, lack of rank-and-file involvement in the negotiating process, and a multi-million dollar legal bill amassed over more than a decade of negotiations. As for the fish, in this zero-sum game the trick will be to share the salmon stocks 'fairly' among all users and still have a sustainable resource. Signs such as the well-below-predicted size of the 2009 sockeye salmon run on the Fraser River do not augur well for a sustainable fish resource.

Mining Industry

The mineral wealth of British Columbia is found in both of the province's physiographic regions—the Cordillera and the Interior Plains. Each has a distinct geological structure. The Cordillera contains a wide variety of minerals, while the Interior Plains is part of the Western Sedimentary Basin, which contains petroleum deposits. Minerals include copper, gold, silver, lead, zinc, molybdenum, coal, and industrial minerals. However, the most exciting play at present is the Horn River shale deposit in northeast BC, which contains vast quantities of natural gas. But where is the market for such an enormous deposit? The answer may lie in Asian markets. Accordingly, a **liquefied natural gas (LNG)** terminal is planned for the west coast of British Columbia that will be connected by pipeline to the Horn River deposit. Specially designed ships with double hulls protect the LNG cargo systems from damage or leaks. In 2007, BC produced

16 per cent of Canadian natural gas output (Table 7.4).

Transportation to world markets poses a challenge for mineral development, yet in 2009 British Columbia had the third-highest value in mineral production among Canada's provinces and territories, at $5.8 billion, just behind Ontario ($6.3 billion) and Québec ($6.2 billion) (Table 7.5). The recession beginning in late 2008 had a huge impact on mineral production across the country, however. BC alone lost over $1.5 billion in production between 2008 and 2009 (Natural Resources Canada, 2010). For BC, location of its mineral deposits, distance to foreign

Table 7.4   BC's Oil and Gas Production, 2007

Commodity	2007	% of Canada
Crude oil (thousands of cubic metres)	1,529.2	2
Natural gas (millions of cubic metres)	30,438.6	16

Source: Statistics Canada (2008).

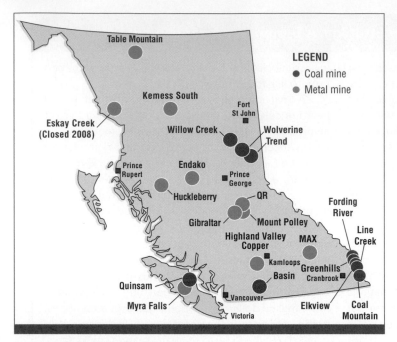

Figure 7.5   Mines in British Columbia.
Source: Adapted from Mineral Resources Education Program of BC (2009).

and rail-to-ship loading facilities provide part of the answer—though at a high construction cost. The vast investment in rail transportation for the Northeast Coal Project in the late twentieth century illustrates this point (Vignette 7.6). Based on the Peace River coal deposits, the Northeast Coal Project and its coal-mining community of Tumbler Ridge are located close to the Willow Creek, Wolverine, and Trend mines shown in Figure 7.5.

Today, the Willow Creek, Wolverine, and Trend mines ship their product by rail to Prince Rupert while those in southeast BC (Fording River, Line Creek Greenhills, Elkview, and Coal Mountain) ship by rail to Roberts Bank just south of Vancouver. Transportation costs have been reduced by unit trains and bulk-loading facilities. Unit trains consist of a large number of ore cars, sometimes over 100, pulled by one or more locomotives. The Roberts Bank terminal was designed as a large bulk-loading facility where coal in rail cars is dumped and then the coal is moved by conveyor belts to the ship's hold. These ships must be moored in deep water, which necessitated building a long causeway to reach the ships. Similar arrangements take place at Prince Rupert.

Think About It

From a geopolitical perspective, there are two views on energy resources and markets: continental and global. First, the US considers energy resources such as the Horn River shale deposits as part of a continental system. Second, Canada, by selling this gas to China, is diversifying its global markets. Which view do you think will prevail?

markets, and proximity to ocean ports are key factors. Unfortunately, many mineral deposits, but particularly coal mines, are located far from ocean ports. An expanded rail system

Table 7.5   Mineral Production by Province and Territory, 2009 ($ millions)*

Province/Territory	Metallics	Non-metallics	Coal	Total	% Share of Production
NL	2,244.1	45.7	—	2,289.8	7.1
PEI	—	3.4	—	3.4	0.0
NS	—	380.1	—	380.1	1.2
NB	749.6	x	x	1,090.4	3.4
Qué.	4,624.4	1,592.7	—	6,217.1	19.3
Ont.	3,790.0	2,540.1	—	6,330.1	19.7
Man.	1,176.8	143.7	—	1,320.5	4.1
Sask.	1,441.2	x	x	5,010.5	15.6
Alta	2.0	951.9	1,061.7	2,015.5	6.3
BC	1,828.4	588.7	3,316.5	5,733.6	17.8
Yukon	245.0	5.9	—	250.9	0.8
NWT	50.1	1,459.5	—	1,509.6	4.7
Nunavut**	—	—	—	—	—
Totals	16,151.6	11,455.6	4,544.4	32,151.5	100

*Preliminary data.
**Nunavut sand and gravel production is included in NWT non-metallics total.
x = confidential.
Source: Adapted from Natural Resources Canada (2010).

Vignette 7.6   Northeast Coal Project: A Failed Megaproject

Back in the 1970s, the vast Peace River coal reserves in the northeast corner of British Columbia represented an enormous untapped energy resource. But distance from an ocean port made these rich coal deposits uneconomic. However, by the late 1970s, global prices for coal reached new highs, making the Peace River coal attractive. With sales secure with Japanese steel companies, the dream of developing the Peace River coal deposits became a reality as the Northeast Coal Project. The Northeast Coal Project comprised Quintette Coal Ltd at Tumbler Ridge, Teck Corporation at Bullmoose, and Gregg River Coal Ltd at Gregg River.

The goals were straightforward: a mega energy project for BC and a stable supply of coking coal for Japan's steel industry. Denison Mines and Teck Corporation were the major Canadian mining companies behind this $4.5 billion construction project (Bone, 2009: 174–5). With a 14-year contract to sell coal to Japanese steel companies, banks and other financial institutions were happy to supply much of the capital (about $2.5 billion) to Denison and Teck. The federal and provincial governments provided $1.5 billion of the $4.5 billion to upgrade the CNR line to Prince Rupert, to extend a BC Rail line to Tumbler Ridge, and to build a coal terminal to load ships that would carry the coal to Japan. The two mining companies contributed about $500 million. Both governments provided another $1 billion to build the coal town of Tumbler Ridge.

Like all megaprojects, the construction phase takes time and production began in 1984. By then, the world demand and price for coal had peaked and a slow but steady decline set in as the global economy slipped into a recession. At that point, the Japanese steel mills no longer needed so much coal and the dream of the Northeast Coal Project collapsed with the bankruptcy of Denison Mines.

Hydroelectric Power

Hydroelectric energy is a renewable energy source dependent on the hydrologic cycle of water, which involves evaporation, precipitation, and the flow of water due to gravity. British Columbia, with abundant water resources and a geography that provides many opportunities to produce low-cost energy, is the second-largest producer of hydroelectric power in Canada. In 2005, BC Hydro produced 60.4 gigawatts of electric energy while Hydro-Québec accounted for 160 gigawatts (Natural Resources Canada, 2006). Within the Cordillera, a combination of elevation, steep-sided valleys, and steady-flowing rivers provides ideal conditions for the construction of hydroelectric dams. The Columbia, Fraser, and Peace rivers offer many excellent hydroelectric sites. In addition, heavy precipitation and meltwater from the mountain snowpack ensure a regular and abundant supply of water. Major hydroelectric developments have taken place on the Columbia and Peace rivers as well as on the Nechako River, a tributary of the Fraser. Minor developments have occurred on Vancouver Island.

In the early days of hydroelectric development, small-scale hydroelectric projects were established near Vancouver and Victoria. These two cities are the major markets in the province for electricity. Hydroelectric dams were also built near smelters that required vast amounts of power to operate. For example, early in the twentieth century, power was generated from privately owned dams on the Kootenay River for the lead-zinc smelter at Trail.

After World War II, public and private companies began to undertake megaprojects. Among these giant industrial construction efforts, three—one on the Columbia River, another on the Peace River, and the third on the Nechako River—had enormous impacts on the economy. They all involved the harnessing of water power from the province's rivers to generate low-cost electrical power, but they also flooded valuable farmland and Indian lands and led to the loss of salmon spawning grounds. Such construction projects were considered major engineering feats. In 1951, Alcan completed the construction of the Kenney Dam across the Nechako River, impounding the water draining from an area more than twice the size of Prince Edward Island. Ten years later, Canada agreed to construct three dams on the Columbia River while the United States constructed a hydro-electric dam at Grand Coulee in Washington. The power was shared, as described in Vignette 7.7. In 1968, BC Hydro built the W.A.C. Bennett hydroelectric dam on the Peace River, creating Williston Lake, the largest freshwater body in the province. From this giant reservoir, vast quantities of water flow through the turbines at the power station, generating much of British Columbia's electrical power and creating a surplus of power for sale in the US.

Tourism: A Growing Sector in BC's Economy

The natural beauty of BC's ocean, mountains, and forest provide the basis of tourism. While many tourists are attracted to the parks and wilderness areas found in the province's forests, most are drawn to the larger urban centres, especially Vancouver and Victoria. Tourism is a key and growing element in the service sector. The 2010 Winter Olympics drew many thousands of tourists to the Games and, with world-class winter sports facilities, tourists are expected to come year after year. Whistler ski resort, a key site for the 2010 Games, will benefit from the global exposure afforded by the Olympics, as will British Columbia generally.

The scale of the industry is revealed in the fact that, in 2005, $9.7 billion was spent by 23 million overnight visitors to the province (BC Stats, 2006). Given British Columbia's scenic landscapes and many parks, tourism has enormous potential for expansion. One example is BC communities tapping into the Alaska cruise ship business. In 2006, cruise ships travelling to Alaska carried nearly a

Vignette 7.7   Alcan in Northwest British Columbia

In the late 1940s, the Aluminum Company of Canada (Alcan) proposed to build an aluminum complex in northwest British Columbia. After obtaining the rights to the waters of the Nechako River for 50 years from the provincial government, Alcan began construction of the first phase of this complex in 1951.[5] Three years later, Alcan built a new town called Kitimat, a power plant at Kemano, and a dam on the Nechako River. The Alcan hydroelectric development in British Columbia illustrates the advantages and pitfalls of megaprojects. Alcan is one of the giants in the world aluminum industry. To keep its production costs low, Alcan is attracted to sites where low-cost hydroelectric power can be developed and then used by its smelting plants to reduce bauxite (a clay-like mineral) into alumina (the chief source of aluminum) and then aluminum. In Canada, water resources are under provincial jurisdiction, so Alcan seeks long-term arrangements with provincial governments to develop the water power for its smelting plants. In turn, provincial governments are attracted to megaprojects because they help to develop a region and employ large numbers of people.

In 1989, Alcan began the second phase of its hydroelectric complex—the Kemano Completion Project. Estimated costs were $1.3 billion. The plan was to enlarge its hydroelectric facility by boring a second tunnel through Mount DuBose, thereby diverting more waters of the Nechako reservoir westward. This time, both the Carrier-Sekani Tribal Council (whose members include the Nechako River as part of their land claim) and environmental groups, spearheaded by the Rivers Defense Coalition, openly challenged the project, claiming that the social and environmental costs were too great. In 1995, the project was half-finished, but the provincial government cancelled its approval for the Kemano Completion Project. Alcan threatened to sue. Later, the British Columbia government agreed to compensate Alcan for the money spent on construction by selling the company electricity at a very low price until 2024 (BC was able to make use of BC Hydro surplus power). The next development came in 2003 when the global economy and the commodity cycle again began to swing upward. With higher prices and growing demand, Alcan (now Rio Tinto Alcan) dusted off its plans for its BC operations. Modernization of Kitimat Works (formerly call the Kemano Completion Project) is underway at a cost of US$2.5 billion, which will result in a leading-edge, highly efficient operation with a completion date within five years (Rio Tinto Alcan, 2010).

Courtesy of Dwight Magee, Rio Tinto Alcan

Photo 7.8

Aerial view of the Kitimat Works smelter.

million tourists (VanderKlippe, 2007b). Most cruises begin in Vancouver and sail north through BC's spectacular Inland Passage where whales and grizzly bears are often sighted. From Vancouver, the cruise ships next stop in Juneau, Alaska. A new and distinctive twist for the cruise ships is to stop along the BC coast to learn about local Indian tribes and their cultures. The world's first Indian-themed cruise ship terminal, the Wei Wai Kum Terminal, opened in 2007, is located near Campbell River on Vancouver Island. This recently constructed terminal is owned and operated by the Campbell River First Nation. It cost $24.5 million, with principal funding from the federal government

and lesser amounts from the province and local governments, including the Campbell River First Nation (Indian and Northern Affairs Canada, 2007). Passengers visit a traditional village complete with totem poles and a Big House, both of which provide a window into Laichwiltach history and culture.

While tourism has grown in importance, BC faces a dilemma. Tourist facilities do not always mesh with preserving the environment. For instance, the expansion of the Sea-to-Sky Highway that winds along Howe Sound from North Vancouver to Squamish and then on to Whistler has met with opposition from environmentalists and residents of West Vancouver. The basic question is: How does

Photo 7.9

A hiking trail on Whistler Mountain with Blackcomb Mountain, with its ski runs, in the background.

Think About It

Did Vancouver make the correct decision to host the 2010 Winter Olympics at a cost of $2 billion or should these funds have been spent on social programs, including housing for the homeless?

such development affect BC's wilderness? The challenge remains how to balance the need for more tourist development and the desire to maintain a natural landscape and an urban environment free of air and water pollution. The answer may lie in making cities more in tune with an 'urban environment' where greens spaces abound, bike paths are widespread, and downtown areas are designed for pedestrians rather than for automobiles.

Key Topic: Forestry

The forest is British Columbia's greatest natural asset. With just over 60 per cent of the province covered by forests, British Columbia contains about half of the nation's softwood timber and leads the nation in export of forest products. With such a dominant position, this region is easily Canada's leading supplier of wood products for construction and

finishing. With effective access to ports, forest firms are ideally located to supply the needs of the US and other global markets. Forests play an important role in the tourist industry with the growing recreational use of forest and wilderness areas by campers, hikers, and whitewater adventurers.

The forest industry remains important, but no longer dominates the province's economy. For instance, 50 years ago forestry alone accounted for 50 per cent of the provincial economy and for most of the employment. By the late 1990s, forestry comprised only 17 per cent of BC's economy and 14 per cent of employment (Barnes and Hayter, 1997: 5). Forestry has slipped from the pedestal and now tourism and knowledge-based businesses have gained the high ground in the province's economy. In the last few years, this shift has been accelerated by the combination of falling sales of softwood lumber and

Table 7.6   Timber Harvest by Species, British Columbia, 2005 and 2008 (million cubic metres)

Species	2005	2008
Lodgepole pine	37.6	28.7
Spruce	11.9	8.5
Douglas fir	10.9	6.4
Hemlock	7.7	5.6
Cedar	6.0	5.0
Balsam	5.1	3.6
Other	4.2	3.2
Total	83.4	61.0

Source: BC Stats (2006, 2009a: 'Forestry', 7).

Vignette 7.8   Access to the Interior Forest

Until the 1950s, the Interior Forest was not accessible because the transportation system was inadequate to permit shipment to world markets. As a result, the forest industry was concentrated along the Pacific Coast. Here, logs were dumped into the Pacific Ocean; rafts of logs formed; and these rafts were pulled by tugboats to sawmills. At that time, Vancouver was a major sawmilling centre. In 1956, the completion of the Pacific Great Eastern Railway from North Vancouver to Prince George radically altered BC forest industry by opening the interior to commercial logging. Prince George became a hub of forest production. Thanks to BC Rail (now CN Rail), sawmills in the interior account for about three-quarters of the lumber production and almost all production is exported.

the burst in tourism associated with the 2010 Winter Olympics. Nevertheless, BC's forest resource endures and, with the higher prices for softwood lumber in early 2010, the industry could rebound in the coming years.

British Columbia's forest consists almost entirely of coniferous **softwood forest**. Within Canada, British Columbia accounts for just over half of the total logging output. The main species harvested are lodgepole pine, spruce, hemlock, balsam, Douglas fir, and cedar (Table 7.6). The processing of timber into lumber, pulp, **newsprint**, paper products, shingles, and shakes supports a major manufacturing industry in British Columbia, which leads in the production of such wood products as lumber and plywood, while Ontario and Québec produce a greater proportion of pulp and paper products. At first, the forest industry was concentrated along the coast where logs could be barged to sawmills and then the lumber exported by ship. The exploitation of the Interior Forest did not take place until the completion of the Pacific Great Eastern Railway (now a CN line) from North Vancouver to Prince George (Vignette 7.8).

Fluctuations in total BC exports by value are affected by forest exports. For example, since 2004 softwood lumber exports have declined, mainly due to falling prices. As a result, the place of the forest industry within the BC economy has diminished. The major factor was the drop in exports, especially to the United States. Natural events added to the economic problems facing the forest industry—the invasion of the pine beetle and several years of massive forest fires have destroyed forest stocks. Finally, the central role of forestry in the BC economy is affected by economic growth in other sectors, especially the construction industry. One indicator of this diminished position is the drop in the value of exports from $14.1 billion in 2005 to $12.3 billion in 2007 (BC Stats, 2006, 2009b). However, once the US housing market recovers and demand for BC lumber and other wood products increases, the forest industry should regain its place in the BC economy.

Forest Regions

Climate and topography have divided this vast forest into two distinct regions—the rain forest of the coast and the boreal forest of the interior. Size, age, and species vary in these two regions. Within the rain forest, the mild, wet climate allows trees to grow to great heights for hundreds, even thousands of years. Old-growth trees that are at least 250 years old are located along the Pacific coast, where, because of heavy precipitation throughout the year, fires are rare. The major species harvested are spruce and Douglas fir. In the interior, the main species logged is the lodgepole pine. Since the climate in the interior is much drier as well as colder in the winter and hotter in the summer than along coast, the lodgepole pine and other trees are smaller and forest stands less dense. Trees have a shorter lifespan (120 to 140 years). While forest fires occur in all parts of BC, the greatest number take place in the dry, hot interior. Here, forest fires are a constant threat and, more recently, pine beetle infestations (discussed above) have become a significant environmental and commercial problem. Another terrain-related issue is clear-cut logging. BC forests often are found in mountainous areas. Clear-cut logging on steeply sloping land leads to erosion and sediment deposition in streams and rivers (photo 7.10). Beyond the environmental concern about clear-cut logging, this harvesting approach is antithetical to the sustainability of the BC forests. Along with this efficient but anti-sustainable logging system, economic cycles of high demand cause companies to accelerate logging in response to economic rather then environmental considerations (Clapp, 2008: 129).

British Columbia's forest is far from homogeneous, largely because of varying climatic zones and the varied relief in the Cordillera. BC's forest lands are divided into two major

Al Harvey/Slide Farm

Photo 7.10

The practice of clear-cut logging keeps costs down but makes the goal of a sustainable industry problematic. Note the logs lying on the steep slope in this photograph. According to a 2005 report by the David Suzuki Foundation, clear-cutting remains the dominant system of harvesting the old-growth trees in British Columbia's rain forest. Two impacts of clear-cutting are blowouts (strong winds knock over isolated trees) and landslides (moderate to steep slopes are subject to mud and landslides during periods of heavy rain).

regions: the Coast Rainforest and the Interior Boreal Forest. Within the Interior Forest are four sub-regions: the Northern Forest, the Nechako Forest, the Fraser Plateau Forest, and the Columbia Forest (Figure 7.6). Within each of these sub-regions there is great variation due to the differences in local growing conditions, which are affected by precipitation and length of growing season. Other factors are elevation, soil conditions, and topography.

The Coast Rainforest is the most luxuriant coniferous forest in Canada. With its wet and mild marine temperatures and abundant rainfall, this is one of the most densely forested areas of North America. The key species are Douglas fir, western red cedar, and western red hemlock. Under ideal conditions, mature cedar and hemlock trees reach 45 m in height and about 1 m in diameter. The Douglas fir is an even larger tree. Mature stands may average 60 m in height and over 1.5 m in diameter. Logs from the mature, old-growth stands are highly valued for lumber and plywood, while logs from immature stands (known as secondary growth) are used for pulp and paper.

Inland from the Coast Rainforest, the climate changes from a wet marine climate to a semi-arid one. At this point, the Interior Boreal Forest begins. However, differences exist within the Interior Forest. The Fraser Plateau sub-region (Figure 7.6) is an open woodland with much smaller trees than those in the Coast Forest. The controlling factor is the dry climate. Trees in the Fraser Plateau must be able to cope with drought conditions. Ponderosa pine and lodgepole pine are the most common species. They often attain heights of 25 m and a diameter of 1 m. These trees are usually converted into lumber.

The Nechako Forest sub-region, which lies to the north of the Fraser Plateau, receives more precipitation and experiences lower summer temperatures. The net result is more moisture for tree growth. Consequently, the forest cover is denser than that in the Fraser Plateau. A common species in the Nechako Forest is the Engelmann spruce, which can attain heights of 40 m and diameters of 60 cm. Timber from this sub-region is processed by pulp and paper mills at Prince George.

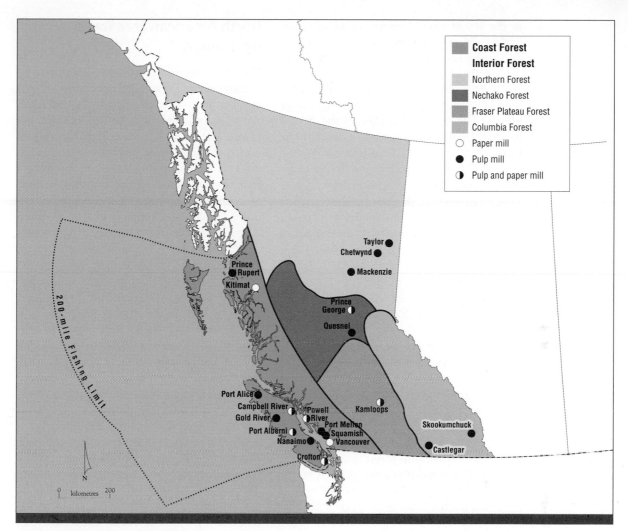

Figure 7.6   Forest regions in British Columbia.
The two principal regions are the Coast Rainforest and the Interior Boreal Forest. The forest lands in the Coast Rainforest are almost entirely in mountainous terrain. The Interior Forest is found in many types of terrain. It is subdivided into four areas that reflect variations in growing conditions. The two main elements are lower temperature towards the north and, because of the warmer summer temperatures, drier conditions.

The Columbia Forest sub-region lies in the easternmost area of southern British Columbia. Again, a dry climate limits tree growth. Because of widely varying terrain, the forest cover is quite varied. Western red cedar and western hemlock are common species, though stands of ponderosa pine and even Douglas fir are also found in this region. Large timber is sent to sawmills, while the smaller logs are used for pulpwood.

The Northern Forest sub-region is the most remote of BC's forests. High transportation costs to ship wood to world markets hinder logging. Tree growth is hampered by cool growing conditions and poorly drained land. Large blocks of land that contain muskeg are either devoid of trees or have trees with little commercial value.

Dependency on the US Market

The small domestic market compels the forest companies to seek foreign buyers. The US market is the major destination for Canadian forest products. One advantage for BC forest exporters is the proximity to the US market. Another is its size. The problem facing the BC forest industry is 'too many eggs in one basket' because fluctuation in US demand can cause havoc. In 2005, BC's timber harvest reached a record high of 83.4 million cubic metres. This huge harvest had a ready market in the booming US house construction industry. The value of this harvest, including processing, was $18.2 billion. Led by lumber and pulp, forest exports were valued at $14 billion. Three years later, the US housing market had collapsed, causing demand for lumber and other forest products to drop sharply. The timber harvest dropped from 83.4 million cubic metres in 2005 to 61.0 million cubic metres in 2008 (Table 7.6). Forest exports, the real barometer of the industry, fell from $14 billion in 2005 to $10.1 billion in 2008 (Table 7.7).

In 2006, the US housing industry took a significant downturn. Consequently, prices for Canadian softwood lumber dropped to half the 2005 prices and value of exports fell by 40 per cent. Plant closures and staff layoffs swiftly followed. This sudden shift from high demand to low demand illustrates the vulnerability of the forest industry, and of single-industry forest communities. A strong Canadian dollar compounds the difficulty of regaining market share in the US.

Think About It

If the high Canadian dollar inhibits exports of softwood lumber to the United States, should the Bank of Canada force the dollar lower?

North American Free Trade Agreement

The US government has imposed a number of countervailing duties against Canadian softwood lumber—in 1982, 1986, 1991, and 2002. NAFTA has reduced trade barriers between the United States and Canada, but it has not prevented trade disputes involving lumber from erupting. Forest exports from British Columbia have already been subjected to the heavy hand of Washington. In each case of countervailing duties, Washington's action has been driven by pressure from the American lumber lobby, the Coalition for Fair Lumber Imports. Most Canadians do not understand why NAFTA has not given Canadian products 'free' access to US markets. After all, was not 'free' access to each other's market the purpose of the original Free Trade Agreement and the subsequent NAFTA? The answer to that question is 'freer' but not 'free' trade. For example, when an American producer is adversely affected by imported products, the US company complains to the American government about the lower-priced products, claiming that such lower prices are a form of unfair trade. The company expects Washington to protect it by creating a trade barrier, which it has done on occasion. The trade agreement has a dispute-settlement mechanism, but resolving trade disputes takes time, during which the Canadian exporter loses sales and profits.

Problems over free and fair trade have been particularly frequent in the forest industry. American lumber production, which operates mainly in the Pacific Northwest and Georgia, can produce a maximum of 15 billion board feet per year. Canadian lumber production nearly doubles the American figure. In fact, lumber production from British Columbia is roughly equal to that of the entire United States. The main difference is that US-produced lumber serves its domestic market, while most Canadian-produced lumber is exported to the United States, Japan, and other foreign customers. Trade disputes over lumber exports can therefore significantly affect BC's forest industry.

When American lumber producers lose market share to their Canadian counterparts, they turn to their lobby organization. The US

Table 7.7   Forest Product Exports from British Columbia, 2005 and 2008 (value in $ millions)

Commodity	2005	2008
Softwood lumber	6,271	3,606
Pulp	2,500	2,878
Paper and paperboard	1,209	1,308
Wood products	1,004	513
Panel products	741	407
Newsprint	624	349
Plywood	460	178
Cedar shakes and shingles	235	203
Other	936	666
Total	13,980	10,108

Sources: BC Stats (2006: 'Forestry', 7; 2009a: 'Forestry', 7).

political system is extremely sensitive to lobby efforts because individual members of the House of Representatives and the Senate have the power to intercede on trade matters. In addition, these elected officials rely on lobby organizations for support at election time. Since 1982, the US Coalition for Fair Lumber Imports has managed to convince the American government to reduce Canadian forestry imports into the United States. In 1986, for example, the US launched a trade action, claiming that Canadian softwood lumber production was subsidized by low provincial stumpage fees. In the following year, the US Department of Commerce ruled that Canadian stumpage rates constituted a countervailing subsidy. Ottawa agreed to impose a 15 per cent export tax on lumber exported to the US. British Columbia and several other provinces raised their stumpage rates to counter the American claim of low stumpage rates and to replace the federal export tax. In 1991, Canada terminated the 15 per cent export tax. In response, the US Coalition for Fair Lumber Imports called for another trade action to restrict the flow of Canadian lumber into the US. In 1992, the US government imposed a countervailing duty on Canadian lumber. Over a two-year period, the US government collected more than $800 million from Canadian exporters. In 1994, a bilateral trade panel (the dispute mechanism created by NAFTA) declared the tax invalid under NAFTA and ordered the US government to reimburse the Canadian companies. Not to be outdone, the US Coalition for Fair Lumber Imports called for new restrictions.

In 1996, the US government insisted on a ceiling for Canadian lumber shipments to the US, and Ottawa and Washington reached a new agreement: a limit of 14.7 billion board feet would be allowed into the US in exchange for five years of non-interference with the lumber trade by the US government. The agreed-to figure was an average of the volume of Canadian lumber exported to the United States over the previous three years. The first 650 million board feet in excess of the figure faced a US tax of $50 per 1,000 board feet. Even with this additional cost, Canadian lumber was still competitive in the US market. Once the 650 million board feet figure was exceeded, the tax jumped to $100 per 1,000 board feet. Even at that tax level, Canadian lumber remained competitive in the US market. All of these efforts by the US lumber lobby and the US government have been attempts to bypass NAFTA in order to protect US lumber interests. This US–Canada agreement ended on 31 March 2001. From a Canadian perspective, the softwood lumber trade dispute has never been about subsidies but rather has served as a means by which Washington protects the US lumber producers. In May 2002, as negotiations to extend the previous agreement reached an impasse, Washington announced a duty of 27 per cent on Canadian softwood lumber entering the United States. Four years later, a new agreement finally was signed in October 2006.

The latest softwood lumber agreement between Ottawa and Washington has ended a bitter trade dispute—at least for the next seven years. The agreement provides protection to the US industry when lumber prices fall. Highlights of the new agreement are:

- The seven-year agreement includes a possible two-year extension.
- Import duties of $4 billion the US charged Canadian companies since 2002 will be returned. The US keeps $1 billion.
- The US is banned from launching new trade actions against lumber imports from Canada.
- Canada is to impose an export tax if softwood lumber prices in the US fall below US $355 per thousand board feet. At that point, Canada must impose a tax on softwood lumber shipped to the United States of at least 5 per cent of the price per 1,000 board feet. The purpose of the export tax is to protect US lumber producers by increasing the price of Canadian lumber and thus give an advantage to US producers.
- Neutral trade arbitrators are to provide final and binding settlements of disputes.

The lumber industry in British Columbia is at a crossroads. The pessimistic interpretation is of an industry on the decline, fizzling out as the resource base itself disappears. The optimistic interpretation, though, is of a kind of industrial renaissance, of a newly fashioned forest products industry that emphasizes high-value products, skilled labour, and leading-edge technology. Industry analysts say that the province has too many pulp mills facing a dwindling supply of wood fibre. The forest industry remains an important part of BC's economy, but in order to protect it, and the economy in general, such issues as those discussed above will have to be addressed. Already the market has caused several pulp and paper mills to close, as well as many more sawmills.

British Columbia's Urban Geography

The most striking aspect of the urban geography of British Columbia is the concentration of people in the southwest corner of the province popularly referred to as the **Lower Mainland**: over 60 per cent of BC's residents are concentrated in this area. Besides the City of Vancouver, the major cities in the Lower Mainland include

Photo 7.11

Back in 2002, the United States invoked a punitive countervailing duty of 27.2 per cent on Canadian softwood lumber. In 2006, a new softwood lumber agreement was reached. During the time of the duty, larger firms were able to increase their efficiency through economies of scale but many small forest operations were hurt by this duty, causing closures and layoffs. By mid-2006, the US housing market had declined sharply, causing lumber prices to drop dramatically. Unfortunately for Canadian producers, the new softwood lumber agreement requires Ottawa to impose an export tax when the price of softwood lumber reaches US$355 per 1,000 board feet.

Abbotsford, Burnaby, Coquitlam, Langley, Maple Ridge, New Westminster, North Vancouver, Richmond, Surrey, and West Vancouver (Figure 7.7). Beyond this population core, a secondary population cluster is found on Vancouver Island, where Victoria forms the second-largest urban population and a series of towns and cities, including Naniamo, stretch northward to Campbell River. In the Interior, Kelowna, Vernon, and Penticton constitute the major urban centres in the Okanagan Valley, while nearby Kamloops is in the Thompson Valley. As well, population centres are located along the north coast around Prince Rupert, in the central interior at Prince George, in the Peace River country where Dawson Creek and Fort St John are located, and along the US border, where Trail and Creston are located.

Canada's third-largest city, Greater Vancouver, has a population of over 2 million. The census metropolitan area of Vancouver dominates the urban geography of British Columbia (Table 7.8). Like the province as a whole, Vancouver's population has increased at a rate well above the national average. Until 1996 it was the fastest-growing large metropolitan area in Canada. Since then, Vancouver's spectacular growth rate has continued, but from 1996 to 2006, Calgary was the fastest-growing city in Canada.

Besides the rapid growth of Vancouver, smaller urban centres in BC are also growing quickly. Since 2001, Kelowna (9.8 per cent), Chilliwack (9.3 per cent), Abbotsford (7.9 per cent), and Nanaimo (7.8 per cent) have exhibited the greatest rates of increase of BC cities with a population of 50,000 or greater (Table 7.9). In comparison, Vancouver and Victoria had increases of 6.5 per cent and 5.8 per cent. At the other end of the scale, the greatest population losses (over 12 per cent) took place in Kitimat and Prince Rupert. The demographic decline of Prince Rupert may have halted since the last census now that its container port is in operation and traffic flows along the CN's much touted **North West Transportation Corridor**. Kitimat's future depends on the completion of the Kitimat Works, which will result in an expansion of aluminum production and, with it, employment. Also in the wind is the prospect for Kitimat to become a transshipment point for Fort McMurray oil going to China.

Unlike Ontario and Québec, each of which has more than one population cluster, BC can claim only the Vancouver area as a veritable population cluster. The Vancouver CMA's 2006 population was 2.1 million. Victoria, the second-largest urban centre, has 330,088 inhabitants (Table 7.8). As the capital of the

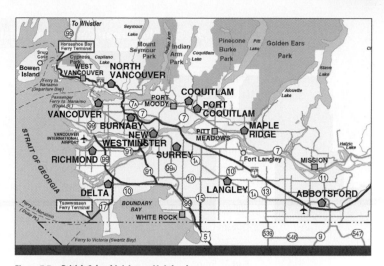

Figure 7.7   British Columbia's Lower Mainland.
The Lower Mainland of British Columbia, with Vancouver as the focal point, dominates British Columbia's urban geography. (British Columbia Adventure Network © 1996–2007 Interactive Broadcasting Corporation). *Source:* Davenport Maps Ltd, at: <www.davenportmaps.com>.

Table 7.8 Census Metropolitan Areas in British Columbia, 2001–6			
Centre	**Population 2001**	**Population 2006**	**Percentage Change**
Abbotsford	147,370	159,020	7.9
Kelowna	147,739	162,276	9.8
Victoria	311,902	330,088	5.8
Vancouver	1,986,965	2,116,581	6.5

Source: Adapted from Statistics Canada publications *Population and Dwelling Counts, 2001 Census,* Catalogue 93F0050XCB2001013, http://www.statcan.ca/bsolc/english/bsolc?catno=93F0050X2001013 and *Population and Dwelling Counts, 2006 Census,* Catalogue 97-550-XWE2006002 http://www.statcan.ca/bsolc/english/bsolc?catno=97-550-XWE2006002.

Table 7.9 Smaller Urban Centres in British Columbia, 2001–6			
Centre	**Population 2001**	**Population 2006**	**Percentage Change**
Kitimat	10,285	8,957	−12.6
Dawson Creek	10,754	10,994	2.2
Prince Rupert	15,302	13,392	−12.5
Squamish	14,435	15,256	5.7
Salmon Arm	15,388	16,205	5.3
Powell River	16,604	16,537	−0.4
Terrace	19,980	18,561	−7.0
Williams Lake	19,768	18,760	−5.1
Quesnel	24,426	22,499	−8.1
Cranbrook	24,275	24,138	−0.6
Fort St John	23,007	25,136	9.3
Port Alberni	25,299	25,297	0
Parksville	24,285	26,518	9.2
Campbell River	35,036	36,461	4.1
Duncan	38,613	41,387	6.6
Penticton	41,564	43,313	4.2
Courtenay	45,205	49,214	8.9
Vernon	51,530	55,418	7.5
Chilliwack	74,003	80,892	9.3
Prince George	85,035	83,225	−2.1
Nanaimo	85,664	92,361	7.8
Kamloops	86,951	92,882	4.4

Source: Adapted from Statistics Canada publications *Population and Dwelling Counts, 2001 Census,* Catalogue 93F0050XCB2001013, http://www.statcan. ca/bsolc/english/bsolc?catno=93F0050X2001013 and *Population and Dwelling Counts, 2006 Census,* Catalogue 97-550-XWE2006002 http://www. statcan.ca/bsolc/english/bsolc?catno=97-550-XWE2006002.

province, Victoria is a 'government' town, as well as an important tourist and service centre. Its mild climate has also attracted retired people, especially from the Prairie provinces. Kelowna and Abbotsford, with populations in 2006 of 162,276 and 159,020, respectively, are the only other CMAs in the province. Four cities—Kamloops, Nanaimo, Prince George, and Chilliwack—have populations between 80,000 and 93,000, and below the 80,000 mark are 17 smaller cities with populations of 10,000 or more (Tables 7.8 and 7.9).

Vancouver

Vancouver has a majestic physical setting. The city, located on the shores of Burrard Inlet, lies across the water from the snow-capped peaks of the North Shore Mountains. The Lions Gate Bridge passes over Burrard Inlet, thereby

linking Vancouver with the North Shore and its main urban centres of West Vancouver and North Vancouver. From West Vancouver, the Sea-to-Sky Highway leads to the Whistler ski resort. To the west is the island-studded Strait of Georgia, while the Fraser River and its deltaic islands (flat, low islands composed of silt and clay near the mouth of the river) mark Vancouver's southern edge. Vancouver has a mild, marine climate, though some find the frequent rain and overcast skies unappealing.

Like Montréal, much of Vancouver's commercial strength stems from its role as a trade centre. Vancouver is the largest port in Canada and one of the largest on the Pacific coast. With most of the world's population located along the Pacific Rim, Vancouver handles $30 billion worth of trade goods each year. In fact, the Port of Vancouver is so

laughingmango/iStockphoto.com

Photo 7.12

Located between Okanagan and Skaha lakes, Penticton lies in the Okanagan Valley of the Interior Plateau. Its hot summer weather and sandy beaches make Penticton a summer tourist hub. Irrigated fruit orchards and vineyards surround the city on the lower slopes of the Okanagan Valley. On the upper slopes, the natural vegetation becomes desert-like, with sagebrush, cactus, and other desert plants.

busy that some shippers may opt for a longer route through the Panama Canal to Halifax.

Much of Vancouver's economic well-being is closely tied to the United States and Pacific Rim countries. Most of BC's exports go to the United States, China, and Japan. In 2005, 80 per cent of exports from British Columbia and 78 per cent of imports involved those three countries (BC Stats, 2006: 12). Forest products made up approximately half of all 2005 exports, of which softwood lumber comprised 44 per cent, pulp 20 per cent, newsprint 5 per cent, and paper and paperboard 10 per cent. The remaining 21 per cent consisted of manufactured wood products such as doors and window frames (ibid., 7). By 2008, the forestry sector accounted for only 30 per cent of the province's exports, though a modest rebound was anticipated for 2010 (Export Development Canada, 2009).

Vignette 7.9   Granville Island

Granville Island houses a public market oasis in the heart of downtown Vancouver. Lying beneath Granville Bridge, the island has turned into a key gathering spot for both locals and tourists. Granville Island was a nondescript sandbar that was converted in 1915 into an industrial core for the growing forest industry. In the 1970s, it began to change into an upscale residential and specialized commercial area centred on the Granville Island Public Market. Other enterprises have widened its appeal—artists' studios and shops, a wide variety of restaurants, and features like the Kids Market, Maritime Market, and Coast Salish Houseposts, a joint endeavour between the Emily Carr College and First Nations. Granville Island is unique to Vancouver and has added another dimension to the wide-ranging tourist attractions in the Greater Vancouver area.

Gunter Marx/Alamy

Photo 7.13

Lying on the north shore of Burrard Inlet, North Vancouver serves as a port where grain, sulphur, and other commodities are shipped to Pacific Rim countries. The freighter is loading sulphur. The North Shore Mountains are in the background.

Sergeibach/Dreamstime/GetStock.com

Photo 7.14

The Lower Mainland is a popular destination for Chinese immigrants and the urban landscape contains Chinese-Canadian 'ethnoburbs' in Vancouver, Richmond, and other cities in the area.

Vancouver is also a transshipment point for resource products from the interior of British Columbia, Alberta, Saskatchewan, and Manitoba; a service centre; and a tourist destination. Unit trains carry coal, grain, lumber, and potash to Vancouver for export. Head offices of private companies (especially fishing, forestry, and mining companies) and federal and provincial public offices are located in Vancouver. While Vancouver (known in 1867 as Gastown) began as a sawmill centre, these air-polluting mills relocated to other centres or were dismantled. False Creek, in downtown Vancouver, had been the prime site for processing logs, but in 1986 False Creek became the site of Expo '86 and then the site of an upscale residential and farmers' market complex known as Granville Island (Vignette 7.9).

The Centralist/Decentralist Faultline in BC

The root of the centralist/decentralist fault-line in BC lies in the province's perceived lack of power within Confederation. On the one hand, BC's population continues to grow rapidly. Yet, this demographic fact has not been fully recognized by political representation in the House of Commons. With BC's steady population increase over the years, its number of members of the House of Commons grew to 36 of the 308 members in the 2008 election. But is that a fair number? According to the 2006 population estimates by Statistics Canada, British Columbia had 13.0 per cent of the national population and therefore should have 40 members. This discrepancy poses an irritant in BC's relations with Ottawa, and reinforces other grievances that tend to increase the centralist/decentralist faultline. Only time will tell whether the addition of seven new seats in the House of Commons (as promised in legislation introduced in 2010), the positive outcomes of the 2010 Winter Olympics, forward movement on land-claim negotiations, and improved markets for British Columbia's natural resources will ease centralist/decentralist tensions. One geographic fact remains irrefutable, however: the **grooves of geography**, in this case the north–south mountain ranges of BC, tend to align this region with the adjacent US Northwest and, despite modern transportation methods, set it apart from the rest of Canada.

Table 3.6, 'Members of the House of Commons by Geographic Region, 2008', page 95, shows that only British Columbia and Ontario are under-represented in Canada's Parliament.

Think About It

Why does Atlantic Canada have an MP in the House of Commons for every 71,000 people, while the proportion for British Columbia is one MP for every 114,000 persons?

SUMMARY

British Columbia, economically and demographically, is a powerful region within Canada. As one of six regions, BC's population continues to increase well above the national rate while its export-oriented economy recently lost some of its momentum, especially as trade with the United States and, to a lesser degree, with Pacific Rim countries declined with the global economic downturn. After years of unprecedented economic growth, British Columbia's economy hit a rough patch. Its first rough patch was the sudden drop in forest exports to the United States when housing construction in the US collapsed. The second stumble was due to the broader global economic crisis. The 2010 Olympics may have signalled better times, but the true test will be a worldwide economic recovery. One reason for optimism is the heavy investment in transportation infrastructure—the superhighway corridor to Vancouver and the expansion of port facilities at Vancouver and Prince Rupert—in the last decade. The expansion of port facilities at Prince Rupert alone has provided another and shorter outlet for natural resources and agricultural products from Western Canada to reach markets in Asia. In addition, the prospects of an energy corridor across the Cordillera to Kitimat and then by ships to Asian countries would greatly strengthen trade. Once the global economy regains its steam, exports to the United States and Pacific Rim countries are anticipated to return to previous or even higher levels. Finally, BC's natural setting makes it a world-class tourist destination. The main challenge facing this upward transitional region is to foster further development of its high-technology industry and to apply some of that technology to the processing of its natural resources, especially its forestry products.

CHALLENGE QUESTIONS

1. Is the appeal of Cascadia rooted in the centralist/decentralist faultline?

2. What new technology allowed natural gas to be extracted from the Horn River shale deposit in the Interior Plains of British Columbia?

3. Is the Aboriginal fishery a treaty right or a temporary political compromise?

4. Why do forest firms prefer clear-cut logging over selective cutting?

5. Why does the US want to limit BC's ability to ship lumber to US markets while the US places no such limits on BC's export of electricity to the Pacific Northwest?

FURTHER READING

Molloy, Tom. 2000. *The World Is Our Witness: The Historic Journey of the Nisga'a into Canada.* Calgary: Fifth House.

On 11 May 2000, the Nisga'a treaty passed into law, marking a historic agreement between this small group of Indians and the rest of Canadian society. In *The World Is Our Witness*, Molloy, who was the chief federal negotiator, describes how this agreement ends the centuries-old colonization by the British and, later, Canadians of the Nisga'a lands and people. Furthermore, this treaty begins the search for a place within Canadian society by the Nisga'a. For the other Indian tribes of British Columbia, this treaty has far-reaching implications. Molloy not only provides an insider's view of the struggle to achieve an agreement, but also explains its significance to Canadians. The Nisga'a treaty represents a compromise between the Nisga'a and other Canadians on how to share the lands and resources found in the traditionally occupied lands of the Nisga'a. Originally, there was no sharing. The early English settlers who established British colonies along the east coast of the United States nearly 400 years ago claimed the land by the right of 'discovery' and acknowledged no Aboriginal right to land. Recognition of Aboriginal rights began with the Royal Proclamation of King George III. In 1763, the Royal Proclamation declared that lands to the west of the Appalachian Mountains were Indian lands. While these lands were lost to the Americans in 1783, the legal precedent laid the foundation for treaty-making across British North America and, later, Canada. The Nisga'a treaty is not only one of the most recent agreements but it has expanded the scope of treaties into the area of resource-sharing.

8

WESTERN CANADA

INTRODUCTION

Western Canada, rich in natural resources, lies in the heart of North America. Geography, in the form of a dry continental climate and distance to ocean ports, has played a dominating role in shaping Western Canada's economy and society. As an exporting region, Western Canada trades mainly with the United States, Pacific Rim countries, and the European Union. Prices for its resource products are critical for companies and farmers. In the last decade, for example, low prices for forest products have stalled this industry while high prices for energy, potash, and canola have had the opposite effect. Canola now exceeds spring wheat in returns to farmers and matches it in sown acres.

Western Canada is not a homogeneous region. Its physical geography reveals two distinct sub-regions: (1) a northern resource hinterland found mainly in the Canadian Shield where the majority of the population is of Aboriginal descent; (2) a more densely populated agricultural/industrial core associated with the Interior Plains. The political geography of this region also is varied, as it consists of the provinces of Alberta, Saskatchewan, and Manitoba.

Western Canada has experienced both economic and population growth in recent years, with Alberta leading the way. This growth is reflected in the major cities—Calgary, Edmonton, Winnipeg, Saskatoon, and Regina. In sharp contrast, rural areas have seen a huge population loss, the disappearance of rural villages, and fewer but larger farms. The 'draining' of rural population is one element associated with ongoing transformation of the region's agricultural economy. We examine this transformation more closely in this chapter's *Key Topic*, 'Agriculture'.

CHAPTER OVERVIEW

The main themes in this chapter are:

- Western Canada's physical and historical geography.
- The basic characteristics of the population and economy of the region.
- Western Canada's population and economy within the context of the region's physical setting and dry continental climate.
- Its increasing economic and demographic power within Canada.
- The transition in Western Canada's agricultural, resource, and manufacturing sectors.

Fields of canola, rye, and wheat in Manitoba. Photo: Mike Grandmaison/All Canada Photos.

Western Canada within Canada

Situated in the vast western interior of North America (Figure 8.1), two physiographic regions dominate the landscape—the Interior Plains where agriculture takes place, and the Canadian Shield, which is noted for mining and logging. Western Canada is composed of three provinces, Alberta, Manitoba, and Saskatchewan. Its geographic location poses two formidable challenges. First, water is a scarce resource and the region's dry continental climate (Vignette 8.1) sometimes results in a water deficit, which, in turn, leads to crop failure. Second, distance to world markets has proven to be a major stumbling block to the region's economic development because of its need to export its surplus energy, food, and mineral resources. Three renewable

resources—the fertile soils of the Canadian Prairies, the boreal forest of its northern lands, and its rivers—offer the prospect of a sustainable economy, if managed properly. As well, the huge deposits of oil sands and potash, while non-renewable, have over a hundred years of reserves and thus might be expected to provide support for the regional economy over that period of time, assuming that commodity prices remain high and unforeseen geopolitical events don't alter the current situation.

In terms of economic output, Western Canada falls behind Ontario and Québec, but it is closing the gap. Overall, Western Canada's economy accounts for 22.6 per cent of Canada's GDP (Figure 8.2). An important measure of its economic well-being is Western Canada's low unemployment rate. In 2006, Western Canada had, at 4.0 per cent, the lowest unemployment rate of the six regions, well below the national average

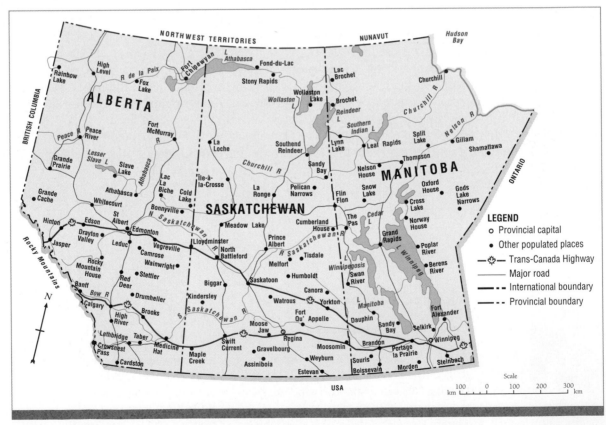

Figure 8.1   Western Canada.

Source: Atlas of Canada, 2003, 'Prairie Provinces', at: <atlas.nrcan.gc.ca/site/english/maps/reference/provinceterritories/prairie_provinces>.

Vignette 8.1   Water Deficit and Evapotranspiration

A dry, continental climate prevails over the interior of Western Canada. Due to a combination of low annual precipitation and high summer temperatures, this zone has a 'water deficit'. The low precipitation is primarily the result of the eastward-moving Pacific air masses losing most of their moisture as they rise over the Cordillera's mountain chains. By the time they descend the eastern slopes of the Rocky Mountains, these moist air masses have turned into dry ones. As a result, the Prairies receive little rain or snow from the Pacific Ocean. This water deficit is often measured in terms of potential **evapotranspiration**, which is the amount of water vapour that can potentially be released from an area of the earth's surface through evaporation and transpiration (the loss of moisture through the leaves of plants). There is a water deficit if the evapotranspiration rate is greater than the average annual precipitation. For example, most of the grassland natural vegetation zone receives less than 400 mm annually, while its evapotranspiration rate exceeds 500 mm. The difference indicates a water deficit of over 100 mm. If a water deficit occurs in a dry year, plants can use water found in the soil from previous years when there was a water surplus. Therefore, soil moisture provides a reserve that plants can draw on, but eventually this reserve is exhausted and plants are damaged or die from a lack of water. The maps of precipitation and natural vegetation in Chapter 2 (Figures 2.5 and 2.7) indicate the geographic location of this water deficit area. If there is a heavy winter snowfall, then the spring melt can add to the soil moisture, thus offsetting the water deficit.

of 6.0 per cent (Figure 8.2), and even following the severe economic downturn beginning in 2008, the three provinces of Western Canada had the lowest unemployment rates across Canada: Saskatchewan, 4.8 per cent; Manitoba, 5.2 per cent; Alberta, 6.6 per cent. Led by Alberta, economic gains in this geographic region are outstripping those of Ontario and Québec, and percentage population growth in Western Canada has been greater than that in Québec over the past decade. Even after the global economic crisis, Western Canada outpaced the other regions of Canada in terms of economic and demographic performance.

Table 5.1, 'Before and After the Crash: Unemployment Rates by Province, 2007 and 2009', page 195, shows how well the provinces in Western Canada has fared relative to the other provinces.

In 2006, the population of Western Canada totalled over 5.4 million. The vast majority live in the regional core, an area of major cities, towns, and villages located in the southern half of the region. Agricultural, industrial, and service activities are well established within this population core. Calgary, Edmonton, Winnipeg, Saskatoon, and

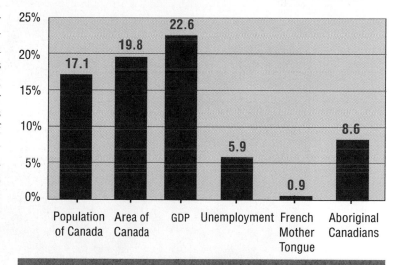

Figure 8.2   Western Canada vital statistics.
Although Western Canada covers nearly 20 per cent of Canada, its population forms only 17 per cent. Yet its economic strength as measured by GDP, at nearly 23 per cent, is greater than that for Québec (19.4 per cent). Significantly, Western Canada—except for the Territorial North—has the highest percentage of Aboriginal Canadians.
Note: Unemployment percentage is for 2009; demographic data are based on the 2006 census.
Sources: Tables 1.1, 1.2, 5.1.

Regina are the leading cities. In the northern half of Western Canada is the regional hinterland, where less than 5 per cent of the region's population lives. While most reside in resource towns, such as Fort McMurray and Thompson, nearly a hundred small

Think About It

Given the past boom-and-bust economic pattern associated with agriculture and resource development, is the author overly optimistic about Western Canada's economic prospects?

Aboriginal settlements are scattered across this boreal landscape.

Western Canada's Physical Geography

Western Canada has two major physiographic regions—the Interior Plains and the Canadian Shield—as well as small portions of two others—the Hudson Bay Lowland and the Cordillera (Figure 2.1). A thin portion of the Cordillera, the Rocky Mountains, forms a natural and political border between southern Alberta and British Columbia. Each physiographic region has a particular set of geological conditions, physical landscapes, and natural resources.

 The four physiographic regions found in Western Canada are discussed in Chapter 2.

The tiny section of the Cordillera, along the eastern flank of the Rocky Mountains in Alberta, provides logging and mining opportunities for Western Canada. However, the main attraction of this slice of the Cordillera is its spectacular mountain landscape. This region has two internationally acclaimed parks, Banff National Park (photo 8.1) and Jasper National Park, which attract visitors from around the world. Calgarians are especially fortunate in having easy access to the Kananaskis Country Provincial Park, where a number of mountain recreational activities—camping, hiking, and skiing—are available.

In the Interior Plains region the sedimentary rocks contain valuable deposits of fossil fuels. By value, the four leading mineral resources are oil, gas, coal, and potash. Most petroleum production occurs in a geological structure known as the **Western Sedimentary Basin**,

Unclejay/Dreamstime.com/GetStock.com

Photo 8.1

Lake Louise, one of Banff National Park's most stunning natural features, lies within the alpine recreation zone of Calgary. The lake's famous turquoise colour is caused by fine rock particles, called 'glacial flour', that are contained in the stream water from alpine glaciers. The steep-sided U-shaped valley behind the lake, known as a glacial trough, indicates the erosional effect of an alpine glacier.

which underlies most of Alberta and portions of British Columbia, Saskatchewan, and Manitoba. However, not all mineral extraction takes place deep in these sedimentary rocks. A few deposits are exposed at the surface of the ground. In southeastern Saskatchewan, brown coal is extracted through open-pit mining and then burned to produce thermal electricity. In northeastern Alberta, the huge petroleum reserves in the **Athabasca tar sands** are exploited by surface mining techniques, such as hydrotransport in which the oil sands are mixed with extremely hot water and transported to an **upgrader** plant by pipeline.

In the Canadian Shield, which extends into much of Manitoba and Saskatchewan and a small portion of Alberta, rocky terrain makes cultivation virtually impossible. While forestry does take place along the southern edge of the Canadian Shield, particularly in southeastern Manitoba, large-scale commercial enterprises are generally limited to mining and the production of hydroelectricity. In northern Saskatchewan, uranium companies produce most of Canada's uranium from open-pit and underground mines. Large-scale hydroelectric dams and generators are located on Manitoba's northern rivers, particularly the Nelson River.

Western Canada's continental climate is a dominating component of the region's physical geography. Far from moderating ocean influences, this climate is characterized by cold, dry winters and hot, dry summers. The resulting range of temperatures is extreme—from lows of −30°C in January to +30°C in July. During the winter, Arctic air masses often dominate weather conditions in the Prairies, placing the region in an Arctic 'deep freeze' (Figure 2.3). The hot, dry summer weather, on the other hand, results from the northward migration of hot, dry air masses from the Southwest US (Table 2.4).

Annual precipitation, whether in the form of snow or rain, is among the lowest in all regions except the Territorial North (Figure 2.6). The region is dry for two main reasons. First, distance from the Pacific Ocean reduces the opportunity of moist Pacific air masses to reach Western Canada. Second, **orographic uplift** of these Pacific air masses over the Rocky Mountains causes them to lose most of their moisture, leaving little precipitation for Western Canada. A combination of strong winds and sub-zero temperatures can produce blizzard-like weather. In southern Alberta, strong winds that become warm and dry as they flow down a mountain slope are known as **chinooks**. On the other hand, the **Alberta clippers** with their strong, frigid winds produce true blizzard conditions, due to severe blowing and drifting snow.

Precipitation increases from west to east. Calgary receives 413 mm annually while Winnipeg gets 514 mm on average per year. Part of the explanation is that southern Manitoba (including Winnipeg) receives summer rainfall from the moist Gulf of Mexico air masses. As a result, southern Manitoba is rarely troubled by droughty conditions compared to the remaining prairie lands in southern Saskatchewan and Alberta. The driest lands are found in **Palliser's Triangle** where less than 400 mm falls each year (Vignette 8.2).

Vignette 8.2　Palliser's Triangle

In 1857, the Palliser Expedition set out from England to assess the potential of Western Canada for settlement. Well documented by Professor Spry (1963), this expedition spanned three years (1857–60). In his report to the British government, John Palliser identified two natural zones in the Canadian Prairies. The first zone was described as a sub-humid area of tall grasses, while the second, located further south, was described as a semi-arid area with short-grass vegetation. Palliser considered the area of tall grasses to be suitable for agricultural settlement. He named this area the Fertile Belt. In Manitoba this belt is south of the Canadian Shield and stretches to the border with the United States. Palliser believed that the semi-arid zone, located in southern Alberta and Saskatchewan, was a northern extension of the Great American Desert. His belief was reinforced by the Great Sand Hills (Vignette 8.8). According to Palliser, these semi-arid lands were unsuitable for agricultural settlement. The area described by Captain Palliser became known as Palliser's Triangle—its area overlaps with, but is slightly larger than, the agriculture zone known as the Dry Belt (see Figure 8.4). Homesteaders called these lands 'heartbreak territory' and most abandoned their attempts at farming because of the frequency of drought-induced crop failures. David Jones's *Empire of Dust* (1987) captures the settling and abandonment of homesteads in the 1930s while Arthur Kroeger's *Hard Passage* (2007) contains a personal recollection of his family's struggle to homestead in Palliser's Triangle and their eventual defeat.

In addition to this spatial variation, precipitation varies from year to year, causing so-called wet and dry years. This annual variation has its greatest effect in the dry lands of Palliser's Triangle. For instance, in the 1930s a series of dry years resulted in the disastrous Dust Bowl that drove thousands of homesteaders off the land.

In Western Canada the natural vegetation and the soil, and therefore the success of commercial crops, are largely determined by the evapotranspiration rate, which is a combination of precipitation and temperature (Vignette 8.1). The evapotranspiration rate varies within Western Canada and therefore creates differences in the region's soil and vegetation. In the northern parts of Western Canada, in the Canadian Shield, cooler weather leads to a lower evaporation rate, resulting in sufficient moisture for tree growth. In this area the boreal forest grows in a podzolic soil (Figures 2.7 and 2.8). The only exception is in the Peace River country, a gently rolling plain where degraded black chernozemic soils are found beneath an aspen parkland. Further south, where the evaporation rate is much higher, the podzolic soil gives way to chernozemic soils, which are more favourable for agriculture.

Indeed, the soil and natural vegetation of the southern part of Western Canada have had a significant impact on the region's role as an agricultural heartland. This southern portion, in the Interior Plains, has two natural vegetation zones (parkland and grassland) and three chernozemic soil zones (black, dark brown, and brown) (Figures 2.8 and 8.4). Just south of the boreal forest, podzolic soil gives way to the black chernozemic soil that supports parkland vegetation. This parkland is a transition zone between the boreal forest and the grassland natural vegetation zone, which is located a little further south and is supported by dark brown and brown chernozemic soils (photo 8.2). Within the grassland, the evaporation rate increases towards the American border, causing this mid-latitude grassland to change from tall grass to short grass as one moves southward. Tall-grass natural vegetation occurs in a wide arc from southwestern

Barrett & MacKay Photography, Inc.

Photo 8.2

Within the Fertile Belt, rich, dark brown chernozemic soils have formed under a tall-grass natural vegetation and they now provide farmers with one of the most fertile soils in Canada. In this photograph taken near Bigger, Saskatchewan, the practice of crop rotation is illustrated by the three types of land use. The dark brown field represents summer fallow where chemicals are used to control weeds, known as **chem fallow**; the yellow field contains a spring wheat crop; and the green field consists of grassland used for pasture.

Manitoba to Saskatoon to Edmonton. Here there are dark-brown chernozemic soils, except in southern Manitoba, where black soils occur because of higher annual precipitation. This tall-grass natural vegetation zone and the parkland to its north make up the area known as the Fertile Belt, where farming has traditionally been more successful. South of this fertile arc is the area of short-grass natural vegetation and brown chernozemic soils known as the Dry Belt (Vignette 8.2). This semi-arid area has presented the highest risk for grain farming, so much of the land is now worked in fallow rotation (a crop is planted every second year) or used for pasture.

Environmental Challenges

Alberta is home to one of the world's largest deposits of oil sands. While Alberta is benefiting from the extraction of these oil sands, this mining operation poses three major environmental challenges to Alberta, Canada, and the world. First, the release of greenhouse gases to the atmosphere is among the largest in Canada. Second, open-pit mining has created a scarred industrial landscape and the reclamation process, if successful, will take time and money. Third, separating the oil from the bitumen requires large amounts of water and the resulting waste product is deposited into large toxic tailing ponds. While each challenge has serious consequences for Alberta and Canada, tailing ponds deserve special attention.

Alberta's oil sands industry produces 1.8 billion litres of toxic water each day. The problem facing industry is what to do with this vast quantity of non-renewable water? Industry's solution is tailing ponds. The problems with tailing ponds are threefold. First, since these toxic waters cannot be released into the local rivers and lakes, they must be stored in large ponds for an indefinite time. Second, the amount of toxic waters is increasing every day, thus either increasing the size of existing tailing ponds or creating new ones. Third, leakage from these ponds has a negative effect on the landscape, groundwater, and surface waters, including the Athabasca River. Native communities downstream from the tar sands development, notably at Fort Chipewyan, have experienced health consequences, including unusually high rates of cancer. Since 1967 when tailing ponds were first established, no pond has been reclaimed. Over 40 years later, the geographic extent of these ponds is enormous and they pose an obvious example of 'dirty oil'. In April 2008, international attention was focused on Syncrude's lake-size tailing pond where a flock of migratory birds landed. Over 400 ducks died in these oily toxic waters. Environmentalists seized on this tragedy as an example of the impact of 'dirty oil' on wildlife. Perhaps because of this highly publicized disaster, the Energy Resources Conservation Board decided to regulate tailing ponds and this provincial agency demanded tailing pond plans from the six companies. The goal is to shorten the time required to reclaim the pond, i.e., contour and cover the site with soil and then plant trees, bushes, and native grasses. However, these requirements are for **tailings** discharged into ponds from 2010 onward but do not apply to ponds constructed before 2010.

Reclamation of the toxic tailing ponds is pushing the oil companies into uncharted waters. While the industry has a number of ideas, they are unproven and, so far, unapproved by the Energy Resources Conservation Board. Pilot projects are set to begin in 2010. One pilot project, called End Pit Lakes, involves burying the toxic sludge into deep pits and covering it with layers of earth, topped off with fresh water that will form a lake.

Western Canada faces other environmental challenges. One is the constant threat of spring floods in the Red River Valley. Efforts to divert the flood waters from reaching Winnipeg by the Red River Floodway have been successful, but other communities and farmland in the valley remain vulnerable. The extent of the April 2009 Red River flood is shown in photo 2.11.

In Chapter 2, 'Extreme Weather Events', page 56, the Red River Valley, spring floods, and the diversion of water around Winnipeg by the Red River Floodway are discussed.

Think About It

Oil from Alberta is a critical element in US geopolitical strategy, which calls for reducing oil imports from the Middle East. Yet, US environmentalists do not want imports of 'dirty oil' from Alberta. Which group do you think will win the day?

Another challenge is the need to remove the radioactive wastes from abandoned uranium mines near Lake Athabasca before the radioactive waste seeps into the lake and eventually spreads throughout the Mackenzie River system. Companies did not remove these radioactive wastes when they ceased operations for two simple reasons: (1) such removal would be very expensive; (2) there was no requirement for the companies to do so. As a result, when small mining operations closed in the 1960s, no effort was made to reclaim the mine sites. The purpose of these mines, which are often described as 'Cold War legacy mines', was to supply the American government with much-needed uranium for its weapons production in the post–World War II period. For years, these dangerous wastes were recognized as a serious hazard to the environment but a decommissioning agreement between the federal and Saskatchewan governments was stalled over the sharing of the cost of a cleanup. In fact, estimating the cost of cleanup was difficult and actual costs may exceed the estimate. In 2007, the federal and provincial governments announced an agreement to share equally the estimated cost of $24.6 million. The final step before work can begin is the approval of an environmental assessment of the proposed cleanup plan. By January 2010, assessments of the various mine sites had been completed and the next step—the actual cleanup—was about to begin.

Intensive cattle and hog facilities pose a serious challenge to the environment. Most large, high-technology operations with thousands of animals are located in southern Alberta. Large-scale and highly specialized livestock factory-like operations are part of the trend towards an industrial agricultural economy in North America. The Lethbridge area contains the heavest concentration of beef-cattle feedlots and hog barns. Waste

Barrett & MacKay Photography, Inc.

Photo 8.3

Pincher Creek, located near the foothills of the Rockies in southwest Alberta, is famous for its livestock industry. More recently, Pincher Creek has become the site of wind energy production, helping to make Western Canada the leading region for wind-produced electricity in Canada. Powerful chinook winds from the Rocky Mountains make southwest Alberta a particularly attractive site for wind turbines.

from cattle and hogs provides a serious threat to water resources. The heavy concentration of manure in a small area poses a danger to surface water through runoff and to groundwater through infiltration. Untreated manure may also contain bacteria such as E. coli and salmonella, viruses, and parasites. The potential of fecal bacteria seeping into the limited water supplies in this dry zone of Western Canada is great and yet provincial governments have been slow to react, preferring to encourage agricultural diversification. Several American states, Québec, and Taiwan have refused to permit additional large-scale hog operations because existing hog farms have caused the contamination of some water resources. Public awareness and pressure from environmental organizations have prompted governments and industry to address the environmental risks associated with intensive livestock operations (Price, 2003: 38). Even so, the so-called 'beneficial management practices' designed to minimize the impacts of animal waste products on water quality are just in the research stage. Even if such research is successful, the likelihood of the livestock industry causing a serious health problem from the contamination of a local water supply along the lines of Walkerton, Ontario, or North Battleford, Saskatchewan, is too great not to invoke the precautionary principle, whereby risk to the environment mandates that certain practices should not be pursued.

Western Canada's Historical Geography

Western Canada's history began long before the three Prairie provinces became part of Canada. In fact, Western Canada's recorded history goes back to the fur-trading days. Beginning in 1670, the Hudson's Bay Company (HBC) administered for 200 years much of Canada's western interior. This area was part of Rupert's Land (all the land draining into Hudson Bay). In 1821, when the company merged with its rival, the North West Company, the HBC acquired control over more land, known as the North-Western Territory (lands draining into the Arctic Ocean). Before 1870, when Canada

was ceded these lands by the British government (Figure 3.4), the HBC used this vast territory exclusively for the fur trade.

The land in Western Canada began to be used for purposes other than fur trading at the beginning of the nineteenth century. In 1810, Lord Selkirk, a Scots nobleman who was concerned with the plight of poor Scottish crofters (tenants) evicted from their small holdings, acquired land in the Red River Valley from the Hudson's Bay Company. The first Scottish settlers arrived in 1812 to form an agricultural settlement near Fort Garry, the principal HBC trading post in the region. This became known as the Red River Settlement. Selkirk's settlers, however, faced an unfamiliar and harsh environment and had great trouble establishing an agricultural colony. Over the years, many gave up and left for Upper Canada and the United States.

At the same time, many former officers and servants of the Hudson's Bay Company, along with their Indian wives and children, settled at Fort Garry. In addition to these English-speaking people were the French-speaking Métis who had worked for the North West Company. Because the Métis were Catholic and spoke French, they formed a separate cultural group within the settlement. After the consolidation of the Hudson's Bay Company and the North West Company, many who worked for the North West Company were no longer employed by the new company. Many Métis, particularly those who were French-speaking, settled at Red River, where they turned their attention to subsistence farming, freighting, and buffalo hunting.

Desert or Arable Land?

By the middle of the nineteenth century, little was known about the suitability of the Canadian Prairies for settlement. What was known came from fur traders and explorers. South of the border, American explorers had dubbed Montana part of 'the Great American Desert'. Did that desert extend into Interior Plains of British North America?

The British government was concerned about the political viability of these lands managed by the HBC. Without settlement, the

British hold on these lands was tenuous, as the loss of the Oregon Territory in 1846 demonstrated. Was history to repeat itself? American settlers had begun to occupy land in the Dakota Territory (land west of the Mississippi River in what is now North Dakota). By 1854, a railway stretched across the United States from New York to St Paul on the Mississippi River and, after 1863, American settlers began to occupy the upper reaches of the Red River Valley. How long would it be before American settlers began to look northward to the unoccupied lands of British North America?

The events that led to the loss of the Oregon Territory in 1846 are considered in Chapter 3, 'National Boundaries', page 85, and Chapter 7, 'British Columbia's Historical Geography', page 282.

Action was needed. In 1857, the British government and the Royal Geographical Society sponsored an expedition into the Canadian West. Their central task was to determine the suitability of the Canadian West for agricultural settlement. John Palliser, an explorer, led the British North American Exploring Expedition. After his party arrived from England, they quickly travelled by rail from New York to St Paul and then by steamboat to the Red River Colony. They travelled by horseback from Fort Garry across the Canadian Prairies to the Rocky Mountains. Palliser reported that fertile land in much of the western interior existed, but that the land in southern Alberta and Saskatchewan near the border with Montana was 'treeless' and therefore far too dry for farming (Vignette 8.2). Palliser believed the **Great American Desert** that American explorers had labelled for the Great Plains extended into the grasslands of southern Alberta and Saskatchewan. Named after him, this semi-arid area is now known as Palliser's Triangle. Another expedition in 1858, organized in Canada West (Ontario) and led by Henry Hind, a geologist and naturalist, confirmed that the parkland (the natural vegetation zone between the grasslands and the boreal forest) offered the best land for agricultural settlement.

During the negotiations with Britain over Confederation, the subject of the annexation of Rupert's Land into the Dominion of Canada arose, and provision was made in the British North America Act for its admission into Canada. In 1869, the Hudson's Bay Company signed the deed of transfer, surrendering to Great Britain its chartered territory for £300,000—with the exception of the lands surrounding its posts and about 1,133,160 ha of farmland. In 1870, Great Britain transferred Rupert's Land to Canada.

In 1869, Canada's new government sent surveyors into the Red River Settlement to prepare a land registry system for the expected influx of settlers. In this **Dominion Land Survey**, the surveyors employed a rectangular grid system known as a township survey. A township consisted of 93 km², or 36 sections. Each section was subdivided into four quarter sections of 65 ha each. Each **homesteader** would receive a quarter section of land and would be required to till the land and build a house on that section. The opening of Western Canada for agricultural settlement officially began in 1870 and continued until 1914, when World War I halted the influx of European immigrants into Western Canada. Following World War I, veterans were encouraged to establish homesteads, especially in the Peace River country, where some arable land remained.

The key attraction for homesteaders in Western Canada was 'free' land, but many settlers, especially those familiar with the mild climate and treed landscape of Western Europe, were ill-prepared for the harsh continental climate and the absence of forests. Unlike the agricultural lands of southern Ontario and Québec, arable land in Western Canada lies much further north. For example, Central Canada's arable lands fall within 42° N and 47° N while Western Canada's are between 49° N and 56° N. Hence, a shorter growing season in the Canadian Prairies affects selection of crops and chances of a successful harvest (Akinremi et al., 2001). Added to the disadvantage of a short growing season were grasshoppers, hail, and untimely frosts, and homesteaders quickly learned that a July bumper crop can turn into a September crop failure. From these experiences of farming in such a physical environment, farmers coined the term, 'Next Year Country' (Vignette 8.3).

But What About the Original Inhabitants?

In the nineteenth century, Plains Indians and Métis formed the population of Western Canada. Both depended on the buffalo and the fur trade. In the last half of the century, commercial hunting of buffalo in the United States and Canada for the buffalo robe trade marked the end of the great buffalo herds. With the virtual extinction of the buffalo, the Plains Indians (Sarcee, Blood, Peigan, Stoney, Plains Cree, Nakota, Lakota, Blackfoot, and Saulteaux) could no longer support themselves. The transfer of Hudson's Bay lands to Canada coupled with Ottawa's plan to settle the arable lands of Western Canada meant that the only option for Plains Indians was to sign treaties and live on reserves (Figure 3.10).[1] The Métis were also confronted by the impact of these historic changes on their way of life. Yet, because they were not semi-nomadic like the Plains Indians and, instead, formed an organized settlement at Red River that was remote from Canada, the Métis were more able to resist Canada's desire to settle the Prairies with farmers.

The rebellions of 1869–70 and 1885 are examined in Chapter 3; see page 107 and page 110.

In 1867, the population at Red River was nearly 12,000, mostly Métis. The arrival of land surveyors and settlers led the Métis, under the leadership of Louis Riel, to mount the Red River Rebellion in 1869. The Métis wanted to negotiate the terms of entry into Canada from a position of strength—that is, as a government—and they obtained major concessions from Ottawa: guaranteed ownership of land, recognition of the French language, and permission to maintain Roman Catholic schools. In 1870, the fur-trading district of Assiniboia became the province of Manitoba. But the rebels' victory was hollow. First, Ottawa sent troops to exert Canada's control over the new province, forcing Riel and

---

Vignette 8.3 'Next Year Country'

The dry continental climate in Western Canada makes farming a risky business. Prairie farmers often describe the land as 'Next Year Country'. The meaning behind these words is simple: Our crops did poorly this year, but we hope that they will do better next year. Often the reason for low yields is insufficient precipitation during the growing period. However, farmers face a whole range of natural hazards: summer frosts, which occur when a cold air mass slides unimpeded from the Canadian Arctic to Western Canada; hail and early snowfall; pests, such as grasshoppers; and diseases, such as stem rust. All of these hazards can have devastating effects—from delaying a harvest or lowering the grade of wheat, to reducing a yield or destroying an entire crop.

---

his followers to flee. Second, settlers began to pour into Manitoba, changing the demographic balance of power and overwhelming the Métis community.

Many Métis left the colony to search for a new place to settle in the Canadian West. Such a place was Batoche, just north of the site where Saskatoon now is situated. Within 15 years, however, settlers would again encroach on the Métis agricultural settlement. In 1885, as before, the Métis, led by Louis Riel, rebelled. This time, the Canadian militia defeated the Métis at the Battle of Batoche and Riel was captured, found guilty of treason, and hanged.

The experiences of Indian tribes during the early period of western settlement were somewhat different. Tribes such as the Blackfoot had roamed across the Canadian Prairies and the northern Great Plains of the United States long before the arrival of European explorers, fur traders, and settlers. The tribes were semi-nomadic and hunted buffalo. By the 1870s, the buffalo had virtually disappeared from the Prairies, leaving the Plains Indians destitute. They had little choice but to sign treaties with the federal government. Between 1873 and 1876, all the tribes (except for three Cree chiefs—Big Bear, Little Pine, and Lucky Man—and their followers) signed numbered treaties in exchange for reserves, cash gratuities, annual payments in perpetuity, the promise of educational and agricultural assistance, and the right to hunt and fish on Crown land until such land was

---

Think About It

Do you think that Palliser's original assessment of the semi-arid areas of Western Canada as being unsuitable for agricultural settlement was correct?

Think About It

If the Canadian Pacific Railway had existed in 1869, do you think that negotiations between Riel's followers and the government of Canada would have taken place?

Library and Archives Canada

Photo 8.4

The Métis search for a place within Confederation began with armed resistance in 1869–70 and 1885. Both rebellions were led by Louis Riel. After losing the Battle of Batoche, Riel surrendered to the Canadian forces. He was tried and convicted of high treason, and on 16 November 1885 Riel was hanged as a traitor. He remains a hero to the Métis to this day.

required for other purposes. In 1882, impending starvation for his people also forced Big Bear to accept Treaty No. 6. Over the next few years, however, the Cree sought other concessions from the federal government. When these efforts failed, Cree warriors supported the doomed Métis rebellion in 1885 by attacking several settlements, including Fort Pitt, the Hudson's Bay post on the North Saskatchewan River near the present-day Alberta–Saskatchewan border.[2]

Treaties with Ottawa offered the Indians prospects for survival and time to find a place in a new economy, but the treaties also

made them wards of the Crown. Living on reserves, Indians were isolated from the evolving Canadian society and became increasingly dependent on the federal government. Further north, the Woodland Cree and Dene (Chipewyan) tribes who lived in the boreal forest were not as affected by the encroachment of western settlers. Although they, too, signed treaties, these northern Indians continued their migratory hunting and trapping lifestyle well into the next century. In the 1950s, their dependency on Ottawa grew with the demise of the fur trade and their subsequent relocation to settlements.[3]

Canadian Pacific Railway

Once treaties ensured the peaceful availability of land for homesteading, the next steps were a land survey in the form of townships and a transcontinental railway. In fact, Prime Minister Macdonald's vision of Canada extending from the Atlantic to the Pacific hinged on a transcontinental railway. The third American transcontinental railway, the Northern Pacific Railway, was just south of the forty-ninth parallel. Without a Canadian counterpart, Ottawa feared that the West would be lost to the Americans. Even if Canada could retain its western territories in the absence of a Canadian transcontinental railway, the north–south transportation pull exerted by the American railways would prevent Ontario's fledgling industrial core from reaching the market in Western Canada, and western settlers would be unable to ship their products to eastern markets. For instance, in 1870, the new province of Manitoba was linked by steamboat to the rail centre of Fargo, North Dakota.

British and Canadian companies were not interested in a risky railway construction project across Canada unless they could obtain substantial financial assistance from Ottawa. Two reasons accounted for their lack of interest: the Canadian Shield and the Cordillera were two formidable (and therefore costly) barriers to overcome in building a railroad. Indeed, physical geography posed a much greater challenge to Canadian railway builders than to their American counterparts. In 1881, Ottawa announced generous terms:

the Canadian Pacific Railway Company was awarded a charter, whereby the company received $25 million from the federal government, 1,000 km of existing railway lines in eastern Canada owned by the federal government, and over 10 million ha of prairie land in alternate square-mile sections on both sides of the railway to a maximum depth of 39 km. The terms were successful—the Canadian Pacific Railway was completed in 1885. As a result, the new Dominion achieved four important nation-forming goals:

- An east–west transportation link united Canada from coast to coast.
- The vast territory west of the Red River Valley was secured for Canada.
- The Canadian Prairies could be settled.
- A rail-based transportation system to eastern ports could be used to ship farm products to the world's major grain market in Great Britain and other European countries.

Settlement of the Land

The settling of Western Canada marks one of the world's great migrations and the transformation of the Prairies into an agricultural resource frontier. Under the Dominion Land Act that Ottawa had passed in 1872, homesteaders were promised 'cheap' land in Manitoba—by building a house and cultivating some of the land, they could obtain 65 ha of land for only $10. Following 1872, an influx of prospective homesteaders began arriving, most coming from Ontario and, to a lesser degree, from the Maritimes, Québec, and the United States. However, settlement did not occur west of the Red River Valley until the Canadian Pacific Railway was completed. Then, homesteaders began to occupy lands in Saskatchewan and Alberta. The first wave of homesteaders came from Ontario, Great Britain, and the United States. The second wave came from continental Europe.

By 1896, the federal government sought to increase immigration by promoting Western Canada in Great Britain and Europe as the last agricultural frontier in North America. The Canadian government initiated an aggressive campaign, administered by Clifford Sifton, Minister of the Interior, to lure more settlers to the Canadian West. Thousands of posters, pamphlets, and advertisements were sent to and distributed in Europe and the United States to promote free homesteads and assisted passages. Prior to 1896, most immigrants came from the British Isles or the United States—these were 'desirable' immigrants. Sifton's campaign, however, cast a wider net to areas of Central and Eastern Europe that were not English-speaking and therefore provided 'less desirable' immigrants. The strategy generated considerable controversy among some English-speaking Canadians who believed in the racial superiority of British people.

Nevertheless, Clifford Sifton's efforts paid off. At the end of the 1880s, the Canadian Prairies had few settlers beyond Manitoba, and most of them had taken land near the Canadian Pacific Railway line. Following the recruitment campaign, a flood of settlers arrived and the land was quickly occupied. Thus began the great migration to Western Canada. After 1896, the majority of settlers—about 2 million—were Central or Eastern Europeans from Germany, Russia, and Ukraine. This large influx of primarily non-English-speaking immigrants led to a quite different cultural makeup in Western Canada from that in Central Canada, where the French and English dominated. Within a remarkably short span of time, cultural and linguistic acculturation had forged a non-British but English-speaking society from the sons and daughters of these immigrants. Some, however, kept separate. For instance, Doukhobor settlers were Russian-speaking peasants whose adherence to communal living made adjustment and acceptance difficult if not impossible. By 1905, Alberta and Saskatchewan had sufficient populations to warrant provincial status. By the outbreak of World War I, the region of Western Canada was settled.

See Chapter 3, 'The Doukhobors', page 112, for an examination of the plight of this pacifist religious sect.

> **Think About It**
>
> Which physiographic regions posed the greatest challenge to John A. Macdonald's dream of a railway stretching to the Pacific Ocean?

Think About It

Do you think that homesteaders from the steppes of the Russian Empire were better prepared for life on the Canadian Prairies than those from Great Britain?

The decision to build the CPR along a southern route (from Winnipeg to Regina to Calgary) meant that much of the land opened to homesteaders was in the driest part of Western Canada.

Life was not easy for homesteaders. Many were ill-prepared for farming, let alone farming in a dry continental environment. Securing supplies of wood and water often posed a problem. Those settlers who could not afford to import lumber were forced to live in sod houses and burn buffalo chips and cow dung for heat. While many members of ethnic and religious groups settled together in the same area, forming communities, isolation still posed a problem for many. The land survey system encouraged a dispersed rural population, with individual farmsteads rather than rural villages. As a result, farm families sometimes did not visit the nearest town or see their neighbours for weeks or even months. Such isolation was particularly hard on farmwives. In spite of these difficulties, the land was settled, towns sprang up, and institutions were created to meet the local and regional needs. In short, a new society was in the making.

By 1921, there were over 250,000 farms in Western Canada. Homesteaders now had to turn to the Peace River country for arable land. The Peace River country, part of the high Alberta Plain, is much further north and its short growing season makes agriculture risky. Even so, this area has a climate and soils that allow for mixed farming (a combination of grain and hay crops with livestock). The section of the Rocky Mountains located to the west of the Peace River country is considerably lower and thus allows more rainfall from Pacific air masses to reach the area. The final settlement of the Peace River country took place after World War I, when returning soldiers were encouraged to settle there.

Geographic Challenges Facing Homesteaders

By the early twentieth century a prairie agricultural economy had replaced the Aboriginal hunting one, but this commercial economy and its transplanted Euro-Canadians faced severe geographic challenges. Few had experienced growing crops at such high latitudes and none had attempted farming in a region undergoing settlement and infrastructure-building. The first challenge was distance to market. While the production of its principal staple, wheat, was a success, getting it to market was a different matter. The high cost of shipping grain long distances by rail to reach ocean ports in Québec and Nova Scotia and then by ocean vessels left little profit for the farmer. An alternative and shorter route to Europe, the Hudson Bay Railway, was constructed with a rail line extending from The Pas, Manitoba, to Churchill, Manitoba (Vignette 8.4; see Figure 8.1). Unfortunately, this route proved ineffective because of the threat of icebergs and the resulting high cost of marine insurance.

The second challenge was the climate. Drought, hail, and frost threatened crops. Farmers in the semi-arid Dry Belt of southern Alberta and Saskatchewan were most vulnerable to dry spells and crop failures. Innovations were sought that would lessen this natural threat. The strategy of dryland farming, where part of the land is left in **summer fallow** each year, reduced the risk of crop failure. In this way, sufficient soil moisture is accumulated over the year, allowing for the seeding of the land every other year.[4] Another challenge is hail, which is associated with severe thunderstorms. Hailstones can reach golf-ball size (and even larger), and can flatten crops within minutes. While hail can occur anywhere, its frequency seems random,

Vignette 8.4   The Hudson Bay Railway

The search for a shorter route to European grain markets caused farmers to call for a railway to Hudson Bay where ocean ships could take their grain to Great Britain and other European markets. After all, why not follow the lead of the Hudson's Bay Company, which used this shorter route to send its furs to London, England? Building a railway, however, is a more onerous and major task in some ways than it is for individual trappers and traders to follow existing water routes by canoe. The Hudson Bay Railway had to cross difficult terrain, including the muskeg and permafrost found in the Hudson Bay Lowland. Completed in 1929, the Hudson Bay Railway stretches from The Pas to Churchill. The railway was never a success and is now operated by an American company, Omni TRAX.

Year	Alberta	Saskatchewan	Manitoba	Western Canada
1971	62,702	76,970	34,981	174,653
1981	58,056	67,318	29,442	154,816
1991	57,245	60,840	25,706	143,791
2001	53,652	50,598	21,031	125,281
2006	49,431	44,329	19,054	112,814
Change 1971–2006	–13,271	–32,641	–15,927	–61,839
Percentage Change	–21.2	–42.4	–45.5	–35.4

Table 8.1   Number of Farms in Western Canada, 1971–2006

Sources: Statistics Canada (1992, 2003, 2007a).

except for areas of relief such as the Alberta foot-hills. Farmers can purchase hail insurance. A third challenge is the short growing season, which leaves crops vulnerable to frost damage. In fact, freezing temperatures have occurred in every month of the summer! Fortunately, most years have frost-free summers. Spatially, the frequency of summer frost is greatest in the higher latitudes where the northern edge of the grain belt is situated. For farmers in the Fertile and Dry Belts, innovation came to the rescue. In 1910, a new strain, Marquis wheat, was developed. Its shorter maturation period overcame the threat of frost.[5] Marquis wheat also extended the growing area for wheat in the Prairies. By 1920, Marquis wheat was the most popular spring wheat in Western Canada. While the Peace River country, located between 54° and 56° N, benefits from long summer days, its northern location and the invasion of cold air masses from the Territorial North make it particularly subject to frost.

Impact of Mechanization

Homesteaders worked the land with oxen and horses. The size of a homestead farm was 160 acres. By 2006, the average farm size had reached nearly 1,500 acres (Table 8.2), a result of mechanization of farm operations and the ensuing economies of scale. The first change was the shift in the farm economy from a labour-intensive operation to a capital-intensive one. At the end of the nineteenth century, many hands were required to successfully deal with the sowing, growing, and harvesting of grain. The introduction of machinery, such as self-propelled steam tractors and threshing machines, changed the way farms were run and reduced the need for farm labour. Further technological changes continued to affect the size of the farm labour force and the size of the farms. For instance, the development of the combine harvester, which can cut and harvest a swath of grain as wide as 15 metres, allowed farmers to harvest hundreds of hectares in a single day. The days of the quarter section of land, the land size obtained by homesteaders, were over.

Consolidating farms into larger and larger units saw the number of farms (and farm population) decline, while the size of farms increased (Tables 8.1 and 8.2). In the Canadian West, this shift began before World War I but accelerated during the Depression (1930s) because of extremely low prices and poor harvests (Figure 8.3). Many farm

> **Think About It**
>
> Since summer fallow conserves soil moisture, why has this technique fallen out of favour?

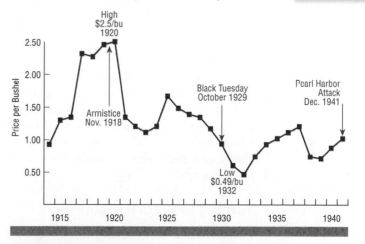

Figure 8.3   Declining world grain prices.
For comparison purposes, US wheat price on May 28, 2010 was $4.57/bu.
Source: US Bureau of the Census, *Historical Statistics of the United States, Colonial Times to 1957.*

Table 8.2	Average Size of Farms in Western Canada, 1971–2006 (acres)			
Year	Alberta	Saskatchewan	Manitoba	Western Canada
1971	790	845	543	726
1981	813	952	639	801
1991	898	1,091	743	911
2001	970	1,283	891	1,048
2006	1,824	1,451	1,000	1,425
Change 1971–2006	1,034	606	457	699
Percentage Change	131	72	84	96

Sources: Statistics Canada (1992, 2003, 2007a).

households fled the land penniless, the bulk from the semi-arid zone known as Palliser's Triangle. Depopulation of rural Western Canada continues. In the 35-year period from 1971 to 2006, the number of farms declined by 35 per cent, while the average farm size doubled from 726 acres in 1971 to 1,425 acres in 2006. Nearly 62,000 farms disappeared, with Saskatchewan alone losing more than half of these. Some lost heart and sold their lands while others were less efficient farmers or lacked sufficient capital. For the surviving grain farmers, larger farms increased productivity and thereby lowered per unit costs.

The move from intensive to extensive agriculture transformed the grain economy of Western Canada. It also transformed the pattern of prairie settlement. In this sense, mechanization of agriculture triggered a demographic movement of farm and rural people to urban settings where jobs and public services ranging from schools to hospitals were available. With fewer people engaged in agriculture, the rural landscape lost much of its population and its villages saw its services drift away to larger regional centres and cities. The ripple effect was that many villages, which had been small service centres, were abandoned and eventually disappeared from the landscape.

Turning to Natural Resources

In the last half of the twentieth century, natural resources became a focus of economic development. Western Canada's northern boreal forest, vast mineral bodies, petroleum deposits, and rivers formed the core of natural resource development that has diversified the region's economy. Commodity prices, too, have played a key role in the pace of development as well as in the flow of royalties to provincial governments. Prices rose following World War II, when a shortage of natural resources in the United States led to a resource boom in Western Canada. One natural resource, oil, skyrocketed in price in the 1970s due to the action of the **Organization of Petroleum Exporting Countries (OPEC)** to control supply, thus forcing prices upward. By the 1980s, the boom cycle for resources had ended and resource prices dropped sharply. At the same time, American demand for oil and gas from Alberta rose significantly; Saskatchewan's potash and uranium extraction grew; and Manitoba became a hub of nickel and hydroelectric production. The forest industry in all three provinces expanded with the rising demand for lumber, pulp, and paper in the United States and other industrial countries.

Rising oil prices in the 1970s ignited an energy boom that had a most dramatic impact on the economy of Alberta. While prices for other natural resources were set by global forces of supply and demand, the OPEC cartel curtailed output and thus caused world prices for oil to rise. This sudden increase had important spinoffs for the Alberta economy, including more jobs and royalties. With higher prices, attention turned to the vast but expensive-to-develop oil sands. Technological advances encouraged by provincial support unlocked a commercial method to separate oil from sand, thus allowing for the commercial extraction of bitumen from the tar sands in northern Alberta. With output increasing, additional pipelines were necessary to carry the oil and gas to markets in Ontario and the United States. During this process, Calgary became the headquarters for major oil and gas companies.

Manitoba and Saskatchewan followed at a much slower pace. In fact, during the decade that followed, low prices for primary products had a greater impact on the economies of these two provinces because they continued to rely heavily on agriculture and resources. By 2006, however, prices for energy and minerals had reached new heights, driven by the expanding economies of Pacific Rim countries, especially China. The economies of Manitoba and Saskatchewan benefited and, with its vast potash resources, Saskatchewan joined Alberta as a 'have' province. Even agricultural prices show signs of improving, leaving only forest products with low prices. Although a downturn in the global economy at the end of 2008 saw commodity prices drop, such a downturn seems unlikely to be of long duration.

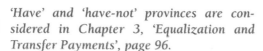

'Have' and 'have-not' provinces are considered in Chapter 3, 'Equalization and Transfer Payments', page 96.

Western Alienation

Western Canada's geography, particularly its unpredictable weather leading to crop failures, the long and expensive rail distance to reach ports for export, the National Energy Policy, and now a possible **carbon tax** on fossil fuels, provides ample ammunition for **western alienation**. In all cases, westerners are faced with a seeming lack of control over their environment, economy, and federal decision-makers. The deep sense of alienation has been an ongoing theme in the history of Western Canada and can now be found in all sectors of Western Canadian society. This negative feeling stems from the peripheral position of Western Canada within Canada and the global economy, which translates into decisions for Western Canada being made by

those outside of the region. For the homesteaders, the targets of their resentment were the eastern banks and railways.[6] Later, for oil producers and provincial governments, the target would become Ottawa's centralist policies. Ottawa is often seen as either an uncaring government that ignores western grievances or a manipulative state power that places the interests of Central Canada over those of Western Canada. For instance, at the time that Alberta and Saskatchewan joined Confederation in 1905, the two western provinces were denied control over natural resources while Ontario, Québec, Nova Scotia, and New Brunswick had obtained this control (and taxing power) when they united to form the Dominion of Canada in 1867. The federal National Energy Program (NEP) of the early 1980s created a chasm between Alberta and the Liberal government. In this case, Ottawa exerted its control over oil prices and levied taxes on oil production. In Alberta's eyes, the NEP was both a 'tax grab' and a political means by which Ottawa favoured energy-deficient Ontario over Alberta, securing low oil prices for Central Canada's manufacturing industry while interfering with Western Canada's resource revenue. The most recent threat—at least as perceived by Westerners—is the dreaded carbon tax. A 2008 report prepared by the Alberta-based Pembina Institute and the BC-based David Suzuki Foundation (Bramley et al., 2008) calls for a drastic reduction in fossil fuel production in Alberta, Saskatchewan, and British Columbia and a transfer of the revenue generated by the carbon tax to other regions of Canada, but most notably to Ontario and Québec. While rejected by the federal government, at least for now, Canada will likely enter into a North American version of the carbon tax when—and if—the United States sets its greenhouse gas emission plan. Not surprisingly, then, western alienation is an expression of the centralist/decentralist faultline.

Sometimes western alienation has expressed itself in political terms. While threats from a few for 'political separation' have gone nowhere, the formation of new political movements is a fact of life because such movements challenge the centralist-oriented

Think About It

Would a carbon tax throw Alberta, Saskatchewan, and British Columbia under the economic freight train and thus reignite the ever-smouldering western alienation?

governments. One such movement was the Co-operative Commonwealth Federation (CCF), a political party formed in 1932 by a coalition of labour and farming interests in an effort to combat the destitution people were experiencing during the Great Depression. (The CCF later became the NDP.) It met with considerable success, and is attributed with leading the way for the creation of a variety of social programs in Canada. Social Credit, another Western-based political party, came into existence at about the same time. The Reform Party of Canada, which became the Canadian Reform Conservative Alliance in 1999 and then amalgamated with the old-line Progressive Conservative Party to form the Conservative Party in 2003, provides the most recent example of a western protest party. Western alienation explains the political fragmentation of Canada's federal parties in the West (British Columbia and Western Canada). Howard Richards (1968), who founded the Department of Geography at the University of Saskatchewan in 1960, argued that isolation from other populated regions and from the political decision-making in Ottawa is at the root of western alienation. While western feelings towards Ottawa ran deep, times have changed—since 2006, the federal government has been led by a Prime Minister from Alberta. Regional tensions remain, but Ottawa now appears to be much more receptive to Western Canada and British Columbia.

Western Canada Today

The economy and population of Western Canada, led by Alberta, continue to grow. While agriculture remains a basic element, most wealth is created by resource industries, especially from oil sands, potash, and uranium. The region is diversifying with a growing trend towards processing its agricultural products and widening its mineral production. As shown in Table 8.3, Western Canada has a relatively high proportion of Aboriginal peoples, a reflection of the region's history, and this demographic reality has consequences for the region and its future; as well, the region has the lowest unemployment rate in

Canada, which underlies its robust economy. Added to these factors, innovations within the knowledge-based economy are accelerating economic change, especially in the field of biotechnology. Most research into plant breeding and seed development takes place in public institutions, though leading global companies such as Bayer and Monsanto now have research facilities in the region. Still, innovative activities extend to the more exotic areas, such as blimp-like transportation for mining and Aboriginal communities in remote areas of northern Canada (Vignette 8.5).

The role of research parks at universities is presented in Chapter 4, 'Economic Structure', page 166.

Alberta's economy, driven by the oil and gas industry, has become the most diversified of the three provinces. Saskatchewan and Manitoba have also diversified but, lacking Alberta's huge petroleum reserves, their economic growth is occurring at a much slower rate. The 1995 cancellation of the **Crow Benefit**, a transportation subsidy that allowed farmers to ship their products by rail at a reduced cost, initiated massive changes in the agricultural sector. In 1995, prices for grain were relatively high and farmers could manage the rail charges. But in the following years, grain prices dropped while rail charges rose, pushing many farmers into bankruptcy. The challenge remains to process more agricultural products within Western Canada and to ship higher-valued processed products in containers by rail and then by ship to world markets.

Alberta is the economic giant of the three provinces. It has over half of the population in Western Canada and accounts for about 63 per cent of the region's GDP (Table 8.3). Both Saskatchewan and Manitoba fall well below this level. Each province has much natural wealth. Saskatchewan, for example, has most of the cropland and is the leading producer of potash and uranium. In addition to having the richest agricultural land in the West, Manitoba produces vast amounts of hydroelectric power from the Nelson River. Even so, Alberta holds the trump resource card—oil and gas.[7] Indeed, by 2015 Alberta is forecast

Think About It

Does the fact that most seats in Western Canada are held by Conservative members suggest that western alienation has withered on the vine, or does it imply a strong distaste for previous Liberal governments and their policies towards Western Canada?

Table 8.3   Basic Statistics for Western Canada by Province

Province	Population (000s)	Unemployment Rate (%)	GDP (%)	Aboriginal Population (%)	Canada's Cropland (%)
Alberta	3,290	6.6	16.4	5.8	26.8
Saskatchewan	968	4.8	3.1	14.9	41.7
Manitoba	1,148	5.2	3.1	15.5	13.1
Western Canada	5,406	5.9	22.6	9.4	81.6
Canada	31,613	8.3	100.0	100.0	100.0

Note: Unemployment percentages are for 2009; demographic data are based on the 2006 census.
Sources: Tables 1.1, 1.2, 5.1.

Vignette 8.5   Reaching for the Sky

Resource development in remote areas is hampered by the high cost of air transportation. Winter roads, such as those to the diamond mines in the Territorial North and to Aboriginal communities, provide one solution but only a seasonal one. Global warming may reduce the winter period and thus winter roads. In spring 2010, ice roads connecting several First Nations settlements in northern Manitoba became impassable much earlier than normal, thus making it impossible for trucks to bring much-needed supplies. Could blimps provide a solution? The concept of blimps has long been touted as a possible solution to northern transportation and Calgary-based SkyHook International Inc. may have the answer with its heavy-lift rotorcraft—a blimp with four helicopter-like rotors underneath that has the capacity to life a 40-tonne load. In 2008, Boeing was awarded a contract to design and build two prototypes by 2014 (photo 8.5).

Courtesy of Boeing Phantom Works Communications

Photo 8.5

Artist's conception of the 'SkyHook' being developed by SkyHook International of Calgary.

to rank as the fifth largest producer of oil in the world (Grauman, 2006).

Back in 1972, when a barrel of oil was worth $2 on the world spot market, Alberta oil was not valuable enough to dominate the western economy. Since then, prices have risen, generating great wealth for Alberta. The price has fluctuated greatly in recent years and even months, reaching its highest point in July 2008 when a barrel of crude oil was worth nearly $150. Prices fell sharply during the global meltdown, and by December 2008 the price was $35/barrel. By late March 2010 the price had recovered to over $80/barrel.

Though Western Canada exports both primary and processed products to other countries—recently increasing its exports to Pacific Rim countries—the petroleum industry is more closely tied to the North American market. Most of the natural gas and oil goes to markets in Ontario and the United States. After World War II, a network of oil and gas pipelines was constructed to serve both the Canadian and American markets. Energy, unlike manufactured goods, had easy access to American markets before the Free Trade Agreement.

World prices hold the key to economic growth for the resource industries. Imagine the impact on the prairie economy if other **primary prices** (prices for primary resources) followed the upward path of oil prices! Imagine, too, if Europeans and Americans dropped their wheat subsidies to Canadian levels! Unfortunately, prairie farmers, miners, and loggers are all too well acquainted with commodity price cycles, which are controlled by global demand. For a long time, prices, except for oil, were depressed. In the first years of the twenty-first century, an upward trend in many commodity prices emerged, driven especially by demand from China.

Economic Structure

Employment by industrial sector in Western Canada reveals the prominence of primary economic activities. While Table 8.4 presents only a generalized picture of the western economy, it illustrates two important aspects. One is the importance of the primary sector. The percentage of people employed in this sector (10 per cent) is over five times the figure for Canada's principal industrial core region—Ontario. The second is the relatively smaller size of the secondary and tertiary sectors. The proportion of employment in the secondary sector is well below that of Ontario while the tertiary sector is just slightly lower (Table 5.3). For instance, percentage of employment in the manufacturing sector is less than half that of Ontario. However, the percentage of workers in the construction industries, a subset of the secondary sector, is higher in Western Canada and this has increased from 2005 to 2008. With oil sands investments alone estimated to reach a staggering $125 billion by 2015, the ripple from the oil sands construction will continue to support a robust mining-oriented manufacturing sector in Edmonton, Calgary, and Saskatoon as well as in industrial centres in Ontario and Québec (National Energy Board, 2006: 11).

Based on scientific breakthroughs, knowledge-based innovations take the form of technical advances and digitized information and they permeate all sectors of the economy. In Western Canada, the application of technological breakthroughs is found in the

Table 8.4	Employment by Industrial Sector, Western Canada, 2005 and 2008			
	2005		**2008**	
Economic Sector	**Workers (000s)**	**Workers (%)**	**Workers (000s)**	**Workers (%)**
Primary	284.3	10.0	311.6	9.9
Secondary	468.6	16.5	553.1	17.7
Tertiary	2,095.3	73.5	2,268.2	72.4
Total	2,848.2	100.0	3,132.9	100.0

Source: Statistics Canada (2009b).

primary sector. For example, oil and gas operations use **horizontal drilling**, which, when combined with **hydraulic fracturing**, has unlocked oil from the Bakken shale formation in the Interior Plains of southern Saskatchewan. Carbon capture and storage is another complex and innovative undertaking. Several companies and universities are engaged in perfecting this technology. Saskatchewan, with its experience in **carbon sequestration** at its Weyburn oil field, is a world leader in this fledgling technology, and the University of Regina is continuing its research in this area. Alberta has provided funds to large coal and oil sands companies to develop carbon capture and storage technology that will meet their specific needs (Vignette 8.6).

The Weyburn carbon sequestration project is discussed in Chapter 2, 'Environmental Challenges', page 61.

Manufacturing

The manufacturing sector in Western Canada, while growing, is still relatively small. The four leading manufacturing centres are Edmonton, Calgary, Winnipeg, and Saskatoon. The explanation for the relatively weak state of manufacturing in the Prairies has been attributed to its small market. Two factors have changed the equation. First, the specialized demands of the expanding resource industries in Alberta and Saskatchewan have created local manufacturers of mining equipment. Second, market size has increased. From 1981 to 2005, a significant increase in Western Canada's domestic market took place as the region's population went from 4.2 million to 5.5 million. Western Canada's portion of total Canadian manufacturing increased to 10 per cent, with most of this occurring in Alberta. Much of the increase stems from the petrochemical, agriculture, and mining industries. In a few cases, global manufacturing companies are turning to Western Canada. Case New Holland recently announced its 2006 global consolidation plan, which called for the closure of several plants in the US Midwest and the relocation of those activities to Saskatoon. The Saskatoon plant will add the production of Case New Holland air-seeders to its planter

and seeder production lines for the North American market. Winnipeg, already a centre for the aerospace industry, has high hopes to receive a substantial portion of the $3.4 billion targeted for four C-17 Globemaster transport planes, which can carry tanks, soldiers, and large equipment around the world without stopping. While the airplanes would be built in the US, Boeing has agreed to spend an amount equal to the $3.4 billion purchase price on projects in Canada.

Key Topic: Agriculture

Agriculture was the driving force behind the settlement and development of Western Canada in the late nineteenth and early twentieth centuries. At that time, most homesteaders grew spring wheat for export to Great Britain. Canola now exceeds spring wheat in returns to farmers and matches it in sown acres, and

Vignette 8.6 Technological Gamble: Carbon Capture and Storage Concept

Climate change is arguably the most serious environmental issue society faces. The heavy reliance on coal and oil sands industries in Western Canada has placed these industries, especially companies developing the oil sands, under the spotlight. In response, these firms are under pressure to reduce their carbon footprints and to deal with the 'dirty oil' label. One potential solution at the top of the industry's research agenda is carbon capture and storage. Applying this technology to the oil sands is not straightforward. Considerable research is necessary and, if successfully applied, it will offer an exceedingly expensive solution. The oil industry has called for public support and Ottawa and Alberta have responded with Ottawa's **Clean Energy Fund** and Alberta's Carbon Capture and Storage Fund. The first international company to receive public funding was Shell's Quest Project, which would focus on reducing greenhouse gas emissions from its Scotford Upgrader at Fort Saskatchewan (just east of Edmonton). Ottawa is contributing $120 million and Alberta $745 million. The estimated total cost is $1.35 billion. The Scotford Upgrader, by turning bitumen from the Athabasca oil sands into synthetic crude oil, is a major emitter of greenhouse gases. Over a 15-year period, the Quest Project is designed to capture around 40 per cent of the CO_2 emissions from the Scotford Upgrader and then inject these greenhouse gases into a stable geological formation 2,300 metres underground.

Think About It

Is it morally correct to consider trade-offs between jobs and the environment? For instance, would you favour exporting more bitumen to upgraders in the US where greenhouse emission would take place as well as create more jobs, rather than employ people in Alberta and refine the bitumen there?

durum wheat and specialty crops are commanding a stronger place in prairie agriculture. At the same time, the area devoted to summer fallow has dropped sharply, thus placing more crop area in production. Cattle and hog production remains on the edge for two reasons. First, exports to the United States are hampered by American border restrictions. Second, disease has affected both consumption and export of beef and pork (Vignette 8.7). For example, the H1N1 flu virus is associated with hogs and has led to a drop in sales. A few years earlier, the cattle industry saw a similar negative impact from bovine spongiform encephalopathy (BSE), commonly known as mad-cow disease. These changes in the mix of prairie agriculture are price-driven. One perennial fact—the high cost of shipping grain by rail—has pushed farmers into other crops. With canola, farmers can deliver their product to local crushing plants where canola oil is produced and the rest of the canola is sold for feed. Another factor is grain subsidies by the European Union and the United States that often price Canadian grain out of export markets.[8]

The combination of these factors, over the years, has forced some farmers into bankruptcy, and by the middle of the twentieth century, agriculture had lost its central place in the economy of Western Canada. Signs of decline, such as deserted communities, abandoned branch rail lines, and the closure of many rural grain elevators, are everywhere in the rural landscape. This decline of rural towns and villages and the loss of farm population illustrate the weak position of the rural economy. The struggle to find its footing has seen a shift from grain to other crops and farm activities. In 2006 and 2007, prices for grain and canola increased because of high oil prices, but in 2008, prices dropped slightly. Higher prices for canola and spring wheat have been triggered by the indirect effect of the expanding demand on the part of European and US ethanol plants for biomass to produce an alternative fuel for automobiles, and this has taken much land out of cultivation for human food. Whether these higher prices will continue depends on three factors. On the demand side, will the growth of the ethanol industry in North America continue and thus reduce the supply of grain? Second, will the demand for agricultural products from Pacific Rim countries, especially China, continue to grow? On the supply side, farmers are quick to respond to price. If the price of grain increases, then more land is seeded to grain the following year and, if weather conditions are favourable, production increases. In this scenario, the price for grain then declines.

Technology plays a large role in agriculture, ensuring greater yields and more efficient farming implements. Innovations in the form of advanced machinery and improved seeds have played a key role in agriculture. Spring wheat, for instance, has evolved over time thanks to the efforts of plant breeding research. Red Fife was brought from Ontario to Manitoba in the late nineteenth century. While these seeds produced high yields, they required a long growing season and so crop loss due to frost was a serious weakness. In 1910, Marquis wheat was developed by federal plant breeders and these seeds had a shorter maturation period. Canola represents a more recent innovation. It is an achievement of Canada's plant breeding community. The initial breakthrough took place at the University of Saskatchewan, where researchers were able to alter rapeseed into a superior product called canola. Today, Western

Vignette 8.7   Negative Impact of BSE on the Livestock Industry

In 2003, the discovery of BSE (bovine spongiform encephalopathy) in Alberta had devastating implications for the livestock industry. Almost immediately, the United States and Japan (the chief export markets) halted the importation of Canadian livestock and meat products. The Canadian livestock industry is concentrated in Western Canada, with Alberta having almost half of Canada's 13.4 million cattle. Beef cattle comprise Alberta's largest agricultural sector, providing about half of farm revenue. Exports are extremely important, making up around 40 per cent of total sales. With the halt of exports, Canadian prices to the ranchers dropped dramatically, though retail prices did not follow. The US lifted the ban on Canadian cattle in 2005. The H1N1 influenza virus or the so-called 'swine flu' of 2009 had a similar impact, causing a sharp drop in pork consumption around the world.

Canada is the global centre for canola research with the major effort taking place in research facilities in Saskatoon, where experimental work is conducted by private seed developers, federal researchers, and university plant breeders. Their combined efforts have increased yields, reduced plant diseases, and improved the quality of the final product. Canola offers another advantage over wheat and other grain crops: farmers can greatly reduce their shipping costs by trucking canola to local processing plants.

Western Canada is blessed with rich black, dark brown, and brown chernozemic soils that are well-suited for growing cereal crops (Figure 8.4). Some crops, such as peas and lentils, require more moisture and are restricted to the black soil zone. Unlike canola, durum wheat thrives in hot, dry weather commonly occurring in the brown soil zone. It also requires a longer growing season than spring wheat and therefore is usually grown south of the fifty-first parallel. Consequently, the agricultural land use varies, forming three distinct agricultural regions: (1) the **Fertile Belt** (black soil associated with parkland and long-grass natural vegetation; (2) the **Dry Belt** (brown soils with short-grass natural vegetation); and (3) the **agricultural fringe** (southern edge of the boreal forest) and the Peace River country (Figure 8.4). These sub-regions have very different growing conditions. The major factors controlling those conditions are the number of frost-free days and the soil moisture. The Fertile Belt provides the best environment for crop agriculture. The Dry Belt occupies semi-arid lands and has become a grain/livestock area. In the agricultural fringe, the short growing season encourages farmers to grow feed grains and raise livestock; in the Peace River country, farmers grow both grain and feed grain for livestock.

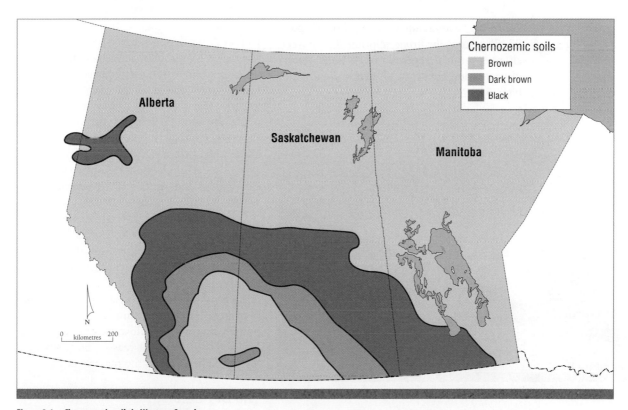

Figure 8.4   Chernozemic soils in Western Canada.
Three types of chernozemic soils are found in the Canadian Prairies: black, dark brown, and brown. The differences in colour are due to the varying amount of humus in the soil, which, in turn, is a factor in the natural vegetation cover—short grass, tall grass, and parkland vegetation. The soil of the Peace River country, formed under an aspen forest, is 'degraded' black soil.

The Canadian Prairies are on the northern margin of crop agriculture where the length of the growing season is short. Slight variations in weather conditions have either a positive or negative impact on crops. Wheat, for example, is one of the more resilient crops that can grow in areas of low precipitation, but wheat harvests can vary widely from year to year in both size and quality. Because of its continental climate, temperature and precipitation vary from year to year so that the risk to the prairie farmer is high compared to that of the southern Ontario farmer. Other weather conditions affecting crop farming include late spring seeding due to cold or wet weather, summer frosts before the crop is mature, and wet weather in the fall that impedes harvesting. All of these weather conditions will reduce the quality of the harvest. Beyond weather conditions, pests, such as grasshoppers, have greatly reduced the size of the harvest in some years.

Table 2.5, 'Canadian Climatic Zones', page 53, illustrates the broad relationship between climate, natural vegetation, and soils; natural vegetation zones are shown in Figure 2.7, page 53.

> ### Think About It
>
> How would global warming affect farming? A longer, warmer growing season would be an advantage, but what about precipitation and the concept of evapotranspiration?

Mark Duffy/World of Stock

Photo 8.6

Foam Lake Terminal, between Yorkton and Saskatoon, is served by the Canadian Pacific Railway. The terminal is now owned by Viterra, which used to be called the Saskatchewan Wheat Pool. The export of grain to foreign markets requires an elaborate transportation and handling system. **Inland terminals** and elevators play a key role by serving as collection points along main railway lines. Here the grain is graded, cleaned, and weighed before being transported by railway cars to Canada's major ports.

The Fertile Belt

The Fertile Belt extends from southern Manitoba to the foothills of the Rocky Mountains west of Edmonton (Figure 8.5). The higher levels of soil moisture, an adequate frost-free period, and rich soils make this belt ideal for a variety of crops and livestock. The most popular crop, since farmers first arrived in the West, has been wheat. In recent years, however, the acreage in grain has declined, while the planting of canola and specialty crops, such as beans, field peas, and sunflowers, has increased. This change was fuelled by rising prices for these crops and declining prices for wheat.

The predominance of wheat in the Fertile Belt is most evident in western Saskatchewan and eastern Alberta. In southern Manitoba and the adjacent parts of eastern Saskatchewan, there is more annual precipitation so farmers there can grow a wider variety of crops as well as keep livestock. As a result, mixed farming is common. Grain and specialty crops (canola, flax, sunflowers, and lentils) are combined with beef, pork, and poultry production. Near Winnipeg, for instance, a livestock industry has developed where cattle are fattened before shipment to meat-packing plants in Winnipeg and Ontario. Feedlots and nearby meat-packing plants are located in other parts of Western Canada, particularly in Brandon, Calgary, Edmonton, Lethbridge, and Saskatoon. Since NAFTA, meat production has increased, especially pork, and most meat products are exported to markets in the United States.

As the urban markets grow, market gardens, dairy farms, and other specialized forms of intensive agriculture are developing near major cities in the Fertile Belt. The demand for specialty-crop products has spawned a number of smaller but very intense production units, including nurseries and greenhouses, to produce flowers and vegetables for local sale. For that reason, farm sizes are much smaller around the major cities.

The Dry Belt

The Dry Belt contains both cattle ranches and large grain farms. It extends from the Saskatchewan–Manitoba boundary to the southern foothills of the Rockies and north

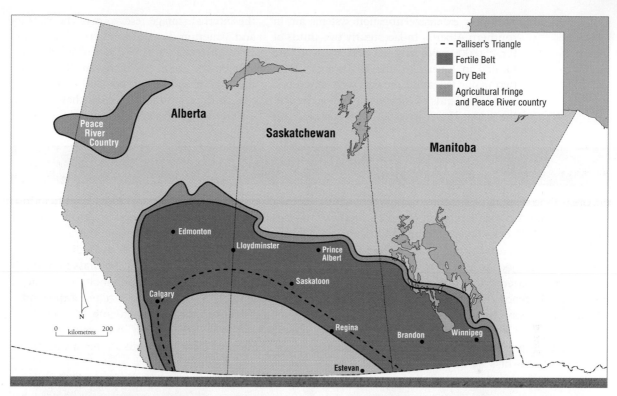

Figure 8.5   Agricultural regions in Western Canada.
Farming in the Prairies can be divided into three areas: the Fertile Belt, the Dry Belt, and the agricultural fringe and Peace River country. Each region has a different type of agriculture because of variations in physical geography.

nearly to Saskatoon (Figure 8.5). However, the driest area, or heart of the Dry Belt, occupies a much smaller area, stretching southward from the South Saskatchewan River to the US border. The arid nature of the Dry Belt is not due to low annual precipitation (it is comparable to other areas of Western Canada) but to longer summers and higher evaporation rates. Within the Dry Belt, feed grain and hay crops are grown to supply winter feed for the cattle ranching that dominates in this area. Cattle ranching began in this area in the 1880s. Today, ranches are large, often many times the size of grain farms, because of the lower productivity of the dry land and the need for huge grazing areas to support a rotational grazing system.

Along the northern edge of the Dry Belt, grain farming is pursued, but the risk of crop failure is high. To conserve soil moisture and control weeds, summer fallowing was a widespread practice, but has been largely replaced with **continuous cropping**. In Saskatchewan, for example, summer fallowing has declined from a high of 15 million acres in 1988 to

4 million in 2009 (Saskatchewan Ministry of Agriculture, 2009a). The purpose of summer fallowing is that two years of precipitation will accumulate sufficient soil moisture to germinate the seed and sustain the young wheat plant. The crop will still require summer rainfall to reach maturity. On average, one-fifth of the arable land in the Dry Belt is kept in summer fallow each year. The drawback to summer fallowing is that the ploughed (fallow) land is exposed to water and wind erosion. In a dry spring, windy weather can result in extensive loss of topsoil with huge clouds of dust stretching for many kilometres across the Prairies. Continuous cropping keeps short, stiff stalks of grain or hay remaining on a field after harvesting, thus protecting topsoil from water and wind erosion. Then, too, advances in technology have allowed for 'one-pass' seeding, spraying, and fertilizing. The expense and time not spent on repeated tilling of a field reduce a farmer's costs and conserve soil moisture.

Irrigation on these semi-arid lands has provided another solution to dry conditions.

The most extensive irrigation systems are in southern Alberta. In fact, nearly two-thirds of the 750,000 ha of irrigated land in Canada are located in Alberta. In the 1950s and 1960s, two major irrigation projects were developed in the dry lands of Alberta and Saskatchewan: the St Mary River Irrigation District is based on the internal storage reservoirs of the St Mary and Waterton dams in southern Alberta, and Lake Diefenbaker serves as a massive reservoir on the South Saskatchewan River. Since the two areas of irrigated land compete for the same potato market, Saskatchewan farmers are at a disadvantage because of the greater

Vignette 8.8   The Great Sand Hills

The Great Sand Hills are situated in a semi-arid climatic zone where short-grass vegetation exists. Located in the centre of Palliser's Triangle, some sand dunes in the Great Sand Hills remain void of natural vegetation, making them subject to wind erosion, while others have been stabilized by a covering of native prairie grasses. Cacti, creeping juniper, and small shrubs like wild rose, saskatoon, chokecherry, and silver sagebrush grow in the Great Sand Hills. This area of southwestern Saskatchewan near the Alberta border covers about 1,900 km², and the hills were formed from wind action causing beach deposits of former glacial lakes to form desert-like sand dunes. Archaeologists and geologists believe that these dunes were active in the late eighteenth century due to wild grass fires and Indians using fire to drive buffalo, thus keeping the grass from encroaching into the sand dunes (Hugenholtz and Wolfe, 2005; Wolfe et al., 2007). With the settling of the land by homesteaders in the late nineteenth century, the natural vegetation slowly began to encroach on the edges of the sand dunes and thus started a stabilization process that continues today.

Ron Erwin/All Canada Photos

Photo 8.7

The Great Sand Hills of southwestern Saskatchewan.

distance (shipping costs) to processing plants in southern Alberta. A second disadvantage facing Saskatchewan irrigators is that they experience a shorter growing season than farmers in southern Alberta. Consequently, their selection of crops is more limited. On the other hand, farmers in southern Alberta are able to grow corn, sugar beets, and other specialty crops that provide a high return per acre.

The Agricultural Fringe and Peace River Country

The agricultural fringe is a narrow transitional strip of forested land located just to the north of the Fertile Belt while the Peace River country lies in the Interior Plateau where the deeply entrenched Peace River flows from the Rocky Mountains and across the High Plains. Grain, livestock, and hay crops are the principal agricultural activities. Specialty crops include legumes and grass seeds. In these higher latitudes (55° to 57° N), crop agriculture is challenged by a short growing season and the threat of frost, hence hay crops are widespread.

As the last agricultural frontier, the Peace River country did not attract settlers until after World War I. By then, the best lands in the Prairies had been settled. Two events laid the groundwork for settlement. First, the signing of Treaty No. 8 in 1899 opened large areas to agricultural settlement. Second, accessibility to export markets improved with the construction of a railway to Peace River with rail connections to Vancouver in 1915; then, in 1955, the Pacific Great Eastern Railway linked Peace River to Prince George, thus allowing wheat and other agriculture products to be shipped to world markets via Prince Rupert.

Agricultural Transition

The liberalization of international trade and the Free Trade Agreement opened up the US and other foreign markets for Western Canadian agricultural products and, at the same time, exposed farmers more directly to global forces of demand and supply. For prairie agriculture, NAFTA was perceived to

Fuss Heini/All Canada Photos

Photo 8.8

In the dry lands of southern Alberta, water is king. Irrigation waters from the Oldman River reservoir supply valuable water to farmers who specialize in growing corn, sugar beets, potatoes, and other vegetables.

provide unfettered access to the US market, thus encouraging farmers and ranchers to expand their production and reap the higher prices for such products in the US market. The liberalization came at a cost, however, and the principal federal subsidy, the Grain Transportation Subsidy, was eliminated by Ottawa in 1995. Prairie farmers soon discovered that free trade was an illusion because their access to the US market could be restricted by Washington and other foreign countries. Such interventions by the US were usually designed to protect US farmers from competition with their Canadian counterparts. Unfortunately, Canadian farmers who expanded production designed for the US market and more recently for Asian markets have ended up with unwanted surpluses. Trade restrictions, in a sense, are indirect. For instance, in 2009, China barred imports of Canadian canola because of the presence of blackleg disease (caused by the fungus *Leptosphaeria maculans*), though that disease has been present in earlier shipments. The real reason is that China has large reserves of canola and is using the ban on Canadian canola to reduce its supply. Other countries, including the United States, have used health/disease concerns when convenient to slow imports.

The First Phase of Transition: Low Prices and Rising Input Costs

The first phase of this agricultural transition was marked by the end of federal subsidies for grain transportation and by rising prices for fertilizers, fuel, pesticides, and other farm inputs. With the end of the Grain Transportation Subsidy, grain farmers had to pay all the freight charges for grain shipments. Worse yet, the price of grain declined from a high in 1995 and stayed low until 2006. During this time, farmers began to grow more and more canola. All the time freight rates increased each year, which again favoured

canola. Therefore, the first phase of this agricultural transition represented hard times for farmers.

While the shipment of feed grain to cattle and hog farms in Ontario and Québec dropped dramatically because of the higher freight rates, the geographic pattern of foreign exports did not alter, with the US remaining the main market for most agricultural exports, including spring wheat, canola, hogs, and cattle. The reason was simple: even with higher freight rates, the railways represented the only route for the export of grain to world markets. Farmers (except those close to the US border who could truck their product to US markets) had three choices: continue to ship to world markets and absorb the additional transportation costs; switch production to another crop or other crops; or sell the farm. Switching to another crop had its drawbacks. Crop rotation might be affected; new crops could require new equipment and certainly new farming techniques; and prices for new crops might be as low as those for wheat.

For Western Canadian farmers, then, trade liberalization was and is a double-edged sword. This new world of so-called open markets has provided export opportunities, but as exports to the US increased, duties were imposed. With prices remaining low and production costs increasing, many farmers went bankrupt or sold their farms. From 2001 to 2006, the number of farms declined by nearly 13,000 (Table 8.1). Trade liberalization also led to the elimination of farm subsidies, such as the transportation subsidy for grain, and it threatens farm marketing boards and farm organizations, such as the Canadian Wheat Board (Vignette 8.9).[9] Canada, being so dependent on exports to the United States, is vulnerable to US trade barriers. While Washington aims at protecting its domestic producers, the cost of this protection is borne by Canadian farmers and ranchers. Perhaps the worst aspect of trade with the United States is the uncertainty—if Canadian producers expand production to service the US market, then their operations are extremely vulnerable to barriers to that market.

Vignette 8.9   The Canadian Wheat Board

In the early years of agricultural settlement, grain farmers felt helpless because they had no control over the prices for their commodities. Private grain buyers, it seemed, made money, but not the farmers. This experience accounts for much of the bitterness that took root in Western Canada. In 1935, the answer to the plight of grain farmers came in the form of a state grain buyer, the Canadian Wheat Board. Prairie wheat farmers can only sell their crops to the Wheat Board. The twin goals of the CWB are to sell as much grain as possible at the best possible prices and to ensure that each producer gets a 'fair' share of the market. The Canadian Wheat Board has been under attack from the United States, which claims that it subsidizes Canadian exports. For that reason, Washington placed duties on Canadian durum and spring wheat. Following a NAFTA decision in 2005, both tariffs were removed by the US government in the following year. The Conservative government, with its free-market perspective, favoured ending the CWB's monopolistic position, but strong support from farmers has caused Ottawa to soften its approach.

The Second Phase: Higher Prices

A second phase of the agriculture transition appears to have begun with global demand outstripping supply. By June 2007, prices for most agricultural products were up substantially over the previous year and the price for wheat had recovered to the highest point since 1995. Two questions are basic for understanding the second phase: (1) Why have prices risen? (2) Are these higher prices, like in the past, just a brief anomaly? Brian Oleson, an agribusiness professor at the University of Manitoba, believes that these higher prices are not anomalous but represent a dramatic shift to permanently higher prices for wheat and other agricultural products. Oleson states: 'We are entering a new era. It's almost as if the platform [for wheat prices] has been raised and we're all dancing on a new dance floor' (Greenwood, 2007). His position is supported by the World Food Price Index, which reached its bottom around 2000 but had jumped sharply by 2007 (Figure 8.6).

The second phase, if real, has been triggered by two international developments. Both spell good news for prairie farmers. Higher prices for grain, canola, and pork can be attributed to (1) demand for more meat, in particular pork, from China and (2) the need of the biofuel industry for corn and other crops. China has become a global economic power and, as a result, a substantial middle class has emerged with money to pay for a better diet, which often includes pork. Consequently, Chinese farmers are increasing their production of hogs and more feed (corn) is needed. In the first six months of 2006, prices for both hogs and corn rose sharply in China. And there is no turning back. As Keith Bradsher (2007) reported:

> Few things are as essential to the Chinese as their pigs. From pork spare ribs and mu shu pork to char siu bao—barbecued pork buns—pork is a staple of the Chinese diet. So in this Year of the Pig, an acute shortage of pork has been national news, as butchers raise prices almost daily and politicians scramble to respond.

Chinese officials offer several reasons for the high pig prices. The cost of animal feed has risen by one-quarter in the last year, partly because more corn is being made into ethanol and partly because more prosperous workers are eating more meat, which requires more animal feed.

As well, the biofuel industry has become a major industry in Europe and the US. For the European Union, the assumption is that ethanol-blended fuel produces cleaner emissions than regular gasoline and thus is an important part of Europe's attempt to reduce the emission of greenhouse gases to the atmosphere. In the case of the United States, the political rationale supporting the ethanol industry is to reduce dependency on foreign oil. In the US, ethanol is made from corn. But corn is also used to feed chickens, hogs, and cattle. As demand for corn increases, its price will also increase, which means a rise in prices for meat, eggs, and dairy products and for other farm produce. While the federal Conservative government committed $2 billion in incentives for the construction of ethanol and biodiesel plants in its June 2007 budget, apart from feasibility studies, progress has been slow: by spring 2010, no major plant had been built. However, US ethanol and biodiesel plants are already operating and their consumption of corn has pushed the US price

Think About It

With rising freight rates, why did farmers prefer canola over spring wheat?

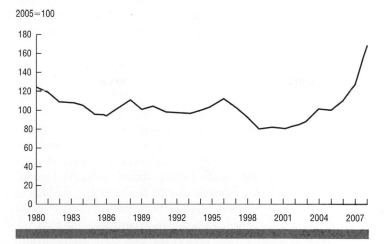

2005=100

Figure 8.6   World Food Price Index upswing: Short-term or long-term?
Note: 2008 data reflects prices as of June 2008.
Source: Agriculture and Agri-Food Canada (2009).

up. Signs indicate that US prices are translating into higher prices for Canadian grain and canola. The large subsidies provided by governments to the biofuel industry represent the only dark cloud on the horizon, i.e., if the subsidies end, then the biofuel industry will likely collapse and grain prices will fall.

A third factor suggesting a permanent turnaround in grain prices is that as cities sprawl into agricultural land—more so in China and India than in Canada—the amount of high-quality land available for food production decreases. Of course, the counter-argument to a smaller arable land base is that technology will allow farmers to produce more on a smaller amount of land. Such an argument, however, is itself countered by the realization that additional technological inputs can and do have negative environmental consequences.

Rural Land-Use Changes

Two major land-use changes have taken place in the last 30 years. First, the amount of land seeded in crops has increased substantially. In Saskatchewan, the increase was a remarkable 13 million acres—from 21.1 million acres in 1981 to 34.1 million acres in 2008 (Saskatchewan Ministry of Agriculture, 2009b: Table 2.3). The explanation lies in a shift from summer fallowing to continuous cropping. Over the same period, the amount of land left in summer fallow dropped by 12.3 million acres—from 16.5 million acres to 4.2 million acres (ibid., Table 2.1).

The second change was driven by high rail transportation costs due to the ending of the Crow Benefit in 1995. Since then, farmers have had to pay the full cost of rail transportation to ship their grain to port.[10] As a result, prairie farmers searched for crops and livestock that can be sold and processed locally, thus reducing their transportation costs. Field grains, for example, use to be shipped to feedlots in Ontario and Québec, but without the Crow Benefit, transportation costs were too high.

The agricultural land use changes fall into two geographic zones. In the *eastern zone*, farmers in eastern Saskatchewan and in Manitoba are growing less grain but more feed grains and specialty crops such as sunflowers, lentils, chick peas, and field peas. Specialty crops can be sold locally, thus keeping transportation costs low. Two new canola crushing plants at Yorkton, Saskatchewan, and the expansion of existing Saskatchewan crushing plants at Nipawin and Clavet will encourage more canola production (probably at the expense of wheat) and keep transportation costs at a minimum. Livestock producers in southern Ontario can no longer afford to import feed grains from Western Canada and a growing livestock industry in Western Canada now supplies beef and pork products to Ontario and Québec. As well, the demand for pork is increasing, opening more markets for Western Canadian producers in both the United States and Pacific Rim countries. To meet the demand, huge hog farms, similar to those in the US, have sprung up across Western Canada. These farms take advantage of modern agro-technology, mass production techniques, and economies of scale. Because of the smell and the risk of contaminating groundwater, large hog barns are located in rural settings far from settlements. Here, the hog barns are providing employment for residents of rural communities, though their waste poses a threat of fecal bacteria seeping into groundwater.

In the *western zone*, farmers in western Saskatchewan and Alberta did not decrease their wheat acreage for two reasons. First, semi-arid growing conditions are ideal for wheat and less suitable for canola. Second, western Saskatchewan and Alberta have lower rail costs for shipping spring wheat to Vancouver. Ranchers, too, benefit from short distances to major markets, with a relatively short rail haul to Vancouver to reach US and Asian markets. Southern Alberta has greatly expanded its cattle- and hog-slaughtering plants, which has resulted in an expansion of the livestock industry and the growth of more hay and fodder crops for winter feed.

Spring Wheat: The Prairie Staple?

Grain production, particularly spring wheat, has been the prairie staple for over 100 years. Grains do well in dry conditions whereas other crops would fail. Low world prices for spring wheat and the loss of the rail transportation subsidy for grain led grain farmers to seek alternative crops.[11] From 1971 to 2001, the acreage seeded to spring wheat in Saskatchewan fell below 10 million acres twice, in 1998 and 2000 (Saskatchewan Ministry of Agriculture, 2009b: Tables 2.3, 2.9). In the same period, the acreage in spring wheat exceeded 14 million acres 18 times (all before 1994). Since 2001, spring wheat acreage has been below 10 million acres and, in 2008, fell to 7.6 million acres. At the same time, canola acreage has jumped from 2.7 million acres in 1971 to a record high of 7.7 million acres in 2008. The same pattern of crop change has occurred in Alberta and Manitoba.

Livestock Industry

The livestock industry is also undergoing change—a combination of restructuring and consolidation of its processing plants. As a result of the Free Trade Agreement in 1989, competition within the North America market has become fierce, forcing Canadian operators to build larger hog-processing plants, to specialize in a single product in each plant, and to demand lower wages from employees. In Western Canada, new hog-slaughtering plants have been built at Brandon, Red Deer, and Lethbridge and another one is planned for Winnipeg. Plant specialization is driven by cost of production—larger plants gain economies of scale, and specialized production lines achieve higher productivity. While it does cost to ship meat from the slaughterhouse to a number of processing plants, the saving gained from efficiencies in processing plants dedicated to a single product offsets the shipping costs. For example, Maple Leaf Foods' new hog-slaughtering operation at Brandon supplies carcasses to its Winnipeg 'ham' plant and its North Battleford 'bacon' plant. Michael McCain, president of Maple Leaf Foods, described competing for a place in the North American market for pork products as 'akin to dancing with elephants. We have to be a little more nimble, a little more agile than our dancing partners' (Bell, 1999: 60).

Western Canada now accounts for 40 per cent of hog production in Canada. Manitoba

Mike Grandmaison/All Canada Photos

Photo 8.9

Spring wheat has been challenged by canola as the principal crop in Western Canada. The reason is not growing conditions but price—canola commands a higher price than wheat. Still, red spring wheat remains a popular crop because of its ability to survive droughty conditions. The farmer is using an old model of swather to cut his wheat while a modern concrete grain terminal with a Saskatchewan Wheat Pool logo occupies the background. Viterra now owns all Saskatchewan Wheat Pool assets.

leads with 3 million pigs, Alberta has 2.1 million, and Saskatchewan follows with 1.2 million. Large-scale hog barns have led the way to this remarkable expansion. Hog barns, like cattle feedlots, produce a great deal of waste that could threaten the local water supply. At a community meeting in 2003 in Foam Lake, Saskatchewan, the issue of economic benefits from a proposed large hog barn was challenged by the possible environmental costs. The proposal included a breeder farrow barn for 5,000 sows, a nursery barn, and three finisher barns. Pig excretion would be placed in lagoons and, in time, would be removed as manure that could be placed on fields as fertilizer. Grain farmer Harry Abtosway had heard enough. He complained that 'Pollution, environment, smell, you name it. I think we are going to ruin our land, our water supply' (Hall, 2003: C8). Large hog barns are not popular with local people, but then, as Terry Markusson, the proponent of the proposed hog barn, put it: 'There is nothing that smells as bad as a dying community.' In March 2003, the Foam Lake municipal council rejected the proposal after residents submitted a petition with over 600 signatures opposing the hog barn plan.

Western Canada's Resource Base

Western Canada's resource economy began to diversify in the 1970s, when oil and gas developments in Alberta benefited from rising prices, uranium mining gained a foothold in Saskatchewan, and Manitoba expanded its hydroelectric facilities on the Nelson River. The dramatic rise in oil prices, resulting from the action of the Organization of Petroleum Exporting Countries in 1973 to control supply, gave an extra boost to Alberta's economy in the 1970s. By 2005, oil, natural gas, electricity, potash, and uranium all benefited from global demand and high prices, but these prices crashed in the 2008–9 economic crisis and are just regaining their strength.

The impact of the global crisis on Canadian exports saw total exports drop from $450.2 billion in 2005 to $369.7 billion in 2009,

with oil and gas exports falling from $66 billion to $58 billion (Statistics Canada, 2006, 2010). However, exports of oil and gas formed a slightly larger percentage of total exports in 2009, at 15.2 per cent compared to 14.7 per cent in 2005, indicating the growing demand for Canadian energy in the United States.

On the other hand, grain continues to suffer from low prices due to increased world production, and the Canadian Wheat Board anticipated low prices for wheat, durum, and barley in 2010 (Jubinville, 2009). Low grain prices explain why farmers are turning to other crops, especially canola and lentils, where prices are stronger. Spring wheat, for instance, accounted for less than $2 billion in 2005. While signs are positive for global demand for agricultural products to increase, reality tells Western Canada that with record prices for fossil fuel and potash deposits, the region's wealth remains with these resources (Figure 8.7), as

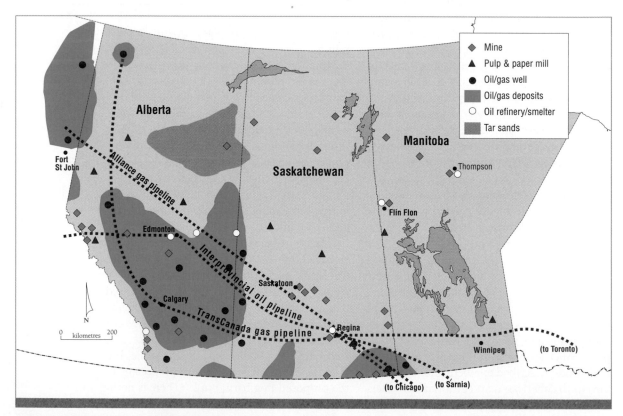

Figure 8.7   Western Canada's resource base, 2010.
The mines, pulp and paper mills, oil and gas wells, and oil refineries and smelters that dot Western Canada's landscape attest to the region's range of natural wealth. The Alliance Pipeline, completed in 2000, is a natural gas pipeline extending 3,000 kilometres from just north of Fort St John, BC, to Chicago, Illinois. Construction should be completed by late 2010 on two bitumen oil pipelines—Alberta Clipper (Figure 8.9) and Keystone—to feed US refineries. Huge petroleum resources are found in Alberta and, to a lesser degree, in Saskatchewan. Saskatchewan has the world's largest reserves of potash and uranium.

well as with hydroelectricity. Manitoba Hydro has three new projects on the drawing board—Wuskwatim (200 MW), Gull/Keeyask (620 MW), and Conawapa (1,380 MW). Power from these projects will flow to Ontario and neighbouring American states.

Energy: Oil and Natural Gas

Oil and gas are Western Canada's most valuable natural resources. Western Canada contains over 70 per cent of Canada's known oil reserves. These oil deposits are located in the southern half of the Western Sedimentary Basin. Alberta has the largest oil reserves (55 per cent of Canada's reserves), followed by Saskatchewan (17 per cent). Manitoba has less than 1 per cent.

Oil production from Alberta's vast reserves of oil and natural gas drives that province's economy. Technological innovations have had an impact on oil and gas production (Vignette 8.10). By 2008, oil from the **Bakken formation** in southern Saskatchewan and Manitoba greatly increased those provinces' production. Western Canada's production of oil comprised 85 per cent of the national figure, with Alberta accounting for 68 per cent and Saskatchewan 16 per cent. (Table 8.5).

In addition to oil and natural gas deposits, vast amounts of oil are contained in the tar sands—oil mixed with sand, known as **bitumen**—of northern Alberta. Extraction takes place in three areas, Fort McMurray, Peace River, and Cold Lake (Vignette 8.11). With most oil sands deposits too deep to mine, an in-situ system is employed, which is similar to conventional oil production. Operations began with open-pit mining because of its low cost per unit of output; more recent operations have had to extract bitumen from much deeper deposits, necessitating the employment of the more expensive method. Suncor Energy, the largest oil sands

Vignette 8.10 Technological Breakthrough: Horizontal Drilling in Oil Shale

Horizontal drilling, often called directional drilling, has revolutionized the way that oil and gas wells are drilled. First, horizontal drilling will produce many times more oil and gas than a vertical well. A vertical well only penetrates a few feet of the oil or gas zone, but a well drilled horizontally may penetrate several thousand feet into this zone. Second, unproductive rock formations, such as the Bakken oil shale, have become productive due to horizontal drilling. But the story does not end here. Since the Bakken oil shale has poor porosity, oil could not flow along these horizontal drill holes. The solution is hydraulic fracturing, which involves pumping fluid and rounded beads to break and keep the fissures in the shale open and thus allow oil to flow. One downside to this technology is that the toxic fluid used in the process can enter the water table and underground aquifers, making the water unfit for drinking.

company, which purchased Petro-Canada in 2009, operates open-pit mines at Millennium and Steepbank and in-situ production at MacKay River and Firebag. In-situ operations call for steam-assisted gravity drainage (SAGD) where parallel pairs of horizontal wells are drilled: one for steam injection and one for oil recovery. The bitumen is then sent by pipeline to Suncor's upgrading facility at Fort McMurray. The other oil sands company now in production is Syncrude Aurora, with operates open-pit mines. Other projects are at the proposal stage (Oilsands Developers Group, 2009).

The economic impact of petroleum on Western Canada, especially Alberta, has been remarkable, and prices, as shown in Figure 8.8, have continued to climb. A combination of investments, demand for labour, and royalties has transformed Alberta into the fastest-growing province in Canada as well as the richest province. In terms of the

Think About It

Given the advantages of trade with the US, is it in Canada's long-term interest to encourage the proposed construction of Enbridge's $5 billion Gateway pipeline from Edmonton to Kitimat on the Pacific coast?

Table 8.5	Oil Production in Western Canada, 2008 (millions m³)				
	Alberta	**Saskatchewan**	**Manitoba**	**Western Canada**	**Canada**
Production	108.7	25.6	1.4	135.7	159.3
Percentage	68.2	16.1	0.9	85.2	100

Source: Statistics Canada (2009a).

core/periphery model, the export of oil reversed a century-old economic relationship—Central Canada was dependent on the western provinces. Ottawa interfered with this new relationship by introducing the National Energy

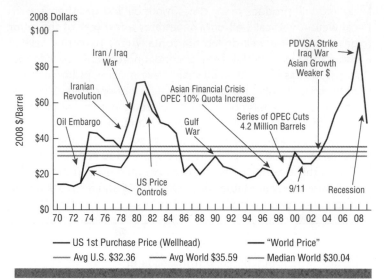

Figure 8.8   Crude oil prices, 1970–2009.
Source: WTRG Economics (2009), www.wtrg.com.

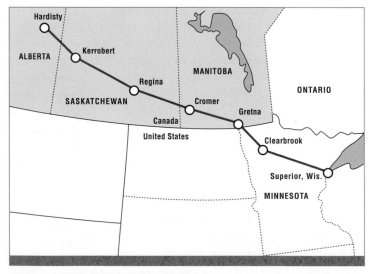

Figure 8.9   Enbridge Pipeline: The Alberta Clipper.
The Alberta Clipper pipeline will transport bitumen from Alberta's oil sands to refineries in the US Midwest by mid-2010. TransCanada's Keystone pipeline, also beginning at Hardisty, Alberta, is planned for completion in late 2010. Keystone will move oil sands bitumen to the US Midwest, via Manitoba and Kansas, and may eventually extend to under-utilized refineries in the US Gulf states. These Gulf refineries now use bitumen from Venezuela.
Source: Enbridge Energy in Schwartz (2009).

Program in 1981 (which was later jettisoned when Brian Mulroney and the Progressive Conservatives took office in 1984). The federal government was seen by Alberta as intruding on provincial rights and attempting to transfer the wealth away from Alberta. The results of this federal program were to widen the political gap between Alberta and Ottawa and to deepen the sense of western alienation and mistrust towards Ottawa and Central Canada.

The petroleum industry has helped to diversify the economy of Western Canada. In 2008, the three Prairie provinces produced 160 million barrels of petroleum, with Alberta accounting for 80 per cent. The production in Western Canada amounted to 85 per cent of the national output (Statistics Canada, 2009a). In Alberta today the oil sands are in the process of rapid development (Vignette 8.11). Demand from the US is growing and other countries, such as China, Japan, and Korea, are investing in the oil sands with the goal of securing a supply of oil in the next decade. To further its ambitions, in 2009 PetroChina invested $1.7 billion to acquire a majority stake in Canada's Athabasca Oil Sands Corporation, which holds properties at MacKay River and Dover, and the same year South Korea's Korea National Oil Corporation bought an oil sands property from Canada's Harvest Energy Trust.

Alberta is faced with two challenges. First, should the pace of development be regulated rather than leaving it to the market? Some, such as former Alberta Premier Peter Lougheed, argue that, by regulating the number of new construction projects in the oil sands, Alberta would benefit from a more stable economy, especially in the construction industry. Second, and related to the first challenge, others believe that the dependency on oil exports, especially unprocessed bitumen, is not in Alberta's interest and perpetuates the image of **'hewers of wood and drawers of water'**. Yet, market conditions have caused two pipeline companies, TransCanada and Enbridge, to build the Keystone and the Alberta Clipper pipelines, respectively, to export Alberta bitumen to US refineries (Figure 8.9). Enbridge also has proposed a pipeline to Kitimat for oil exports to China and other Pacific Rim countries.

Vignette 8.11   Alberta Oil Sands

Canada's oil sands deposits represent one of the world's largest oil reserves. Because of the oil sands, Canada ranks second in the world in terms of proven oil reserves. Oil sands are found in three areas of northern Alberta—Fort McMurray, Cold Lake, and Peace River (Figure 8.10). Alberta oil sands production reached 267 million barrels in 2007, up by almost five times over its 1997 output level.

Development of the oil sands is not without environmental consequences. The mining process requires huge amounts of water to extract the tar-like bitumen and turn it into synthetic crude oil. According to the Pembina Institute (2007), the ratio of water to bitumen is 2 to 4.5 cubic metres of water to 1 cubic metre of bitumen. During this process, the water taken from the Athabasca River becomes highly toxic and, thus, cannot be returned to the river but is stored in tailing ponds. Production is expected to double again in the next five years, which means that the demand for water will grow, as will the size of the tailing ponds, while water levels downstream and in Lake Athabasca will continue to fall. Suncor, one of the major oil companies, has accelerated its plans to restore mature tailing ponds back to a natural state.

Mining in Western Canada

Like the oil and gas industry, the mining industry has helped to diversify the economy of Western Canada. The variety and value of mineral production in Western Canada are enormous. Although the Canadian drop in value of mineral production from 2008 to 2009 was 31.5 per cent, from $47 billion to $32.2 billion, Western Canada, led by Saskatchewan's production value of $5 billion, was the leading Canadian region in mineral production for 2009 at $8.3 billion (Natural Resources Canada, 2010). Mining companies produce a full range of mineral products, including metals, non-metals, structural materials, and fuels. The geology of each province differs sufficiently to produce three distinct types of mining. Alberta contains rich coal reserves along the eastern slopes of the Rocky Mountains. Coal mining began in 1872 just west of Lethbridge. By the 1990s, Alberta produced more coal than any other province, which contributed to Canada's position as the fourth-largest coal exporter in the world. Today, coal mining takes place in the East Kootenay and Peace River coalfields. Like other resource industries, these Alberta mines depend on world demand and prices. For instance, in 1998 the Asian economic slump resulted in a drop in demand and price for Alberta coal, and, again in 2009, the global

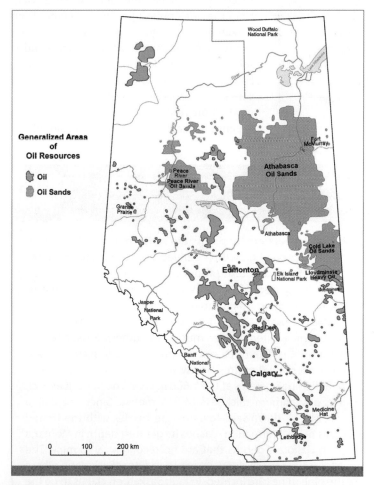

Figure 8.10   Alberta's hydrocarbon resources: Oil sands and oil fields.
Source: Alberta (2007) Energy Resources Conservation Board/Alberta Geological Survey.

economic crisis saw a drop in demand and price. By early 2010, however, the coal industry was on the road to recovery.

Table 7.5, page 298, shows production values for minerals in 2009 by province and territory.

Potash, uranium, and diamonds are the major mineral deposits in Saskatchewan, though only potash and uranium have been developed. The **potash** deposit lies approximately 1 km below the surface of the earth, reaching its thickest extent around Saskatoon, where six of the nine potash mines are located. Canada is the world's largest producer and exporter of potash. Saskatchewan has the largest and highest-quality deposit in the world, and except for a potash mine in New Brunswick, all of Canada's potash production comes from Saskatchewan (Vignette 8.12). Potash is used to produce potassium-based fertilizers and is sold to firms in the United States and Asia, especially China and Japan. World demand varies, but Canadian potash producers adjust production volume to maintain price—at the expense of their workers, who are laid off.

Think About It

How can potash companies vary production to keep prices high while farmers cannot vary production?

Vignette 8.12   Potash: Saskatchewan's Underground Wealth

Potash is a general term for potassium salts. Potassium (K), a nutrient essential for plant growth, is derived from these salts. Roughly 95 per cent of world potash production goes into fertilizer, while the remainder is used in a wide variety of commercial and industrial products, ranging from soap to explosives.

In Saskatchewan, potassium salts are found in the Prairie Evaporite, which extends over much of southern Saskatchewan at varying depths. The three potash mines near Saskatoon operate at just over 1,000 metres below the surface while Belle Plain mine near Regina operates at the 1,600 m level. The deposit at each mine has a maximum thickness of 210 metres. Since the Prairie Evaporite slopes downward towards the border with the United States, this potash formation reaches its greatest depth in Montana and North Dakota—depths that are not economical to mine. The more accessible potash deposits are found near Saskatoon where six of the nine mines are located. In fact, Saskatoon claims to be the 'Potash Capital of the World'.

Uranium mining also takes place in Saskatchewan. Production began in 1953 on the northern shores of Lake Athabasca near Uranium City. Since the late 1970s uranium mining has shifted south to the geological area of the Canadian Shield known as the Athabasca Basin. Currently, five mines are operating—Rabbit Lake, Eagle Point, Key Lake, McArthur River, and McClean Lake.

Manitoba has two major mineral deposits—copper-zinc and nickel—both located in the Canadian Shield. Mining for copper and nickel in Manitoba is relatively expensive. As well, the high cost of shipping the processed product to distant markets adds to their economic disadvantage. Copper and zinc ore bodies are found near Flin Flon, a mining and smelter town in northern Manitoba that began production in 1930, shortly after a rail link to The Pas was completed in 1928. At one time, Flin Flon produced most of Canada's copper and zinc but today it is an aging resource town with a declining population. Thompson, located some 740 km north of Winnipeg, is a nickel-mining town. In 1957, after a rail link to the Hudson Bay Railway was completed, the mine facility, smelter, and town were constructed. Unlike Flin Flon, Thompson was a specially designed resource town with a complete array of urban amenities. The first nickel was produced in 1961. During the 1960s, Thompson's population soared to 20,000. Since then, the population of Thompson and of many other resource towns has dwindled: greater mechanization in the mining process results in a smaller labour force, and this, in turn, leads to a contracting service sector. In 1991, Thompson had a population of 14,977, but by 2006 it had fallen to 13,446. For similar reasons, Flin Flon's population dropped from 7,119 in 1991 to 5,594 in 2006.

Forest Industry: On the Skids

While the boreal forest stretches across the northern part of Western Canada, the forest industry depends on exports to the United States. As with other regions, Western Canada's forest industry has hit the skids. Plants have closed and little logging is taking place. Aboriginal workers and businesses have been particularly hard hit because they depend heavily on this industry for their livelihood.

For discussion of forestry exports to the US, see Chapter 7, 'Dependency on the US Market', page 305, and 'North American Free Trade Agreement', page 306.

In 2007, logging in Western Canada amounted to 15 per cent of Canada's total timber harvest (Table 8.6). Approximately 75 per cent of this production takes place in Alberta.[12] Alberta's advantage comes from its large northern forest stands. The forest-covered Interior Plains region, which extends beyond the North Saskatchewan River to the border with the Northwest Territories, is the largest forest stand in Western Canada. This, together with the timber in the foothills of the Rocky Mountains, means that Alberta has by far the largest commercial forest of the three Prairie provinces.

The federal government's Green Transformation Program is discussed in Chapter 9, 'Forest Industry: A Weak Sister', page 387.

Each of Western Canada's three provinces has a diversified timber operation, ranging from sawmills to pulp and paper plants. Most pulp and paper mills are located on the southern edge of the forest, but with sales to the United States falling sharply, several mills have closed, included the mill at Prince Albert. In Alberta, mills are located in Athabasca, Grande Prairie, Hinton, and Peace River. In Manitoba, The Pas is the site of a major pulp and paper mill. By 2009, the plight of these mills was apparent and the federal government, under its Green Transformation Program, provided financial assistance for them to invest in capital projects that improve their environmental performance and economic efficiency.

Western Canada's Population

Western Canada's population has undergone significant changes since settlers first came to this region, and these changes have affected

Table 8.6 Forest Harvest, Western Canada, 2007

Province	Area (ha)	Volume (000s m³)
Manitoba	13,648	2,000
Saskatchewan	13,000	2,412
Alberta	54,981	20,513
Western Canada	81,629	24,925
Percentage Canada	11.1	15.3

Source: Natural Resources Canada (2009).

and been affected by many aspects of prairie life. In 1921, Western Canada had a population of nearly 2 million. At that time, it comprised 22 per cent of Canada's population. By 2006, its population had increased to 5.4 million but accounted for only 17 per cent of the national population.

Two migrations transformed Western Canada. Just 100 years ago, land-hungry homesteaders poured into the Prairies, creating a rural landscape of small farms and villages. No one could have foreseen that this rural landscape would lose most of its population in a remarkably short time. In the second migration, the so-called rural-to-urban migration, rural people moved from the countryside to towns and cities in Western Canada and to urban centres in British Columbia, Ontario, and the United States. The push factors of this migration were the mechanization of the agricultural sector, the consolidation of farms, and the shrinking need for farm labour that resulted. Pull factors included the employment opportunities and greater amenities that existed in towns and cities. The rural-to-urban migration resulted in a redistribution of people within Western Canada. Today, most people live in the five major cities of the region: Calgary, Edmonton, Winnipeg, Saskatoon, and Regina (Figure 8.1).

Within Western Canada, Alberta has the largest population, at 3.3 million. Manitoba has 1.1 million and Saskatchewan just under 1 million. Over the last 25 years, the rate of population increase has been highest in Alberta, followed by Manitoba. Recent population changes between the 2001 and 2006 censuses confirm that this trend continues, with Alberta growing at a rate of 10.6 per cent over this five-year period followed

Think About It

The Indian Posse is only one street gang. The larger question is, why has the gritty world of street gangs, which cuts across ethnic lines, become a common feature of Canadian cities?

by Manitoba at 2.6 per cent. Although Saskatchewan's population decreased by 1.1 per cent over the 2001–6 period (Statistics Canada, 2007b), the province has rebounded due to its post-2006 economic boom. By 1 July 2008, Saskatchewan's population was estimated to have surpassed one million for the first time (Statistics Canada, 2009c).

Another demographic feature of Western Canada is the size of the Aboriginal population. By 2006, just over half a million Aboriginal people lived in Western Canada, forming nearly 10 per cent of Western Canada's population. The high natural rate of increase of Aboriginal peoples accounts for the fact that both Indian reserves and Métis communities and the urban Aboriginal population are increasing. The attraction of cities is due to the pull of urban amenities and opportunities and the push of insufficient employment opportunities and social problems on reserves and in Native communities. While an Aboriginal middle class has emerged in cities, poverty-driven social problems take the form of street gangs, drug and alcohol abuse, and

Al Harvey/Slide Farm

Photo 8.10

Saskatoon, situated on the South Saskatchewan River, is Saskatchewan's largest city with a population of nearly 234,000 (2006). Known as the 'Bridge City', Saskatoon has witnessed rapid population growth over the last 25 years. During that period, many Indians and Métis have settled in Saskatoon. By 2006, the Aboriginal population comprised over 10 per cent of Saskatoon's residents.

prostitution. Winnipeg's North End is a troubled neighbourhood where the Indian Posse holds sway. The 2004 film, *Stryker*, by Noam Gonick, captures the brutal turf war between two street gangs in Winnipeg's North End.

Western Canada's Urban Geography

The process of urbanization in Western Canada has lagged behind that of Ontario, British Columbia, and Québec. Even so, nearly three-quarters of Western Canada's residents live in urban centres, and most of these reside in the five major cities—Regina, Saskatoon, Winnipeg (Vignette 8.13), Edmonton, and Calgary. A second order of urban centres includes Airdrie, Lethbridge, Red Deer, Grande Prairie, Medicine Hat, Wood Buffalo (Fort McMurray), Brandon, Prince Albert, and Moose Jaw. In 2006, these cities had a combined population of 3.7 million (Table 8.7). From 2001 to 2006, the rate of urban growth varied considerably for the towns and cities of Western Canada. This variation reflects differences in local economic growth and in the pace of consolidating populations into regional centres. From 2001 to 2006, the fastest-growing cities in Western Canada were in Alberta: Airdrie at a remarkable 41.8 per cent, Wood Buffalo at 23.6 per cent, Grande Prairie at 22.3 per cent, and Red Deer at 22 per cent. The fastest-growing cities in Saskatchewan and Manitoba were Saskatoon at 3.5 per cent and Brandon at 4.3 per cent (Table 8.7).

Calgary–Edmonton Corridor

The Calgary–Edmonton Corridor has emerged as the most urbanized region in the province and one of the densest in Canada. With major universities and colleges in its cities, it is a hub for the knowledge-based activities where high-tech industries thrive and cutting-edge research takes place. Over the past five years, this urban corridor has had a high rate of population growth, exceeding 10 per cent.

Anchored by Calgary and Edmonton, the 400-kilometre corridor includes the cities of Airdrie and Red Deer and a host of

Vignette 8.13   Winnipeg

Winnipeg, the capital and the largest city in Manitoba, is located at the confluence of the Red River and the Assiniboine River—the location of Lord Selkirk's early nineteenth-century settlement for Scottish crofters that became the Red River Settlement with a predominantly Métis population. With the building of the Canadian Pacific Railway, Winnipeg became known as the 'Gateway to the West'. By the turn of the twentieth century, Winnipeg was the principal city in Western Canada, controlling the grain trade and also acting as the administrative, financial, and wholesale hub for Western Canada. Today, Winnipeg dominates the economy of Manitoba, but its role within Western Canada has been considerably diminished by the growth of Calgary and Edmonton and, to a lesser degree, by Saskatoon and Regina.

After the completion of the Panama Canal in 1914, Alberta farmers and some Saskatchewan farmers began to ship their wheat through Vancouver instead of through Winnipeg. Next, service industries began to emerge in Edmonton, Calgary, Saskatoon, and Regina. Each city captured some of the trade previously held by Winnipeg merchants. Winnipeg's stranglehold over the sale and distribution of agricultural machinery and products to the farmers of Western Canada was broken.

After World War II, Winnipeg still remained the largest city in Western Canada. However, Calgary and Edmonton grew rapidly in the following years, partly because of developments in the oil and gas industries. In 1951, however, Winnipeg remained the largest city in Western Canada, with a population of 357,000 compared to 177,000 for Edmonton and 142,000 for Calgary. By 1981, both of these cities had surpassed Winnipeg in population. This demographic trend continued. By 2006, Calgary and Edmonton's populations had jumped to just over one million. Winnipeg, in sharp contrast, saw its population slowly expand to 706,900. Winnipeg, however, is attracting many Aboriginal peoples. These urban Aboriginals make up 10 per cent of the city's population.

Table 8.7   Major Urban Centres in Western Canada, 2001–6

Centre	Population 2001	Population 2006	% Change
Airdrie	20,382	28,892	41.8
Moose Jaw	33,519	33,360	−0.5
Prince Albert	41,460	40,766	−1.7
Brandon	46,273	48,256	4.3
Wood Buffalo	42,581	52,643	23.6
Medicine Hat	61,735	68,822	11.5
Grande Prairie	58,787	71,868	23.3
Red Deer	67,829	82,772	22.0
Lethbridge	87,388	95,196	8.9
Regina	192,800	194,971	1.1
Saskatoon	225,927	233,923	3.5
Winnipeg	676,594	694,668	2.7
Edmonton	937,845	1,034,945	10.4
Calgary	951,494	1,079,310	13.4
Total	3,444,614	3,760,392	9.0

Source: Statistics Canada (2007b). Adapted from Statistics Canada publication *Population and Dwelling Counts, 2006 Census,* Catalogue 97-550-XWE2006002 http://www.statcan.ca/bsolc/english/bsolc?catno=97-550-XWE2006002.

Table 8.8   Aboriginal Population* by Province, Western Canada

Province	Population 2001	Population 2006	Change (%)	Regional (%)**
Alberta	156,225	188,365	20.3	16.1
Manitoba	150,045	175,395	16.5	15.0
Saskatchewan	130,185	141,890	9.0	12.1
Western Canada	436,455	505,650	13.2	43.2
Canada	976,305	1,172,790	20.1	100.0

*Population based on responses to the identity (not ancestry) question in the 2006 census. **Percentage of total Aboriginal population
Source: Statistics Canada (2008).

smaller centres. Calgary has become one of Canada's key corporation headquarters, especially for the oil and gas industry. British Petroleum, EnCana, Imperial Oil, Suncor Energy, Shell Canada, and TransCanada are headquartered in Calgary, as is Canadian Pacific Railway. With its proximity to Banff, the Rocky Mountains, and Kananaskis Country, Calgary offers easy access to mountain recreation. Edmonton, besides being the provincial capital, is a petrochemical industrial node. Known as the 'Gateway to the North', Edmonton serves as a staging centre for the oil sands and for diamond mining in the Northwest Territories.

Faultline: Aboriginal/ Non-Aboriginal Populations

Western Canada has the highest proportion of Aboriginal population of the five regions in southern Canada. Not surprisingly, many Aboriginal/non-Aboriginal achievements, challenges, and problems are found in this region. These issues range from employment to housing and health care. Demography helps explain this focus.

Think About It

Do urban places offer a solution to the economic and social issues facing First Nations peoples?

First, most Aboriginal peoples reside in Western Canada (Table 8.8). Second, cities and towns in Western Canada contain the highest percentages of Aboriginal peoples (Statistics Canada, 2009d). Winnipeg leads the way with 68,380 Aboriginal peoples, representing 10 per cent of city's total population. Edmonton and Calgary, with 52,100 and 26,575, have the next largest urban Aboriginal populations, comprising 5 per cent and 2 per cent of their respective populations. Saskatoon (21,535) and Regina (17,105) also have large Aboriginal populations, making up 9 per cent of these cities' populations. The highest percentages are found in smaller cities near large First Nations reserves. Thirty-six per cent of the population of the northern mining town of Thompson, Manitoba, is Aboriginal, while Prince Albert, Saskatchewan, the gateway to northern Saskatchewan, has 34 per cent.

Further discussion of Aboriginal demography is found in Chapter 4, 'The Expanding Aboriginal Population', page 158. Table 1.2, page 8, explains the Statistics Canada distinction between Aboriginal identity and Aboriginal ancestry.

SUMMARY

Geography has dealt Western Canada a full hand of resources, but also two deuces—distance and climate. The long haul to ocean ports makes its export products more expensive while its dry, continental climate makes crop agriculture uncertain. Price plays a critical role. For example, low prices for forest products have stalled this industry while high prices for energy, potash, and canola have had the opposite effect. In the last decade, canola has exceeded spring wheat in returns to farmers and has matched spring wheat in sown acres.

Exports are the lifeblood of this region, and the emergence of Asian industrial powerhouses and their burgeoning middle classes have created a strong demand for many of Western Canada's resources, including foodstuffs, potash, and uranium. Added to this growing export market, the United States remains the principal importer of oil, gas, and hydroelectricity. Alberta's oil sands, which contain the second-largest oil reserves in the world, supply the US with over 10 per cent of its oil needs and, with further expansion, could supply more than 20 per cent. From a geopolitical perspective, Alberta's oil sands play a key role in US energy security and this strategic fact means that the construction of additional oil sands plants, upgraders, and pipelines will continue and thus anchor Western Canada's economic boom.

The transformation of Western Canada's economy is clear—more processing of agricultural products, a larger service sector, and growing urban centres where knowledge-based research clusters focus on innovations in agriculture, oil sands, and mining. If the upward-moving trend line for global prices continues for the region's underground wealth, a robust economy is secure for the short and middle terms. These positive signs raise the question: Has Western Canada reached a turning point where the promise of Next Year Country finally arrived? The only dark clouds on the horizon are:

- the threat of lower global prices;
- failure of carbon capture and storage to solve the 'dirty oil' issue;
- global warming and climate change, which could result in a shortage of water for agriculture, industry, and cities and create havoc for trucking goods along winter roads.

CHALLENGE QUESTIONS

1. Palliser's Triangle is one of the driest areas in the Canadian Prairies. Besides low precipitation, what other factor causes evapotranpiration to be so high in this area?

2. Distance to global markets translates into high transportation costs for farmers. In the 1920s, farmers called for an 'on to the Bay' solution based on the old Hudson's Bay Company route to England. The attraction of this route is that the length of the rail line is much shorter than the line to Montréal. But why was the Hudson Bay Railway route not successful?

3. Nature plays a critical role in prairie agriculture. Crop yields vary from year to year. Besides low soil moisture, what other natural factors affect size and quality of the harvest?

4. Wheat is no longer king. Why are more and more farmers switching from wheat to canola and specialty crops?

5. Why doesn't Alberta insist that all bitumen produced in its oil sands be refined in the province, thus reversing the image of Canada as a nation of 'hewers of wood and drawers of water'?

FURTHER READING

Casséus, Luc. 2009. 'Canola: A Canadian Success Story', *Canadian Agriculture at a Glance*. Ottawa: Statistics Canada Catalogue no. 96–325–X.

In the 1970s, canola (an abbreviation for 'Canadian oil') was developed by plant breeders in Saskatchewan and Manitoba to yield food-grade oil. Canola is often grown in rotation with Canada's traditional cereal crops of wheat, oats, and barley. The area seeded with canola varies according to prevailing economics, among other factors, but the trend is to increasing area. Wheat, on the other hand, has been showing the opposite trend in planted area. Increasing yields, improved marketing, higher-quality crops, and increases in both canola prices and quantity sold have helped boost Canada's farm cash receipts for canola. By 2005 the crop had surpassed wheat to become the most valuable field crop in Canada. In 1976, canola accounted for only 4.9 per cent of total crop receipts; by 2006, this percentage had climbed to 17.2 per cent. Over the same period, wheat receipts dropped from 35.5 per cent to 15.0 per cent of total crop receipts. Farm cash receipts for wheat totalled $2.2 billion in 2006, while those for canola totalled $2.5 billion.

9

ATLANTIC CANADA

INTRODUCTION

Atlantic Canada, located on the eastern margins of the country, is far from the centres of economic and political power in Central Canada. Yet, its proximity to Europe allowed French settlers to reach its shores as early as the seventeenth century; English and Scottish settlers would follow later. Fish and timber were among the first exports, but harvesting extended beyond sustainable levels and gradually Atlantic Canada became an 'old' resource hinterland.

For some time, Atlantic Canada has lagged far behind the other Canadian regions in terms of economic and population growth. Heavily dependent on equalization payments, Atlantic Canada is classified as a 'have-not' region. Not surprisingly, then, the region suffers from out-migration of its younger members, thus accelerating the greying of its population and creating a shortage of skilled workers. For rural Atlantic Canada, the failure of the northern cod stocks to regain their former prominence has seriously hurt the local fishing economy and has resulted in the decline and even closure of many coastal communities. For many people and many communities, the cultural loss—the loss of a way of life—has been at least as devastating as the economic difficulty.

Yet, Atlantic Canada shows signs of economic rejuvenation, driven largely by offshore oil and gas developments, its role as an energy supplier to New England, and its position as a gateway for container traffic into Central Canada and the US Midwest. Under these positive circumstances, the region's economic pace shows promise, but virtually all has taken place in the major urban centres, leaving rural Atlantic Canada in a seemingly relentless decline. Still, the task of revitalizing Atlantic Canada's economy remains. Until then, the region's dependency on Ottawa and the outflow of its younger population to the faster-growing areas in Canada will undercut its rejuvenation. In this chapter, the *Key Topic*, the fishing industry, demonstrates both its declining role in Atlantic Canada's economy and its negative impact on coastal communities.

CHAPTER OVERVIEW

Important issues and topics examined in this chapter include:

- Atlantic Canada's physical and historical geography.
- The basic characteristics of Atlantic Canada's population and economy as seen from two perspectives: an 'old' resource hinterland and a rejuvenating one.
- The population and economy within the context of Atlantic Canada's physical
- setting, especially its fragmented geography.
- The offshore energy resources of the region.
- The cod fishery as an example of the 'tragedy of the commons'.

St. John's, Newfoundland is alive with bright colours while its harbour provides safe haven. Photo: Sean White/All Canada Photos.

Atlantic Canada within Canada

Stretching along the country's eastern coast, the region of Atlantic Canada consists of two parts: the Maritimes (Nova Scotia, Prince Edward Island, and New Brunswick) and Newfoundland and Labrador[1] (Figure 9.1). In spite of these two geographic parts, Atlantic Canada is united by a rich sense of place that has grown out of the region's history, its British and French settlements, and its geographic location on the Atlantic Ocean. Through all of this history and geography, the Atlantic Ocean has dominated the history of this region from the early days of the fishery to the 'Golden Age of Wooden Ships and Iron Men', to the dream of an Atlantic Gateway. Offshore petroleum developments, the energy corridor to New England, and the wealth returning to Atlantic Canada from its commuters to Alberta's oil sands have added new dimensions to the regional economy. Still, the image of Atlantic Canada remains the small fishing community like Diamond Cove (photo 9.1).

In Chapter 1, 'Canada's Geographic Regions', page 6, the rationale for Canada's six geographic regions, including Atlantic Canada, is elaborated.

Dale Wilson/All Canada Photos

Photo 9.1

The island of Newfoundland still has hundreds of small outports, like Diamond Cove, along its rocky and indented coastline. Since the loss of the cod fishery, most outports no longer have an economic base and their future is in doubt.

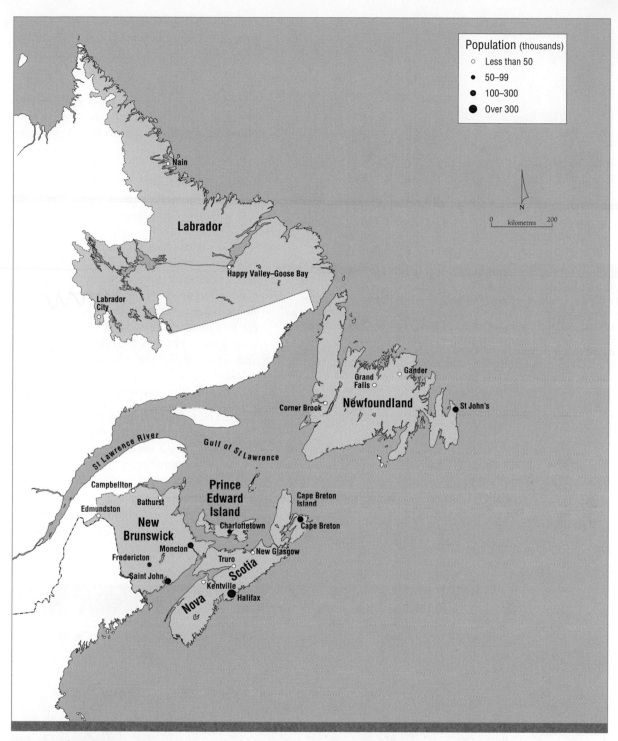

Figure 9.1   Atlantic Canada.
Atlantic Canada contains four provinces, Nova Scotia, New Brunswick, Newfoundland and Labrador, and Prince Edward Island. These four provinces have the smallest populations and, historically, the weakest economies of the 10 Canadian provinces, although its offshore oil resources recently have made Newfoundland and Labrador a 'have' province. The largest city is Halifax, with a population of just under 400,000, ranking thirteenth among Canada's census metropolitan areas.

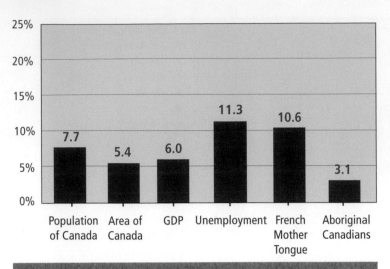

Table 1.4, 'National Version of the Core/ Periphery Model', page 19, describes the four regions of Friedmann's model.

Figure 9.2   Atlantic Canada vital statistics.
Except for its oil and gas sector, Atlantic Canada has a weak economy. All four provinces, until 2009, have been considered 'have-not' provinces and have received equalization payments. However, Newfoundland and Labrador did not receive equalization payments for 2009–10. Atlantic Canada has 7.2 per cent of Canada's population but produces only 6.0 per cent of the country's GDP. The high rate of unemployment is another indication of its poor economic performance. Acadians account for the high number of French-speaking Canadians in Atlantic Canada, which has the second-highest percentage by region in Canada. In 2006, Aboriginal peoples, however, formed only 3.1 per cent of the region's population.
Note: Unemployment percentage is for 2009; demographic data are based on the 2006 census.
Sources: Tables 1.1, 1.2, 5.1.

time, the region was troubled by past exploitation of its renewable resources and the exhaustion of its most accessible and richest non-renewable resources. Atlantic Canada's troubles are summed up by the subpar performance of its economy relative to the rest of Canada and the seemingly unstoppable out-migration to faster-growing regions of Canada.

One measure of Atlantic Canada's overall economic performance is reflected in the region's per capita gross domestic product figures and its level of unemployment. In 2006, Atlantic Canada's GDP per capita was the lowest in Canada, while the region's 2006 unemployment rate was the highest (Figure 9.2 and Table 9.1). The primary reasons for Atlantic Canada's weak economic performance include the following:

- The political division of Atlantic Canada into four provinces discourages the emergence of an integrated economy in which economies of scale might occur and increases the cost of government.
- Atlantic Canada has been exploiting its resources for a long time and some

Yet, as Canada's oldest hinterland, Atlantic Canada has experienced both growth and decline over the last 200 years. Atlantic Canada has been, in Friedmann's regional version of the core/periphery model, a downward transitional region. During that

Table 9.1	Basic Statistics for Atlantic Canada by Province				
Province	**Population (000s)**	**Population Density**	**% of National GDP**	**Unemployment Rate (%)**	**% Aboriginal**
Prince Edward Island	136	23.9	0.3	12.0	1.3
Newfoundland and Labrador	506	1.4	1.9	15.5	3.8
New Brunswick	730	10.2	1.7	8.9	2.5
Nova Scotia	914	17.3	2.1	9.2	2.7
Atlantic Canada	2,286	4.9	6.0	11.3	3.1
Canada	31,613	3.5	100.0	8.3	3.3

Note: Unemployment percentages are for 2009; demographic data are based on the 2006 census.
Sources: Tables 1.1, 1.2, 5.1.

of these, such as coal and iron, have been exhausted, while its renewable resources, such as the northern cod, have been overexploited.

- Atlantic Canada's population is widely dispersed and, with the exception of Halifax, consists of small markets.
- Distance from national and global markets has stifled its manufacturing base.

All these factors have made it extremely difficult for economic development to flourish in the region. Furthermore, over the years Atlantic Canada has become heavily dependent on Ottawa for economic support through equalization payments and social programs. Yet, Atlantic Canada has received a second chance with the discovery of offshore oil and gas deposits, its potential as an Atlantic Gateway, and its position as an energy corridor to New England. By taking full advantage of these new opportunities, Atlantic Canada may well become a more resilient and prosperous region and thus shed its moniker as a 'have-not' region.

Atlantic Canada's Physical Geography

Atlantic Canada contains two of Canada's physiographic regions: the Appalachian Uplands and the Canadian Shield. The Appalachian Uplands are found in the Maritimes and the island of Newfoundland while Labrador is part of the Canadian Shield. In terms of geologic time, the Appalachian Uplands represents the worn-down remnants of an ancient mountain chain. Formed in the Paleozoic Era, the Appalachian Mountains have been subjected to erosional forces for some 500 million years. As photo 2.8 illustrates, streams have cut deeply into the Cape Breton Highlands of the Appalachian Uplands, resulting in rugged, hilly terrain. In Labrador, the most prominent feature of this portion of the Canadian Shield is the uplifted and glaciated Torngat Mountains (photo 9.2). Unlike the rest of the Canadian Shield, the Labrador portion was subjected to the mountain-building process (**orogeny**) in which the rocks were folded and faulted some 750 million years ago. More recently in geologic time, these mountains were covered with glaciers, which, as the glaciers slowly moved down slope, carved the mountain features and eventually reached the sea, where they created a fjorded coastline.

See photo 2.2, page 37, which illustrates Labrador's fjorded coastline.

The climate of Atlantic Canada is quite varied because of the meeting of continental air masses with marine air masses. The flow of continental air masses from west to east brings

Think About It

Given its offshore oil and gas resources, might Atlantic Canada be reclassified as an upward transitional region?

Vignette 9.1 Consequences of the Geographic Fragmentation of Atlantic Canada

According to Macpherson (1972: xi), Atlantic Canada is not a homogeneous region but 'a region of geographical fragmentation' consisting of three sub-regions: the Maritimes, Newfoundland, and Labrador. Macpherson argues that these physical divisions have impacted all aspects of life in Atlantic Canada, hindering the emergence of a common political will, a common regional consciousness, and an economic union among the four Atlantic provinces. Whether or not an economic union is a desirable political arrangement for Atlantic Canada is uncertain, though the topic of union has long been debated. As early as 1949, Joey Smallwood, Premier of Newfoundland, proposed a union of the four eastern provinces to enable them to increase their bargaining power in Ottawa. In the 1970s, the Deutsch Commission on Maritime Union recommended a 'full political union as a definite goal'. However, there was no support within Atlantic Canada, and before the 1970s had ended Nova Scotia Premier Gerald Regan pronounced the proposal for a Maritime Union 'as dead as a doornail'. For the residents of Atlantic Canada, the advantages of a union are not readily apparent, and Atlantic premiers have not shown much enthusiasm for this proposal.

Alexandra Kobalenko/All Canada Photos

Photo 9.2

The Torngat Mountains, a national park reserve since 2005, stretch along the fjorded coast of northern Labrador. One such fjord is Ramah Bay (with an iceberg floating in its waters). The mountains were recently, in geologic time, subjected to alpine glaciation, resulting in extremely sharp features, including arêtes, cirques, and horns. These mountains, including Mount Caubvik (also known as Mont D'Iberville), straddle the Québec/Labrador boundary. They attain heights of 1,652 m (5,420 ft) above sea level and are located near the sixtieth parallel. For both reasons—high elevation and high latitude—the Torngat Mountains lie beyond the tree line.

warm weather in the summer and cold weather in the winter, yet, with no part of Atlantic Canada more than 200 km from the Atlantic Ocean, its moderate, marine-type weather is tempered by the cold waters of the Labrador Sea and the warmer waters in the Gulf of St Lawrence and the North Atlantic Ocean. The result is generally unsettled weather. Still, Atlantic Canada, especially Labrador, has a strongly continental aspect to its climate. This continental effect is due to the movement of air masses across Canada. During the winter, for instance, cold, dry air frequently enters Atlantic Canada from the interior of Canada. Occasional winter storms are caused by the meeting of cold Arctic air with warmer, humid air from the south. In summer, occasional incursions of hot, humid

air from the Gulf of Mexico take place, but the dominant weather is cool and rainy. In the winter, influxes of moist Atlantic air produce mild snowy weather.

Annual precipitation is abundant throughout Atlantic Canada. It averages around 100 cm in the Maritimes and 140 cm in Newfoundland, but gradually diminishes further north in Labrador. Much precipitation comes from **nor'easters**—strong winds off the ocean from the northeast—that draw their moisture from the Atlantic Ocean. Atlantic Canada but especially the Maritimes and the island of Newfoundland have foggy weather. Thick, cool fog forms in the chilled air above the Labrador Current when it mixes with warm, moisture-laden air from the Gulf of Mexico. With onshore winds these banks of fog move far inland, but the coastal communities experience the greatest number of foggy days. Both St John's and Halifax, for instance, experience considerable foggy and misty weather.

With such varied weather conditions, Atlantic Canada has three climatic zones—Atlantic, Subarctic, and Arctic zones. The great north–south extent of this region is one reason. For example, the distance from the southern tip of Nova Scotia (44° N) to the northern extremity of Newfoundland and Labrador (60° N) is over 2,000 km. Then, too, Atlantic Canada is the meeting place of Arctic and tropical air masses and ocean currents, resulting in wet, cool, and foggy weather. In addition, close proximity to the Atlantic Ocean exerts a moderating effect on the region's climate.

The Arctic zone of Labrador is found in northern Labrador, in the Torngat Mountains and along the coast. An Arctic storm track funnels extremely cold and stormy weather along the Labrador coast while the **Labrador Current** brings icebergs from Greenland to the Labrador and Newfoundland coastlines. The Arctic zone is associated with tundra vegetation as the summers are too cool for tree growth. The Subarctic climate zone exists over the interior of Labrador. The interior of Labrador experiences much warmer summer weather and this area is associated with the boreal forest. The Atlantic zone includes the Maritimes and the island of Newfoundland. For most of the year, this more southerly area falls under the influence of warm, moist air masses that originate in the tropical waters of the Caribbean Sea and the Gulf of Mexico. Only in the winter months does the Arctic storm track dominate weather conditions. The coastal areas of the Maritimes and Newfoundland can be affected by tropical storms in the late summer and fall.

The main air masses affecting the region originate in the interior of North America and from the Gulf of Mexico and the North Atlantic Ocean. Consequently, summers are usually cool and wet, while winters are short and mild but often

---

Vignette 9.2   Weather in St John's

St John's has one of the most varied but mild climates in Canada. Weather is characterized by foggy days, overcast skies, and stormy, wet weather. Fog is common from April to September. Throughout the year, a salty smell is in the air. While most of the year has mild temperatures, strong northwest winds result in heavy winter snowfalls. Jutting into the Atlantic Ocean at the eastern tip of the island of Newfoundland, the marine climate of St John's is affected by the westerly circulation of the atmosphere and by the waters of the North Atlantic. As David Phillips (1993: 155), Canada's well-known weather pundit, put it, 'Of all major Canadian cities, St John's is the foggiest, snowiest, wettest, windiest, and cloudiest.'

---

associated with heavy snow and rainfall. Most precipitation falls in the winter, and temperature differences between inland and coastal locations are striking. Temperatures are usually several degrees warmer in the winter near the coast than at inland locations. During the summer, the reverse is true, with coastal areas usually several degrees cooler.

Along the narrow coastal zone of Atlantic Canada, the climate is strongly influenced by the Atlantic Ocean. The summer temperatures of coastal settlements along the shores of Newfoundland are markedly cooled by the very cold water of the Labrador Current (a cold ocean current in the North Atlantic Ocean), which brings with it icebergs as far south as Newfoundland. Besides filling St John's harbour with pack ice, icebergs are often sighted offshore as late as June. Another effect of the sea occurs in the spring and summer—fog and mist result when the warm waters of the northeastward-flowing **Gulf Stream**, which originates in the Gulf of Mexico, mix with the cold, southerly flowing Labrador Current. In the winter, the clash of warm and cold air masses sometimes results in severe winter storms characterized by heavy snowfall: St John's, for example, had a record 648.2 cm of snow (almost 22 feet) in the

Vignette 9.3   The Annapolis Valley

The Annapolis Valley is a low-lying area in Nova Scotia. At its western and eastern edges the land is at sea level, but it rises to about 35 m in the centre. The area is surrounded by a rugged, rocky upland that reaches heights of 200 m and more. The fertile sandy soils of the Annapolis Lowlands originate from marine deposits that settled there about 13,000 years ago. After glacial ice retreated from the area, sea waters flooded the land, depositing marine sediments that consisted of minute sand and clay particles. Isostatic rebound then caused the land to lift and slowly these lowlands emerged from the sea. In the seventeenth century, the favourable soil of the Annapolis Valley attracted early French settlers, the Acadians, who built dikes to protect parts of this low-lying farmland from the high tides of the Bay of Fundy and Minas Basin. Today, the Annapolis Valley's stone-free, well-drained soils and its gently rolling landforms provide the best agricultural lands in Nova Scotia. In photo 9.3, at high tide, waters from Minas Basin (seen in the background) extend into the low, wet land in the foreground. Land use is changing with vineyards replacing apple orchards.

Barrett & MacKay/All Canada Photos

Photo 9.3

Nova Scotia's Annapolis Valley, just north of Wolfville near Cape Blomidon. In the foreground is a small apple orchard, for which Annapolis Valley is famous; in the middle is a tidal stream; in the background are the waters of the Minas Basin. Like other apple-growing areas in Canada, vineyards are replacing fruit trees.

winter of 2000–1, and a vicious winter storm hit Nova Scotia and Prince Edward Island on 19–20 February 2004, leaving behind snowfall accumulations of 50–70 cm (Conrad, 2009: 36–7).

Environmental Challenges

Without a doubt, the major environmental disaster to strike Atlantic Canada has been the collapse of the cod fishery. But this disaster was due more too technological advances that enable much larger catches, coupled with federal mismanagement, than to natural factors such as the warming of the Atlantic waters or the expanded seal population consuming vast quantities of cod. Clearly a case of a 'tragedy of the commons', this ecological collapse of one of the world's greatest fish stocks is documented later in this chapter. Here, our attention is focused on an unwanted remnant of Cape Breton's glory days as a major iron and steel centre in Canada—the so-called tar ponds.

By the end of the twentieth century, Sydney, Nova Scotia, not only had lost its industrial sector but the community was saddled with an environmental disaster—the Sydney Tar Ponds, which are composed of tar and a wide variety of chemicals that are dangerous to human beings. Arsenic, lead, and other chemicals are found in the tar ponds. These toxic wastes came from the operation of the Sydney Steel Co., or Sysco, especially its coke ovens, where coal was heated at extremely high temperatures. Waste products from coke ovens include tar and toxic gases. These toxic wastes included benzene, kerosene, and naphthalene.

In the case of Sysco, these wastes were discharged into a nearby stream and gradually seeped into Muggah Creek. Since the 1980s, residents living near the tar ponds complained of orange goo seeping into their basements, while others found that dust from the tar ponds was blowing into their yards and houses. When Health Canada stated that those living closest to the tar ponds were in greater danger than those living further away, Ottawa and Halifax began to pay attention. Action began after a cancer specialist concluded that there was a higher risk of dying from cancer in Whitney Pier and Ashby, communities closest to the tar ponds, than the national average. At that point, the federal and provincial governments joined forces with the local government to fund a cleanup of the tar ponds.

The Sydney Tar Ponds are the site of the biggest environmental cleanup project in Canadian history. In 1998, Ottawa and Halifax agreed to a $62 million cost-sharing agreement to clean up the tar ponds and the site of the coke ovens. Since 2001, the tar ponds, coke ovens site, and other areas within the Muggah Creek watershed have been the scene of intense activity to repair the environmental damage. Since the cleanup began, the community is seeing improvements on a number of fronts: the construction of a sewer interceptor that will divert the tonnes of raw sewage that now flow daily into the tar ponds; the demolition and removal of derelict structures on the coke ovens site; and the closure and capping of the old Sydney landfill. Still, much remains to be done.

Atlantic Canada's Historical Geography

Atlantic Canada was the first part of North America to be discovered by Europeans. The Vikings arrived first, establishing for a brief time around AD 1000 a settlement at the northern extreme of the Northern Peninsula in Newfoundland. The first documented discovery of land in North America was made by John Cabot. The Italian navigator, employed by the King of England, reached the rocky shores of Atlantic Canada (the exact location, Cape Bonavista, Newfoundland, or Cape Breton Island, Nova Scotia, is in dispute) on 24 June 1497. Like Columbus far to the south in 1492, Cabot was searching for a sea route to Asia. In England, Cabot's report of the abundance of **groundfish**—cod, grey sole, flounder, redfish, and turbot—in the waters off Newfoundland lured European fishers to make the perilous voyage across the Atlantic to these rich fishing grounds, though some, especially the Basques, possibly had been fishing these waters at an earlier date. In any event, Canada's Atlantic coast quickly became a popular area for European fishers and though landings on shore took place—for drying the fish and establishing temporary habitation during the fishing season—permanent settlements were slow to take hold in this part of North America.

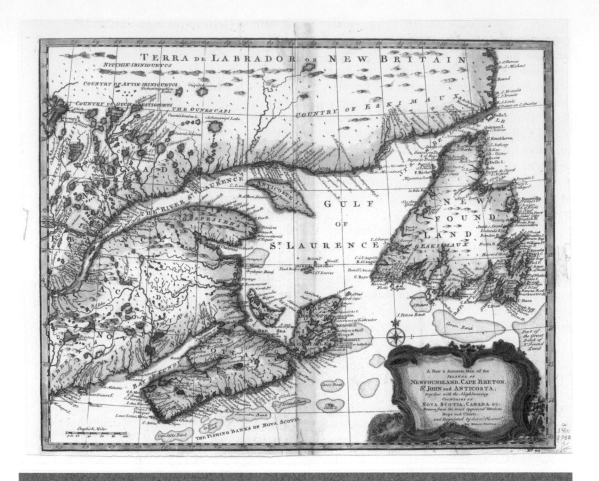

Figure 9.3   Atlantic Canada in 1750.

In 1605, a handful of French settlers, at **Port Royal** on the Bay of Fundy coast of present-day Nova Scotia, established the first permanent European settlement in North America north of Florida. During the seventeenth and part of the eighteenth centuries, French settlers spread into the Annapolis Valley and other lowlands in the Maritimes. By 1750, French-speaking Acadians numbered over 12,000. These French settlements, united by culture, language, and a common economy, became known as Acadia. In the coming decade, Acadians were deported to various parts of British North America.

Source: Bibliothèque nationale du Québec <www.nsexplore.ca/maps1/ns1750bowen.htm>.

While French settlers had established themselves in the Maritimes by the early seventeenth century, English settlers began arriving in 1610 in another part of Atlantic Canada—at Cupids in Conception Bay on Newfoundland's Avalon Peninsula. By the 1750s, there were over 7,000 permanent residents, mostly English, living in hundreds of small fishing communities along the shores of the island of Newfoundland. The French Shore is a coastal area along the west coast of the island and much of the northern coast where French fishers enjoyed treaty rights to fish. These rights were first granted by the British in the 1713 Treaty of Utrecht, when the French surrendered any territorial interest in Newfoundland, and though the Treaty of Versailles in 1783 cut back on the northern and eastern extent of the French Shore, it replaced what was taken away by extending the French Shore to the southwest corner of the island at Cape Ray. These French rights did not end until 1904. Today, several communities along Newfoundland's southwest coast are populated by descendants of early French settlers.

At the dawn of the eighteenth century, British possessions in Atlantic Canada were little more than names on a map. On the ground, the French colony of l'Acadie and their allies, the Mi'kmaq, were the most numerous inhabitants of the Maritimes, while French and English settlers occupied coastal settlements in Newfoundland with the **Beothuk** still occupying the interior of Newfoundland.

Over the first half of the eighteenth century, war between the two European colonial powers in North America—England and France—was almost continuous. During that time, the French forged an alliance with the Mi'kmaq and Maliseet, drawing them into the conflict with the English and their Iroquois allies.[2] Under the terms of the 1713 Treaty of Utrecht, France surrendered Acadia to the British. However, many French-speaking settlers, the Acadians, remained in this newly won British territory, which was renamed Nova Scotia. During the previous century, the Acadians had established a strong presence in the Maritimes with settlements and forts. Most Acadians lived along the Bay of Fundy coast, tilling the soil. Until the mid-1700s, Britain made little effort to colonize these lands, leaving the Acadians to till the land in this British-held territory. The Mi'kmaq remained close to the Acadians, but, as the British power took hold, they became strangers in their own land. Between 1725 and 1779, the Mi'kmaq signed a series of peace and friendship treaties with Great Britain. However, the Mi'kmaq remained close to the Acadian settlers. Events turned against the Mi'kmaq and Acadians with the founding of Halifax in 1749. The expulsion of the Acadians in 1755 from Nova Scotia, and in 1758 from Île Royale (Cape Breton Island) and Île Saint-Jean (Prince Edward Island), eliminated the Mi'kmaq's ally and relations between the British and the Mi'kmaq deteriorated. British rangers were unleashed to harass the Natives, to destroy their villages, and to drive them far beyond the British settlements. After the British defeated the French, the Treaty of Paris in 1763 ceded all French territories in North America to the British except for the islands of Saint-Pierre and Miquelon near the southern coast of Newfoundland.

The next event to influence the evolution of Atlantic Canada was the American Revolution. Following victory by the American colonies, approximately 40,000 Loyalists made their way to Nova Scotia and New Brunswick where they occupied the fertile lands of the recently departed Acadians and the prime hunting lands and fishing areas of the Mi'kmaq. With its superb harbour for ships of the British navy, Halifax became known as the 'Warden of the North'. Over the next 100 years, more and more British settlers came to the Maritimes. Nova Scotia alone received 55,000 Scots, Irish, English, and Welsh. Most Scots went to Cape Breton and the Northumberland shore. The driving forces pushing them from the British Isles were the **Scottish Highland clearances** and the **Irish famine**, which resulted in large influxes of migrants with Celtic cultural roots. These immigrants helped to define the dominantly Scottish character of Cape Breton and the Irish character of Saint John. The cultural impact of these Celtic peoples still resonates, and people of Scottish descent are still the largest ethnic group in Nova Scotia ('New Scotland').

For more details on Loyalist migrations to Canada, see 'The Loyalists' in Chapter 3, page 116.

By the time impoverished settlers from Britain arrived in the Maritimes, the old world of the Mi'kmaq had collapsed. Denied a place in the new economy, the Mi'kmaq were forced into a state of subsistence in which they soon became outcasts in their own land. It would not be until some 200 years later that the Supreme Court of Canada ruled that the Mi'kmaq have a right to fish commercially in waters where they once ruled (Vignette 9.4).

See Vignette 5.6, 'Historical Timeline of the Caledonia Dispute', page 194, for an outline of another modern-day confrontation between Aboriginal Canadians and non-Aboriginal Canadian law and society.

Think About It

Did the Six Nations of the Iroquois Confederacy, who were allies of the British during the American Revolution, fare better than the Mi'kmaq, who earlier aligned themselves with the French?

Vignette 9.4   Mi'kmaq Become Lobster Fishers

In 1999, the Supreme Court of Canada, in its landmark *Marshall* decision, acquitted Donald Marshall Jr of a federal Fisheries Act violation for catching fish (eels) for sale outside of the regulated season. In effect, the ruling meant that the Mi'kmaq should have a share of the commercial fishery. This led, in 2000, to confrontations between non-Aboriginal lobster fishers and Mi'kmaq at Burnt Church, New Brunswick, when the Mi'kmaq—as their Aboriginal and treaty right—began to set lobster traps in Miramichi Bay at a time when the non-Aboriginal fishery was closed. The non-Native fishers took it upon themselves to destroy the Native traps. Fisheries and Oceans officials, in their attempt to impose Fisheries Act regulations, rammed and capsized a Mi'kmaq fishing boat, and as the trouble escalated, the Supreme Court took the unprecedented step of seeking to qualify its earlier decision. Former Ontario Premier Bob Rae was brought in by the federal government to negotiate a solution among the involved parties, and in October 2000 the Mi'kmaq pulled their traps from the water earlier than they had planned.

In the typical Canadian style of resolving political confrontation, compromise was reached, though not without some residual bitterness on the part of the involved parties. Five years later, the Mi'kmaq had gained access to about 3 per cent of the federal lobster allocation by receiving about 320 commercial fishing licences. When operating at full capacity, these licences should allow the Mi'kmaq fishers to generate around $25 million worth of lobster each year. 'Further Readings', at the end of this chapter, offers a summary of Ken Coates's book on the *Marshall* decision.

CP/Jacques Boissinot

Photo 9.4

A Mi'kmaq lobsterman.

Historic Head Start

In the early nineteenth century, the harvesting of Atlantic Canada's natural wealth increased. This frontier hinterland of the British Empire exploited its rich natural resources—the cod off the Newfoundland coast and the virgin forests in the Maritimes—and became heavily involved in transatlantic trade of these resources. Furthermore, the availability of timber and the region's favourable seaside location provided the ideal conditions for shipbuilding in Atlantic Canada. By 1840, Atlantic Canada entered the 'Golden Age of Sail', with Nova Scotia and New Brunswick becoming the leading shipbuilding centres in the British Empire. Exports from Atlantic Canada were primarily cod and timber, while imports were manufactured goods from England and sugar and rum from the British West Indies. Several world events in the mid-nineteenth century added strength to this economic boom in the Maritimes. Demand for its exports rose, especially for timber, which was used to build British merchant ships and warships. Overseas trade with other British colonies in the Caribbean, as well as with Britain, formed the cornerstone of Atlantic Canada's trade and prosperous economy. In the first half of the nineteenth century, trade with New England was relatively limited because both the Maritimes and New England had almost identical resource-oriented economies. However, after the American Civil War, New England would industrialize, leading to greater trade between the two regions. In addition, Britain's move to free trade in 1849 would mean the loss of Atlantic Canada's protected markets for its primary products, resulting in even greater interest in the American market, especially in the Maritimes. Just before Confederation, however, external events dampened Atlantic Canada's resource-based economy. Iron was replacing wood in shipbuilding; the three-way trade with Britain and its Caribbean possessions collapsed, largely because a world glut of sugar caused prices to drop; and the end of the Reciprocity Treaty in 1866 cut off access to the Maritimes' natural trading partner, New England. These events resulted in the deterioration of Atlantic Canada's economic position among the Atlantic trading nations and colonies.

Confederation

The provinces of Atlantic Canada joined Canada at different times and for different reasons. Nova Scotia and New Brunswick joined at the time of Confederation; Prince Edward Island followed in 1873; Newfoundland reluctantly joined Canada in 1949. None saw much advantage to joining Canada. Each had to be pushed into Confederation. For the Maritime colonies, the decision to join Canada was not an easy one. In spite of their historic head start and their excellent access to sea and world markets, they were on the margins of the new country and still looked to New England as their natural market. So strong was the opposition in Nova Scotia that its provincial

government argued that Confederation was no more than the annexation of the province to the pre-existing province of Canada. A motion passed by the Nova Scotia House of Assembly in 1868 refusing to recognize the legitimacy of Confederation. Newfoundland, on the other hand, stood to gain from social programs available to Canadians—two were particularly popular. First, mothers received monthly family allowance payments for each child and that payment provided a reliable source of cash. Second, unemployment insurance payments to seasonal workers like fishers and loggers greatly stabilized family income and, through spending, supported rural business.

Yet, Confederation had a downside because Ottawa wanted to create a manufacturing base in Central Canada. Atlantic Canada had the basic resources for heavy industry—coal in Nova Scotia and iron in Newfoundland. An iron and steel complex on Cape Breton Island did compete in the Canadian marketplace, but eventually distance from markets and declining resources ended this experiment. Instead, the focus was on Central Canada where the bulk of the Canadian population was located.

See 'The Territorial Evolution of Canada' in Chapter 3, page 82.

The federal government's National Policy of 1879 reinforced Central Canada's natural advantage, and, in doing so, placed the Maritimes' fledgling manufacturing base in danger. To stimulate Canadian manufacturing, Canada imposed duties on imported goods. In retaliation, Americans increased their trade barriers on Canadian manufactured goods. Canada even placed a tariff on foreign-produced coal to protect the high-cost coal mines of Nova Scotia. In effect, Canada had created a 'closed' economy that facilitated the industrialization of Central Canada by relegating the rest of Canada to the dual roles of domestic market for Canadian manufactured goods and resource hinterland for Central Canada, the United States, and the rest of the world.

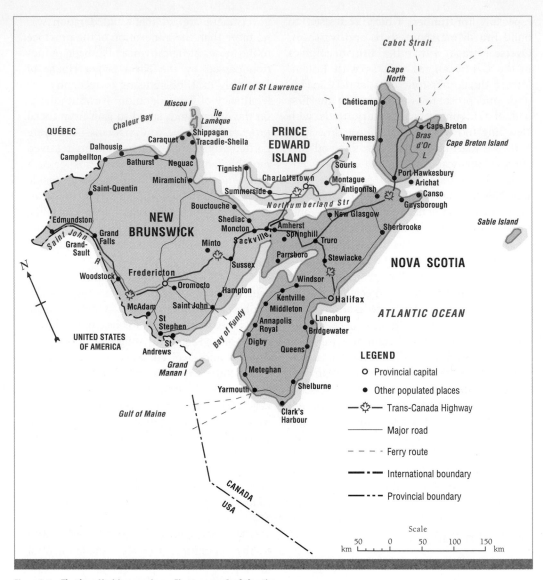

Figure 9.4   The three Maritime provinces: First to enter Confederation.
New Brunswick and Nova Scotia joined the Province of Canada (Québec and Ontario) to form the new Dominion of Canada in 1867 while Prince Edward Island entered Confederation four years later.
Source: Atlas of Canada, 2003, 'Maritime Provinces', at: <atlas.nrcan.gc.ca/site/english/maps/reference/provinceterritories/maritimes>.

What Atlantic Canada needed was access to the market of Central Canada. Ottawa's answer was the Intercolonial Railway (completed in 1876). Built to fulfill the terms of Confederation, the Intercolonial was never a commercial success, but it did stimulate economic growth in the Maritimes. The Intercolonial was operated by the federal government: freight rates were kept low to promote trade, and substantial annual deficits were paid by Ottawa. With low-cost transportation to the national market and the possibility of firms achieving economies of scale, a general economic surge occurred in the Maritimes. Cotton mills, sugar refineries, rope works, and iron and steel manufacturing plants were established or expanded to serve the much larger national market and to grab a share of the expanding western market. The railway boom in the

Canadian Prairies had a positive impact on Nova Scotia in the form of heavy investment in the province's steel mills, where steel production was directed to manufacturing steel rails and locomotives. By taking advantage of Cape Breton's coalfields and iron ore from Bell Island, Newfoundland, the steel industry in Sydney prospered, accelerating Nova Scotia's economic growth well above the national average. However, in 1919, the Maritime economy suffered a deadly blow when federal subsidies for freight rates were eliminated. Immediate access to the national market became more difficult and, with the loss of sales, many firms had to lay off workers, while others were forced to shut down. Even before these troubled times, the Maritimes economy was unable to absorb its entire workforce, leading many to migrate to the industrial towns of New England and Central Canada. From then on, out-migration was a fact of life in the Maritimes and, after 1949, in Newfoundland. For the rest of the twentieth century, this region languished because of its scattered and relatively small population, its narrow resource base, the high cost of transporting goods to the national market, and barriers to trade with New England. Even during periods of national affluence—for instance, the 1950s and 1960s—Atlantic Canada experienced limited economic growth; numerous federal loans and grants to firms in Atlantic Canada made little difference; and the forlorn struggles of its iron and steel industry typified the problem of manufacturing in Atlantic Canada. In 1968, Professor David Erskine at the University of Ottawa drew a rather dismal picture of Atlantic Canada as a hinterland:

> The region is, in the Canadian context, one of 'effort' rather than of 'increment'. Small scale resources once encouraged small scale development, but only large scale resources encourage modernization. The small scale and lack of concentration of its resources makes the region one in which government investment is easily dispersed without bringing about growth. Low levels of professional services and low levels of education result from the high taxes from low incomes; thus, still further retardation of economic growth occurs. (Erskine, 1968: 233)

Steel, Iron, and Coal—The Rise and Fall of Nova Scotia's Industrial Base

For over 100 years, coal and iron mining provided the basis for the Nova Scotia iron and steel industry. The iron and steel industry in Cape Breton Island near Sydney, Nova Scotia, was for a long time the heavy industrial heartland of Atlantic Canada. During that time, Sydney was the principal city on Cape Breton Island and the second-largest city in Nova Scotia.[3] Sydney's fate was closely tied to its major industrial firm, the Sydney steel mill, and to local coal mines. The mill and mines prospered in the early part of the twentieth century, and both the town of Sydney and steel production expanded with much steel exported as rails for the construction of railways in Western Canada. Two dark sides to this iron and steel complex existed. One was the loss of life in the coal mines. While the pay was good, danger went with the job and, for some, the cost was their lives. In 1956 and again in 1958, the Springhill mine near the head of the Bay of Fundy was the site of two underground explosions that took the lives of nearly 100 miners. In 1992, the Westray mine in Pictou County, Nova Scotia, was the site of another mine disaster that took 26 lives. The second dark side was the environmental degradation caused by seepage of toxic fluids from the steel mill, which resulted in the Sydney Tar Ponds discussed above.

A major turning point took place after World War II when demand for steel dropped and the size of the labour force was reduced. This process of deindustrialization, while delayed by federal and provincial subsidies, eventually saw the steel mill closed. By 2001, the dream of a heavy industrial base in Nova Scotia was gone. Without the steel and coal industry, Sydney and Cape Breton fell on hard times and their economic future is uncertain at best.

Atlantic Canada Today

Atlantic Canada contains a natural beauty and captivating cultural roots that continue to foster a quality of life for Atlantic Canadians. Yet, young people of the region continue to seek economic opportunities in other parts of the country. The prospect for strong economic growth remains elusive and the challenge facing Atlantic Canada is to translate its fossil fuel resources, minerals and metals, forests, and marine and freshwater resources into a strong sustainable economy.

Barriers to such a sustainable economy are formidable, but opportunities do exist. First of all, Atlantic Canada has the handicap of being an old resource hinterland. As such, many of the region's most valuable natural resources have already been harvested, and by the twenty-first century the first-growth forests, coal, and the cod stocks had been depleted. Second, Atlantic Canada's attempt at industrialization has been troubled by a small local market, distance to larger markets, and, until the Free Trade Agreement, trade barriers to its natural market in New England. Third, Atlantic Canada has a fractured geography, which divides its population into a series of small provincial markets and thereby creates four provincial governments—this leads to very high per capita cost of government. Fourth, rural Atlantic Canada has fallen into a steep decline with many coastal villages disappearing. Rural Atlantic Canada is confronted with a mismatch of resources and population. Single-industry villages and towns dependent on the cod fishery and forestry are leading the rural retreat. The principal factor underlying this rural retreat is the loss of jobs in the primary sector; logging jobs, for example, are disappearing because US trade restrictions hamper access to Atlantic Canada's principal market. Fifth, urban Atlantic Canada has only one major city—Halifax—with the population size to form a regional capital of the Maritimes. To a lesser degree, St John's plays the same role in Newfoundland and Labrador. Lastly, the out-migration of Atlantic Canada's workforce to more prosperous parts of Canada drains away many of the more ambitious,

younger, and talented people. These six factors prevent Atlantic Canada from making use of economies of scale, which is necessary to keep production costs low, and its entrepreneurs from adventuring into new and often risky enterprises. The region's predicament is summed up by its subpar economic performance, high unemployment rates, and the seemingly unstoppable out-migration to faster-growing regions of Canada.

Yet, Atlantic Canada has a second chance. The first opportunity has been the discovery and exploitation of offshore oil and gas deposits. Most petroleum deposits are located 200 to 300 km east of Newfoundland in waters of the Grand Banks. A secondary deposit, mainly natural gas, is located offshore of Nova Scotia near Sable Island. The Atlantic Canada petroleum industry now includes the export of natural gas to New England. Liquid natural gas (LNG) facilities have been constructed at Saint John, New Brunswick, which is ideally situated as an energy hub for exporting regasified (warmed) LNG to its natural state for transport through pipeline networks to utilities in New England. Goldboro, Nova Scotia, may also become an LNG site. A recent agreement with the state of Maine adds to Saint John's role as an energy hub. This agreement calls for the establishment of a **Northeast Energy Corridor** to facilitate the flow of electricity and natural gas from Saint John to markets in New England. The second opportunity stems from the mining of the huge nickel deposit at Voisey's Bay, Labrador, and the **hydrometallurgy** processing of nickel ore at the Long Harbour facility now under construction near St John's on the island of Newfoundland. The combination of these two projects maximizes the long-term economic gain for Newfoundland and Labrador.

Political agreements might provide the basis for a third opportunity to achieve further economic gains in Atlantic Canada. One agreement, the 2004 Atlantic Energy Accord, offers a greater share of resource royalties and equalization payments to the governments of Nova Scotia and Newfoundland and Labrador.

Photo 9.5

Artist's sketch of the Long Harbour nickel processing and docking facility on Placentia Bay, south west of St John's.

Actual payments are shown in Table 9.2 and the controversy about the parallel equalization systems introduced in 2008–9 is discussed in Vignette 9.5

For more on equalization payments, see Chapter 3, 'Equalization and Transfer Payments', page 96.

A second political agreement with potential economic impact is the Northeast Energy Corridor agreement between New Brunswick and Maine. Following on its heels was the bold initiative of Hydro-Québec to purchase the power facilities in New Brunswick and create an energy corridor, although this latter deal fell through at the last moment. The aim

Table 9.2	**Equalization Payments to Atlantic Canada Provinces ($ millions)**					
Year	NL	PEI	NS	NB	Atlantic Canada	Percent Canada
2006–7	632	291	1,386	1,451	3,760	33.3
2007–8	477	294	1,308	1,477	3,556	27.9
2008–9	197	310	1,294	1,492	3,293	25.5
2009–10	0	340	1,571	1,689	3,600	25.0

Source: Canada, Department of Finance (2009) 'Federal Transfers to Provinces and Territories'. Reproduced with permission of the Minister of Public Works and Government Services, 2007. <www.fin.gc.ca/fedprov/eqp-eng.asp>.

Vignette 9.5 The Atlantic Energy Accord: Jackpot or Not?

In 2004, Premiers Danny Williams and John Hamm signed the Atlantic Energy Accord with Prime Minister Paul Martin, thus allowing both Newfoundland and Labrador and Nova Scotia to receive all of their offshore oil and gas revenues without reducing their equalization payments for eight years. In 2007, Ottawa announced a political compromise between honouring the Atlantic Accord and accepting the recommendations of the Expert Panel on Equalization and Territorial Formula Financing. This compromise resulted in two versions of equalization calculations. On the one hand, Ottawa kept the equalization system that excluded natural resource revenue in its calculations, thus honouring the Atlantic Accord. On the other hand, Ottawa accepted the recommendation of the Expert Panel that 50 per cent of natural resource revenues become part of the calculations for equalization payments. The rub was that Ottawa increased the amount of equalization payments from 2007–8 to 2008–9 by 5.4 per cent and that increase was not available to provinces opting for the equalization calculations. The Newfoundland and Labrador government remained in the original equalization calculations while Nova Scotia selected the new formula. Newfoundland and Labrador is gambling that future oil and gas production and prices will provide more revenue than the lost annual increase in equalization funds. In 2009–10, Newfoundland and Labrador saw its equalization payments clawed back while Nova Scotia jumped by $1.2 billion. When the Atlantic Energy Accord ends in 2012, each province will know if it made the right choice.

Michael de Adder/Artizans.com

Photo 9.6

Jackpot! Premiers Williams and Hamm celebrated in this cartoonist's rendering of the outcome of the Atlantic Energy Accord, but it remains to be seen who the real winner is.

of this corridor was to connect hydroelectric facilities in northern Québec and Labrador to the huge markets in New England (see Vignette 9.6). This agreement also called for the modernization of the New Brunswick system. It was on this issue, in March 2010, that the New Brunswick–Québec agreement came apart, as Hydro-Québec found that the upgrading and refurbishing of New Brunswick facilities could be too risky and costly (Bissett, 2010; CBC News, 2010). The failure of the agreement, however, does not alter its basic aims: more direct and less costly transmission of Québec power to New England, and a unified and more efficient energy system in Atlantic Canada.

Economic gains for the region also can be driven by the region's major cities. Halifax, St John's, Moncton, and Saint John are benefiting from an ever-widening tertiary base to their urban economies. Most knowledge-based businesses, however, are located in Halifax, the de facto the regional capital of the Maritimes. In addition, with its deep harbour, Halifax is ideally suited to participate in global container trade.

Finally, even a positive spin can be put on out-migration. The movement of workers to Alberta's oil sands has had one positive impact on Atlantic Canada, especially its smaller communities—the transfer of wealth back home. Many of these workers commute to the oil sands in a cycle of 20 days at work followed by eight days back home, and thus bring home much of their wages. Others send money back to their parents and close relatives. Some have

---

Vignette 9.6   Can Hydro-Québec Mend Atlantic Canada's Fractured Geography?

The physical setting of Atlantic Canada has prevented this region of Canada from achieving a cohesive economic and political unit. The island of Newfoundland stands alone, separated from the rest of Atlantic Canada by the Cabot Strait and from Labrador by the Strait of Belle Isle. These water barriers hamper Newfoundland from having close economic links with the Maritimes. At the same time, Labrador's geographic connection to Québec has had profound economic implications and has drawn Labrador into Québec's economic orbit. Examples include: (1) the transmission of Labrador hydroelectric energy across Québec; (2) the shipment of Labrador's iron ore to the port of Sept-Îles, Québec; and (3) the Labrador–Québec highway that runs from Labrador's largest town, Happy Valley–Goose Bay, to Baie-Comeau, thereby connecting to Québec's provincial highway system. Connections also exist along the border between New Brunswick and Québec, a shared culture and language prominent among them.

In October 2009 Hydro-Québec struck a deal with New Brunswick Power to purchase transmission and distribution systems as well as New Brunswick's power stations. Then, in early 2010, New Brunswick withdrew distribution and transmission from the agreement. The basic deal, however, remained intact—Québec would purchase New Brunswick power plants for $3.2 billion. Finally, on 24 March 2010, when Hydro-Québec sought to change the terms after it found that refurbishment costs would be significantly greater than expected, New Brunswick's Premier Shawn Graham backed out (Bissett, 2010).

Is this the end of the story? Probably not. The dissolution of the agreement was notably amicable, but political pressure against it within New Brunswick and in some other Atlantic provinces had been intense. Newfoundland and Labrador Premier Danny Williams, for one, opposed the initial deal because it would essentially give Hydro-Québec control of electricity transmission corridors between Atlantic Canada and the US markets and thus block Williams's plan to transmit his province's electric power via an undersea cable to Nova Scotia, New Brunswick, and then to market in New England. Given the geopolitical manoeuvring of Hydro-Québec, might it purchase the power system in Prince Edward Island and Nova Scotia, and perhaps even the system in Newfoundland and Labrador? Or how long might it be before a new deal between Québec and New Brunswick is on the table? The individual players in such geopolitical games inevitably change, as do the social, political, and economic contexts in which they operate. The fact remains that economies of scale—and the profits to be gained—will continue to push those players towards a unified system of energy generation and transmission.

made their fortunes and have returned to Atlantic Canada. Unfortunately, the size and impact of this transfer of wealth from Alberta to Atlantic Canada are difficult to estimate. Anecdotal stories abound, such as that Newfoundland and Labrador's second-largest city is Fort McMurray. Yet, does economic opportunity in distant places trump sense of place?

See Chapter 1, 'Sense of Place', page 15, for discussion of this powerful cultural bond to a 'home' region.

Industrial Structure

Atlantic Canada's primary resources are fish, forests, minerals, and petroleum deposits, but the region's economic future lies in its tertiary sector, especially high-technology industries

that focus on ocean technology. The rapid growth of ocean technologies in Atlantic Canada is fuelled in part by the strong growth of the offshore oil and gas and the defence/marine security industries. Atlantic Canada's highly developed skills in cold-water engineering are in full view with the $1 billion Confederation Bridge that spans the Northumberland Strait between New Brunswick and Prince Edward Island (photo 9.7).

Employment statistics provide a picture of the basic economic structure of Atlantic Canada. In 2008, employment in primary activities accounted for 5.4 per cent of the labour force; secondary employment constituted 16.4 per cent; and tertiary employment was 78.2 per cent (Table 9.3). Compared with the figures for 2005, the greatest percentage decline took place in the primary sector while the largest increase occurred in the secondary sector.

Photo 9.7

The Confederation Bridge connects Prince Edward Island with New Brunswick. As an integral part of the Trans-Canada Highway network, the 12.9-km Confederation Bridge is the longest bridge over ice-covered waters in the world. Opened in 1997, the economic impact on Prince Edward Island has been significant in four areas: increased tourism; a real estate boom; expanded potato production and potato-based processed foods; and greater export of time-sensitive and high-priced seafoods. These economic gains help to account for the province's increased population.

Barrett & MacKay/All Canada Photos

Table 9.3   Employment by Economic Sector in Atlantic Canada, 2008

Economic Sector	2005		2008	
	Workers (000s)	Workers (%)	Workers (000s)	Workers (%)
Primary	62.5	5.8	59.6	5.4
Secondary	171.7	16.0	181.5	16.4
Tertiary	841.6	78.2	868.8	78.2
Total	1,075.8	100.0	1,109.9	100.0

Source: Statistics Canada (2009b).

Significant variations among the provinces' employment figures exist. Newfoundland and Labrador, as expected, ranks first in employment in the fisheries, but it also ranks first in mining, primarily because of the large iron-mining operations in Labrador and the offshore oil extraction. Employment in manufacturing is much stronger in Nova Scotia and New Brunswick than in Newfoundland and Labrador and Prince Edward Island, mostly because there is greater processing of local raw materials in these two provinces—from the processing of agricultural, fish, and wood products to the processing of mineral products. However, Prince Edward Island has expanded its food-processing industries, especially french fries and potato chips. Newfoundland and Labrador has the highest unemployment in Atlantic Canada and consequently many workers have turned to commuting or relocating to labour-short Alberta.

Atlantic Canada has exceedingly high rates of unemployment, especially in rural and coastal communities. From 1996 to 2009, unemployment figures decreased (Table 9.4) but the gap with other provinces, particularly Alberta, helps to explain the out-migration from the region. While progress is occurring slowly, the region continues to adjust its economy to twenty-first-century realities. Many rural towns and villages, for example, have their roots in a fishing industry that can no longer support many of them. Great Harbour Deep is one such isolated community that has disappeared from the map of Newfoundland (Vignette 9.7). On the other hand, economic growth and high rates of employment are concentrated in the major urban centres, especially Halifax and St John's. Ironically, Atlantic Canada, and especially Newfoundland and Labrador, has a mismatch between the demand for skilled workers and the education/training level of its labour force. The shortage of skilled workers is partly due to the exodus of these workers to such locations as Alberta's oil sands, where wage levels are much higher and overtime is readily available.

Table 9.4   Unemployment Rates in Alberta and in Atlantic Canada by Province, 1996, 2006, and 2009

Province	1996	2006	2009
Newfoundland and Labrador	25.1	14.8	15.5
Prince Edward Island	13.8	11.0	12.0
New Brunswick	15.5	8.8	8.9
Nova Scotia	13.3	7.9	9.2
Alberta	7.2	3.4	6.6
Canada	10.1	6.1	8.3

Sources: Statistics Canada (2007b, 2009a).

Vignette 9.7   The End of Great Harbour Deep

For 400 years, people fished cod in the waters off Great Harbour Deep. Located on the Northern Peninsula, Great Harbour Deep was one of the most isolated communities in Newfoundland. Access to the outside world was by a three-hour boat ride in the summer and by a ski-equipped aircraft in the winter. While residents cherished their independence, the need for outside amenities, including health services, had become more and more pressing. In the 1990s, the loss of the inshore fishery and the closure of the town's fish plant led to the economic collapse of this outport. Some left, while others worked on the large trawlers for months at a time. By 2003, the population had dropped to 180 residents. After a series of public meetings, community residents decided to accept the government's $5 million package and leave their community for good. In the summer of 2003, the 50 families of Great Harbour Deep left their community, leaving behind memories and empty homes but hoping to start a new life in a new community.

Photo 9.8

The sheltered bay of Great Harbour Deep in winter.

Key Topic: The Fishing Industry

Nature has given Atlantic Canada a vast continental shelf that provides an excellent physical environment for fish: the warm ocean currents from the Gulf of Mexico and the cold Labrador Current create ideal conditions for fish reproduction and growth. The continental shelf extends almost 400 km offshore (Figure 9.5). In some places, where the continental shelf is raised, the water is relatively shallow. Such areas are known as banks. The largest banks are the Grand Banks off Newfoundland's east coast and Georges Bank off the south coast of Nova Scotia (Vignette 9.8).

The diversity that characterizes Atlantic Canada extends to the fishery as well. Although each province relies on the fishery, there are striking differences (particularly between Newfoundland and Labrador and the Maritime provinces) in the type of catch and the location. Before the restrictions on cod fishing, Newfoundlanders depended mostly on cod fishing, while Maritimers have harvested a

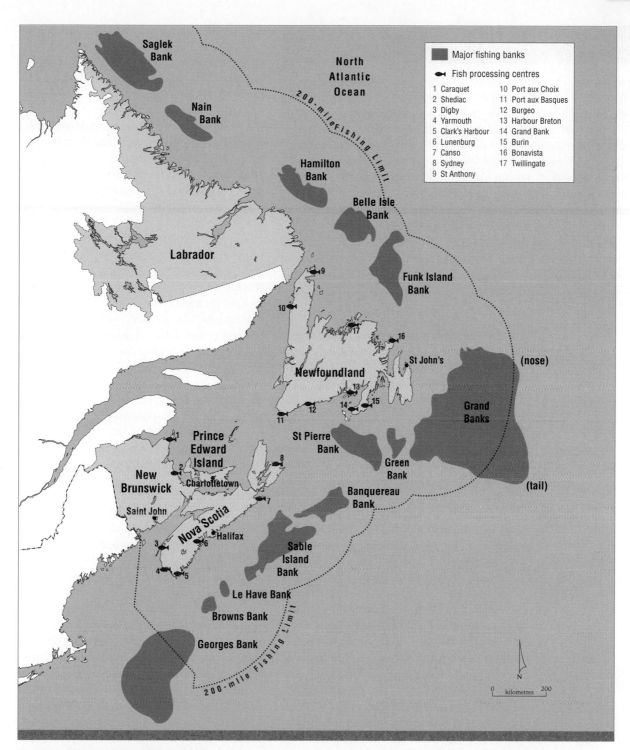

Figure 9.5   Major fishing banks in Atlantic Canada.
The Atlantic coast fishery operates within a vast continental shelf that extends some 400 km eastward into the Atlantic Ocean, southward to Georges Bank, and north to Saglek Bank. Within these waters are at least a dozen areas of shallow water known as 'banks'. The Grand Banks of Newfoundland is the most famous fishing ground while Georges Bank contains the widest variety of fish stocks. Scallops, for instance, are harvested in beds on Georges Bank and Browns Bank.

Vignette 9.8   Georges Bank

As part of the Atlantic continental shelf, Georges Bank is a large, shallow-water area that extends over nearly 4,000 km². Water depth usually ranges from 50 to 80 m, but in some areas the water is 10 m or less. Georges Bank is one of the most biologically productive regions in the world's oceans because of the tidal mixing that occurs in its shallow waters. This brings to the surface a continuous supply of regenerated nutrients from the ocean sediments. These nutrients support vast quantities of minute sea life called plankton. In turn, large stocks of fish (such as herring and haddock), as well as scallops, feed on plankton. Cod also feed in these waters, but their numbers have dwindled due to overfishing, especially in international waters. Properly managed, Georges Bank can sustain annual fishery yields of about 400,000 tonnes.

variety of sea life, including cod, lobsters, and scallops. Fishers from the Maritimes ply the waters of Georges Bank and smaller banks just offshore of Prince Edward Island and Nova Scotia. In these waters, Maritimers find cod, grey sole, flounder, redfish, and shellfish. Newfoundlanders, on the other hand, historically fished in the waters of the Grand Banks and, especially, in the inshore fishery around the island of Newfoundland and along the shore of Labrador, where cod and other groundfish congregate in the shallow areas of the continental shelf. Since the collapse of the cod fishery, the Newfoundland fishing industry has diversified and shellfish have become the most important and valuable species.

The value of the Atlantic fisheries reached $1.6 billion in 2008. Shellfish, led by lobster, crab, and shrimp, made up 85 per cent of the total value. A similar pattern exists in Newfoundland and Labrador. The value of its landings totalled $528 million, with shellfish accounting for 82 per cent. However, as the best lobster grounds are far from Newfoundland, the leading shellfish for Newfoundland and Labrador were shrimp and crab (Fisheries and Oceans Canada, 2009). With the decline of the northern cod stocks, lobster became the most valued species, making up the bulk of the total value of the Atlantic fishery. Lobster was not always considered a valuable species and, in the

eighteenth century, was only eaten by poor families—and as a last resort—and was used to fertilize potato mounds. The disdain for lobster slowly waned, and by the nineteenth century the poor man's chicken became highly desired—and high-priced.

While lobster fishing grounds are widespread within a narrow coastal zone surrounding the Maritimes, the most productive grounds are near Yarmouth on the southern tip of Nova Scotia. The opening of the lobster season in November 2009 saw more than 1,500 fishing boats crewed by an estimated 5,000 fishermen ready to set thousands of traps on the fishing grounds just off the tip of southwest Nova Scotia. Traps are set on the seabed either individually or in groups of up to eight on a line. The traps are hauled to the fishing vessel by powered winches, emptied, re-baited, and lowered again. The catch is transported live to harbour where it is often kept in seawater-permeated wooden crates for later sale. Since most are exported to New England, the US demand for lobster sets the price. With the US economy in trouble, lobster prices have dropped sharply, thus greatly reducing the income of fishers.

The fishing industry consists of an inshore fishery, an offshore fishery, and fish-processing plants. Traditionally, Newfoundland and Maritime fishers used a flat-bottomed skiff called a dory and stayed close to the shore. Only the larger skiffs ventured to the Grand Banks and other offshore banks. Today, inshore fishers still use smaller boats to fish close to the shore. During the winter, rough waters and ice prevent the smaller craft from leaving port. Offshore fishers travel in steel stern trawlers to the Grand Banks and other offshore fishing banks where they fish for weeks. Upon returning to port, they unload hundreds of tonnes of iced groundfish at a fish-processing plant. Offshore vessels, often much larger than 25 tonnes, are equipped with radar, radios, sonar detectors, and other sophisticated equipment that enable them to operate in winter as well as summer.

The modernization of the Canadian fishing industry, with its larger trawlers and more efficient nets and gear, was accompanied by huge investments by fish companies and individuals and an expansion of the fishing fleet in number and size. Atlantic Canada's fishery, particularly in Newfoundland and Nova Scotia, was able to increase its catch and its income. Now, fishing companies usually own the ships, so fishers no longer share in the value of the catch but receive a salary. However, modernization had many downsides:

- Fewer fishers were required by this more efficient fishing system.
- The new fishing system based on technological advances and huge dragger nets created the circumstances for overfishing.
- Overfishing and the collapse of the northern cod stocks forced inshore fishers to turn to government assistance to survive.

Management of Fish Stocks

The expansion and modernization of the world fishing fleets have threatened world fish stocks. With no management of fish stocks in international waters, pressure on selected fish stocks grew. By the 1970s, the world fishing fleets had greatly increased their take of groundfish on the Grand Banks. As is the case in British Columbia with the salmon, the issue of fishing limits in Atlantic Canada was crucial in trying to manage the fish stocks in a sustainable way. Management of the international fish stocks in Atlantic Canada was so important that, in 1977, Ottawa claimed the right to manage the fisheries within a 200-nautical-mile zone off the east coast of Atlantic Canada. Washington also extended its fishing zone to the 200-mile limit, making

Table 9.5 Cod Landings for the Atlantic and Newfoundland and Labrador Fisheries, 1990–2008 (metric tonnes live weight)*

Year	NL	Atlantic Canada	% NL
1990	245,896	395,266	62.2
1991	178,687	309,031	57.8
1992	75,138	187,804	40.0
1993	37,068	76,644	48.4
1994	2,292	22,719	9.8
1995	863	12,438	6.9
1996	1,147	15,541	7.4
1997	12,317	29,899	41.2
1998	22,764	37,894	60.0
1999	38,663	55,527	69.6
2000	30,216	46,177	65.4
2001	23,774	40,440	58.8
2002	21,083	35,741	59.0
2003	14,290	22,768	62.8
2004	14,534	24,729	58.8
2005	16,257	26,156	62.2
2006	17,050	27,307	62.4
2007	17,784	26,593	66.9
2008	17,594	26,783	65.7

*Fisheries and Oceans Canada has two operational regions for Atlantic Canada, the Maritime region and the Newfoundland and Labrador region. The department reports its statistics by fisheries that correspond to its operational regions. However, it includes the Maritime and Newfoundland fisheries within the larger, non-operational region of Atlantic fisheries.

Note: Since 1992, a modest cod catch was allocated to inland fishers.

Source: Adapted from Fisheries and Oceans Canada (2009).

Georges Bank an area of dispute between the two countries.

Fishers from both the Maritimes and New England harvest the fish stocks on Georges Bank, which is located at the edge of the Atlantic continental shelf between Cape Cod and Nova Scotia. In the 1970s, fish stocks, especially herring, dropped to dangerously low levels due to overfishing. A boundary decision made by the International Court at The Hague in October 1984 allocated five-sixths of Georges Bank to the United States. However, the easternmost one-sixth that was awarded to Canada is rich in groundfish and scallops. The Nova Scotia fishers have greatly benefited from this decision. With the scallops now under Canadian management, sustainable harvesting of the scallops and groundfish has brought a measure of stability to this sector of the fishing industry.

However, overexploitation of international fish stocks remains a worldwide problem. The root of the problem is that common, i.e., public, resources are decimated by the selfish actions of individuals who have no regard for the well-being of the resource. Such mismanagement is known as the 'tragedy of the commons'. Such ecological disasters occur when no one is responsible for ensuring the proper management of resources—or when those entrusted with management do a poor job. Northern cod stocks suffered huge losses from overfishing and mismanagement (Vignette 9.9). Mismanagement was based on three factors: (1) the estimates of cod stocks by the federal Department of Fisheries and Oceans (DFO) were overly optimistic and too often disregarded the anecdotal local ecological knowledge of fishers, and consequently DFO set the quotas too high; (2) strong pressure for high cod quotas came from Newfoundland politicians and from the major fish processors whose trawlers worked the Grand Banks, as well as from some outport

Photawa/Dreamstime/Getstock.com

Photo 9.9

Neil's Harbour is a small fishing village on the northern tip of Cape Breton. The protected harbour provides an ideal place to moor small fishing boats.

Vignette 9.9   Ecological Crises and the Resource Cycle

The decline in groundfish stocks in Atlantic Canada is an ecological crisis. But history tells us that many natural resources have suffered such a crisis. Professor Robert Clapp (1998: 129) examined this issue and concluded that 'the long record of failure to sustain the yield of biological resources suggests that such restriction is possible only in theory.' Clapp offers the resource cycle as an explanation for ecological crises, in other words, what begins as a rich resource leads to overexploitation and the collapse of the resource.

The cause of overexploitation of the northern cod is well known. Much is due to the use of draggers by the Canadian and foreign fishing fleets. The attraction of this form of fishing is its cost-efficiency. The Canadian Atlantic Fisheries Scientific Advisory Committee revealed in mid-1992 that stocks had declined sharply since the 1960s, when the **biomass** (the volume of fish aged three years or older) was estimated at over 3 million tonnes. By 1991, the biomass had dropped to between 530,000 and 700,000 tonnes, the lowest level ever calculated. **Spawning biomass** (generally seven years or older) represents the reproductive part of the fish stock. In 1991 it amounted to between 50,000 and 110,000 tonnes. Thirty years before, the spawning biomass had been as high as 1.6 million tonnes, and even in 1987 it was estimated at 400,000 tonnes. Faced with such a serious crisis, the Canadian government announced in 1992 a moratorium on cod fishing in the waters of Atlantic Canada. By the summer of 2006, cod stocks had not yet recovered sufficiently to allow the reopening of the cod fishery. As shown in Table 9.5, the last large catch occurred in 1991. Since then, cod fishing has been restricted, on a limited basis, to inshore fishers.

communities; and (3) Canada had no control over foreign fishing along the nose and tail of the Grand Banks. The circumstances and consequences of the demise of cod stocks in Atlantic Canada can lead to a greater appreciation of the need for sustainable harvesting in the fishing industry.

Overfishing and the Cod Stocks

The Atlantic cod (*Gadus morhua*) is the major natural resource in Atlantic Canada. While the Atlantic cod (sometimes called the northern cod) is found as far north as the southern tip of Baffin Island and as far south as Cape Hatteras, cod stocks are concentrated off Newfoundland and along the Labrador coast. The largest single concentration of northern cod, historically, was on the Grand Banks.

The Atlantic cod has been one of the world's leading food fishes for centuries. Cod is sold fresh, frozen, salted, and smoked. Salted cod was very popular in the past, prior to quick freezing and refrigeration, but now most of the cod sold is frozen. During the sixteenth century, the catch was small, probably averaging less than 25,000 metric tonnes a year. By the end of the seventeenth century, however, the annual catch may have reached 100,000 metric tonnes. Annual cod landings during the nineteenth century were about 150,000 to 400,000 metric tonnes. As fishing technology advanced, catches of cod rose to nearly 1 million metric tonnes in the 1950s. By the 1960s, the annual catch reached a peak of almost 2 million metric tonnes. European trawlers accounted for most of this catch.

With the emergence of a modern fishing fleet, fish stocks around the world were overfished. Several fish stocks, including the California sardine, Peruvian anchovy, and Namibian pilchard, collapsed in the 1980s from overfishing—not so much from local fishing operations but from the technology-enhanced fishing fleets that scan the world's oceans for fish. The simple hook-and-line fishing system did not have the capacity to overfish, but foreign and Canadian fishing fleets that employ sophisticated fishing equipment and that drag huge, weighted nets across the ocean floor indiscriminately trapping all groundfish regardless of age, species, and value do have that capability. Known as draggers, these fleets create enormous waste because 'non-commercial' fish—the bycatch—are simply discarded. In addition, the draggers

Think About It

In the late 20th century, did politicians, both federal and provincial, support the high quotas for the northern cod because they believed that full exploitation of the cod fishery would lead to a more prosperous Atlantic Canada?

destroy fragile ocean-floor ecosystems, including coral reefs and breeding habitat. In spite of the many excuses for the collapse of the northern cod stocks, including changing ocean temperatures and an increase in seal population, the primary cause has been the multinational seafood products companies that ran the technologically advanced fishing fleets with factory ships that could stay at sea for extended periods of time. These fleets ruthlessly exploited the cod and other fish of the sea, to the point where a study by 14 international scholars published in late 2006, which analyzed all existing datasets, forecast that all of the world's seafood fisheries will collapse by the year 2050 (Physorg.com, 2006).

In the case of cod, Ottawa's ability to manage the fish stocks was complicated because, as noted above, the nose and tail of the Grand Banks extend into international waters. This made it difficult for Canadian fisheries officials to monitor and enforce catches in these areas. An agreement was reached with the countries belonging to the European Common Market, but monitoring the actual catches proved difficult. Some European fishing vessels, particularly those from Spanish ports, were exceeding their fish quotas and were catching undersized cod by using nets with a small mesh, which is illegal. During the 1980s, groundfish stocks, especially the northern cod stocks, dropped significantly—perhaps by as much as 60 per cent (Cashin, 1993: 19). This collapse translated into a decline in fish landings and processing. Cod landings for Newfoundland were reduced by over half from 1991 to 1992. In 1992, Ottawa was forced to impose a moratorium on the cod fishery. The fear was that the northern cod stocks had collapsed and further fishing would make their recovery more difficult. While fisheries landings have remained small, the lowest point occurred in 1995 (Table 9.5). By 2007, scientists were still unsure as to how long it will take for the northern cod stocks to recover, if indeed they ever will. Cod must reach the age of seven before the fish can reproduce, so recovery to a point to make the cod fishery even marginally viable is a long-term proposition at best.

The impact of the cod moratorium on Atlantic Canada, particularly Newfoundland, was severe. The cod has been the mainstay of Newfoundland's economy. Quite justifiably, the Atlantic cod was called 'Newfoundland currency'. In the past, the total catch by Newfoundland fishers was close to four-fifths groundfish. By comparison, the groundfish would constitute only about half of the catch in Nova Scotia and about one-third of the catch in Prince Edward Island and New Brunswick.

Before the moratorium, most jobs were in the fish-processing plants. Often these jobs lasted for only a short period, sometimes less than a month. However, employment at fish-processing plants provided two highly valued items—much-needed cash and enough weeks of work to qualify for unemployment insurance. In May 1994, fish-processing employees became eligible for financial support from The Atlantic Groundfish Strategy and were included in the extension to TAGS announced in 1998. In Newfoundland, the fish-processing sector was dominated by two firms, Fisheries Products International and National Sea Products. Prior to the moratorium on cod fishing, these two firms accounted for about half of the fish-processing jobs or nearly 8 per cent of the total number of employed people in Newfoundland.

The fishing industry in Atlantic Canada has suffered under the July 1992 moratorium on cod fishing. Since 1995, however, Newfoundland's fishing industry has recorded higher and higher production values because of high prices for crab, shrimp, lobster, clams, turbot, and other stocks. In this highly managed industry, only a few fishers have licences for these special kinds of catches. Only about 1,200 licences to catch crab are issued to Newfoundland fishers each year. On the other hand, before the cod moratorium, approximately 7,000 licences were issued for groundfish. Without cod, wealth from the fishery is concentrated in fewer and fewer hands. By 1997, small catches of cod by inshore fishers were permitted, but, in spite of strong pressure from fishers, Ottawa is still not prepared to allow the cod fishery to reopen. In Atlantic Canada, fishing is not only an important economic industry, it is a way of life. By 2010, cod stocks showed no signs of recovery and the inshore fishing way of life based on out-port settlements is slipping away.

Atlantic Canada's Resource Wealth

Though Atlantic Canada's early settlement and subsequent development revolved very much around the harvest of the sea, the land mass of the four provinces has, in the past, bolstered the region's economy through the harvesting of its natural resources, including forestry, mining, and agriculture. Today, the offshore petroleum deposits have sparked a boom focused on St John's (Figure 9.6).

Forest Industry: A Weak Sister

Beyond the sea, Atlantic Canada's most important renewable resource has been its forests. The rugged Appalachian Uplands in Atlantic Canada contain 22.9 million ha of forest land. Yet, the best days for this industry are long gone. In fact, the forest industry in Atlantic Canada (and across Canada's boreal forest) is contracting due to diminishing demand in the US and in the rest of the world for forest products. Other factors contributing to the weak state of the forest industry are the appreciating Canadian dollar against the US dollar, making forest products more expensive for American buyers; declining demand for newsprint due to a slumping newspaper industry; and rising electricity costs in Atlantic Canada, which constitute a large portion of operating costs in pulp and paper. A few figures support the weak nature of the forest industry. In 2001, for instance, the value of forest products was approximately $4 billion; by 2008, the value had sunk to less than half this figure. Another measure of decline lies in the smaller number of hectares harvested for logs. In 2003, just over 180,000 hectares were logged, but by 2007 this total had dropped to 135,000 hectares (Natural Resources Canada, 2009a, 2009b). An analysis by the Atlantic Provinces Economic Council in 2007 reinforces this dismal news: from 2004 to 2006, the

value of output fell by 9 per cent, the number of forest firms decreased by 11 per cent, and, worst of all, employment dropped by 24 per cent. This downward cycle has continued in the current global economic crisis and the collapse of the US housing market. All forest operations have suffered: logging, sawmilling, and pulp and paper plants. Pulp and paper mills have been hit hard. In 2002, nine pulp plants operated in New Brunswick, three in Nova Scotia, and three in Newfoundland. From 2004 to 2009, five mills have permanently shut down—three in New Brunswick and two in Newfoundland and Labrador—with the loss of more than 1,000 direct jobs. Clearly, Atlantic Canada's forest industry has become a weak sister. Until the US economy recovers, the forest industry will remain in a slump. Modest help from Ottawa is underway for the forest industry under a greenhouse gas reduction ploy—the $1 billion **Canada Pulp and Paper Green Transformation Program** will allow Canadian pulp and paper companies to receive federal funding by using a liquid by-product (black liquor) to generate renewable energy and thereby avoid a US claim of a forestry subsidy that would result in retaliation.

The bulk of the forest industry is located in New Brunswick. In 2007, 53 per cent of the logged area, including logs from private woodlots, was in New Brunswick, followed by Nova Scotia at 33 per cent. Most logs are sold to local pulp mills. Despite the closure of some pulp mills, other mills are upgrading their operations, including the New Brunswick mills at Edmundston, Nackawic, and Atholville, the latter two of which had previously been mothballed.

Because Atlantic Canada was settled first, its hardwood and softwood forests have been harvested for a very long time—over 300 years. In Nova Scotia and New Brunswick, some areas have been logged three or four times. Logging is an important occupation in Atlantic Canada. Unlike in the rest of Canada, where forest land is usually Crown land (public), the proportion of private timberlands to Crown lands in the Maritimes is extremely high. Private timberlands make up 92 per cent of the commercial forest in Prince

Think About It

In spite of great natural wealth, Atlantic Canada, unlike Western Canada, remains unable to shift from a resource economy to a more diversified one. Does this mean that Innis's staples thesis, described in Vignette 1.3, is flawed?

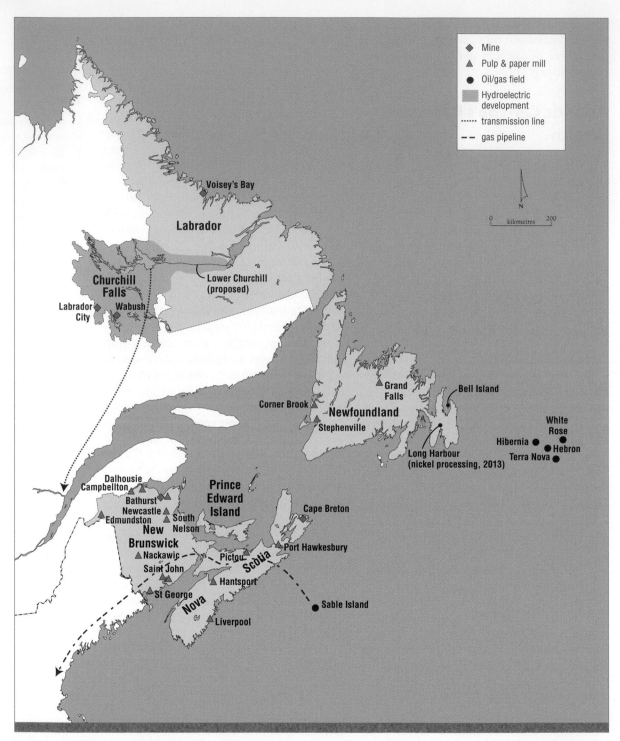

Figure 9.6   Natural resources in Atlantic Canada.
Besides fish, Atlantic Canada contains important natural resources. Enormous iron and nickel deposits are located in Labrador, huge oil and gas reserves are found below the continental shelf, and one of Canada's largest hydroelectric facilities is situated on the Churchill River. Especially important for Newfoundland and Labrador is that, in late August 2007, Premier Danny Williams negotiated a stake in the Hebron–Ben Nevis offshore project, including a 4.9 per cent equity share at a cost of $110 million. Depending on the price of oil, this share could amount to $16 billion over the 25-year life of the project (McCarthy, 2007).

Photo 9.10

The AV Nackawic mill is a joint venture of India's Aditya Birla Group and Québec's Tembec, and produces rayon from wood fibre for the Asian market. Like many other mills, it is on the financial edge. The mill closed in 2004, but the present ownership group took over and reopened the mill in 2006 with millions of dollars of provincial loans and loan guarantees. In April 2008 the New Brunswick government fronted the owners with another $10 million loan, and then added another $10 million loan in July 2009 to keep the mill running and preserve 300 jobs.

Edward Island, 70 per cent in Nova Scotia, and 50 per cent in New Brunswick. In Newfoundland and Labrador, like the rest of Canada, private ownership makes up only 2 per cent of the forested area. On average, the rate of logging on private lands is very high, sometimes exceeding the annual allowable cut estimated by the province. Exceeding the allowable cut impedes sustainable harvesting practices in the region. The high rate of logging often takes place on farms where timber sales are an important source of income.

A new issue in the forest industry in Atlantic Canada relates to Aboriginal people's right to timberlands. Aboriginal Canadians are demanding greater access to timber. Most Crown land in Atlantic Canada has been leased to large companies, mainly to the Irving Forest Corporation. In 1998, a dispute erupted over cutting rights on Crown land. The dispute went to court. Following a provincial court ruling in 2003, the right of Mi'kmaq Indians to harvest timber on Crown lands claimed by these Indians was confirmed. However, in 2005, the Supreme Court of Canada, in a decision described as a setback for Native rights, ruled that the Mi'kmaq of Nova Scotia and New Brunswick need permits if they want to cut down trees on Crown land for commercial gain. The Mi'kmaq are still seeking a compromise with the provincial governments whereby some Crown land currently leased to large companies would be reallocated to them. Just how this ruling of the Supreme Court will translate into a negotiated settlement remains unclear at this time.

The Future: Mining and Petroleum Industries

Atlantic Canada is endowed with mineral and petroleum deposits. The Canadian Shield found in Labrador has rich deposits of iron ore and nickel while the sedimentary strata of the Appalachian Uplands contain many minerals, including coal and potash, while the sedimentary rocks underlying the Grand Banks are a rich source of petroleum. Historically, coal mining on Cape Breton Island was extremely important and was the basis of Nova Scotia's iron and steel industry. But those days are gone and the last coal mine was closed in 2001.[4]

In 2008, mineral production, led by iron and nickel, was valued at nearly $6 billion, while oil and gas production totalled $13.6 billion (Table 9.6). Never before has Atlantic Canada's underground wealth come close to $20 billion! In 2002, for example, the total was $7 billion. In a relatively short time, oil production in the offshore waters of Newfoundland and Labrador has expanded and the royalties have propelled that province from a 'have-not' to a 'have' province. Nova Scotia has also benefited, though to lesser extent, with its Sable Island natural gas deposit. In 2008, the flow of natural gas from Nova Scotia's Sable Island field was valued at close to $1.3 billion.

The recession at the end of 2008 meant a huge drop in the value of mineral production for 2009. See Table 7.5, 'Mineral Production by Province and Territory, 2009', page 298.

The petroleum deposits off Canada's east coast were identified in the late twentieth century, but the high cost and technological challenges prevented full-scale exploitation. The oil deposits consist of a **light sweet crude**. These oil fields are situated in sedimentary basins found off the east coast of Newfoundland on the Grand Banks. Within the Jeanne d'Arc Basin, oil and natural gas deposits have been discovered and three oil projects—Hibernia, Terra Nova, and White Rose—are now operating.

These megaprojects have added a new dimension to Newfoundland and Labrador's economy. In 1997, the Hibernia oil project began producing oil; Terra Nova in 2002; and White Rose in 2005. All three oil fields are located some 300 km east of St John's, in water ranging from 80 to 120 m deep. The oil deposits extend another 2,500 to 4,000 m below the seabed. These projects required huge capital investments. In turn, they generated construction booms by creating a high demand for workers, especially skilled tradesmen, and for a variety of products and services. Once operational, employment in the construction industry dropped sharply and the number of permanent workers in the oil production is relatively small. Hibernia (Vignette 9.10) uses a fixed platform while the Terra Nova and White Rose oil fields employ floating production storage and offloading (FPSO) vessels. The FPSO facility is a ship-shaped vessel with integrated oil storage from which oil is offloaded onto a shuttle tanker.

The Hibernia drilling site has an annual output of about 30 million barrels and adds greatly to Newfoundland's mineral output. Based on an average price of $60 per barrel, the annual value of production is $1.8 billion. Over the next 20 years with the same price and production, the total value of production is estimated at $36 billion. Oil production began in 1998. By 2000, Hibernia accounted for 12 per cent of Canadian oil production. Oil is exported to American and other foreign

Table 9.6	Mineral and Petroleum Production in Atlantic Canada, 2008 ($ millions)				
Product	**NL**	**PEI**	**NS**	**NB**	**Atlantic Canada**
Minerals	4,133	3	340	1,367	5,843
Petroleum	12,283	0	1,276	0	13,559
Total	16,416	3	1,616	1,367	19,402

Source: Natural Resources Canada (2009c); Canadian Association of Petroleum Producers (2009).

Vignette 9.10   The Hibernia Platform

The Hibernia oil project is located 315 km east of St John's, Newfoundland, on the Grand Banks. To tap the estimated 615 million barrels of oil from the Hibernia deposit, an innovative offshore stationary platform was needed. About 4,000 workers built a specially designed offshore oil platform that can withstand the pounding storms of the North Atlantic and crushing blows from huge icebergs. The massive concrete and steel construction sits on the ocean floor, with 16 'teeth' in its exterior wall designed to absorb the impact of icebergs.

The 111-metre-high Hibernia platform, which includes oil-storage units, weighs over 650,000 tonnes. In the summer of 1997 the platform was placed on the ocean floor just above the oil deposits. The depth of the water at this point is about 80 m, leaving the oil platform approximately 30 metres above the ocean surface. This structure is designed as a platform for the oil derricks, and houses pumping equipment and living quarters for the workers, as well as storage facility for the crude oil. The rig extracts oil from the Avalon reservoir (2.4 km under the seabed) and from the Hibernia reservoir (3.7 km deep). The crude oil is then pumped from the Hibernia storage tanks to an underwater pumping station and then through loading hoses to three 900,000-barrel supertankers for shipment to foreign refineries.

Source: Adapted from Cox (1994)

(c) Toronto Star/Peter Power

Photo 9.11

The Hibernia platform has a massive concrete base that supports its drilling and production facilities as well as the workers' accommodations. Since the platform was positioned on the ocean floor in 1997, the province has joined the ranks of other oil-producing provinces. The government of Newfoundland and Labrador at first received half of the royalties generated by these offshore developments. With a new agreement with Ottawa, in 2006 the province received all of the oil revenues. By 2009, oil revenues had made Newfoundland and Labrador a 'have' province.

refineries. According to the owners, the Hibernia Consortium, production should continue for approximately 20 years.

Despite the lure of megaprojects, they often come at a cost in human life. In February 1982, the *Ocean Ranger*, a semi-submersible mobile drilling rig involved in exploratory drilling on the Grand Banks east of St John's, sank in a violent storm. Its crew of 84 died. More recently, in March 2009, a helicopter crashed in bad weather while transporting workers to oil platforms off Newfoundland's east coast, killing 17.

Megaprojects also present problems for regional development. First, they are capital-intensive undertakings. During the construction phase a large labour force is required, but in the operational phase relatively few employees are needed. Second, megaprojects in resource hinterlands lose much of their spinoff effects to industrial areas. As a consequence, economic benefits related to the manufacture of the essential parts for building a megaproject go outside the hinterland, as does the processing of the resource once the project is up and running. Interventions by the government of Newfoundland and Labrador to address this classic problem have had mixed results. The biggest success story comes not from the petroleum industry but from the agreement with the developers of the Voisey's Bay nickel mine whereby the government of Newfoundland

and Labrador obtained a commitment from the company, Vale Inco,[5] to process the ore at Long Harbour in Placentia Bay.

In 2008, the annual value of mineral production in Atlantic Canada was $5.8 billion (Table 9.6). Most production came from iron and nickel mines in Labrador, with nickel valued at $1.9 billion and iron ore at $1.5 billion (Natural Resources Canada, 2009c). Much smaller production by value comes from the lead-zinc mine operated by Brunswick Mining and Smelting Corp. near Bathurst, New Brunswick. In 2008, its production came to $43 million.

The most exciting and promising mineral development is taking place at the huge nickel deposit, the Ovoid deposit, at Voisey's Bay, Labrador. This deposit lies close to the surface. It consists of 32 million tonnes of relatively rich ore bodies—2.8 per cent nickel and 1.7 per cent copper. Another deposit, Eastern Deeps, though larger (about 70 million tonnes), has a lower grade of nickel and copper (similar grade to that found at the Sudbury mines in Ontario). The Voisey's Bay deposit has three attractive features: high-grade nickel, a surface deposit with the

potential for open-pit mining, and proximity to ocean shipping. These characteristics may make it the lowest-cost nickel mine in the world. In fact, there is concern that production from this mine might force higher-cost producers, such as the nickel mines in Sudbury, to close their operations. For Newfoundland and Labrador, the Voisey's Bay mining and smelting operation could employ over 1,000 workers. In late 1996, after Inco Ltd had purchased the Voisey's Bay deposit for $4.3 billion, the company announced that it would build a $1 billion nickel smelter and refinery at Argentia on the southeastern coast of Newfoundland with approximately 750 workers employed at the smelter and refinery. Later, Inco proposed shipping the nickel ore to its Sudbury smelter. Newfoundlanders were outraged and their government refused to sign the surface lease that would allow Inco to move forward with its plans for a mine. In 2002, the government of Newfoundland and Labrador concluded an agreement with the company so that it soon will process ore in the province. In April 2009, after experimental trials, construction began on a US$2.2 billion nickel processing plant at Long Harbour, a

CP/Andrew Vaughan

Photo 9.12

Discovered in 1993, the Voisey's Bay nickel deposit lies along the coast of Labrador approximately 350 km north of Happy Valley–Goose Bay. In 1996, Inco Ltd acquired the rights to the Voisey's Bay property. Once in full operation, the mine and concentrator are expected to employ 400 people. Since the deposit lies within the land claims of both the Labrador Innu and Inuit, Inco was obliged to reach an Impacts and Benefits Agreement with each of the Aboriginal groups.

Vignette 9.11   Lower Churchill, Churchill Falls, and Hydro-Québec

At a cost of $6.5 billion and at least 10 years of construction, the Lower Churchill hydroelectric project involves a dam on the Churchill River plus generation stations at Gull Island and Muskrat Falls. Together, these stations would produce 3,074 megawatts of energy. The government of Newfoundland and Labrador has two choices when it comes to sending the power to the New England market. First, St John's could negotiate a deal to use Hydro-Québec's high-voltage transmission system, but the past agreement for Churchill Falls power still does not sit well in Newfoundland and Labrador. The 1969 contract was initially for 40 years, from 1976—when Churchill Falls power came on stream—to 2016. However, a 'controversial' 25-year extension was added with the price of Churchill Falls power fixed at 1960 levels (Feehan and Baker, 2005). The second choice calls for an undersea cable to carry Labrador's electric power to the island of Newfoundland and then another undersea cable to transmit it to Nova Scotia. From Nova Scotia, the power would move through above-ground lines to New Brunswick and then to New England. In 2009, another wrinkle emerged with the Tshash Petapen Agreement (New Dawn Agreement) between the Labrador Innu and the provincial government. The agreement provides land and compensation for the Churchill Falls project and for the proposed Lower Churchill project (Moore, 2009).

short distance north of Argentia. This plant is expected to be fully operational by February 2013 (Roberts, 2009).

Resource Development and the Aboriginal/Non-Aboriginal Faultline

The Voisey's Bay mining development has sparked a renewed interest in land-claim settlements in Labrador. With such agreements, resource developments are more assured of a trouble-free working environment. This point was made back in 1973 when the courts forced Hydro-Québec to reach an agreement with the Cree and Inuit of Québec who claimed that their respective areas of northern Québec were their traditional homelands. Such lands have given rise to a legal claim known as Aboriginal title. With an unresolved claim of Aboriginal title, there is an uncertainty over ownership of the land in dispute. The Québec government (as did all other provincial governments of the day) considered that Aboriginal people could use Crown lands for hunting and trapping at the pleasure of the Crown and, therefore, no special arrangement with Aboriginal residents was necessary when the provincial government granted a lease or sold the land to a developer. How wrong they were!

Without such an agreement, construction of the James Bay Project could have been halted. By the twenty-first century, however, companies can proceed with their developments before a land-claim agreement is achieved if they reach a special arrangement known as an Impact and Benefits Agreement with the concerned Aboriginal group(s).

The James Bay Project is examined in Chapter 6, 'Hydro-Québec's James Bay Project', page 252.

By 2003, Inco had achieved Impact and Benefits Agreements with both the Labrador Inuit and the Labrador Innu. Yet in all of the negotiations, the Innu—whose traditional homeland includes the site of the Voisey's Bay mining project, who are without treaty, who have never achieved a comprehensive land-claim agreement, and who continued to suffer from incredibly high suicide rates and substance abuse problems—were drawn into the vortex of a huge construction project and forced to answer the question, 'What do you want in exchange for development?' (Samson, 2003: 96–102). While the Innu, like the Labrador Inuit, reached an Impact and Benefits Agreement with Inco to provide specific benefits such as employment opportunities, some Innu (as well as many Inuit) have only a limited command of the English language, and most lacked the basic education level (high school diploma) or employment experience to take advantage of opportunities provided by the construction of the Voisey's Bay nickel mine.

Then, too, the issue of a comprehensive land claim is more complicated for the Labrador Innu than for the Labrador Inuit because the Innu are considered Indians by the federal government and therefore they must be dealt with under the Indian Act. The Labrador Inuit began negotiations in the 1990s and, in August 2003, reached a land-claim agreement.[6] In 2005, the three levels of government (federal, provincial, and Inuit) signed this agreement, thus bringing it into law. The agreement, which includes provision for self-government, is the first of its kind in Atlantic Canada. On the other hand, progress on the Innu land claim has proceeded very slowly because Ottawa insisted that the Labrador Innu first register as Indians under the Indian Act, the benefits of which they had been denied for more than 50 years since Newfoundland joined Canada. This would mean that they must first negotiate for and be given small reserve lands, and only then seek a comprehensive land-claim settlement that would involve their traditional lands (lands that would extend far beyond their reserves) and that may then remove them from the Indian Act they had to adhere to before seeking a comprehensive agreement in the first place (Canadian Human Rights Commission, 2003). By the spring of 2010, the Innu had reached an Agreement-in-Principle with the province on issues under provincial jurisdiction, such as education; negotiations at the federal level dragged on.

Agriculture

Agriculture is limited by the physical geography in Atlantic Canada. Arable land constitutes less than 5 per cent of the Maritimes. Arable land is even rarer in Newfoundland and Labrador, making up less than 0.1 per cent of its territory. Though limited in size, agricultural production significantly contributes to the economy of Atlantic Canada. In 2005, the value of agricultural production in

Barrett & MacKay Photography, Inc.

Photo 9.13

The rich, red soils of Prince Edward Island are famous for growing potatoes, which are the primary cash crop in the province. Prince Edward Island is Canada's leading potato province, responsible for almost one-third of Canadian production. Its potatoes are grown for three specific markets: seed, table potatoes, and processing. Seed potatoes are sold to commercial potato growers and home gardeners to produce next year's crop; table potatoes go to the retail and food service sectors; and processing potatoes are manufactured into french fries, potato chips, and other processed potato products.

the region was about $1 billion. Specialty crops, especially potatoes, contributed heavily to the value of this production.

Atlantic Canada has nearly 400,000 ha in cropland and pasture. Almost all of this farmland is concentrated in three main agricultural areas—Prince Edward Island, the Saint John River Valley in New Brunswick, and the Annapolis Valley in Nova Scotia. Potatoes and tree fruit are important cash crops, though vineyards are gaining ground in the Annapolis Valley. In all three agricultural areas, dairy cattle graze on pasture land. The dairy industry in Atlantic Canada has benefited from the orderly marketing of fluid-milk products through marketing boards.

Prince Edward Island is the leading agricultural area in Atlantic Canada. It has almost half of the arable land in the region. Most of Prince Edward Island's 155,000 ha of farmland are devoted to potatoes, hay, and pasture, with the principal cash crop being potatoes. Since the 1980s, most potato growers have had contracts with the island's major potato-processing plants—Irving's processing plant near Summerside and McCain's plant at Borden–Carleton now dominate the potato industry on the island. The second major agricultural area, the Saint John River Valley, is in New Brunswick. Its 120,000 ha of arable land make up about one-third of Atlantic Canada's farmland. The Saint John River Valley has the best farmland in New Brunswick. Nova Scotia has nearly one-quarter of Atlantic Canada's farmland, with 105,000 ha. Nova Scotia's famous Annapolis Valley, the region's third agricultural area, is the site of fruit orchards and market gardens. The valley's close proximity to Halifax, the major urban market in Atlantic Canada, has encouraged vegetable gardening. In both New Brunswick and Nova Scotia, potatoes are a major cash crop. Almost all potato farmers in these two provinces seed their potatoes under contract to McCain Foods, a multinational food-processing corporation based in New Brunswick. The company has benefited from NAFTA after the removal of tariffs on its food products, especially french fries and potato chips, for export to the United States. Newfoundland has the least amount of farmland—just over 6,000 ha.

Atlantic Canada's Population

Since Confederation, Atlantic Canada's population has increased but at a rate well below the national average. But in the 10-year period from 1996 to 2006, Atlantic Canada's demography hit rock bottom when its population declined by nearly 50,000 (Table 9.7). By far the greatest loss—nearly 46,000—took place in Newfoundland and Labrador. New Brunswick did not escape and lost just over 8,000 residents. Only Nova Scotia and Prince Edward Island saw their populations increase and these increases were very modest. For Newfoundland, many took advantage of high-paying jobs in Alberta. Since 2003, many relocated while others joined the **Big Commute**, working in Alberta but keeping family and home in Newfoundland (Vignette 9.12). Professor Keith Storey (2009) notes that commuting became so popular that some 5,000 workers in St John's lined up for jobs in Alberta, and that by 2008, about 7 per cent of the Newfoundland and Labrador workforce was employed outside of the province but living in the province.

Table 9.7	Population Change in Atlantic Canada, 1996–2008				
Province	1996	2006	2008	Change 1996–2006	Change 2006–8
PEI	134,557	135,851	139,818	1,294	3,967
NL	551,792	505,469	507,895	−46,323	2,426
NB	738,133	729,997	747,302	−8,136	17,305
NS	909,282	913,462	938,310	4,180	24,848
Atlantic Canada	2,333,764	2,284,779	2,333,325	−48,985	48,546

Sources: Statistics Canada (2002, 2007a, 2009d).

Vignette 9.12   The Big Commute and Transfer Payment

Although an untold number of Atlantic Canadians have moved to Alberta for work in the oil fields and, more generally, in construction, many others commute from their homes on the east coast on a cycle of 20 days of work in the oil patch and eight days back home. Estimates suggest that as many as 10,000 Newfoundland trades workers regularly commute to the Alberta oil sands to work for salaries that start above $100,000 a year, not including overtime. Their air travel is paid for by companies that have regularly scheduled commuter flights from Newfoundland to Fort McMurray, and once there they are fed and housed at company expense (CBC News, 2007, 2009; Storey, 2009). The earnings they bring back are believed to amount to hundreds of millions dollars—for Newfoundland and Labrador, a hidden economic boost for a recently anointed 'have' province, and a monetary infusion that is keeping some communities viable, and at the same time dependent on the fortunes of an industry practically at the other end of the country. The hidden costs—to families, to social structure, to individual lives and values—are perhaps even more difficult to discern.

Think About It

Which has the higher social costs for a family, unemployment or cross-country commuting?

The main demographic factors accounting for this decline were a falling birth rate and a steady death rate, which resulted in a very low rate of natural increase; little in-migration; and a massive out-migration, especially from Newfoundland. In Newfoundland and Labrador, the unemployment rate was the highest of the 10 provinces, thus providing a push to seek work in other parts of Canada

Then, unexpectedly, in only two years, according to Statistics Canada estimates, Atlantic Canada rebounded with a population gain of 48,546 that virtually matched the loss of the previous 10 years. Nevertheless, Atlantic Canada appears to have turned the corner and is now enjoying population growth. The reason for this change appears to be a much more positive economic situation since 2006, especially in Nova Scotia, so that workers and their families are staying in Atlantic Canada and others, mainly Newfoundlanders, are returning home. In addition, a construction slowdown in the Alberta oil sands had a 'return-home effect'.

Rural Atlantic Canada

In the 2006 census, Atlantic Canada had the greatest percentage of its population classified as rural. The contrast with the rest of Canada is striking: In 2006, nearly six million people, almost 20 per cent of Canadians, were living in rural areas, but the percentage of rural residence in Atlantic Canada was more than twice as high, at 46 per cent (Table 9.8).

Geography and history explain why Atlantic Canada has such a large rural population, many of whom live in small coastal communities. In the seventeenth and

Table 9.8   Rural Atlantic Canada

Province	2006 Rural Population	2006 Rural Percentage
Newfoundland and Labrador	213,370	42.2
Prince Edward Island	74,683	55.0
New Brunswick	357,072	48.5
Nova Scotia	406,537	44.5
Atlantic Canada	1,051,662	46.0
Canada	6,216,135	19.7

Note: Statistics Canada's definition of rural population is farms and small communities with less than 1,000 persons.
Source: Statistics Canada (2008).

eighteenth centuries, when the first British and French families arrived in Newfoundland and Nova Scotia, these newcomers searched for small, sheltered harbours suitable for mooring boats that would provide quick access to cod fishing grounds. Often these outports consisted of no more than three or four families. While many coastal villages still remain, they are anachronisms in the twenty-first century. Delivering basic public services, such as education and health care, in these coastal places is expensive because populations are small and isolated, and both because of this and for economic reasons, small coastal fishing communities are disappearing. The rural-to-urban migration, long underway in other regions of Canada, is slowly grabbing hold in Atlantic Canada. While the settlement context is different (fishing rather than farm communities), the result is the same—relocation to larger towns and cities. In 1945, nearly 1,400 tiny outports with populations under 200 dotted the coast of Newfoundland (Staveley, 1987: 257); by 2001, there were fewer than 400 outports with populations under 200. This decline was the result of two forms of relocation—voluntary and government-assisted. In short, people, particularly young people, no longer wanted to live in these tiny places where life revolved around the fishery. Outport living could not survive a depressed fishing industry and the growing desire/need for urban amenities. The abandonment of outport villages began in 1946 and by 1954, 49 small communities had been abandoned without government assistance. From 1955 to 1966, 110 communities were abandoned and the

Canaport LNG

Photo 9.14

With the city of Saint John in the background, the strategic location of Canaport LNG facility, in this artist's conception, is ideally located for access to the huge New England energy market. Liquefied natural gas arrives from a variety of locales, including the Caribbean and Middle East. Here, the liquefied gas is stored in containers at −162°C, then is regasified and sent by pipeline to New England markets. The first shipment of Qatar natural gas arrived in December 2009 in the *Q-Flex*, the world's second-largest LNG tanker.

residents resettled in larger centres of their choice. In 1967, the federal government joined forces with Newfoundland. The federal–provincial Newfoundland Resettlement Program added new conditions to relocation. Migrants had to relocate to one of 77 designated 'growth centres'—selected communities that offered more social and economic opportunities. Another stipulation was that at least 75 per cent of the households must agree to move for them to receive public relocation funds. Under the joint program, another 150 communities were abandoned from 1967 to 1975. After this program ended in 1976, few communities were abandoned until the late 1990s, after the cod fishery was closed. Following the 1992 cod moratorium, many communities saw their fish plants close, causing people to seek work elsewhere. Such was the fate of Great Harbour Deep, a tiny outport of 180 inhabitants on the Northern Peninsula that was founded some 400 years ago. In April 2003, the residents accepted the provincial government's offer of financial help if all agreed to leave their community. Each household received up to $100,000 to help cover the costs of relocation (Vignette 9.7).

Think About It

If the cod fishery returned, could Newfoundland's outports survive?

Russ Heinl/All Canada Photos

Photo 9.15

When the British founded Halifax in 1749, they were attracted by its magnificent harbour. The high hill overlooking the harbour offered a perfect location for a fortress to defend the new town and its naval base. Named the Halifax Citadel, this fortress is an impressive star-shaped masonry structure complete with defensive ditch, earthen ramparts, musketry gallery, powder magazine, garrison cells, guard room, barracks, and school room. The Citadel is now a National Historic Site.

Atlantic Canada's Urban Geography

The urban geography of Atlantic Canada is characterized by few large cities and, as we have seen, many small coastal villages. Statistics Canada's definition of an urban place requires a minimum population concentration of 1,000 persons and a population density of at least 400 persons per square kilometre (Statistics Canada, 2009c). It is not surprising, then, that Atlantic Canada in 2006 was the least urbanized region of Canada with just over half of its population living in urban centres. In comparison with other regions, the difference is both striking and an indicator of how much more urban growth (or rural decline) is likely. For instance, in Ontario and British Columbia the urban population accounts for 85 per cent of total population. As well, Atlantic Canada has none of Canada's largest cities: Halifax ranks thirteenth in population and St John's is twentieth (see Table 9.9). Equally significant, Atlantic Canada's fractured geography results in these two cities being the dominant metropolitan centres in the region. Halifax serves as the urban focal point for the Maritimes, while St John's fills a similar role for Newfoundland and Labrador. Halifax's advantage is its deep, ice-free harbour, its role as a naval base, and its relatively large population/market. Halifax serves as a major container port and has the

Table 9.9 Urban Centres in Atlantic Canada, 2001–6

Urban Centre	Population 2001	Population 2006	% Change
Labrador City	7,744	7,240	–6.5
Gander	9,651	9,951	3.1
Bay Roberts	10,531	10,507	–0.2
Grand Falls–Windsor	13,340	13,558	1.6
Campbellton	18,820	17,888	–5.0
Edmundston	22,173	21,442	–3.3
Kentville	25,172	25,969	3.2
Corner Brook	26,153	26,623	1.8
Bathurst	32,523	31,424	–3.4
New Glasgow	36,735	36,288	–1.2
Truro	44,276	45,077	1.8
Charlottetown	57,234	58,625	2.0
Fredericton	81,346	86,688	5.3
Cape Breton*	109,330	105,928	–3.1
Saint John	122,678	122,389	–0.2
Moncton	118,678	126,424	6.5
St John's	172,918	181,113	4.7
Halifax	359,183	372,858	3.8

*Cape Breton includes Sydney, North Sydney, Glace Bay, and other Cape Breton Island municipalities.

Source: Statistics Canada (2007a). Adapted from Statistics Canada publication *Population and Dwelling Counts, 2001 Census,* Catalogue 93F0050XCB2001013, Released 16 July 2002, http://www.statcan.ca/bsolc/english/bsolc?catno=93F0050X2001013.

potential of becoming the hub of the Atlantic Gateway. On the other hand, St John's today is focused on offshore oil, the fishing industry, and government services, and may become a centre for Arctic marine research and resupply. Within the Maritimes, Saint John, NB, is ideally situated as an energy hub to New England. The natural gas shipped to the United States comes from foreign sources and is shipped to Saint John in a liquefied state (see photo 9.14).

See Chapter 4, 'Urban Population', page 136, including Table 4.5, 'Percentage of Urban Population by Region, 1901–2006', for further understanding of Atlantic Canada's lagging urbanization.

Vignette 9.13 Halifax

Halifax, the capital of Nova Scotia and the largest city in Atlantic Canada, was founded in 1749. By 2006, Halifax had a population of almost 373,000. As in the past, its strategic location allowed Halifax to play a major role on the Atlantic coast as a naval centre, an international port, and as a key element in the Atlantic Gateway concept. Along the east coast of North America, its deep, ice-free harbour is ideally suited for huge post-Panamax ships. Yet, because of its relative distance from the major markets in North America and its reliance on indirect transferring of goods between ships, trains, and trucks, Halifax cannot provide lower transportation costs than New York. The economic strength of Halifax rests on its defence and port functions, its service function for smaller cities and towns in Nova Scotia, and its provincial administrative functions. Halifax also has a small manufacturing base and a growing service sector, as well as a small but growing high-technology industry.

SUMMARY

Atlantic Canada lies on the eastern rim of Canada. For decades, Atlantic Canada was Canada's economically troubled region with high unemployment figures and record out-migration of workers. Yet, while unemployment and out-migration continue, Atlantic Canada, and especially Newfoundland and Labrador, has a second chance due to its offshore petroleum resources, its rich nickel deposit at Voisey's Bay, and the promise of hydroelectric power from the proposed Lower Churchill project. Newfoundland and Labrador has already shaken off the mantle of a 'have-not' province. Even the booming economy of Alberta has helped through the Big Commute, which transfers to Atlantic Canada much needed wages. But Atlantic Canada's cities are growing and they play a critical role in the region's second chance.

Thus, the future for Atlantic Canada in the twenty-first century looks promising, but the prospects for sustainable economic growth remain problematic. On the one hand, mega-resource developments, especially offshore oil, have the potential to transform the regional economy, and the possibility of an Atlantic Gateway—a trade corridor connecting Atlantic Canada with North America and the rest of the world—if realized, would be a tremendous boom. For Newfoundland and Labrador and Nova Scotia, the expansion of offshore oil and gas is stimulating their economies while New Brunswick is banking on serving as an energy hub for New England. On the other hand, two nagging question persists. First, the new economy is taking place in the major cities, leaving rural Atlantic Canada stranded. Second, does energy production hold the key to economic rejuvenation and sustainability? The answer seems more positive today than it did in the past because the energy-producing provinces keep 100 per cent of the royalties. As the global economy recovers, higher prices for its energy resources could propel Atlantic Canada's economy into a more prosperous state—one in which unemployment rates decline, business opportunities increase, and out-migration slows and even reverses. Such an economic state may create that elusive balance between economic growth and a 'down East' way of life.

CHALLENGE QUESTIONS

1. What economic and political developments provide Atlantic Canada with an opportunity to break the downward economic spiral depicted by Friedmann's downward transitional region?
2. Why does Georges Bank contain such large and rich fish stocks?
3. Why did draggers damage the cod stocks and other marine life?
4. Why does Atlantic Canada have such a high percentage of its population classified as rural?
5. Compared to the other regions, how would you rank Atlantic Canada's sense of place?

FURTHER READING

Coates, Ken S. 2000. *The Marshall Decision and Native Rights*. Montréal and Kingston: McGill-Queen's University Press.

On 7 September 1999 the Supreme Court of Canada ruled in a case involving Donald Marshall Jr that the Mi'kmaq could earn a 'modest income' from the fishery in the Maritimes. In one swoop, Atlantic Canada woke up to the Aboriginal desire and right to participate in the fishery, especially the lucrative lobster fishery. This is easier said than done because sharing of natural resources, such as lobsters, means taking away from those who already have the right to harvest such resources. The lobster solution involved the federal government obtaining lobster fishing licences from non-Aboriginal fishers and allocating them to the Mi'kmaq fishers.

Coates places the *Marshall* decision, which changes the relationship between the Mi'kmaq and other Maritimers, within the larger Canadian context where a search for a new relationship between Aboriginal Canadians and other Canadians is underway.

10

THE TERRITORIAL NORTH

INTRODUCTION

The Territorial North is Canada's last frontier, but it is also a homeland for Aboriginal peoples who form the majority of its population (Table 1.2). Located in the highest latitudes of Canada, the Territorial North, remote and permafrost-affected, remains a paradox—rich in natural resources but slow to develop. Two other challenges unique to Canada face this most northerly region: finding a more secure place for Aboriginal peoples within the unfolding modern version of territorial society and its economy; and dealing with Arctic sovereignty issues. With these challenges, plus global warming, land-claims settlements, and increasing world demand for its resources, what does the future hold for the Territorial North?

In Friedmann's regional scheme of the core/periphery model, the Territorial North would be described as a resource frontier (Table 1.4). In a developing region in a remote part of the country, economic and social development involves huge capital investments and substantial public funding through transfer payments from Ottawa. Like developing frontiers around the world, the Territorial North's economic performance is limited to non-renewable resources, which make the region vulnerable to sharp fluctuations in global demand for its exports. For these reasons, the *Key Topic* explores megaprojects.

CHAPTER OVERVIEW

Topics and issues examined in this chapter include the following:

- The Territorial North's physical and historical geography, including the search for the Northwest Passage, the birth of Nunavut, and the impact of modern land claims.
- The dualistic nature of the region's population and economy.
- The environment, Arctic sovereignty, and control of the Northwest Passage, as seen in the context of global warming.
- The region's changing economic and political position within Canada, the circumpolar world, and the emerging global economy.
- The role of megaprojects in the North's development.

Employed for communication and survival purposes across the Arctic, inukshuk means 'in the likeness of a human'. This inukshuk is found on the rock-strewn Victoria Island, Nunavut. Photo: Wayne Lynch/All Canada Photos.

The Territorial North within Canada

The Territorial North has the largest geographic area of the six regions, but the smallest population and economy (Figure 10.2). Its population of just over 100,000 persons is spread over nearly 4 million km², making the Territorial North one of the world's most sparsely populated areas. Its narrowly based mining economy is dependent on global markets and prices, making it extremely vulnerable to **boom-and-bust cycles**.

The Territorial North's demographic features have been shaped by three factors. The first, its small population, is due to the limited

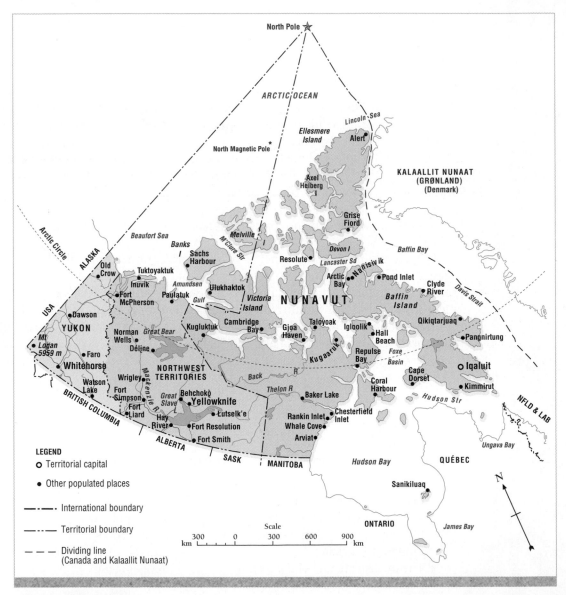

Figure 10.1 The Territorial North.

The Territorial North consists of three territories. Bordering several foreign countries, its boundaries are fixed with a few exceptions. Two areas where international boundaries are unclear are the sea border between Yukon and Alaska and the unclaimed international waters of the Arctic Ocean.

Source: Atlas of Canada, 2006, 'The Territories', at: <atlas.nrcan.gc.ca/site/english/maps/reference/provincesterritories/northern_territories>.

capacity of the land to support people. The second is the Aboriginal population—especially the Inuit, whose very high birth rate and extremely low death rate account for the population growth in the Territorial North. Third, the non-Aboriginal population tends to move to job opportunities so that when the North's economy stalls, non-Aboriginal residents are more likely than Aboriginal residents to move to southern Canada. However, as the education level and job skills have improved for many younger Aboriginal northerners, some are beginning to respond to opportunities outside of the Territorial North, whether for post-secondary education and training or more challenging jobs that better match their educational achievement.

The discussion in Chapter 4, 'Population Density', page 131, provides insight into the issue of limited capacity of the land to support people.

As a **resource frontier**, the Territorial North has an economy based on the exploitation of its energy and mineral resources. Such industrial activities greatly disturb the natural environment and wildlife. Exploitation of non-renewable resources has spurred economic growth in the region, but this type of economic development lacks stability because it is entirely dependent on exports to national and global markets. Variation in demand makes the Territorial North's economy subject to a boom-and-bust cycle. This cycle is caused by the finite nature of non-renewable resources, which results in mine closures, and by fluctuations in world prices. Downward price movements have caused temporary shutdowns of mining operations.

Two Visions

The people of the Territorial North have two powerful and seemingly contradictory visions—one is of a **northern frontier**, while the other is of a **homeland**. The traditional image of the northern frontier is one of great wealth just waiting to be discovered. For example, during the Klondike gold rush (1897–8), prospectors flooded the Yukon to

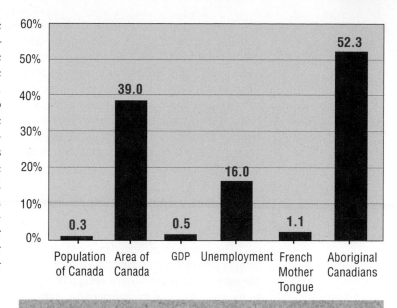

Figure 10.2   The Territorial North, 2006.
Though the region is the largest in Canada, its population and economy are smallest. The Aboriginal population continues to increase, jumping from 51.7 per cent of the total population in 2001 to 52.5 in 2006. The Territorial North suffers from high **unemployment** as well as **underemployment**, i.e., when no jobs are available, individuals do not seek employment and hence are not classified as unemployed.

Sources: Tables 1.1, 1.2.

pan for gold along the Klondike River and its tributaries. A more contemporary version of this image comprises large multinational corporations with their vast capital and advanced technology undertaking megaprojects—mining for gold, diamonds, lead, and zinc, and drilling for oil and gas. **Megaprojects** are large-scale resource developments financed and managed by multinational corporations designed to meet global needs for primary products. Such projects create an economic boom during the construction period, but in their operational phase fewer employment opportunities are available and economic spinoffs for local businesses are limited. Because of the risks associated with developing resources in a frontier—from overcoming physical barriers unique to the Territorial North to coping with downturns in world prices for resources—such projects usually are undertaken by large corporations. In return, these corporations reap large profits and supply the industrial cores of the world with raw materials and energy.

Think About It

If population density were to be measured by physiological density or the carrying capacity of the land to sustain life, how would the Territorial North compare with the other five geographic regions?

Northerners, particularly Aboriginal peoples, see the North as a homeland. This perception is based on a special, deep commitment to the North, which cultural geographers often attribute to a sense of place. Local people have a strong appreciation for natural features, cultural traits, and the political and economic issues affecting their homeland. A sense of place evokes a feeling of belonging as well as a commitment to a particular place.

Sense of place is discussed further in Chapter 1, page 15.

The concepts of homeland and resource frontier are not always compatible. For instance, the interests of multinational companies relate to the profitability of the northern frontier's untapped natural wealth. The interests of northerners, however, especially Aboriginal northerners, are best served when the long-term environmental and social well-being of their communities is considered. 'Frontier' economic activities often bring jobs and investment, but the Aboriginal community has had minimal involvement, partly because of a mismatch between the needs of the companies and the skill sets of Aboriginal workers. Nevertheless, on balance, positive change is taking place with these two visions interacting to place a 'duality' stamp on the regional character of the Territorial North. The very interplay between actual events representing visions indicates a capacity to change over time, largely due to acculturation and accommodation processes. This dynamic produces a powerful social current described in this chapter, especially in comprehensive land-claim agreements and the formation of Nunavut.

In the Territorial North, the concepts of homeland and regional consciousness have resulted in the devolution of political power from the federal government to the territorial

Stephen J. Krasemann/Valan Photos

Photo 10.1

The South Nahanni is one of the world's great wild rivers. Located in the boreal wilderness of Nahanni National Park Reserve in the southwestern part of the Northwest Territories, this untamed river is seen surging through the steep-walled First Canyon. Downstream, its waters rush past hot springs, plunge over a waterfall twice the height of Niagara, and cut through canyons more than one kilometre deep.

governments. Elected governments exist in the three territories and six land-claim agreements with First Nations have been concluded, with the most recent being the Tlicho Final Agreement (2003). Significantly, the Tlicho Final Agreement included provisions for self-government for the Dogrib who live between Great Bear and Great Slave lakes in the Northwest Territories. Future comprehensive land-claim agreements are likely to include a section on self-government. Aboriginal self-government received an enormous boost in 1992 when the provision for the territory of Nunavut was placed in the Nunavut Land Claims Agreement under the Nunavut Political Accord. Unlike Yukon and the Northwest Territories, Nunavut is an expression of 'ethnic' regional consciousness and yet Nunavut is a public government and therefore is different from First Nations' 'ethnic' self-governments. The next challenge facing the Territorial North, but especially Nunavut, is to generate sufficient economic growth and to create a labour force that can take advantage of such growth. While a tall order, if achieved, then the Territorial North's economic dependency on Ottawa would diminish. At the same time, the Territorial North, while blending Western and Aboriginal ways, would create a homeland accepted by all.

Physical Geography of the Territorial North

The Territorial North extends over four of Canada's physiographic regions: the Canadian Shield, the Interior Plains, the Cordillera, and Arctic Lands (including the Arctic Archipelago) (Figure 2.1). As a result, the region encompasses a more varied topography than the other five regions. Its topography ranges from mountainous terrain and forest stands in the west, to barren plains, ice-covered islands, and Arctic seas further east and north.

One special feature is the aurora borealis or **northern lights**. While not unique to the Territorial North, these illuminations occur regularly in the long winter nights, causing spectacular displays of shifting or streaming coloured light in northern skies. These northern lights are caused by the interaction of charged solar particles with gases in the atmosphere under the influence of the earth's magnetic field, thus they coincide with geomagnetic disturbances that can disrupt communication and electrical systems.

The vegetation in this region is quite varied and includes a small portion of the boreal forest in the southwest, the tundra with its mosses and lichens further north, and a polar desert in the highest reaches (Figure 2.7). The region is known for the several rivers that wind through it, the many lakes that dot its landscape, and the Arctic Ocean that supports a range of aquatic wildlife.

The physical geography of the Territorial North is governed not so much by physiography as by a cold environment. Cold persists throughout most of the year and in many ways affects human activities. The cold environment includes permafrost (Figure 2.9) and long winters with sub-zero temperatures. Except for the northern half of Yukon, the Territorial North was subjected to glaciation. For that reason the rich Klondike placer deposits were not affected by the scouring effects of the Cordillera ice sheets that affected southern Yukon. The region's main climate zones, the Arctic and the Subarctic (Figure 2.6), are characterized by very short summers.[1] In the Arctic climate, summer is limited to a few warm days interspersed with colder weather, including freezing temperatures and snow flurries. The Subarctic climate has a longer summer that lasts at least one month. During the short, but warm, summer the daily maximum temperature often exceeds 20°C and sometimes reaches 30°C.

Arctic air masses dominate the weather patterns in the Territorial North. They are characterized by dry, cold weather and originate over the ice-covered Arctic Ocean, moving southward in the winter. The Arctic zone has an extremely cold and dry climate. Distinguished by long winters and a brief summer, the Arctic climate is normally associated with high latitudes and lower levels of solar energy. The ice-covered Arctic Ocean and continuous permafrost keep summer temperatures cool even though the sun remains above the horizon for most of the summer. These cool

Think About It

Does Nunavut represent a merging of the two visions of the Territorial North?

summer temperatures, which Köppen defined as an average mean of less than 10°C in the warmest month, prevent normal tree growth. For that reason, the Arctic climate region has tundra vegetation, which includes lichens, mosses, grasses, and low shrubs. In the very cold Arctic Archipelago, much of the ground is bare, exposing the surface material. As there is little precipitation in the Arctic Archipelago (often less than 20 cm per year), this area is sometimes described as a 'polar desert'.

Beyond 70° N, growing conditions become very limited. With lower temperatures and less precipitation than in the lower latitudes of this climatic zone, tundra vegetation cannot survive. The Arctic climate, however, does extend into lower latitudes in two areas: along the coasts of Hudson Bay and the Labrador Sea. These cold bodies of water chill the summer air along the adjacent land mass. In this way, the Arctic climate extends along the coasts of Ontario, Québec, and Labrador well below 60° N, sometimes extending as far south as 55° N.

The Arctic Ocean was called the 'Frozen Sea' by early explorers. This extensive ice cover, known as the **Arctic ice pack** or polar pack ice, drifts in a clockwise motion in the Beaufort Sea. Because of the extent and thickness of the Arctic ice pack, few ships can navigate these waters without the assistance of icebreakers. Two former mining operations in the High Arctic—the Nanisivik mine on northern Baffin Island and the Polaris mine on Little Cornwallis Island—stored their ore until the summer navigation season, when ships reinforced against ice, sometimes aided by icebreakers, transported the ore to European and other world markets. However, if the impact of global warming continues at its present pace, more open water will appear in the short summer, making the Northwest Passage a reality.

The geology of the Territorial North provides much of its wealth. For example, the sedimentary basins of the Interior Plains and Arctic Lands contain large deposits of oil and natural gas (Vignette 10.1). Even the sedimentary strata

Vignette 10.1   Petroleum Basins

The Territorial North has many sedimentary basins, some of which contain petroleum deposits. The three main petroleum-containing basins are the Western Sedimentary Basin, the Mackenzie Basin, and the Canadian Arctic Basin. The Canadian Arctic Basin contains several smaller basins, including the Sverdrup Basin. According to Indian and Northern Affairs Canada, oil reserves in the Territorial North total 1,899 million barrels and gas reserves amount to 32 trillion cubic feet (2009: Table 1). In the same report, INAC (2009a: 9) states that: 'A current snapshot of oil and gas resources in Canada based on the new

estimates puts 33 per cent of Canada's remaining conventionally recoverable resources of natural gas . . . and 35 per cent of remaining recoverable light crude oil . . . in Canada's Northwest Territories, Nunavut, and Arctic offshore (in regions under the administration of the Minister of Indian and Northern Affairs).' Based on the market value of discovered oil and gas at current prices (2008), these energy resources are valued at over $300 billion (calculated from INAC, 2009a: Tables 1 and 2). They remain trapped in remote areas, but if Arctic navigation becomes a reality, the market value of these petroleum resources turns into real money.

Table 10.1   Discovered Petroleum Resource Inventory, Territorial North, 2008

Region	Crude Oil (10^6 m³)	Million Barrels	Natural Gas (10^9 m³)	Trillion Cubic Feet
Northwest Territories and Arctic Offshore	250.5	1,576.3	462.2	16.4
Nunavut and Arctic Offshore	51.3	322.9	449.7	16.0
Total	301.6	1,899.2	911.9	32.4

Note: Compiled and integrated from several published sources that may underestimate or overestimate actual field resources. Volumes and distribution should be regarded as approximate and reflect the opinion of the author. Numbers may not add due to rounding.
Source: Adapted from Drummond (2009: 60, Table 1).

beneath the Arctic Ocean just beyond Canada's current jurisdiction hold vast energy deposits. The Cordillera and Canadian Shield of the Territorial North have already yielded some of their mineral wealth to prospectors and geologists. These minerals include diamonds, gold, lead, uranium, and zinc.

Environmental Challenge: Global Warming

The Territorial North is faced with a major environmental challenge—the warming of its lands and waters. Global warming is expected to reach its maximum temperature increase in the Arctic because of the albedo effect, whereby greater solar warming of the land and water will occur because of the reduction of ice and snow cover.

As discussed in Chapter 2, global warming is the increase over time of the earth's average surface temperature. Several factors are involved in greater temperature increases occurring in the Arctic, such as heat transfer from lower latitudes to higher ones, but the primary factor is the albedo effect. Light from the sun takes the form of short-wave radiation while energy emitted from the earth's surface takes the form of long-wave radiation. Long-wave radiation is more readily absorbed by the atmosphere and thus warms the atmosphere whereas short-wave radiation escapes into outer space without warming the atmosphere. Currently, the Arctic has a high albedo because of its cover of snow and ice, which means most solar energy is reflected back into outer space without warming the atmosphere. However, as ice and snow cover decreases in the Arctic, its albedo will shift from high to low, meaning that the solar energy reaching the Arctic will be more effective in warming its atmosphere and thus raising temperatures.

The impact of global warming in the Arctic is expected to have both positive and negative impacts on the wildlife and northern peoples. By the end of the twenty-first century, global warming is expected to have created an ice-free Arctic Ocean and Hudson Bay,

Bill Terry/Take Stock Inc.

Photo 10.2

The Yukon River Valley at Dawson City. For Aboriginal peoples as well as for fur traders and prospectors, this long and winding river has been an important transportation route in the history of the Territorial North.

Vignette 10.2   Toxic Time Bombs: The Hidden Cost of Mining

Mining brings jobs and wealth to the North but it also leaves behind toxic wastes. The short lifespan of most mines—less than 20 years—results in a geography of toxic time bombs. But why don't companies accept the social responsibility for cleaning up their mess? The answer is that they wish to skirt the high cost of cleanup. New and more stringent regulations have corrected this situation—except in cases of bankruptcy, which are not uncommon in the mining industry. For example, gold mining near Yellowknife has ended but hidden costs remain. The refining of gold at the Giant gold mine at Yellowknife left residues of arsenic. Now closed, this mine has left a toxic time bomb of 237,000 tonnes of arsenic trioxide, a by-product of gold refining. Since the last mining company (Royal Oak Mines) declared bankruptcy, the cost of the cleanup, estimated at a quarter of a billion dollars, is left to the federal government—and that means the Canadian taxpayer (Bone, 2009: 227; Danylchuk, 2007).

Historical Geography of the Territorial North

European Contact

At the times of initial contact with Europeans, seven Inuit groups and seven Indian groups belonging to the Athapaskan language family (also known as Dene) occupied the Territorial North. The Inuit stretched across the Arctic: the Mackenzie Delta Inuit lived in the west; further east were the Copper Inuit, Netsilik Inuit, Iglulik Inuit, Baffinland Inuit, Caribou Inuit, and Sadlermiut Inuit. Inuit also lived in northern Québec and Labrador. By the early twentieth century, two groups (most of the Mackenzie Delta Inuit and all of the Sadlermiut Inuit) would succumb to diseases that European whalers brought to the Arctic. The Indian tribes that resided in what is today the territorial Subarctic were the Kutchin, Hare, Tutchone, Dogrib, Tahltan, Slavey, and Chipewyan. More recently, these tribes are known in the aggregate as the Dene.

These Aboriginal peoples had developed hunting techniques that were well adapted to two cold but different environments. The Inuit employed the kayak and harpoon to hunt seals, whales, and other marine mammals, which enabled them to occupy the Arctic coast from Yukon to Labrador. As a result, the Inuit depended extensively on marine mammals and fish. The Indians hunted and fished in the northern coniferous forest, where the birchbark canoe, the bow and arrow, and snowshoes enabled them to hunt in summer and winter. The Dene tribes relied heavily on big game like the caribou, the Chipewyans often following the caribou to their calving grounds in the northern barrens of the Arctic. Both the Inuit and the Dene moved across the land in a seasonal rhythm, following the migratory patterns of animals. Operating in small and highly mobile groups, these hunting societies depended on game for their survival. Cultural traits, such as the ethic of sharing, developed from this dependency on the land and sea for food.[2]

thus allowing for ocean transportation of its petroleum reserves and the reduction of water transportation costs for other mineral deposits in the Arctic. Land transport, with the melting of permafrost, will be another matter. Wildlife will be affected. Already scientists have noted negative impacts on polar bears but a positive effect on seal populations. The huge migrating caribou herds that have their calving grounds in the Arctic could be affected. The reduction of the size of calving grounds would have a negative impact on preferred space for reproduction. The Dene and Inuit communities that rely on these herds for much of their country food may have to purchase more of their food from local stores. On the other hand, more open water would permit a return of large numbers of bowhead whales, which used to sustain the Thule. Perhaps their descendants, the Inuit, could again harvest these huge mammals. The cultural and nutrition implications are large. Caribou meat forms an important part of the culture of Dene and Inuit while the superior nutritional value of all country food compared to store-bought frozen meat and poultry is well recognized.

The Time of European Exploration

Though the Vikings were the first to make contact with northern Aboriginal peoples around 1000, little is known of those encounters.[3] At that time, the Arctic Ocean had much less ice cover because of a warmer climate. About five centuries later, in 1576, Martin Frobisher, in searching for a Northwest Passage to the Far East, reached Baffin Island. Unfortunately for Frobisher, his expedition took place at the height of the 'Little Ice Age', and his ships met with heavy ice conditions in Davis Strait (which separates Baffin Island from Greenland).

The Little Ice Age is discussed in Vignette 2.11, 'Fluctuations in World Temperatures', page 68.

Frobisher reached Baffin Island where he encountered a group of Inuit, some on land, others in their kayaks. During a skirmish between Frobisher's men and the Inuit, Frobisher was wounded by an arrow. In the exchange, five of his men were lost and three of the Inuit were captured. The Inuit and one kayak were taken back to England, as often was done in the early years of European exploration, as proof of Frobisher's discovery. All three of the captives soon succumbed to illness. Over the next three centuries, the search for a Northwest Passage through Arctic waters led to misadventure for various European explorers, including John Franklin, whose famous last expedition ended in disaster (Vignette 10.3). However, cultural exchange between Europeans and the original inhabitants of these lands remained limited until the nineteenth century, when the trade in fur pelts and whaling peaked in North America.

Whaling and the Fur Trade

Whaling, the first commercial venture in the Arctic, began in the late sixteenth century in the waters off Baffin Island. During those early years of whaling, whalers had little opportunity or desire to make contact with the Inuit living along the Arctic coast. The Inuit probably felt the same, particularly those who had heard stories of the nasty encounter with Frobisher's men. During early summer, whaling ships set sail from British, Dutch, and German ports for Baffin Bay, where they hunted whales for several months. By September, all ships would return home. In the early nineteenth century, the expeditions of John Ross (1817) and William Parry (1819) sailed farther north and west into Lancaster Sound. Their search for the Northwest Passage had limited success but opened virgin whaling grounds for whalers. These new grounds

Think About It

Photo 3.2, page 79, illustrates a skirmish between Frobisher's men and the Baffin Island Inuit. Do you think that similar skirmishes occurred with the Vikings?

Vignette 10.3   The Northwest Passage and the Franklin Search

In 1845, Sir John Franklin headed a British naval expedition to search for the elusive Northwest Passage through the Arctic waters of North America. This British naval expedition took place at the end of the Little Ice Age, meaning that ice conditions would have been much more challenging than those occurring today. He and his crew never returned. Their disappearance in the Canadian Arctic set off one of the world's greatest rescue operations, which was conducted on land and by sea and stretched over a decade. The British Admiralty organized the first search party in 1848. Lady Franklin sent the last expedition to look for her husband in 1857. These expeditions accomplished three things: (1) they found evidence confirming the loss of Franklin's ships (the *Erebus* and *Terror*) and the death of their crews; (2) one rescue ship under the command of Robert McClure almost completed the Northwest Passage; and (3) the massive rescue effort resulted in a greater knowledge and mapping of the numerous islands and various routes to the north and west of Baffin Island in the Arctic Ocean. The exact sequence of events that led to the Franklin disaster is not known. However, archaeological work, conducted in the early 1980s on the remains of members of the expedition, revealed that lead poisoning, caused by the tin cans in the ships' food supplies, probably contributed to the tragic demise of the Franklin expedition.

were of great interest as improved whaling technology had reduced the whale population in the eastern Arctic. In fact, the period from 1820 to 1840 is regarded as the peak of whaling activity in this area. At that time, up to 100 vessels were whaling in Davis Strait and Baffin Bay.

As whaling ships went further to find better whaling grounds, it became impossible to return to their home ports within one season. By the 1850s, the practice of 'wintering over' (that is, allowing ships to freeze in sea ice along the coast) was adopted by English, Scottish, and American whalers. This allowed whalers to get an early start in the spring, providing for a long whaling season before the return trip home at the onset of the next winter. Wintering over took place along the indented coastlines of Baffin Island, Hudson Bay, and the northern shores of Québec and Yukon. Permanent shore stations were established at Kekerton and Blacklead Island in Cumberland Sound, at Cape Fullerton in Hudson Bay, and at Herschel Island in the Beaufort Sea. Life aboard whaling ships was dirty, rough, and dangerous, and many sailors died when their ships were caught in the ice and crushed.

Despite the early unfortunate encounters with Europeans, the Inuit soon welcomed the whaling ships because of the opportunity for trade. The Inuit were attracted to shore stations and often worked for the whalers by securing game, sewing clothes, and piloting the whaling ships through difficult waters to promising sites for whale hunting. Some Inuit men signed on as boat crew and harpooners. In exchange for this work, the Inuit obtained useful goods, including knives, needles, and rifles, which made domestic life and hunting easier. While this relationship brought many advantages for the Inuit, there were also negative social and health aspects, including the rise in alcoholism and the spread of European diseases among the Inuit (Vignette 10.4). Perhaps the most devastating result of this trade relationship for the Inuit was the unexpected end of commercial whaling, which represented the loss of access to highly valued trade goods. Just as the twentieth century began, demand for products made from whales—whalebone corsets, lamp oil—decreased sharply, halting the flow of whalers, and thus trade goods, that were sailing into Arctic waters. By now, the Inuit depended on trade goods for their hunting activities. Somehow, they had to find other means of obtaining these useful goods.

Fortunately for the Inuit, the fur trade had been expanding northward into the Arctic, thereby providing a replacement for whaling. The fur trade had already been

Vignette 10.4   European Diseases

Whalers, fur traders, and missionaries introduced new diseases to the Arctic. As the Inuit had little immunity to measles, smallpox, and other communicable diseases such as tuberculosis, many of them died. In the late nineteenth century, the Sadlermiut and the Mackenzie Delta Inuit were exposed to these diseases. According to Dickason (2002: 363), in 1902 the last group of Sadlermiut, numbering 68, died of disease and starvation on Southampton Island, 'a consequence of dislocations that ultimately derived from whaling activities'. The Mackenzie Delta Inuit, whose numbers were as high as 2,000, almost suffered the same fate but managed to survive.

The Mackenzie Delta Inuit occupied the northwestern Arctic coast, in present-day Yukon, the Northwest Territories, and part of Alaska. Herschel Island, lying just off the Yukon coast, was an important wintering station for American whaling ships. Whalers often traded their manufactured goods with the local Mackenzie Delta Inuit, who became involved with the commercial whaling operations. Through contact with the whalers, the Inuit were infected by European diseases. By 1910, only about 100 Mackenzie Delta Inuit were left. Gradually, Inupiat Inuit from nearby Alaska and white trappers who settled in the Mackenzie Delta area intermarried with the local Mackenzie Delta Inuit, which secured the survival of these people. Today, their descendants are called Inuvialuit.

successfully operating in the Subarctic for some time—a relationship between European traders and the Subarctic Indian tribes was established through the trade of fur pelts, especially beaver. However, by the beginning of the twentieth century, the fashion world in Europe had discovered the attractive features of the Arctic fox pelt. Demand for Arctic fox pelts rose, which led the Hudson's Bay Company to establish trading posts in the Arctic. Soon the Inuit were deeply involved in the fur trade. The working relationship between the Hudson's Bay Company and the Inuit was based on barter: white fox pelts could be traded for goods.

Until the 1950s, the fur trade dominated the Aboriginal land-based economy. It lasted for less than 100 years in the Arctic and for over three centuries in the Subarctic. Did the fur trade, as well as Arctic whaling, create a form of dependency whereby Indians and Inuit could not survive without trade goods? The answer is a qualified 'yes'. At first, Aboriginal peoples had a form of partnership with European traders and whalers. Each side had power—for instance, the European traders needed the Indians to trap beaver and, often, to show them how to survive in the harsh climate, and the Indians needed the traders to obtain European goods and technology. Gradually, however, the power relationship shifted in favour of the European traders. By the nineteenth century, the fur companies controlled the fur economy. Fur-trading posts dotted the northern landscape. Indians, who had long ago integrated trade goods into their traditional way of life—including their hunting techniques and their migration patterns—were therefore heavily dependent on trade. In fact, when game was scarce, tribes relied on the fur trader for food. Ironically, by securing game for the traders, Indians reduced the number of animals that would be available for their own sustenance. In the Territorial North, game became scarce around fur-trading posts from overexploitation.

The problems of a growing dependency on European goods and a changing way of life for northern Aboriginal peoples were compounded after the arrival of Anglican and Catholic missionaries in the 1860s and the North West Mounted Police (NWMP) in the 1890s. The Indians and Métis were confronted with the full force of Western culture in the late nineteenth century, as were the Inuit in the early twentieth century. Western ideas and rules introduced by the missionaries and police who now lived near the trading posts had a profound impact on Aboriginal culture. The NWMP (which added 'Royal' to its name in 1904 and, in 1920, was renamed the Royal Canadian Mounted Police) imposed Canada's system of law and order on Aboriginal peoples, while the missionaries challenged their spiritual values and encouraged the Inuit, Indians, and Métis to remain in the settlements. Also, on behalf of the Canadian government, both Anglican and Catholic missionaries placed young Aboriginal children in church-run residential schools, where they were taught in either English or French. In this attempted assimilation, most children learned to read and write in English or French, but they were inadequately prepared for northern life. As they lost the opportunity to learn from their parents about how to live on the land, they became trapped between the two very different worlds of their Aboriginal communities and the Euro-Canadians. Under these circumstances, many lost their indigenous language, animistic beliefs, and cultural customs. Fur traders opposed many of these induced Western cultural adaptations because they needed the Aboriginal peoples on the land to trap. Nevertheless, the influence of the churches, the power of the state, and the number of non-Aboriginal residents in the North increased in the twentieth century, placing Aboriginal cultures under siege and crippling their land-based economy. However, political and social changes were occurring at this time that would lead to territorial governments, then to relocation of Aboriginal peoples to settlements, and most recently to land-claim agreements and self-government.

From the Land to Native Settlements

Perhaps the greatest single impact on Aboriginal peoples living in the Subarctic and Arctic was the relocation to settlements in the

Think About It

If you were the Minister of Northern Affairs and Natural Resources in the 1950s, what policy would you propose for northern peoples?

1950s and 1960s. The initial impact was an enormous cultural shock where the freedom of living on the land was exchanged for a regulated life in settlements. The relocation from the land to these tiny settlements marks the beginning of a new way of life with some good aspects and some bad ones. Advantages included food security, access to medical services, and public education. Food security eliminated hunger and starvation but it also meant more store foods and less country food in their diet. Unfortunately, relocation had many negative impacts, including destroying the traditional social hunting/trapping unit from a family-based one to a male one.[4] Children were required to attend schools so mothers stayed in the settlements. In fact, the much-needed family allowance payments to the mothers, introduced in 1945, only took place if the children remained in school. Finally, being based in settlements, it became virtually impossible for the people to follow the seasonal cycle of wildlife movements.

Relocation continues to be a controversial subject in northern history. Williamson (1974), Elias (1995), Marcus (1995), and Rowley (1996) provide different perspectives. However, leaving the Aboriginal peoples in what was seen to be a failing hunting/trapping economy was not an option. Living off the land was, from time to time, a challenge but the shortage of cash/credit from trapping to purchase goods was critical. One option might have been to subsidize the hunting/trapping economy by providing the necessary cash to Aboriginal families as well as more time to adjust to relocation, as is now done in northern Québec where Cree hunters and trappers are paid for living on the land and acquiring country food. When relocation to settlements became the order of the day, few of the people had a full command of English (or French in Québec), which would become so necessary for participating in the affairs of settlement life. Nevertheless, the apparent political urgency of the day caused Ottawa to relocate Aboriginal peoples; but these settlements were ill-prepared. Few had 'extra' housing, schools, or nursing stations and none had a sufficiently vigorous economic base to offer much employment.

Federal officials saw relocation from two perspectives. First, it was seen as a necessary step in protecting northern peoples from the hardships of living on the land, such as life-threatening food shortages. Second, concentrating Aboriginal people in settlements allowed Ottawa to provide a variety of services, including schooling for their children. These two perspectives were part of the overall strategy of 'modernization' of northern Aboriginal peoples. One could argue that, since few Aboriginal families abandoned settlement life to return to the land, the attractions of settlement life outweighed the disadvantages.

Yet, how serious was the hardship of living on the land? The search for game (food) might not always be successful and so hardship and hunger were characteristic of their ancient culture. Most of the time, however, Aboriginal peoples found their hunting, fishing, and trapping lifestyle most satisfying. Yet, when game was scarce, hardships and hunger were severe, and these troubles were not an accepted or acceptable part of the culture of modern Canadian society. Some areas were more risky for hunting peoples. The **Barren Grounds** of the central Arctic were a particularly challenging place to live off the land because of the heavy dependence for sustenance on the migrating caribou herds. Failure to find the caribou translated into hunger and even starvation. Reports of deprivation and even cases of death by starvation among the Caribou Inuit had reached Ottawa before, but no action was taken. In the early 1950s, the Canadian media reported that about 60 Caribou Inuit starved to death. How could people living in a modern country like Canada starve to death? This sad event pushed the government into action, leading to a relocation policy. By 1958, Ottawa made the decision to relocate the Caribou Inuit to settlements, such as Baker Lake and Eskimo Point, but by then starvation had taken its toll—the Caribou Inuit population had dropped from 'about one hundred and twenty in 1950 to about sixty in 1959' (Williamson, 1974: 90). At the same time, Ottawa extended this relocation program to coastal Inuit and to Indians and Métis in the Subarctic who also lived off the land. However, Ottawa was unprepared for the economic, psychological, and social consequences of settlement life for hunting peoples.

After 60 years of settlement life, what has happened? Some impacts are negative but others are positive. On the one hand, access to store food and medical services resulted in a population boom. By 2006, the Aboriginal population formed a clear majority. On the other hand, the increased population did not match the availability of public housing and jobs, resulting in overcrowding and chronic underemployment. As well, Aboriginal communities face deep-rooted social dysfunctions, including extremely high suicide rates among their young people. The cause probably lies in the fact that Native communities have no solid economic base resulting in heavy dependency on government for the impoverished and few opportunities for young people. The two principal sources of income are wages and various forms of government payments, including social assistance. The major employer is the government. As well, lower-income households, especially the elderly, have limited access to country food, which comes mainly through sharing, and insufficient income to purchase store food. Another alarming trend reported by Chan (2006) is that the increase in the consumption of store food rich in carbohydrates, particularly by younger generations, has already caused obesity and diabetes. Perhaps the most positive outcome is the emergence of a more educated population. From these ranks, Aboriginal leaders have had a hand in transforming the Aboriginal society and economy in new directions through successful negotiations for comprehensive land-claim agreements, for the first Aboriginal territory within Canada, namely Nunavut, and for their international involvement in the Arctic Council.

Comprehensive land-claim agreements are discussed in Chapter 3, 'From Hunting Rights to Modern Treaties', page 101.

Alexandra Kobalenko/All Canada Photos

Photo 10.3

Pangnirtung is a small but fast-growing hamlet on the coast of Baffin Island. Like most Inuit communities, its natural increase far outstrips other Canadian urban centres. Economic development may hinge on federal funds to construct a modern harbour, which would support the region's turbot-fishing industry.

Territorial Expansion: Rupert's Land, the Arctic Islands, and the Arctic Seabed

The Territorial North fell under Canadian jurisdiction in three stages. First, the transfer of Rupert's Land to Canada by Britain took place in 1870. Second, Great Britain transferred the Arctic Islands to Canada in 1880. Third, in 1985, Canada declared a 200-mile economic zone that extended its control over the Arctic Ocean. In addition, in the same year Canada announced its Arctic Waters Pollution Prevention Act. Still, the last remaining territory that may become part of Canada consists of the waters and seabed of the Arctic Ocean that now lie in international waters. Canada's claim must be submitted to the United Nations by 2013.

When Canada was formed in 1867, much of what is now Canada remained a British possession. Britain had claimed British North America on the basis of settlement, trade, and exploration. In the Territorial North, the British declared ownership of Rupert's Land as a result of early discovery and exploration, and in 1670 this wide territory had been granted by Charles II to the newly formed Hudson's Bay Company, which had maintained a presence in the region from that time. The claim to the Arctic islands was based on the British Navy's efforts to find the Northwest Passage, including the search for the missing Franklin expedition. In the rest of the Territorial North, beyond Rupert's Land, the Hudson's Bay Company, after its 1821 amalgamation with North West Company, had established and maintained a number of fur-trading posts in the forested lands of the Mackenzie Basin and the Yukon. By extending its fur-trading economy over this area, the British government claimed these Subarctic lands.

Forgotten Frontier: Confederation to 1939

Canada never paid much attention to the Territorial North. In fact, the region was a forgotten part of Canada for several reasons. First, it had little value for agricultural settlement or, because of its remote location, for resource development. Second, Ottawa had its hands full with the provinces where almost all Canadians lived. Third, the fur trade in the North depended on the Aboriginal peoples' living on the land. In short, the Territorial North was not a 'priority' region and thus received minimum attention from Ottawa. With the exception of the Klondike gold rush in the Yukon, the North's economy was left in the hands of the nomadic Dene and Inuit who hunted and trapped, moving seasonally with the wild animals, such as the caribou. Ottawa had adopted a laissez-faire policy to minimize federal expenditures, leaving the fur traders and missionaries to deal with the food and health needs of a hunting society.

Strategic Frontier

With the outbreak of World War II, the Territorial North became a strategic frontier. Military investments and activities included military bases, highways, landing fields, and radar stations. While the nature of its strategic role changed over time, the Territorial North served as a buffer zone between North America and the Soviet Union for over 50 years. This role ceased with the collapse of the Soviet Union and the end of the Cold War in 1991.

For the Americans, Canada's North provided a secure transportation link to the European theatre of war and, in 1942, to Alaska, where the threat of Japanese attack was real. The air routes consisted of the Northwest Staging Route and Project Crimson. Each consisted of a series of northern landing strips that would enable American and Canadian warplanes to refuel and then continue their journey to either Europe or Alaska. In the Northeast, Project Crimson involved constructing landing fields at strategic intervals to allow Canadian and American airplanes to fly from Montréal to Frobisher Bay (now Iqaluit) and then to Greenland, Iceland, and, finally, England. In Canada's Northwest, American aircraft came to Edmonton and then flew along the Northwest Staging Route to Fairbanks, Alaska, where their major military base was located. The Alaska Highway, built at the same time, provided road access to the various landing fields and to Alaska. The US Army command had decided that the oil needed by the American armed forces in Alaska must be made secure by increasing the oil production at Norman Wells in the NWT and sending it by pipeline across several mountain ranges to Whitehorse and then northward to the military facilities at Fairbanks. Known as the Canol Project, the oil pipeline was completed in 1944, but with the disappearance of the Japanese threat it was closed within a year.

After World War II, the geopolitical importance of northern Canada changed. The North's new strategic role was to warn of a surprise Soviet air attack. The defence against such an attack was a series of radar stations that would detect Soviet bombers and allow sufficient response time for American fighter planes and (later) American missiles to destroy the Soviet bombers. In the 1950s, 22 radar stations, called the Distant Early Warning line, were constructed in the Territorial North along 70° N. Before the end of the Cold War, these radar stations were abandoned and replaced with more sophisticated methods of detecting incoming Soviet planes or missiles. With the collapse of the Soviet Union, Ottawa withdrew its military establishment at Inuvik, did not proceed with its plans for a military base at Nanisivik, and downsized its operation at Alert.

American military investment did expand the Norman Wells oil fields, along with a pipeline to Whitehorse, but most of their investment went into improving the transportation system, such as the construction of the Alaska Highway. Private resource development moved along much more slowly as it responded to world demand.

Arctic Sovereignty: The Arctic Ocean and the Northwest Passage

In the twenty-first century, Arctic sovereignty has taken on a fresh urgency. Ottawa has responded with the Prime Minister declaring 'the first principle of Arctic sovereignty is use it or lose it' (BBC News, 2007). What prompted this sense of urgency is pressure from Russia and other **circumpolar countries** that are actively staking their claims to the Arctic seabed and from shipping nations that want the Northwest Passage to be defined as lying in international waters. National boundaries have yet to be set for the Arctic Basin and global warming appears to have opened a summer ice-free shipping route through the Northwest Passage. In addition, the Yukon/ Alaska offshore boundary, shown as a blue dashed line in Figure 10.3, has not been resolved. Consequently, does Canada or the US control management of fish resources, deal with environmental issues, and authorize deep-sea drilling? This disputed area will likely be resolved by negotiations, possibly under the United Nations Convention on the Law of the Sea (UNCLOS).

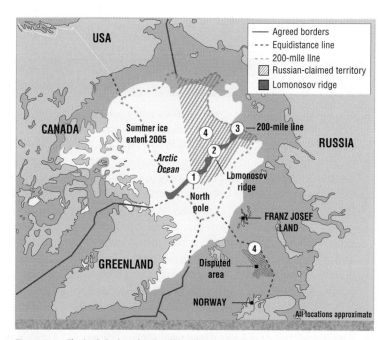

Figure 10.3 **The Arctic Basin and national boundaries.**
The sea claim rush directly involves five countries: Canada, the United States, Russia, Norway, and Denmark (Greenland). The four designated areas shown on the map are: (1) North Pole: in 2007 Russia planted its flag on the seabed, 4,000 m (13,100 ft) beneath the surface, as part of its claim for oil and gas reserves. (2) **Lomonosov Ridge**: Russia argues that this underwater feature is an extension of its continental territory and is looking for evidence. (3) 200-nautical-mile (370-km) line indicates how far countries' agreed economic areas extend beyond their coastlines; this often is set from outlying islands. (4) Russian-claimed territory: the bid to claim a vast area is being closely watched by other countries. Some could follow suit.

Source: BBC News (2007). Based on 'Maritime jurisdiction and boundaries in the Arctic region' by International Boundaries Research Unit, Durham University.

Think About It

Should only those nations bordering the Arctic Ocean be able to lay claims to the remaining international waters or should other nations, such as other members of the Arctic Council, be allowed to make a claim?

For Canada, the Arctic Basin possibly is its last territorial acquisition. Ottawa recognizes that:

- Vast quantities of petroleum deposits lie beneath the floor of the unclaimed zone of the Arctic Ocean.
- Global warming may turn the frozen Arctic Ocean into a commercial ocean route.
- Canada's international position within the Arctic Council and, by extension, within the Circumpolar World is at stake (Vignette 10.5).

While Canada did step to the plate in the Circumpolar World by proposing the Arctic Council, Ken Coates and others question if Canada is ready to lay claim to seabed deposits and to control the Northwest Passage, or if Ottawa will continue to treat Arctic sovereignty as a 'zombie' of its foreign policy. Prominent Canadian scholars question Canada's readiness. In a provocative book, *Arctic Front: Defending Canada in the Far North*, Coates et al. (2008: 1) probe Ottawa's past, present, and hopefully more vigorous future efforts to claim Arctic waters and the resources below:

Arctic sovereignty seems to be the zombie—the dead issue that refuses to stay dead—of Canadian public affairs. You think it's settled, killed and buried, and then every decade or so it rises from the grave and totters into view again. In one decade the issue is the DEW Line, then it's the American oil tanker *Manhattan*, streaming brazenly through the Northwest Passage, then the *Polar Sea* doing the same thing. In August 2007, a Russian submarine planted a flag at the North Pole. Or perhaps it was under the North Pole, as the UK *Daily Telegraph* reported, raising an image of a striped pole floating in the ocean, with the devious Russians diving underneath it. Perhaps the flag did land on the pole, though good luck

---

Vignette 10.5   The Arctic Council and the Circumpolar World

The Circumpolar World is an enormous area, sprawling over one-sixth of the earth's landmass and spanning 24 time zones. The Arctic Council focuses its attention on this massive land area, its environment, and its peoples. In 1996, Canada played a key role in the establishment of the Arctic Council, which is designed to serve as a high-level intergovernmental forum to provide a means for promoting co-operation, co-ordination, and interaction among the Arctic states and peoples, with the involvement of the Arctic indigenous communities and other Arctic inhabitants on common Arctic issues—issues of sustainable development and environmental protection in the Arctic. The member states are Canada, Denmark (including Greenland and the Faeroe Islands), Finland, Iceland, Norway, the Russian Federation, Sweden, and the United States of America. The various national organizations of indigenous peoples are represented as permanent participants, and some European nations, including the UK and Germany, have observer status. Canada has a number of permanent participants on the Arctic Council, including the Athabaskan Council, the Gwich'in Council, and through its participation in the International Inuit Circumpolar Council. The Arctic Council provides information to its member states, especially in six areas—Arctic contaminants; Arctic monitoring and assessments; conservation of Arctic flora and fauna; prevention and response to environmental problems; protection of the Arctic marine environment; and sustainable development.

But all is not warm and cordial in Arctic political affairs. When Canada convened a meeting near Montréal of the foreign ministers representing the 'Arctic Five' coastal states in late March 2010, other participants in the Arctic Council were miffed by their exclusion, and US Secretary of State Hillary Clinton left the meeting early, after severely criticizing her Canadian hosts for setting a conference that did not include all interested parties (Boswell and O'Neill, 2010). Some commentators wondered if the Harper government was backing away from its broader commitment to the Arctic Council and to environmental issues and, instead, was primarily concerned with narrow economic and security interests.

with that, since the pole is a point with no size at all, so the Russians likely missed it. However it was, they are up there, and the zombie has come to life once more.

With the exception of Hans Island, a barren rock that Denmark and Canada both claim, the international community recognizes Canada's ownership of the islands in the Arctic Ocean. However, the ownership of Arctic waters between the islands in Canada's archipelago remains unclear. The United States considers the Northwest Passage as an international sea route and claims ownership of a narrow strip of the Beaufort Sea—some 21,436 km². While this route is not used for commercial purposes because of the difficulty of navigating through ice, a number of 'unknown' countries have sent their nuclear submarines under the ice cover, and the Americans have made several trips through the Northwest Passage on the surface. In 1969, an American tanker, the SS *Manhattan*, made a voyage through the Northwest Passage without asking Canada's permission. It was an attempt to prove the passage was a viable route for shipping oil from Alaska's Prudhoe Bay oil fields. Ottawa did provide a Canadian icebreaker to escort the *Manhattan*. In 1970, the *Manhattan* made another trip through the passage. In 1985, the US Coast Guard icebreaker *Polar Sea* transited the passage—once again, without asking the Canadian government for permission. The political fallout over what was considered the most direct challenge to Canada's sovereignty in the Arctic led to the signing of the Arctic Co-operation Agreement in 1988 by Prime Minister Brian Mulroney and US President Ronald Reagan. The document

Gary Clement, *National Post*, 28 July 2005. © Gary Clement

Photo 10.4

Whose island? Hans Island is a small, barren rocky knoll located along the border between Canada and Greenland (Denmark). While Hans Island has no known value and is uninhabited, its importance lies in two areas. First, the underlying continental shelf may contain oil and gas deposits. Second, its location in the centre of Kennedy Channel of Nares Strait marks the beginning of the Northwest Passage and therefore the ownership of this island may have implications for the Northwest Passage and offshore petroleum deposits.

states that the US is to refrain from sending ice-breakers through the Northwest Passage without Canada's consent; in turn, Canada will always give consent. However, the issue of whether the waters are international or internal was left unresolved.

Over the years, Canada has sought to legalize its sovereignty over the Arctic. In 1907, Canada first announced the 'Sector Principle', which divided the Arctic Ocean among those countries with territory adjacent to the Arctic Ocean. Accordingly, Canada could claim title over a wedge of the Arctic Ocean north of its territory between 60° W and 141° W longitudes and extending to the North Pole. While the Soviet Union supported this principle, the United States and other countries have not.

More recently, Canada has looked to environmental legislation as a means of exercising its sovereignty over Arctic waters. In the age of supertankers and container ships, the threat of toxic spills is more likely than ever before. Ottawa has two concerns. First, how would it ensure sovereignty over Arctic waters in order to manage such ocean traffic? Second, how would Canada protect its waters and adjacent islands from toxic wastes discharged from foreign ships passing through the Northwest Passage? In 1985, Ottawa's response took the form of the Arctic Waters Pollution Prevention Act, which gives Canada the right to control navigation in its sector of the Arctic Ocean and to manage its ocean environment.

In 2003, Canada ratified the UN Convention on the Law of the Sea, which specifies that coastal countries have the right to control access to their coasts. This access zone is 12 nautical miles (22.2 kilometres) wide. While the Convention on the Law of the Sea supports Canada's claim to Arctic waters, it leaves some grey areas, including the fact that some of the islands in the Arctic Archipelago are separated by more than 90 kilometres of water. For that reason, the greatest threat to Canada's sovereignty claims remains the practice of foreign countries passing through the Northwest Passage without obtaining permission from the federal government.

Think About It

Can you think of any reasons why fly-in, fly-out air commuting is more attractive to mining companies than building permanent communities?

The Territorial North Today

The Territorial North remains a resource frontier far from the Canadian ecumene. Surprisingly, almost everyone lives in a settlement, town, or city. Rural communities and farms, as found in southern Canada, do not exist. Another surprising fact is that mining sites, rather than being associated with resource towns, often involve workers being flown to and from the mine. The three diamond mines provide examples. Each is located in an isolated area but the companies house their workers in camps and fly them to and from Yellowknife and other centres rather than build a permanent community where the workers' families would live.

A difference from the rest of Canada is that most Arctic urban centres are very small and isolated from one another (Figure 10.4). From Statistics Canada's perspective, these settlements do not qualify as **urban areas** because their populations are less than 1,000. In 2006, for example, approximately three-quarters of these centres had populations under 1,000, and more than 40 per cent of the Territorial North's population lived in three cities: Whitehorse (22,898), Yellowknife (18,700), and Iqaluit (6,184). In practical terms as measured by function, centres in this region fall into two principal categories: Native settlements and regional service centres. Most Aboriginal people reside in Native settlements, where they form more than half the population but where there are few job opportunities. This mismatch between Native settlements and jobs has its roots in the relocation of Aboriginal peoples to settlements in the 1950s. On the other hand, Native communities are more closely linked to traditional ways of life and therefore provide a cultural link with the land and the past. A growing number, however, have become regional centres, particularly the three capital cities where employment opportunities in the public sector are greatest. As well, a small but growing number of Aboriginal families have relocated to cities in southern Canada.

Population

From 2001 to 2006, the population of the Territorial North increased by nearly 10 per

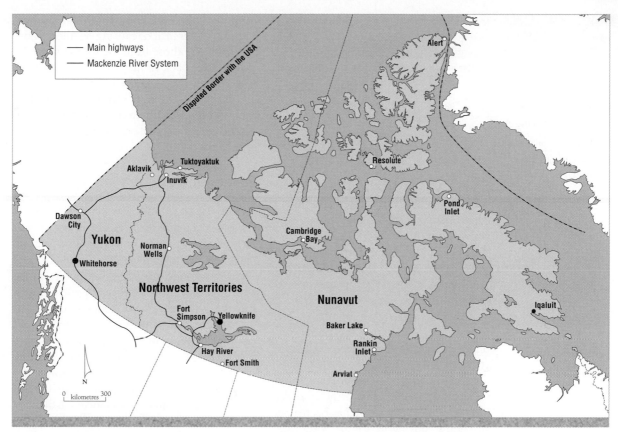

Figure 10.4 Major urban centres in the Territorial North.
The major cities are the territorial capitals, Whitehorse, Yellowknife, and Iqaluit. With most government jobs in these cities, underemployment is much less a problem than in many smaller centres, especially Native settlements. Alert, located at the northern tip of Ellesmere Island, remains a military base from the Cold War. Resolute, on the other hand, soon will become a naval base as part of Canada's effort to exert more control over its Arctic waters, including the Northwest Passage.

cent, reaching a figure of 101,575 (Table 10.2). This population growth is due exclusively to a high rate of natural increase (Table 10.3). Overall, Aboriginal peoples make up 52 per cent of the northern population. However, the percentage of Aboriginal peoples varies widely between the three territories. Nunavut has the highest percentage at 85 per cent, followed by the Northwest Territories at 50 per cent and Yukon at 23 per cent (Table 10.2).

In 2007–8, the rate of natural increase in the Territorial North was approximately 1.5 per cent. Within the three territories, the highest rate (2.1 per cent) occurred in Nunavut. In comparison, the national figure was only 0.3 per cent. The reason for this significant difference is the much higher birth rate in the Territorial North (18 births per 1,000 compared to the national figure of 11.1 births per 1,000) and a lower death rate (4.8

deaths per 1,000 compared to the national figure of 7.2 deaths per 1,000), the result of a younger population. While birth and death rates are not collected by ethnicity, given the proportion of Aboriginal peoples in each territory, the association of high birth rates with

Table 10.2 Population and Aboriginal Population, Territorial North

Territory	Population 2006	% Change 2001–6	Aboriginal Population 2006	% Aboriginal of Total 2006 Population
Yukon	30,190	15.8	7,580	22.8
Northwest Territories	41,060	0.2	20,635	50.1
Nunavut	29,325	10.2	24,915	85.0
Territorial North	100,575	9.7	53,130	52.3

Source: Statistics Canada (2008).

Table 10.3   Components of Population Growth for the Territories, 2007–8

Demographic Event	Canada	Yukon	NWT	Nunavut
Births/1,000 persons	11.1	10.9	16.0	25.2
Deaths/1,000 persons	7.2	5.8	4.4	4.4
Natural rate of increase (%)	0.3	0.5	1.2	2.1
Net interprovincial migrants		221	–805	–426

Source: Statistics Canada (2009a).

a high percentage of Aboriginal populations indicates the highest birth rates are among Aboriginal peoples, especially among the Inuit in Nunavut (Table 10.3).

The population of the Territorial North also is affected by migration. Migration to the North normally occurs when economic expansion creates jobs. Since World War II, many southerners have moved north to take jobs in the mining industry, the public service, and the business sector. However, since many newcomers remain for only a few years, a rapid turnover in the non-Aboriginal population occurs. Since the end of the resource boom in the early 1980s, there has been a net flow of people moving to southern Canada. As shown in Table 10.4, the total migratory growth rate for the three territories varies widely from year to year. Three basic reasons account for this migration. The primary factor is that many are economic migrants. When

Table 10.4   Total Migratory Growth Rate (per 1,000 population), 1981–2007

Year	Yukon	NWT	Nunavut*
1981	–51	86.1	—
1986	7.4	–33.0	—
1996	8.9	–13.6	–9.6
2001	–7.3	1.2	–5.7
2002	–0.4	5.5	3.6
2003	10.2	10.1	–7.7
2004	–4.2	–9.8	–6.9
2005	3.4	–19.7	–3.7
2006	–10.6	–18.0	–9.2
2007	4.6	–6.0	0.0

*Nunavut is included under NWT prior to 1996.
Source: Statistics Canada (2009b).

northern jobs end, they move to southern Canada. Other factors causing people to move south include those from the south who eventually move back home; some families move south when their children reach school age; and those seeking post-secondary education/ training or health care only available in southern hospitals relocate, perhaps only temporarily to the south.

Economic Structure

As a northern frontier, the Territorial North's economy depends heavily on private investment in its primary industries and transfer payments for its public sector. Added to these purely economic matters, the region is caught up in the race for control over the Northwest Passage, the unclaimed zone of the Arctic Ocean, and the ongoing adjustment of its Aboriginal population to the market economy. All require strong financial support from Ottawa. In sum, the Territorial North is a high-cost area for economic development, social programs, and geopolitical challenges. With a limited tax base and without the right to collect royalties from resource companies, the three governments of the Territorial North must depend on Ottawa. All of this translates into the simple fact: Canadians in other parts of the country will be called on to invest in the country's last frontier for decades to come. Canadians should take solace from the territorial version of equalization payments called Territorial Formula Financing—transfer payments and the cost of sovereignty are essential costs of nation-building. In a broader sense, perhaps, this is the true meaning of the Prime Minister's phrase, 'use it or lose it'.

See Chapter 3, 'Equalization and Transfer Payments', page 96, for more on this subject.

Energy and mining are the principal commercial elements of the northern economy. In terms of employment, the primary sector in the Territorial North is much larger than the same sector in other geographic regions. For example, approximately 15 per cent of the workers are in the primary sector compared to only 2.7 per cent in Ontario. The Territorial

North has approximately 83 per cent of its workforce allocated to the service sector, making it the dominant sector. With a small tax base, large transfer payments from Ottawa to the three territorial governments play a major role in their budgets, allowing them to have similar levels of education, health, and social services to those found in the provinces. As a result, the contemporary economic structure of the Territorial North mirrors these two forces—resource extraction and transfer payments—making the tertiary and primary sectors of greatest significance. The secondary sector, manufacturing, is almost non-existent in the Territorial North (Table 10.5).

Most workers are employed by one of three governments: federal, territorial, or local. Local governments are settlement councils, band councils, and other organizations funded by a higher level of government. The reason the tertiary sector is so large in the Territorial North is that geography demands such an investment of people and capital to ensure the delivery of public services. Territorial governments must spend more money per resident than do provincial governments to provide basic services. Much of the cost differential is attributed to overcoming distance in the Territorial North and hiring staff for small communities. The drawback is that economies of scale are difficult to achieve in small communities where teachers and nurses often have a relatively small number of students and patients compared to those working in larger centres. However, the social importance of the public service sector goes beyond the number of employees. The wide geographic distribution of public jobs across the North is a major social benefit to those in small communities. As a result, employment opportunities, while concentrated in the three capital cities, exist in every community. In small, remote communities, where unemployment rates are often as high as 30 per cent, virtually all jobs are associated with one or more levels of government. A second social benefit is the governments' ability to implement social policies in their hiring practices. For example, Nunavut's government seeks to employ mainly Inuit in its civil service.

Table 10.5 Estimated Employment by Economic Sector, Territorial North, 2006

Economic Sector	North Workers (%)	Ontario Workers (%)	Difference (percentage points)
Primary	13	2.7	10.3
Secondary	2	24.6	−22.6
Tertiary	85	72.7	12.3
Total	100.0	100.0	

Source: Author's estimate.

Government Structure

The Territorial North consists of three territorial governments: Yukon, the Northwest Territories, and Nunavut. The capital cities of these three territories are Whitehorse, Yellowknife, and Iqaluit. Territorial governments have fewer powers than provincial governments and, in this sense, they are political hinterlands. For example, the federal government retains power over natural resources and collects a substantial amount of tax revenue from companies extracting natural resources, so this important revenue source is not available to territorial governments. Instead, they depend on Ottawa for transfer payments. Without these transfer payments, the territorial governments could not offer the basic services available in southern Canada. For example, Nunavut receives 90 per cent of its revenue from Ottawa. The level of fiscal dependency is somewhat lower for Yukon and the Northwest Territories. The Northwest Territories is challenging Ottawa over the issue of resource royalties. With more diamond and natural gas projects on the drawing board, the Northwest Territories is demanding a share of resource revenues.

The Territorial North is changing, especially politically. One recent political change was the division of the Northwest Territories to create a new territory, Nunavut. The second political change is being triggered by the resolution of outstanding land claims between Aboriginal peoples and Ottawa through comprehensive land-claim agreements that transfer land, cash, and administrative powers to northern Aboriginal organizations. The most recent comprehensive land-claim agreement—finalized in 2003 with the Dogrib First Nation and known as the Tlicho Final Agreement—includes provisions for self-government.

The Territory of Nunavut

The new territory of Nunavut was made possible through a land settlement agreement between Canada and the Inuit of the eastern Arctic in 1993. The terms of the agreement included the use of Crown lands for the Inuit to hunt, fish, and trap, and the transfer of part of the land to the Inuit, with a portion of this area involving rights to subsurface minerals. The same year the land agreement was reached, the federal government made the commitment to create a new territory by passing the Nunavut Act. This Act, which provided the legal basis for the creation of a distinct territory and territorial government, also allowed for a six-year transition period, giving the Inuit time to form their government, recruit their civil servants, and select a capital city. By means of a plebiscite, Iqaluit was selected as the capital of the new territory. Following an election in February 1999, the 19 members of the Nunavut Assembly took office on 1 April 1999. Unlike First Nations, the Inuit created a public form of government, meaning that every resident has the same political rights.

The creation of a separate territory for the Inuit brought hopes for a brighter future. Through an Inuit government, a sustainable economy was thought to be achievable within 20 years. The Bathurst Mandate (Nunavut, 1999) gave voice to that hope:

> As Nunavummiut we look to support ourselves and contribute to Canada through the potential of our land, the responsible development of our resources and the contributions of our peoples and our cultures.

The word 'Nunavut' means 'our land' in Inuktitut. The vast majority of the approximately 30,000 people who reside in Nunavut are Inuit. A primary goal of the Nunavut government is to reflect the people's aims and aspirations. In addition to expanding economic development and increasing the number of jobs in the new public service for residents of the highly decentralized administration of Nunavut, the new government strives to promote Inuit culture and the Inuktitut language. For many Inuit, Nunavut is a dream come true where their cultural survival within Canadian society is assured. The dream also brings with it the challenge of building a northern economy that can provide jobs for its growing labour force and thereby reduce its financial dependency on Ottawa. However, the challenges facing the Inuit government are formidable. As Légaré (2008a: 367) wrote:

> For now, though, the urgent socio-economic plight of Nunavut does not bode well for the future. The vision of a viable Nunavut society by the year 2020, as expressed through the Bathurst Mandate, seems to be, at least for now, an illusion.

Aboriginal Economy

Aboriginal peoples participate in both the land and wage economies. While the 'mix' varies among Aboriginal peoples, harvesting wildlife from the land and sea tends to be highest in Native settlements where traditional practices are strong and job opportunities are few and far between. Trapping, hunting, and other land-based activities persist in the North not so much because of their commercial value but largely because of their cultural importance. For instance, hunting produces food for the family and country food remains a core cultural feature among northern Aboriginal families. While trapping and hunting sometimes go hand in hand, interest in trapping has diminished because of low prices for furs and because of the effective lobbying of the European Union by animal rights groups. Since peaking in the late 1980s at a value of just over $6 million for the Northwest Territories, fur production has declined significantly; the value of fur production in 2006–7 in the Northwest Territories was $1.3 million, and in some recent years it has been below $1 million. With low prices and rising costs, few trappers can make a profit and most supplement their trapping activities from other income.

Think About It

What is holding Nunavut back from achieving a viable economy?

Efforts of animal rights groups have focused on the seal hunt on the ice floes near the island of Newfoundland where their case is based on 'inhumane' killing practices. However, with the ban on importing seal pelts, the Inuit have been affected as well. In the past, the seal provided both food and cash for Inuit families but without a use for the pelt the incentive to hunt seal has diminished, pushing Inuit to rely more on store food.

The Aboriginal economy took a new turn with comprehensive land-claim agreements. These agreements provided the structure and capital to participate in the market place. The Inuvialuit, for example, are major shareholders in the Aboriginal Pipeline Group.

Comprehensive Land-Claim Agreements

Aboriginal land claims are based on the peoples' long-time use of the land for hunting, fishing, and trapping. In Canada, Aboriginal land claims are settled by treaty: the Aboriginal tribe surrenders its claim to all the land in exchange for title to a smaller amount of land, a cash settlement, and, in most instances, usufructuary right to a larger territory owned by the Crown for hunting and fishing. Other issues involved in land-claim negotiations are self-government, management of wildlife resources, and preservation of language and culture. In simple terms, land claims are an attempt by Aboriginal peoples and the federal government to resolve the issue of Aboriginal rights.

For further discussion of land claims, see Chapter 3, 'From Hunting Rights to Modern Treaties', page 101.

During the latter part of the twentieth century, Indian, Inuit, and Métis organizations in the Territorial North made land claims on behalf of their members. In the 1970s, three major land claims were made in the region: Yukon First Nations (Yukon), Dene/Métis (Denendeh), and Inuit (Arctic). Over the next two decades, the Inuit land claim was split into the western Arctic (Inuvialuit) and the eastern Arctic (Inuit/Nunavut), and the Dene/Métis land claim for Denendeh was divided into five

separate land claims (Gwich'in, Sahtu/Métis, North Slavey (also referred to as Dogrib or Tlicho), South Slavey, and Deh Cho).[5]

Until the late 1970s, Ottawa did not recognize traditional claims to Crown land. The evolution of claims policy and objectives has been greatly influenced by jurisprudence. In the *Calder* case in 1973, the Supreme Court of Canada acknowledged the possible existence of Aboriginal title in Canadian law, that is, some claims to landownership might be valid but such claims had to prove earlier occupancy. The Supreme Court ruling caused Ottawa to reconsider its position. Between 1973 and 1985, the rethinking of the federal government's position on land claims resulted in the comprehensive and specific land-claims policy, which was formally announced in 1986. Prior to this time, negotiations with the Inuvialuit followed the guidelines of the new policy and, in 1984, the Inuvialuit reached a final agreement with Ottawa. They were followed by the Gwich'in in 1992, and, in the next year, the Sahtu/Métis, Inuit, and Yukon First Nations. In 2003, the Dogrib or Tlicho Agreement was completed, leaving only two outstanding claims—Deh Cho and South Slavey.

Figure 3.10, 'Modern treaties', page 104, shows the geographic extent of modern land-claim agreements in the Territorial North.

As with all comprehensive land-claim agreements, the Inuvialuit Final Agreement (IFA) was based on the Inuvialuit's traditional use and occupancy of lands in the western Arctic, that is, the land that they and their ancestors used for hunting and trapping. The Inuvialuit's land claim was originally part of the Inuit land claim to the entire Arctic. However, the Inuvialuit broke away from the Inuit and, as a result, there are two large **settlement areas** in the Arctic, the Inuvialuit and the Nunavut settlement areas.[6] The goals of the Inuvialuit were to preserve their cultural identity and values within a changing northern society; to enable themselves to be equal and meaningful participants in the northern and national economy and society; and to protect and preserve the Arctic wildlife, environment, and biological productivity (Canada, 1985: 1). Canada's goal, on the other hand, was to extinguish the Inuvialuit claim to the western Arctic.

Though the JBNQA (1975) represents the first modern treaty, Ottawa, as pointed out above, instituted a new negotiating system composed of specific and comprehensive agreements some 10 years later. As the first comprehensive land-claim agreement in Canada, the IFA served as a model for subsequent ones. Since 1984, as shown in Table 10.6, five more comprehensive land-claim agreements have been reached. Each agreement is very similar to the IFA. One common feature is that each agreement has created economic and environmental administrative sectors. In the case of the IFA, the economic sector is the Inuvialuit Regional Corporation, which manages and invests the cash settlement received as part of the agreement.[7] The second sector, the Inuvialuit Game Council, is responsible for environmental issues that affect their hunting economy. The creation of these two sectors was the Inuvialuit's attempt to straddle two worlds; their old world was based on harvesting game from the land, while their new world is part of the global industrial economy. However, the IFA was silent on one important element so necessary for Aboriginal peoples—self-government. Over time, the issue of self-government became part of comprehensive land-claim agreements. The Nunavut and Tlicho final agreements do spell out the specific nature and unique structures of self-government for the Inuit and the Dogrib.

The Yukon First Nation agreement, signed in 1993, was different in that it only established the basic elements of final agreements for each of the 14 Yukon First Nations. Known as the Umbrella Final Agreement, this 1993 arrangement provided the basic framework within which each of the 14 Yukon First Nations (Carcross/Tagish; Champagne and Aishihik; Dawson; Kluane; Kwanlin Dun; Liard; Little Salmon/Carmacks; Nacho Nyak Dun; Ross River Dena; Selkirk; Ta'an Kwäch'än Council; Teslin Tlingit Council; Vuntut Gwitchin; and White River) can conclude a final claim settlement agreement. So far, five First Nations have achieved a final agreement: Teslin Tlingit Council (1993), Selkirk First Nation (1997), Little Salmon/Carmacks (1997), Ta'an Kwäch'än Council (2002), and the Kluane First Nation (2003).

For the Inuvialuit, the agreement was an important step towards defining their place within Canadian society. In exchange for cash and land, the Inuvialuit gave up their Aboriginal rights, including their claim to the vast lands of the western Arctic. In exchange, they received legal title to about 90,650 km² of land, which is slightly less than 20 per cent of the settlement area (Canada, 1985: 6). The land to which the Inuvialuit gained title lies within the Inuvialuit settlement area, which extends from the Arctic mainland into the Arctic Ocean (Figure 3.10). A number of islands, including Banks Island, Prince Patrick Island, and the western parts of Melville Island and Victoria Island, are situated here. Of the total settlement land, 12,950 km² include surface and subsurface mining rights for the Inuvialuit. The Inuvialuit enjoy exclusive rights to hunting, trapping, and fishing over the remaining 77,700 km² and the offshore waters (as noted in Figure 10.3, the sea boundary with Alaska is not yet resolved).

Country Food

Country food remains a key element of Aboriginal culture and practices. While Aboriginal cultural identities vary across the Territorial North, certain core elements contain the glue that maintains their distinctness. These core elements include a strong attachment to the land, to country food, and to the ethic of sharing. **Country food** is food obtained from the land, a preferred source of

Table 10.6   Comprehensive Land-Claim Agreements in the Territorial North

Aboriginal People	Date of Agreement	Cash Value*	Land (km²)
Inuvialuit	1984	$45 million (1977)	90,650
Gwich'in	1992	$75 million (1990)	22,378
Sahtu/Métis	1993	$75 million (1990)	41,000
Inuit (Nunavut)	1993	$580 million (1989)	350,000
Yukon First Nations	1993	$243 million (1989)	41,440
Dogrib (Tlicho)	2003	$152 million (1997)	39,000

*By dollar for the stated year.

Sources: Canada (1985: 6, 31; 1991: 3; 1993a: 3; 1993b: 81, 215; 2004); *Globe and Mail* (1993).

meat and fish. As equivalent store-bought foods are expensive, most Aboriginal northerners keep their food costs low by consuming country food. While it is true that there are substantial costs expended in harvesting country food, to some degree this cost is offset by the pleasure and spiritual rewards of being on the land and participating in hunting and fishing. Sharing also remains an important component in the harvesting and distribution of country food among family members, relatives, and close friends. Respect for cultural diversity is a hallmark of Canadian identity. In May 2009, the issue of respect took on a particular nasty face, namely, the negative reaction in some quarters to Governor General Michaëlle Jean eating a slice of seal heart at an Inuit festival in Rankin Inlet, Nunavut. By participating in this traditional Inuit practice, the Governor General showed respect and implied solidarity with their hunting of seals and the selling of seal pelts (photo 10.5). The negative reaction from animal rights groups and the European Union reflects a form of cultural relativism. Who sets the standards for social conduct? Mary Simon, president of Inuit Tapiriit Kanatami, Canada's national Inuit organization, applauded Jean for her public support of traditional Inuit culture. 'Hunting and eating a seal is not a political act, nor is it "bizarre" or "disgusting" as the anti-sealing lobby have commented', she said. 'To us, this kind gesture is an acknowledgement by the Governor General of our culture and our dependence upon our wildlife as an important resource for our communities today' (O'Neil, 2009). In March 2010, Canadian parliamentarians followed the example of the Governor General when the parliamentary restaurant, for the first time, featured various seal dishes.

The European Union, formerly a major importer of seal pelts, has banned their import, with economic consequences for the Inuit that are hurtful. The issue is complicated because of the connection between seal as a food and seal pelts as a commercial product. For centuries, European nations have imported furs and pelts from Canada. Ironically, Britain and France originated the fur trade, which has entailed the killing of seals.

Photo 10.5

Governor General Michaëlle Jean, centre, helps an Inuit elder skin two seals during a community feast in Rankin Inlet, Nunavut, 26 May 2009.

Respect for the diverse cultures in Canada is discussed in Chapter 4, 'Multiculturalism', page 151.

Figures on the cost, size, and value of harvesting country food are not available. However, estimates suggest that the 'substitute value' of country food (the value of equivalent store-bought food) was about $40–$50 million per year in the mid-1980s (Usher and Wenzel, 1989). A visit to an Aboriginal community quickly reveals the importance of wildlife harvesting to the daily diet. While some store-bought foods can be found on the tables of Aboriginal families, the preference is for game and fish from the land and waters.

Country food is so important to Aboriginal northerners that they have made wildlife an essential issue in land-claim negotiations. The agreements establish co-management committees that administer the environment and wildlife. For example, these committees approve (or reject) proposals for new industrial projects and determine the total allowable harvest of wildlife based on biological principles and **traditional ecological knowledge**. In the case of the Inuvialuit Final Agreement, the members of the co-management committees are named by the Inuvialuit Game Council and the three governments (Canada, Northwest Territories, and Yukon). The members of the Inuvialuit Game Council are selected by the Hunters and Trappers Associations found in each of the six Inuvialuit communities. Under the direction of these committees, professional scientists record and monitor the state of the environment and wildlife. As a result of comprehensive land-claim agreements, power to control the use of the environment and wildlife now rests, in part, with most Aboriginal peoples in the Territorial North. Still, resistance to traditional knowledge remains alive and well in a few corners of academe (Widdowson and Howard, 2008).

Early contacts and European hopes of discovering wealth in the North are discussed in Chapter 3, 'Initial Contacts', page 75, and Vignette 3.1, 'Contact between Cartier and Donnacona', page 76.

Resource Development in the Territorial North

Since the sixteenth century, explorers—now replaced by multinational companies—sought to exploit the North's natural wealth, particularly its minerals. The search for precious metals began soon after explorers reached North America. In Mexico and South America, the Spanish and Portuguese were particularly successful in obtaining vast quantities of gold and silver. In North America, English and French explorers searched for similar wealth. Martin Frobisher, on three separate expeditions in the 1570s, returned to England with hundreds of tons of 'sandstone flecked with mica' in the hope he had discovered gold—his shiploads of Arctic rock ended up being used in Elizabethan-era road-building; years earlier, Jacques Cartier had returned to France with iron pyrites (fool's gold) (Bumsted, 2010: 56, 54). Northern Indians trading at the Hudson's Bay posts often wore copper ornaments and had copper tools. Richard Norton, the factor or chief trader at Fort Prince of Wales, ordered one of his men, Samuel Hearne, to find the source of

this copper and assess its commercial worth to the company. In one of the great overland journeys of all time, Hearne travelled with Chipewyan leader Matonabbee's tribe from Fort Prince of Wales on the west coast of Hudson Bay (present-day Churchill, Manitoba) to the mouth of the Coppermine River at Coronation Gulf on the Arctic Ocean. In 1771, Hearne found the source of copper, but this natural copper occurred only at the surface in small amounts. Others seeking mineral wealth in the North were more fortunate. In 1896, George Carmack, an American prospector, and his Indian brothers-in-law, Skookum Jim and Tagish Charley, found gold nuggets in a small tributary (later renamed Bonanza Creek) of the Yukon River. When word of their find reached the outside world, the Klondike gold rush sparked the greatest gold rush on record. Dawson City was the centre of this frenzy. Thousands of prospectors flooded the area to scour the sandbars of the rivers and pan for gold. Within two summers, the more accessible placer gold was gone and by 1900 many left for a new gold find in Alaska. The only remaining gold was deep in the frozen terraces found along the sides of the Klondike River. Permafrost is widespread at latitudes above 60° N. Companies with both capital and technology took over gold mining using hydraulic and steam-mining techniques to thaw the ground and force the sand, gravel, and fine gold particles into separating devices commonly known as sluice boxes. The commercial, large-scale gold mining signalled the birth of megaprojects.

Resource Development and Transportation

Until Prime Minister Diefenbaker's 'Northern Vision' in the late 1950s, Ottawa paid little attention to the Territorial North. With the Development Road Program, Ottawa began to invest funds into highway construction. Still, the North is so vast and so sparsely populated, the case for spending large sums on road-building is difficult to make—the cost of highway construction is extremely expensive, which accounts for the paucity of highways in

Photo 10.6

BHP Billiton's EKATI diamond mine has produced on average over three million carats per year of rough diamonds over the last three years with production transitioning from predominantly open-cut to a mix of open cut and underground mining. EKATI has development options for future open-cut mines. Annual sales from EKATI represent approximately two per cent of current world rough diamond supply by weight and approximately six per cent by value. Most rough diamonds from EKATI are sold to international diamond buyers through BHP Billiton's Antwerp sales office with a smaller amount sold to Canadian manufacturers based in the Northwest Territories.

the Territorial North. One reason for the high cost is the presence of permafrost; another is the distance between places, which translates into extra expenses to assemble road-building equipment and materials. The federal government plays a major role in financing highway construction. Mining companies lobby Ottawa for roads to their mining sites because the costs of air transportation are either impractical and/or too expensive. For the three diamond mines in the Northwest Territories, winter roads provide relatively low-cost trucking routes.

The Territorial North has few highways. Nunavut has none that connect to the national highway system. The most prominent ones are: Yukon's Alaska Highway and the Dempster Highway; and several in the Northwest Territories, including the Mackenzie Highway, Yellowknife Highway, Fort Smith Highway, and

Liard Trail. Winter roads greatly extend the road system, making the trucking of heavy equipment, building supplies, and other goods possible in the winter. The construction of the Deh Cho Bridge (photo 10.7) will greatly reduce the cost of trucking goods to Yellowknife and other centres along the Mackenzie River served by road. CN operates on a single track to Hay River, Northwest Territories, where it connects with the river barge system on the Mackenzie River.

The pattern of mining developments in the North clearly indicates the importance of water transportation and winter roads. In 1935, the first major gold discovery in the Northwest Territories took place near Yellowknife on Great Slave Lake. Water transportation was critical to the opening of the Con-Rycon gold mine, which was near the shore of Great Slave Lake and therefore had

Courtesy of Infinity Engineering Group

Photo 10.7

A drawing of the proposed Deh Cho Bridge, which will cross the Mackenzie River near Fort Providence. The bridge will replace the ferry system in the summer and the ice road in the winter.

excellent access to the Mackenzie River transportation system. Mines located far from water transportation, such as the now-closed Lupin gold mine (about 250 km northeast of Yellowknife), required the construction of **winter roads**. The need for roads in mining is not so much for transporting minerals to external markets as for bringing equipment and supplies to the mine site. Mines along the Arctic coast can take advantage of ocean transportation. Three examples are the former mine at Rankin Inlet on the shores of Hudson Bay, the abandoned lead-zinc mine at Nanisivik near the northern tip of Baffin Island, and the now-closed Polaris lead-zinc mine on Little Cornwallis Island in the Arctic Archipelago (see Figure 10.5).

In assessing the cost of transportation, the value of the mineral is taken into account. For example, copper, lead, nickel, and zinc are low-grade ores (much of the ore has no commercial value). Even after the first-stage separation of some of the waste material from the valuable mineral, the enriched ore still remains a bulky product, with significant waste

material remaining that can be removed only through smelting. Shipping such a low-value commodity is very expensive. Ideally, such ore is transported to a smelter by ship. Railways are the second-most effective transportation carrier for low-grade ore. The critical nature of transportation for such mines is clear in the example of the lead-zinc deposit at Pine Point in the Northwest Territories. Discovered in 1898 by prospectors heading overland to the Klondike gold rush, the Pine Point deposit was not developed until 1965, when a railway was extended to the mine site. With a means of transporting the ore to a smelter, the company, Cominco, could send massive amounts of ore by rail to its smelter at Trail in British Columbia. The mine closed in 1983, leaving Pine Point a ghost town.

In the past, companies built resource towns near their mining sites—Pine Point, for example. However, the life expectancy of resource towns tends to be short because when the mine closes, no economic support remains for the community. Some companies have avoided this problem by turning to **air commuting**, which

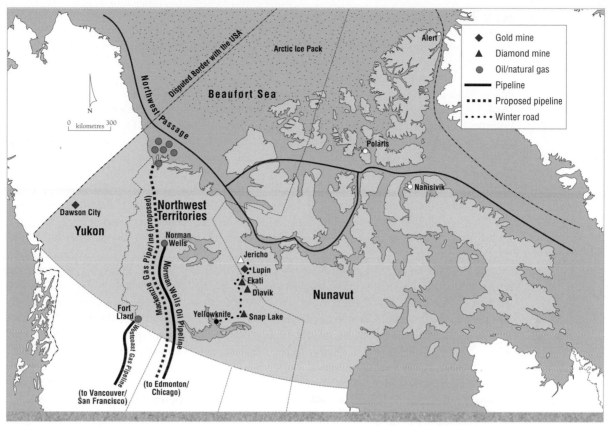

Figure 10.5 Resource development in the Territorial North.
The mineral wealth of the Territorial North lies mainly in the Northwest Territories. Diamonds, gold, natural gas, and oil drive this territory's resource economy. In contrast, the resource economies of Yukon and Nunavut are much smaller. Petroleum exploration in the Beaufort Sea is complicated by the border dispute between Canada and the United States. Washington claims a narrow strip of the Canadian section of the Beaufort Sea—an area of 21,436 km². Mines often have a short life-span. Jericho, Polaris, and Nanisivik are abandoned mines.

may be an effective way to obtain skilled southern workers for remote resource projects but does have drawbacks for the North. Southern-based air commuting systems hurt the North's economy because: (1) workers spend their wages in their home communities in southern Canada, thereby stimulating provincial, not territorial, economies; and (2) workers who reside in a province but work in the territories pay personal income tax to provincial rather than to territorial governments, thereby depriving territorial governments of valuable personal income tax. For Aboriginal communities, air commuting is advantageous because it provides access to high-paying jobs in the mining industry and, at the same time, allows workers to keep their families in their Aboriginal communities. Known as 'cultural commuting', the fly-in and fly-out work schedule has another attraction by allowing workers time to hunt and fish on their week(s) off.

Vignette 10.6 Sea Transportation on the Arctic Ocean

Arctic transportation takes advantage of nature. In late summer when the shore ice melts along the western Arctic, a narrow stretch of water opens between the shore and the polar pack ice. Small ships take advantage of this open water to bring supplies to communities located along the coast of the Beaufort Sea. Most of these supplies are transported by barge northward along the Mackenzie River. In the eastern Arctic, after the shore ice has retreated, ocean-going ships bring fresh supplies to Arctic communities. In other places, most of the surface of the Arctic Ocean is permanently covered by floating sea ice known as the Arctic pack ice. Here, only specially reinforced ships and icebreakers can traverse the ice-covered waters. How thick is the ice? One-year ice is about 1 m thick while older ice can reach 5 m thick. Such voyages by icebreakers in older ice are not common. In August 1994, two icebreakers, the American *Polar Sea* and the Canadian *Louis S. St Laurent*, ploughed through thick ice on a scientific voyage to the North Pole.

Key Topic: Megaprojects

The Territorial North has entered a new phase of resource development characterized by megaprojects controlled by multinational companies. Megaprojects usually cost more than $1 billion and require several years to complete the construction stage. Megaprojects have integrated the Territorial North's economy into the global economy, thereby firmly locking the North into a resource hinterland role in the world economic system. The most recent megaprojects in the Territorial North have involved diamond mining. The first diamond mine to come into production was the Ekati mine. As part of the NWT Diamonds Project (MacLachlan, 1996), preliminary construction of buildings, roads, and mines began in the early 1990s, but the Ekati mine was not completed until 1998. Two other mines, Diavik and Jericho, are now in production, while the Snap Lake mine will begin production in 2008. Supplies for these remote mines are trucked along a winter road (Figure 10.5). All are located in the Slave geological region that straddles the border between the Northwest Territories and Nunavut.

Proponents of resource development describe megaprojects as the economic engine of northern development, though others challenge this assumption, claiming that they offer few benefits to the region and, more particularly, to the Aboriginal communities (Bone, 2009: 168–9). These large-scale ventures are designed for the export market. By injecting massive capital investment into the construction of giant engineering projects, megaprojects create a short-term economic boom. However, most construction expenses are incurred outside hinterlands because the manufactured equipment and supplies are produced not in hinterlands but in core industrial areas. This reduces the benefits of megaprojects to the hinterland economy and virtually eliminates any opportunity for economic diversification. As well, since all megaprojects in the Territorial North are based on non-renewable resources, these developments have a limited lifespan. At the end of these projects, the local economy suffers a collapse. Three examples are the closure of mines at Faro, Yukon; Pine Point, Northwest Territories; and Rankin Inlet, Nunavut.

However, megaprojects can more readily inject much-needed capital and create development in the region. Megaprojects in resource hinterlands are high-risk ventures. Multinational companies can reduce their risks in three ways. First, they can create a consortium of companies and thereby spread the investment risk among several firms. Second, they can arrange for long-term sales of the product at a fixed price before proceeding with construction. Third, they can obtain government assistance, which often takes the form of low-interest loans, cash subsidies, and tax concessions. Three megaprojects are discussed in the following sections: the Mackenzie Valley Pipeline Project: 1970s and 2000s; the Norman Wells Oil Expansion and Pipeline Project; and the NWT Diamonds Project. Except for the 2000 version of the Mackenzie Valley Pipeline Project, each project was approved after an environmental impact assessment. The Norman Wells Project received its approval in 1980, while the NWT Diamonds Project was approved in 1996.

The Mackenzie Valley Pipeline Project: 1970s

By 1970, the plans formulated by oil and pipeline companies to bring natural gas from Prudhoe Bay, Alaska, to major markets in the United States, primarily in the Midwest, were submitted to Ottawa.[8] Economic conditions at the time included a high demand for natural gas in the continental United States and the fact that natural gas was a by-product of oil production at Prudhoe Bay. One plan was to build a pipeline, known as the Mackenzie Valley Pipeline Project, which would transport natural gas from the Arctic coast of Alaska to the Mackenzie Delta and then south along the Mackenzie Valley, eventually reaching Alberta and then the United States. Shortly after these plans were announced, world oil prices increased sharply as a result of joint action to limit supply by the Organization of Petroleum Exporting Countries (OPEC). OPEC's control of world oil production changed the economic landscape for energy and added another economic reason for the Mackenzie Valley Pipeline Project. At these higher prices, the Mackenzie Delta deposits were no longer too expensive to develop, providing its natural gas could be added to the proposed pipeline by means of a short spur pipeline.

In Canada, this project was subjected to the Mackenzie Valley Pipeline Inquiry (also known as the Berger Inquiry). The Berger Inquiry began in 1974 and was concluded in 1976. Led by British Columbia Justice Thomas Berger, who a few years earlier had been chief counsel for the Nisga'a in their landmark land-claim case (*Calder*), this federal investigation examined the potential environmental, social, and economic impacts of the proposed gigantic construction project. The Berger Inquiry established a new precedent by holding community hearings in numerous remote Native settlements, and its final report, *Northern Frontier, Northern Homeland*, issued in 1977, recommended that a natural gas pipeline from the Mackenzie Delta to Alberta was feasible but should only proceed after further study and after settlement of the Dene land claims. To complete these two tasks, Berger recommended a 10-year delay in the construction of such a pipeline. At the same time, he rejected the construction of a pipeline from Prudhoe Bay to the Mackenzie Delta because of the danger of harming the delicate Arctic environment and wildlife found along the North Slope of Alaska and Yukon.

For discussion of the Calder *case, see Chapter 3, 'From Hunting Rights to Modern Treaties', page 101.*

Without the link to Prudhoe Bay and faced with rising construction costs, the proposal lost its economic attraction and the pipeline never was built. With new supplies of Mexican and Albertan natural gas reaching the United States in the 1980s, prices declined in the North American market. However, by 1998, this surplus had disappeared and prices began to rise. From 1998 to 2003, natural gas prices more than quadrupled. Consequently, deposits in the Territorial North again commanded the attention of the major oil and gas companies servicing the North American market. In 2000, for instance, natural gas from northeastern British Columbia (just south of the Northwest Territories) began to flow through the Alliance Pipeline to the Chicago market, underlining the continental system of energy production and marketing in North America.

The Mackenzie Valley Pipeline Project Revisited

The anticipated development of the natural gas deposits in the Mackenzie Delta has brought about a revisiting of the old Mackenzie Valley Pipeline Project. Imperial Oil, Gulf Canada Resources, Shell Canada, and Exxon Mobile, all of which hold considerable natural gas reserves in the western Arctic, are poised to begin the construction of a natural gas line along the Mackenzie River Valley. This proposed megaproject includes the development of an estimated six trillion cubic feet of natural gas in the three largest onshore fields discovered in the Mackenzie Delta and the building of a gas and liquids pipeline system. The Mackenzie Valley gas pipeline would have a capacity of 1.2 billion cubic feet per day and could be expanded to accommodate gas from other fields in the future. The first step in obtaining Ottawa's approval is to complete a statement of the project called a Preliminary Information Package (PIP), which is then reviewed by the National Energy Board (now completed) and by the Joint Review Panel, which submitted its report on environmental and social issues in December 2009. Thirty years ago, Aboriginal peoples strongly opposed the project for two reasons. First, the Dene insisted that their land claims be settled before the construction of a Mackenzie Valley pipeline. Second, the Dene had concerns about the impact of this construction project on their lands and people. More specifically, they feared that wildlife would become less plentiful, causing great harm to their hunting/trapping life. Other concerns also were identified, such as the Dene fear that their communities would not gain a fair share of jobs and business opportunities, and that the expected increase in non-Aboriginal population

Think About It

What reasons can you suggest for holding community hearings in Aboriginal communities as part of a public inquiry? Why do you think this hadn't been done prior to the Berger Inquiry?

would generate negative impacts on Dene culture and social well-being.

By 2003, these concerns were no longer seen as major obstacles. Several land-claim agreements have been concluded, and the Dene workforce and business community are more developed and better prepared to take advantage of such a mega-construction project. The Aboriginal Pipeline Group has obtained a one-third interest in the pipeline. When project details were submitted to Ottawa in 2003, the cost of such a pipeline was approximately $5 billion. Nellie Cournoyea, chairwoman and chief executive officer of the Inuvialuit Regional Corporation, believes that the Aboriginal Pipeline Group 'signals our capacity to work in the economy of the twenty-first century' (Northern Gas Pipelines, 2009).

The attention focused on the Mackenzie Delta gas fields is clearly related to the rising price of natural gas, a result of the growing demand for natural gas in the United States. By 2015, the demand is projected to soar to 31.3 trillion cubic feet a year in North America, well above the current production of 24 trillion cubic feet, possibly pushing prices well over current rates. By then, existing production in southern Canada and the United States is expected to diminish—perhaps to 20 trillion cubic feet—leaving a possible gap of 11 trillion cubic feet. One of the two proposed Arctic pipelines should be completed by that time, possibly making up as much as half of this expected shortfall.

The race to build the first Arctic pipeline to North American markets pits Prudhoe Bay against the Mackenzie Delta. At the moment, the Mackenzie Delta is favoured because of two main advantages: (1) a much lower construction cost because of a shorter distance to market and because the diameter of its pipe is much smaller (and therefore less costly) than that proposed for the Alaskan pipeline; and (2) a much shorter construction time (six years) compared to the proposed Alaskan pipeline (10–12 years). However, the US government may subsidize the construction of the much more expensive Alaskan pipeline (estimated in 2007 at over $30 billion compared to nearly $8 billion for the all-Canadian pipeline). The stakes are high, causing much

lobbying in Washington. Alaska has two strong points: (1) Alaska's reserves are much larger than those in the Mackenzie Delta (Alaska holds known reserves of 35 trillion cubic feet and an estimate of an additional 100 trillion cubic feet, compared to about 9 trillion cubic feet of known reserves in the Mackenzie Delta, plus estimated reserves similar in size to those in Alaska); and (2) American security concerns call for the development of American petroleum deposits, though the Prudhoe Bay pipeline would follow the Alaska Highway and therefore must pass through Canadian territory (Yukon, British Columbia, and Alberta) to reach the main markets in the United States.

The Mackenzie Gas Project proposes to build a 1,220-kilometre pipeline system along the Mackenzie Valley. It would link northern natural gas sources to southern markets. The main Mackenzie Valley pipeline would connect to an existing natural gas pipeline system in northwestern Alberta. The proposed project crosses four Aboriginal regions in Canada's Northwest Territories (NWT): the Inuvialuit Settlement Region, the Gwich'in Settlement Area, the Sahtu Settlement Area, and the Deh Cho Territory. A short segment will be in northwestern Alberta near the NWT border. The Deh Cho, the only directly affected Aboriginal group without a land-claim agreement, have been adamant that no pipeline can cross their territory in the Fort Simpson area until they have a settlement, but the Conservative government in Ottawa has pressed ahead with the pipeline project regardless. The natural gas exploration and development companies involved in the Mackenzie Gas Project have interests in three discovered natural gas fields in the Mackenzie Delta—Taglu, Parsons Lake, and Niglintgak. Together, they can supply about 800 million cubic feet per day of natural gas over the life of the project. Other companies exploring for natural gas in the North also are interested in using the pipeline. In total, as much as 1.2 billion cubic feet per day of natural gas could be available initially to move through the Mackenzie Valley Pipeline.

But will the project go ahead? Natural gas prices have sunk, reaching record lows by 2009 because of fresh supplies in the continental United States and northern British Columbia—from natural gas extracted from shale deposits. Construction costs have increased, too. In 2003, they were estimated at $5 billion. In May 2007, however, Imperial Oil announced revised figures for the costs to build the Mackenzie Gas Project. The total Imperial estimate rose to $16.2 billion, with $7.8 billion for the construction of the pipeline, $4.9 billion for the development of the gas fields, and $4.9 for the gas-gathering system. Could such a jump in estimated costs be a ploy to obtain concessions from Ottawa?

Imperial Oil and the Northwest Territories government see federal financial involvement as crucial. For Imperial, federal support would reduce the risk of low gas prices and help ensure a 'double-digit' rate of return (CBC News, 2007). The NWT government sees the project as in the 'national interest' because, like the CPR, it will open up the last major frontier in Canada's North.

The Norman Wells Oil Fields, 1982–2010

The Norman Wells oil fields were discovered in 1920. Until the pipeline to southern Canada was built, production was limited to providing for local communities and mining sites along the Mackenzie River. By 1982, geological investigations had revealed the quantity of proven reserves to be as high as 100 million m³ (Bone, 2009: 178–81). The challenge was to market Norman Wells oil in the United States. The purpose of the Norman Wells Project was to increase the oil production, build a pipeline, and ship the oil to southern markets. Prior to the pipeline, annual output was less than 180,000 m³. With the completion of the pipeline in 1985 and the expansion of oil production by a factor of 10, the Norman Wells oil fields became a major player in the Territorial North. Since 1985, annual output has exceeded 1 million m³ and reached almost 2 million m³ several times.

The Norman Wells Project, which cost almost $1 billion in 1996 dollars, was hailed by industry as a model megaproject for the North. Both the federal government and Esso Resources Canada believed that the project would not harm the natural and social environments in the construction impact zone, which stretched from Norman Wells to Fort Simpson and then to the Alberta border. Aboriginal organizations, such as the Dene Nation, and environment groups, such as the Canadian Arctic Resources Committee, did not agree.

Plans for this major oil and pipeline construction effort by Esso Resources Canada began in the late 1970s, when the price of oil was rising sharply. Once the proposal was approved by the federal government, construction began, running from 1982 to 1985. The petroleum pipeline route begins at the resource town of Norman Wells in the Northwest Territories and ends at Zama in northern Alberta, the northern terminus for the national oil pipeline system. Much of the terrain north of 60° N lies in the zone of discontinuous permafrost. As a result, special construction measures were necessary. Limiting construction to the winter months prevented excessive damage to the terrain and vegetation by trucks and other heavy equipment, and efforts to re-establish vegetation along the pipeline route following construction were designed to minimize soil erosion and the warming of ground temperatures by solar radiation. The danger from higher ground temperatures in the summer is that this could accelerate the melting of ground ice, leading to the subsidence (sinking) of the ground and possible rupturing of the buried pipeline. Such projects in future could be compromised, too, by our warming climate: the seasonal span for winter roads appears to getting shorter and shorter.

The Norman Wells Project represents a successful and profitable operation. In 2008, just over one million cubic metres of oil were produced, valued at $639 million (Northwest Territories, 2009). For 25 years, Esso has produced and transported crude oil to markets in southern Canada and the United States. During that time, while some subsidence of the ground along the pipeline route has

Terry A. Parker/Viewpoints West

Photo 10.8

A drilling rig near Norman Wells (second island from the left). The Norman Wells Project allowed for a significant increase in oil production and supply to southern markets.

occurred, the pipeline has not been damaged and there have been no major oil spills. Equally important, the operation of the Norman Wells Project has not triggered social problems in the communities along the pipeline route. From the oil industry's perspective, the Norman Wells Project demonstrated that pipelines can be built in areas with permafrost. However, from the Dene Nation's perspective, land claims should have been settled before the project was approved. Under such circumstances, the Dene would have become business partners in the construction and ensuing operation. In 2001, efforts to kick-start negotiations resulted in the Framework Agreement, but that exploratory effort failed to produce results (INAC, 2009b). As of April 2010, Dene land claims and a resolution of

the route of the proposed Mackenzie Valley Pipeline remain in limbo.

The NWT Diamonds Project

Canada is now the third-largest producer of diamonds in the world, behind Botswana and Russia, and accounts for 15 per cent of the world supply, thanks in large part to the three diamond mines in the Northwest Territories—Ekati, Diavik, and Snap Lake. In 2008, the Northwest Territories accounted for $2.1 billion of diamonds (Table 10.7).

How did this remarkable development come about? Until 1991, geologists believed that the Canadian Shield was not a geological structure where diamond could be formed. Two prospectors, Charles Fipke and

Stewart Blusson, proved them wrong and turned conventional thinking on its head when they discovered diamond-bearing kimberlite near Lac de Gras in the Northwest Territories. Following extensive drilling, their find proved to have commercial viability, and BHP Diamonds Inc. decided to develop a mine. The company applied to Ottawa to secure approval, and in 1996 the federal government approved the NWT Diamonds Project, which entails the operation of the Ekati diamond mine in the Lac de Gras area of the Northwest Territories for about 25 years. Five diamond-bearing **kimberlite pipes** are being mined, four of them located within a few kilometres of each other. All five kimberlite pipes are located under lakes that had to be drained to facilitate mining operations.

During the operation of the Ekati diamond mine, between 35 and 40 million tonnes of waste rock are removed from the kimberlite pipes each year. Since only a small proportion of that rock contains diamonds, the rest of the rock is placed in piles near the mines. Recovery of diamonds from the ore takes place in a processing plant where the ore is crushed and diamonds are separated. Final sorting is done using X-rays to separate the diamonds from the remaining waste material. BHP then sells rough-cut diamonds at international diamond markets. The mining operation employs about 800 workers with an estimated annual wage bill of $40 million. BHP uses an air-commuting system to bring workers from Yellowknife to its mine site on a two-weeks-in and two-weeks-out rotation.

Diamond mining has quickly become the backbone of the mining industry in the Territorial North. By value of production, number of employees, and spinoff effects, BHP Diamonds is the leading mining company in the North. In 2006, the two mines, Ekati and Diavik, produced $1.6 billion worth of diamonds (Table 10.7). Their effect on the economy of the Northwest Territories has been profound. One spinoff was the designation of Yellowknife as the pickup point for miners who commute by air on a rotational scheme to the mine site. This reverses the usual trend

Table 10.7	Mineral Production in the Territorial North, 2006 ($ millions)			
Mineral Product	Yukon	NWT	Nunavut	Territorial North
Metals	37.7	55.6	0	93.3
Non-metals	55.7	1,573.3	29.2	1,658.2
Fuels	0*	497.0*	0	497.0*
Total	93.4	2,125.9	29.2	2,248.5

*Estimate.

Source: Adapted from Natural Resources Canada (2007); Indian and Northern Affairs Canada (2007).

of cash and tax flowing out of the territories. A second spinoff was the location of three diamond-cutting and -polishing businesses in Yellowknife.

Another diamond mine, Jericho (Nunavut), began production in 2006 but closed the following year because of high operating costs. Snap Lake began production in 2008 but a drop in global demand caused De Beers to cut production in the following year. Other diamond mines also cut production in 2009. When world prices dropped in 2008, diamond production slowed and was scheduled to halt in late 2009. But then, as the global economy showed signs of recovery, diamond prices rebounded and mining continued.

How important is the diamond industry to the Northwest Territories? The figures are dazzling—diamond production has gone from zero in 1997 to $2.1 billion in 2008. The estimated total value of the diamond reserves at the three mines is around $23.3 billion (Statistics Canada, 2004). Furthermore, diamond production came on stream just as gold mining was ending. With the closing of the Giant gold mine at Yellowknife, the last gold was minted in 2004 but closing down operations lasted until 2005.

Megaprojects: Achilles Heel?

Megaprojects in the Territorial North are based on non-renewable resources, that is, petroleum and mineral deposits. Forests, water, and wildlife are classified as renewable resources. The value of non-renewable resources in the North far outweighs the economic value derived from renewable resources. In 2006, mineral production in the Territorial North was valued at $2.25 billion, which was almost 2.5 per cent of Canada's mineral output (Table 10.7). Several mines have closed: the Nanisivik and Polaris lead/zinc mines closed in 2002; Con gold mine in 2003; and Giant gold mine in 2005; Jericho diamond mine closed in 2007. The dominant role of the Northwest Territories is illustrated in Table 10.7, which shows that it contributed 95 per cent of the value of mineral production in the Territorial North in 2006. The leading minerals by value were diamonds, tungsten, oil, and natural gas. The major mines were located at Ekati, Diavik, and Snap Lake (diamonds); and Fort Liard (natural gas). Placer gold, valued at $50 million per year, is obtained from the valleys of the Yukon River and its tributaries. Norman Wells is the only oil-producing site in the Territorial North, but there are much larger oil and gas deposits in the Mackenzie Delta, Sverdrup Basin, and the Beaufort Sea (Vignette 10.1).

Non-renewable resource development, because of the limited lifespan of such projects, subjects the Territorial North to a boom-and-bust economic cycle. Since mineral and petroleum deposits are in finite quantities, they do not offer long-term economic stability for the region. This economic cycle is the Achilles heel of northern development. In addition, because resource development is based on world demand, production fluctuates with this demand, creating great instability in resource communities. Fluctuations in mineral production provide a measure of this instability, and although production has been on the upswing in recent years, especially with the development of diamond deposits, it fell precipitously in 2009 with the global recession—Canadian diamond production by value fell nearly 30 per cent from 2008 to 2009 (Natural Resources Canada, 2010: Table 2). Individual communities in the past have been devastated by mine closings and the same, inevitably, will occur in the future.

Table 7.5, 'Mineral Production by Province and Territory, 2009', page 298, includes statistics for the three northern territories.

Negative economic and social effects in the larger society and globally are deeply felt in resource towns. Mines even suspend their operations during periods of low prices. In 1997, for example, the Faro mine was closed for the fourth time in its short history. Again, low metal prices made the zinc, lead, and silver-mining operation unprofitable. The Faro mine had been the principal mining activity for 30 years in Yukon. Each time the mine closed, the value of mineral production in Yukon dropped drastically. For example, in 1992 Yukon accounted for nearly half a billion dollars worth of minerals. After the Faro mine closed in 1993, production dropped to $81 million in 1994.

Megaprojects have made important contributions to the northern economy. Often, they are touted in the press as 'the engines of economic growth'. Certainly, megaprojects give the regional economy a boost, but is that boost enough to lead to regional diversification in the Territorial North? True, megaprojects generate the bulk of the North's GDP, expand the northern transportation infrastructure, and, during the construction phase, provide high wages for large numbers of employees. The diamond industry holds out some hope of even greater spinoffs, such as the processing of diamond gems in Yellowknife, but like other fledgling businesses, the global economic downturn saw a loss of business and the closure of Arslanian Cutting Works in June 2009 (CBC News, 2009). Megaprojects, however, have so far failed to diversify the Territorial North's economy in a significant manner. While generating profits for the corporations, megaprojects have been unable to broaden the northern economy or solve the massive underemployment problem in Native settlements, and they have increased the region's economic vulnerability to sudden changes in world demand for its primary products. Worse, all northern megaprojects are based on non-renewable resources, which, having a fixed lifespan, cannot provide the basis for a stable economy over the long term. From this perspective, megaprojects are not the engine, but the Achilles heel of the northern economy.

Given the problem of **economic leakage**, should governments intervene in the marketplace to ensure northern benefits? The Norman Wells Oil Expansion and Pipeline Project is one the federal government 'encouraged'. Imperial Oil and Inter Provincial Pipelines designed contracts suitable for small northern firms, including businesses owned by Dene entrepreneurs and by businesses operated by First Nations in the Mackenzie Valley. Both measures significantly increased economic benefits to the North. Some 25 years later, First Nations and their communities are exploring innovative ways to participate in megaprojects and thus secure an annual flow of income. The Mackenzie Gas Project, while still not off the ground, has attracted interest.

At the turn of the twenty-first century, the Aboriginal Pipeline Group obtained a one-third interest in the pipeline for an initial investment of $4 million. Another proposal around the same time was for communities along the pipeline route to seek 'tax revenue' from the company. Imperial Oil's initial response was to reject this demand and to offer a one-time payment for access, which is the normal easement arrangement made with landowners south of 60°. However, with the Mackenzie Gas Project 'stuck in the mud', according to Fred Carmichael (Ebner, 2009), chairman of the Aboriginal Pipeline Group and a junior partner in this project, pre-construction activities have ground to a halt.

The potential downside of megaprojects relates to Innis's staples thesis. See Chapter 1, Vignette 1.3.

Think About It

Why cannot megaprojects solve the massive underemployment found in Arctic communities?

SUMMARY

The geography of the Territorial North consists of four dominating factors—the Territorial North is huge, sparsely populated, Canada's last resource frontier, and the homeland for numerous Aboriginal peoples. These facts have dominated and will continue to dominate the course of events in the region. At present, Arctic sovereignty once again looms large on the northern political landscape because circumpolar nations are laying claim to the seabed of the Arctic Ocean that lies beyond traditional national boundaries.

As Canada's last frontier, it serves as a resource hinterland with a tiny local market and a great distance to world markets, which means that resource companies must depend on external demand for their products. These characteristics contrive to make the Territorial North's economy extremely sensitive to fluctuations in world prices for its resources. While three levels of government—federal, territorial, and local— seek sustainable and robust economies, the geographic realities make those goals virtually impossible to achieve without substantial financial support. Added to these structural characteristics are weaknesses in the northern labour force and economy. For instance, few skilled and professional workers are in the northern labour force,

forcing companies and governments to seek such workers from other regions of Canada. Then, too, the weak northern economy results in economic leakage. This raises the question: *Are these circumstances an indication of a staple trap?*

In the next decade, megaprojects will continue to draw the Territorial North more closely into the global economy as a resource frontier. While diamond production is now the leading industry, if the Mackenzie Gas Project proceeds, natural gas will become the leading resource by value. The implication of such resource development, based on its non-renewable nature, is that stable economic growth and economic diversification remain elusive goals. Without a more diversified economy, the Territorial North will continue to suffer from boom-and-bust economic cycles and the economic instability thus created. Even if ownership of natural resources is transferred to territorial governments, these governments will remain financially dependent on Ottawa because of the 'have-not' character of the region's geography. One question remains: Does Canada recognize that this financial dependency represents the cost of incorporating the Territorial North and its Aboriginal peoples into the rest of Canada?

CHALLENGE QUESTIONS

1. How have comprehensive land-claim agreements equipped Aboriginal peoples to chart a new and brighter future?

2. Why would the British naval expedition led by Sir John Franklin stand a better chance of navigating the Northwest Passage today than in 1845?

3. Are there any weaknesses in Canada's case for claiming ownership of the Northwest Passage?

4. Megaprojects are the driving force behind the economy of the Territorial North. What is it about these projects that prohibits sustainable growth?

5. Should the federal government financially support the proposed Mackenzie Gas Project because it is in the 'national interest' to open the North to development?

FURTHER READING

Coates, Ken, P. Whitney Lackenbauer, William Morrison, and Greg Poelzer. 2008. *Arctic Front: Defending Canada in the Far North.* Toronto: Thomas Allen.

The authors of *Arctic Front* make the provocative statement that 'Arctic sovereignty seems to be the zombie—the dead issue that refuses to stay dead—of Canadian public affairs.' They argue that only when a crisis emerges—such as the rush to claim the Arctic Basin seabed—does Ottawa respond. And the crisis is huge—the last large land grab about to take place beneath the unclaimed waters of the Arctic Ocean where 30 per cent of the world's untapped oil and gas deposits are found. Added to that fact, the prospect of a commercial sea route across the Arctic Ocean makes the stakes very high. Besides Canada, four other nations, Denmark, Norway, Russia, and the United States, are seeking to lay claim to the potentially great storehouse of mineral wealth and natural gas under the Arctic Ocean, and shipping countries, notably Japan and Britain, are deeply interested in the Northwest Passage being recognized as an international waterway.

CANADA: A COUNTRY OF REGIONS

INTRODUCTION

'Canada is big—preposterously so', wrote geographer Kenneth Hare (1968: 31). The sheer size of Canada has meant that the country spans a diverse set of physical and cultural geographies, thus transforming the nation into a country of regions. As in any country, tensions emerge; but because of Canada's size and the distribution of populations across its breadth and width, sometimes those tensions threaten the fabric of the country. Though regional stresses have arisen because of this diversity and no doubt will continue to arise in the future, Canadians have recognized that differences cannot be obliterated by might. What is important—beyond the heated words—is the existence of a political process that addresses these regional differences even though the agreements may be temporary and incomplete. Final solutions are not the basis of democracies, but negotiation, compromise, and resolution keep populations flexible and open to ongoing solutions to problems as they arise or alter. Canada's future, while full of challenges and uncertainty, contains one truth: Canada's strength lies in its ability to reconcile cultural and regional disputes. In John Ralston Saul's view (1997: 8–9), Canada is a 'soft' country 'because the classic nation-state is hard—hard in the force of its creation and its maintenance'. In this concluding chapter, four critical topics are addressed: Canada's regional character, spatial structure, faultlines, and future.

Trans-Canada Highway at sunrise through Banff National Park, Alberta. Photo: John E Marriott/All Canada Photos.

Regional Character

Canada's strength—you might even say what makes it interesting—is its regional diversity and ensuing political struggle to balance regional interests with national ones. As a country of regions, the character of Canada flows from its six geographic regions—Ontario, Québec, British Columbia, Western Canada, Atlantic Canada, and the Territorial North. These geographical regions have been shaped by history. Canada evolved from a small country consisting of four former British colonies to the second-largest country in the world—with a strong possibility of further territorial gains in the unclaimed Arctic Ocean.

Each region has a special regional character with strengths and weaknesses. *Ontario*, for instance, has the most favourable natural conditions and location for agriculture, economic growth, and industrialization in Canada. Complementing these natural advantages were some federal policies, such as the National Policy in 1879 and the Auto Pact in 1965, designed to encourage a manufacturing base. In spite of these advantages, Ontario's automobile manufacturing industry has witnessed a substantial contraction, thus weakening its manufacturing. Even though the automobile industry has already regained a portion of its output, its long-term future is problematic because of the threat of low-cost imports and the longer-term threat of oil supplies diminishing and oil prices rising. Without a vigorous automobile industry, will an unstoppable hollowing-out of Ontario's manufacturing base weaken Canada's economy?

Québec remains Canada's second core region. La Belle Province has been favoured by the St Lawrence River, which has provided

Jessica K. Robertson, US Geological Survey

Photo 11.1

Researchers aboard the Canadian Coast Guard icebreaker *Louis St. Laurent* and the US Coast Guard cutter *Healy* engaged in a joint Arctic survey mission in 2009. Their objective was to explore largely unknown parts of the Canada Basin, north of the Beaufort Sea. By combining their efforts in this joint mission, the results will provide a much more convincing database and a stronger submission to the United Nations for part of the unclaimed waters of the Arctic Ocean. Both countries will use data from this mission to prove their continental shelves extend beyond their existing 200-nautical-mile economic zones under the United Nations Convention on the Law of the Sea. At stake are areas rich in resources and, if polar ice melts, international ocean routes.

a transportation link to the Great Lakes and to the rest of the world. Nature also has blessed Québec with extensive hydroelectric resources, most of which were obtained by land acquisitions from the federal government after Confederation. Its culture, shaped by its historical relations with France and by its French-speaking inhabitants, represents a cultural anomaly in North America. Despite its apparent locational advantages, Québec remains the recipient of the largest equalization payments, suggesting that economically all is not well.

British Columbia, cut off by the Cordillera, has a distinctly Pacific coast frame of mind—one that Garreau (1981) calls **Ecotopia**. With its trade future linked to the Pacific Rim, British Columbia continues to expand its population and economy. Yet, the principal industry, forestry, remains stalled because its major market is the depressed US housing market. A forestry rebound must wait for an upturn in the US construction sector.

Western Canada, with its fertile soil and vast underground resources, also benefits from greater trade with Pacific Rim countries. Then, too, China, Japan, and South Korea are investing heavily in those resources needed to keep their economies going: one target is Alberta's oil sands while another is Saskatchewan's potash. Locked in the heart of the continent, Western Canada must factor into its finances the long transportation routes to its markets. Western Canada craves respect within Confederation. Whether its objections are real or perceived, Western Canada, historically, has felt ignored by Ottawa and exploited by the Québec–Ontario heartland. This feeling fuels tensions between Western Canada and Ottawa. For instance, Western Canadians are now asking themselves if a proposed carbon tax, like the National Energy Program, will suck resource revenue from Alberta to Ontario and Québec.

The relation of Atlantic Canada to the sea continues to play a paramount role in its regional character, though now more attention goes to its offshore petroleum resources than to its fishery. Atlantic Canada has a location advantage in its proximity to New England. With the growing demand for clean energy, New England has turned to Québec for hydroelectricity. Paradoxically, a proposed electric transmission route through New Brunswick, but owned and operated by Hydro-Québec, may have benefited Québec more than Atlantic Canada, one reason perhaps why the sale of NB Power to Hydro-Québec fell through in March 2010.

The regional character of the *Territorial North* has three prominent features. First, it is the largest region. Second, its cold environment is a dominating geographical feature—though global warming may affect the Territorial North more than any of the other five regions. Third, its Aboriginal population forms a majority, which has political and social implications for the region. Unlike the other five regions, a decision at the United Nations may greatly increase its geographic size and offshore energy reserves. These are but a few of the regional characteristics that define Canada's six regions.

Vignette 1.4, 'Is Continentalism Inevitable?', page 24, discusses Joel Garreau's classic book, The Nine Nations of North America, *and includes a map of Garreau's nine North American regions.*

Regional character changes with time. As each region has grown and matured in its unique way, populations have developed attachments to their regions and a strong sense of place now exists and continues to adapt to new circumstances. Time is the dynamic element of regional geography. The thoughtful study of regional geography takes into consideration static factors of geographic space by considering and by seeking out the forces of change. The broad sweep of Canadian geography has been elegantly expressed in various ways, but always at the heart of understanding is the concept of the region. American geographers de Blij and Murphy (2006) have described geography as 'destiny' while British geographer Sir Gordon Conway (2009: 221) views the discipline as one that looks 'at what happens in a particular place and why'. Next it asks: 'how does it link with other localities?' John Ralston Saul

(1997: 69) places even more weight on the philosophical shoulders of geography:

> Because, in spite of intellectual claims to the contrary, not religion, not language, not race but place is the dominant feature of civilizations. It decides what people can do and how they will live.

See Chapter 1, 'Geography as a Discipline', page 4, for more on the nature of regional geography.

The economic performance of Canada's six regions varies. Some regions have grown swiftly while others have lagged behind. Ottawa's intervention through equalization payments has played a critical role in reducing the economic gap between regions (Table 3.7). 'Have-not' regions use these payments to ensure a degree of parity in terms of education and social services. For many years, Ontario led Canada's regions in economic growth. However, if Ontario's recent economic decline is not temporary, then the well-being of the federal equalization program may be in jeopardy. Ottawa has depended heavily on tax revenue from Ontario and its citizens to fund the equalization program.

In summary, Canada's regional character is reflected in its six regions. In turn, the core/periphery model has cast them into two distinct categories—core and periphery, but with a further differentiation of the periphery into upward transition, downward transition, and resource frontier. The original purpose of this model was to provide an understanding of relationships between regions. However, since some regions have grown more quickly than others, one needs to ask if relationships have changed enough to affect the model? Or, is the classification of Ontario and Québec into core regions more a factor of history than of geography? With these thoughts in mind, we now turn our attention to a critical review of the spatial framework.

Canada's Spatial Structure

The core/periphery spatial framework places Canada's regional geography in a particular theoretical perspective, revealing complex economic and political relationships between the manufacturing heartland in Central Canada, the resource hinterlands, and the federal government that must attempt to balance regional interests. An important concept for understanding Canada's spatial structure is the concept of 'power of place', which stresses resource differences and geographic location as key factors in regional economic differentiation. Further, the concept of 'sense of place' adds a social dimension to each region, and over time, creates layers of cultural history in each region.

'Power of place' and 'sense of place' as geographic concepts are discussed more fully in Chapter 1.

This spatial framework consists of: core regions (Ontario and Québec); and three types of hinterlands—upward transitional regions (British Columbia and Western Canada); a downward transitional region (Atlantic Canada); and a resource frontier (the Territorial North). The danger of theoretical frameworks is their apparent rigidity; one must remember that reality is not rigid and models must adjust. For instance, three watershed events have transformed our reality—global restructuring, the 2008 global financial crisis emanating from the United States, and the increasing demand for commodities from the rapidly industrializing East. All have led to an expansion of core regions in the global economic structure and, in turn, to an unravelling of what had been seen as Canada's 'spatial structure'. In the last decade, Canada's manufacturing base in Central Canada lost its momentum and, with the global crisis, its central pillar—the automobile industry—fell into bankruptcy with General Motors and Chrysler requiring huge public investments to stay alive. At the same time, commodity prices reached new levels in response to strong demand from Asian countries, especially China, causing boom-like conditions in resource-rich regions. Canada's regions are witnessing radical changes in their economic circumstances and relationships that produce challenges to the core/periphery theory. The original core/periphery theory, as

Think About It

If Ottawa were to discontinue its equalization program and use those funds to pay down the national debt, which provinces would suffer the most?

described by Innis (1930) many years ago and as applied to the world system by Wallerstein (1979, 1998), assumed that manufacturing prices always would increase at a greater rate than foodstuffs and commodities; hence the manufacturing centres of the world would gain increasing wealth. If the so-called super commodity price shift or super cycle theory (see Greenwood, 2007; Parkinson, 2009; Agriculture and Agri-Food Canada, 2009), proves enduring, then the consequences for Canada's resource-rich regions are most favourable—while the resources last—and the core/periphery model requires adjustment.

Yet, the utility of this spatial framework has not been lost and its revitalization is pertinent to two crucial matters:

- *Revitalization of the core.* Ontario and Québec have to regain their economic strength, i.e., manufacturing in the core has to reinvent itself and its knowledge-based economy needs to grow within a global context.
- *Strengthening of the East–West connection.* Internal trade and cultural exchanges are among Canada's strengths and reduce regional tensions.

Canada's Faultlines

As a country of regions, tensions naturally arise. Four faultlines stand out. Disputes between regions often draw Ottawa into the political fray. The tension between centralists and decentralists was rooted in the struggle for regional power and continues to take the spatial form of core regions against hinterlands. The second faultline, between non-Aboriginals and Aboriginal peoples, has existed since Europeans arrived on what is now Canadian soil. For centuries, Aboriginal Canadians have struggled with non-Aboriginals over such issues as land, rights, and the environment. The third faultline, between English-speaking and French-speaking Canada, flared up in the late twentieth century and saw the country nearly broken into two.[1] The fourth faultline—between 'newcomers' and 'old-timers'—reflects a gap in values and expectations between recent immigrants to the country and those

who were born and raised in Canada and whose families immigrated earlier.

Canada's principal faultline, between centralists and decentralists, stems from one of Canada's most prominent features—regional diversity. Regional diversity has led to regional disparity—some regions are endowed with both political and economic power while other regions lack both. This disparity has generated a tension between those regions having power, as well as the federal government, and those regions struggling for more power. Though federalism already affords a sharing of power between the central government and regional (provincial and territorial) governments, some argue that power is not shared equally. Because of this sentiment as well as other issues, such as the unilateral cuts in transfer payments to the provinces by Ottawa in the 1990s, provincial governments have been demanding a greater share of power, including more taxing power. With trade liberalization and the federal government's reduced ability to inject money into depressed regions of the country, centralist/decentralist tensions will only intensify. Compromises will have to be reached if provinces are to meet their socio-economic responsibilities, while still maintaining a level of power for the central government that ensures the preservation of national standards among the services available to Canadians in all regions.

The next eruption of the regional faultline may result from regional variations in carbon footprints. A much-needed North American greenhouse-gas emissions program pits oil-rich provinces against Ontario and Québec. Ottawa's plan involves setting CO_2 equivalent emissions targets by provinces. In such a North American based cap-and-trade system of CO_2, a critical question is whether provinces with surplus production of tonnes of CO_2 equivalents should be forced to buy carbon offsets from provinces with deficit production of CO_2 equivalents. Such an arrangement could spark a fierce political battle among oil-producing and oil-consuming provinces, and, in doing so, threaten national unity. Already in the post-Copenhagen period, relations between energy-producing and

Think About It

Does the core/periphery spatial structure, even if bent or broken, illustrate the value of a theoretical framework in regional geography?

Michael de Adder/artizans.com

Photo 11.2

Ottawa, as cartoonist Michael de Adder suggests, plans to ride on the coattails of the US in establishing a North American cap-and-trade system for greenhouse gases. The US bill (American Clean Energy and Security Act) sets a national cap on greenhouse gases and allows polluters to buy and sell allowances (cap-and-trade system). The proposed US legislation was passed by the House of Representatives in June 2009 but remains under consideration by the US Senate. One provision in this bill is to assist energy-intensive, trade-vulnerable industries—which could include Alberta's oil sands producers—by giving them lower emissions targets.

energy-consuming provinces have turned frosty. Ottawa is looking for a compromise solution and, by following the US lead, may merely exacerbate tense relations between provinces. However, such a plan to harmonize emissions targets for North America might arouse fears in Ontario and Québec that the oil sands producers will receive lower emissions targets.

What drives the regional faultline in Western Canada? The forces behind western regionalism stem from the continuing shift westward of Canada's population and economic centre of gravity, and resource differences, i.e., the fact that Alberta has oil while Québec has hydroelectric power, influence their views of public policy, notably solutions to greenhouse gas emissions. The overwhelming factor is that Western Canada and British Columbia remain the sparkplugs for Canadian growth. In fact, the most recent figures, as shown in Figure 11.1, indicate that all four western provinces have had population growth rates above the national average, while Ontario, Québec, and the provinces of Atlantic Canada fell below the national average. In 2008–9, Alberta led the country with a population growth rate of 2.5 per cent, followed by British Columbia and Saskatchewan, both at 1.6 per cent. Such population growth fuels the desire for more political power in

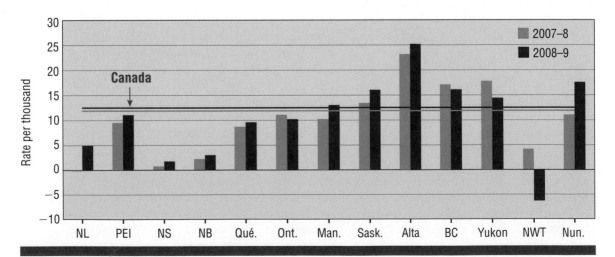

Figure 11.1   Population growth rates by component, Canada, provinces, and territories, 2007–8 to 2008–9.
Source: Adapted from Statistics Canada (2009). <www.statcan.gc.ca/pub/91-215-x/2009000/ct002-eng.htm>.

these provinces, i.e., more seats in the House of Commons. Figure 11.2 provides similar evidence for a greater rate of economic growth in those two western regions.

The three remaining faultlines have their roots in social tensions that often find a regional expression. In the Territorial North, the Aboriginal/non-Aboriginal faultline still resonates. For a majority of Native peoples, their political weight has more power in the Territorial North than in Ottawa. Consider, for example, the continued opposition to the Mackenzie Gas Pipeline by the Deh Cho Nation. Centred around Fort Simpson, the Deh Cho Nation is the only one of five Aboriginal governments along the pipeline route that has not signed a land-claim agreement and the Deh Cho have repeatedly insisted that their claim must be settled before construction of the pipeline begins. Within Québec, the French/English faultline has cooled down while the old-timer/newcomer faultline has heated up. Québec is changing. Québecers already have decided to remain in Canada (at least for now), have embraced the global economy, have moved away from the Church-oriented French-Canadian identity towards a more secular one, and have opted for immigration to sustain their demographic and economic growth.

Yet, many Québecers are not happy with all aspects of the new cultural elements arriving from abroad, especially those from Muslim countries. In part, the problem lies in the rapid changes in Québécois society that may give many a feeling that that their way of life is fading. The pervasive influence of one cultural anchor, the Church, has diminished. Other cultures, including those associated with Islam, have emerged. The vitriolic debate over 'reasonable accommodation', or *accra* as it's known in French shorthand, was sparked by the realization Québec society had moved into uncharted waters. A flashpoint came in 2007 with Hérouxville's 'code of standards', which was designed 'to inform the new arrivals that the lifestyle that they left behind in their birth country cannot be brought here with them and they would have to adapt to their new social identity' (Hamilton, 2007: A7). The

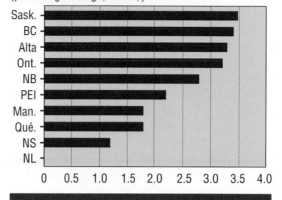

Real GDP by Province, 2010
(percentage change; 2002 $)

Figure 11.2   Percentage change in gross domestic product from 2002 to 2010.
Source: The Conference Board of Canada and Statistics Canada in Scoffield (2009).

danger of letting these issues fester rather than addressing them could drive newcomers further away from mainstream Québec society, leading to core fragmentation of its society and geographic isolation of its population into ghettos. Québec's answer to the backlash against Hérouxville's 'code of standards' was the Consultation Commission on Accommodation Practices Related to

Photo 11.3

Northern lights near the Victor diamond mining camp in the Hudson Bay Lowland of Ontario.

Cultural Differences headed by Gérard Bouchard and Charles Taylor. One of their principal recommendations to achieve 'reasonable accommodation' was that secularism must provide a balance between its four main components, i.e., freedom of conscience, the equality of citizens, the reciprocal autonomies of church and the state, and the neutrality of the state. In doing so, the rights of minorities and majorities would be served. Finding this balance is more difficult. One recommendation of Bouchard and Taylor to promote secularism within the cultural context of Québec society was to remove the crucifix from the National Assembly, but this suggestion was rejected by Premier Jean Charest. In another instance, a Muslim woman taking a French-language class for immigrants refused to remove the veil (niqab) from her face and was asked to leave the class, setting off more debate over the lengths to which public institutions should go to accommodate Muslims (Perreaux, 2010).

 For further discussion of cultural change, see Vignette 1.2, 'The Challenge of Cultural Change: Québec and Muslim Canadians', page 13.

> ## Think About It
>
> 'Reasonable accommodation' is not only a vague concept but one difficult to enforce. Where are the boundaries?

Vignette 11.1   Canadian Identities

Saul's concept of a 'soft' Canada is based on its heterogeneous nature, which fits nicely with the concept of Canada as a country of regions. This identity quandary facing Canada is due to the absence of a unified, powerful Canadian identity. **Pluralism**, which is hardly a national identity, prevails. Each region has its own character, though 'Canadian' values thread their way through regional identities. Québec may be the exception to this rule because its values centre on Québécois culture. Unlike some other parts of Canada, Québec sees multiculturalism as a threat because its people do have a clear and relatively unified identity, though the issue of accommodation to Canadian values, as we have seen, remains to be resolved. The identity quandary across the country, however, has helped make Canada freer of prejudice than is the case in some other countries, and, at the same time, more of a place where immigrants can feel comfortable.

The Future

Regional change is inevitable. The recalibrated global economy is revising the nature and structure of Canada's regional geography. Powerful Asian countries led by China have caused resource-rich regions to prosper and manufacturing regions to suffer. Not only has this global tidal wave overturned the basic principle of the core/periphery model but, in practical terms, it has benefited Western Canada and British Columbia and hurt Ontario and Québec. Added to their economic strength, population growth in Alberta (five new seats) and British Columbia (seven new seats) soon will be rewarded with more seats—and political clout—in the House of Commons. However, Ontario, according to legislation proposed in April 2010, would receive the most new seats—18—thus enhancing its place as the political centre but weakening Québec and Atlantic Canada (Ibbitson, 2010).

By 2020, the face of Canada and its regions will have adjusted to these new circumstances. Already, global events have set the stage for more adjustments in Canada's economy and its regions. By then, Canada and the rest of the world may have agreed to 'hard' emission targets, which could result in a system of cap-and-trade and possibly to a halt in the expansion of open-pit mining of Alberta's oil sands. How would a cap-and-trade system affect Canada's regions? Would such a system inflame regional tensions and threaten national unity because of a transfer of wealth to Ontario and Québec? At the same time, the territorial size of Canada likely will have increased, but by how much? The answer to that question depends on how the United Nations, through UNCLOS, divides the unclaimed area of Arctic Ocean among the five Arctic coastal countries.

By the end of the second decade of the twenty-first century, the shape and strength of Canada's regions and their structure should have stabilized. Canada will continue to have six regions, although the Territorial North likely will extend further into the Arctic Ocean. The implication of such a territorial expansion is that the means to administer these additional waters will be required, especially if the polar ice retreats or even disappears. In the latter event, however, cataclysmic geographical changes may force Ottawa to place most of its attention elsewhere. Other changes are less certain and depend on an answer to the question: Were the changes of the last few years an anomaly or a more lasting shift in regional economic power and population? More specifically, will the manufacturing base of Ontario and Québec regain its former size and strength or will Canada surrender its manufacturing operations to China and other countries where wages are low? Second, what about the dream of a resurgence into a post-industrial era where Canada, led by Ontario and Québec, would gain a place on the world stage for the export of its knowledge-based products and services? Then, too, will predictions of a super commodity price shift that would move all commodity prices to a higher level than those in the first decade of the twenty-first century become a

reality? Resource-oriented regions—but also Québec with its vast hydroelectric resources and its firm grip of the lucrative market of New England—would benefit from higher commodity prices. Finally, will the construction of the Mackenzie Gas Pipeline signal a welcome new beginning for the Territorial North, or an unfortunate end to old ways?

Submerged within this resurgence and affecting regional adjustments are:

- a greener urban Canada where a cultural environment to attract the **creative class**, as proposed by Richard Florida (2002, 2005, 2008), flourishes and where Aboriginal residents find their footing in Canada's economy and society;
- a more comfortable pluralistic society where tensions subside between newcomers and old-timers;

Ken Davies/Masterfile

Photo 11.4

One dream of resurgence sees Toronto as a leading global financial centre and as Canada's leading centre for knowledge-based activities (Greenwood, 2009). As the headquarters for Canadian banks—seen on the world stage as reliable, stable, and profitable—plus several important universities, a creative class, and **ethnoburbs** (Wei, 2009), Toronto is well positioned to take the next step as a world financial hub and centre for knowledge-based firms. But will it?

- a more diversified export effort that would take advantage of fast-growing economies in Asia without decoupling from our long-standing trade relationship with the United States;
- a greater urban resurgence that would lead regional growth deeper into the knowledge-based economy as well as foster a more pluralistic Canadian fellowship.

In sum, geographic regions have and will endure. While regional power—as measured by population and economic performance—is shifting westward, Canada will remain a country of regions, still led by Ontario and Québec. Another truth is that tensions will continue to flare up and the well-tested Canadian 'ad hoc' process of seeking adjustments leading to compromises so necessary in a 'soft' democratic country will still be needed. Finally, Canada is not a homogeneous nation-state, but a pluralistic country where regional differences are cherished by those living in these regions. Canada's survival will depend on the ability of its citizens to accept the country's geographic structure together with its regional identities (Frye, 1971: i–iii). In the next decade, Canada needs to build cultural as well as economic bridges between its regions and, in so doing, strengthen national identity and unity. Such a national identity, however frail or fractured it sometimes appears, is the linchpin holding Canada and its six regions together; if strengthened it can empower economic and cultural bonds that will secure the future of Canada and its regions.

The future, while full of uncertainty, contains two certainties: Canada will remain a large, diverse country and, as such, will continue to experience regional tensions. By engaging in open and vigorous debate, compromises—not solutions—will follow. Compromise works at the heart of the exercise of democratic power in a diverse society by demanding a balance between individual and collective rights, between regional and national interests, and between those with power and those in need of it. The ability to strike this balance is critical to Canada's well-being and regional harmony.

CHALLENGE QUESTIONS

1. 'Canada is big—preposterously so', wrote geographer Kenneth Hare in 1968. But how, when, and where can its territory enlarge?

2. Given the current emphasis on knowledge-based industries, would the future of Canada's economy have been better served if Ottawa had loaned $4 billion to Nortel rather than to Chrysler and General Motors?

3. With increasing immigration from non-Christian countries, it seems inevitable that the demand for public funding for all religious schools will become a major political issue. How would you resolve this religious/political issue and thus prevent rumblings from becoming an earthquake on the fault-line between newcomers and old-timers?

4. The original core/periphery model was based on exploitation of poor regions by rich regions. Do federal programs, such as equalization and transfer payments, repudiate the notion of regional 'exploitation'?

5. Has John Ralston Saul fallen either into postmodern relativism—a recent trap for social thinkers—or into environmental determinism—an old trap for geographers—when he writes that 'in spite of intellectual claims to the contrary, not religion, not language, not race but place is the dominant feature of civilizations. It decides what people can do and how they will live'?

FURTHER READING

Saul, John Ralston. 2009. *The Collapse of Globalism and the Reinvention of the World*, 2nd edn. Toronto: Penguin Canada.

John Ralston Saul's *The Collapse of Globalism* brings a fresh argument to the question: has globalization spread wealth to all nations through international trade? No one doubts the world is a 'richer' place, but the bumps along an unregulated road have displaced lesser economies. His argument has implications for Canada's regions. Essentially, Saul calls for more regulations of capitalism: 'Are political decisions meant to be made in deference to the economy and markets, or are political institutions meant to shield us from harsh realities of globalization?' As a case in point: should Alberta allow the marketplace to determine where its bitumen is upgraded if the processing is done outside of Alberta and Canada?

GLOSSARY

Aboriginal ancestry Statistics Canada defines the Aboriginal ancestry population as those people reporting an Aboriginal identity as well as those who report being Aboriginal to the ethnic origin question, which focuses on the ethnic or cultural origins of a person's ancestors.

Aboriginal identity According to Statistics Canada, those persons identifying with at least one Aboriginal group, that is, North American Indian, Métis, or Inuit, and/or those who report being a treaty Indian or a registered Indian, as defined by the Indian Act, and/or those who report they were members of an Indian band or First Nation.

Aboriginal peoples All Canadians whose ancestors lived in Canada before the arrival of Europeans; includes status and non-status Indians, Métis, and Inuit.

age dependency ratio The ratio of the economically dependent sector of the population to the productive sector, arbitrarily defined as the ratio of the elderly (those 65 years and over) plus the young (those under 15 years) to the population of working age (those 15 to 64 years); the old-age dependency ration is similar to the age dependency ratio except that it focuses only on those over 64.

agricultural fringe Agriculture at its physical limits. Along the southern edge of the boreal forest and in the Peace River country, farmers have cleared the land, but the short growing season prevents most crops from maturing, thus, many farmers turned instead to cattle.

air commuting Travel to a work site in a remote area, such as a mine, by aircraft owned or hired by the company. Until the 1970s, companies built resource towns to house workers and their families. Employees remain at the work site for a week or longer, working long shifts (often 12 hours per day), then have a week or two at home. The company pays for the air transportation and for food and lodgings at the work site.

air pollution Any chemical, physical, or biological agent that modifies the natural characteristics of the atmosphere.

Alaska Panhandle A strip of the Pacific coast north of 54° 40′ N latitude that was awarded to the United States in 1903 following what is known as the Alaska boundary dispute.

albedo effect Proportion of solar radiation reflected from the earth's surface back into the atmosphere.

Alberta clipper A low pressure system that begins when warm, moist winds from the Pacific Ocean come into contact with the Rocky Mountains and then the winds form a chinook in southern Alberta; winter storms occur over the Canadian prairies when it becomes entangled with the cold Arctic air masses. Eventually, the storm reaches Ontario and Québec. Also, 'Alberta Clipper' is the name of a natural gas pipeline that runs from Alberta to the US Midwest.

allophones A term used by Statistics Canada to identify those whose mother tongue is not English, French, or one of the Aboriginal languages.

alpine permafrost Permanently frozen ground found at high elevations.

anglophones Those whose mother tongue is English.

Arctic Circle An imaginary line that signifies the northward limit of the sun's rays at the time of the winter solstice (21 December). At a latitude of 66° 32′ N, the sun does not rise above the horizon for one day of the year (the winter solstice). Except for a short period of twilight, darkness prevails for 24 hours. At the summer solstice (21 June), the sun's rays do not fall below the horizon, providing constant daylight for 24 hours.

Arctic ice pack Floating sea ice in the Arctic Ocean that has consolidated into an ice pack, with an extent of over 10 million km^2. New sea ice (less than one year in age) is often about 1 m thick; old sea ice can reach 5 m in thickness. Ice ridges are formed, reaching 20 m in thickness. Scientists have found that higher temperatures are reducing the geographic extent and thickness of the Arctic ice pack.

arêtes Narrow serrated ridges found in glaciated mountains. Arêtes form when two opposing cirques erode a mountain ridge.

Asia-Pacific Gateway and Corridor A system of transportation infrastructure, including British Columbia's Lower Mainland and Prince Rupert ports, road, and rail connections, that reaches across Western Canada and into the economic heartlands of North America, including major airports and border crossings.

Athabasca tar sands The largest reservoir of crude bitumen in the world and the largest of three major oil sands deposits in Alberta; also known as the oil sands.

backward linkages Investments in production of infrastructure, such as railways, to move products from source to market.

Bakken formation A geological structure containing large quantities of oil trapped in shale. It lies in Williston Basin in southern Saskatchewan and Manitoba and extends into North Dakota, Montana, and South Dakota.

Barren Grounds The area of tundra stretching from the west coast of Hudson Bay to Great Slave and Great Bear lakes and northward to the Arctic Ocean. The barren ground caribou use this region for calving each summer before migrating to the boreal forest.

basins Structural depressions in sedimentary rock caused by a bending of sedimentary strata into huge bowl-like shapes. Petroleum may accumulate in sedimentary basins.

Beothuk   Before the arrival of fishing boats from Europe, the Beothuk Indians, who probably spoke an Algonkian language, hunted and fished on the island of Newfoundland. Relations with fishers and settlers often resulted in conflicts, which confined the Beothuk to the inland. With access to coastal resources cut off and under attack by settlers, the Beothuk struggled to survive in the resource-poor interior. In 1829, the last of the Beothuk died.

Big Bear   The last of the great chiefs, who had a vision to unite the Plains Cree to stand together against the impending wave of settlers and to find a way to sustain their culture. Dissatisfied with Treaty No. 6 and the other numbered treaties, he wanted a new treaty with one huge reserve for all Plains Indians. His people facing starvation, Big Bear, like Poundmaker, could not control his warriors at the time of the 1885 Northwest Rebellion. Big Bear was tried for treason and sentenced to prison for three years; released from prison early, he died within months of his release.

Big Commute   Air travel by Newfoundland trades workers to and from the Alberta oil sands, on a cycle of 20 days in Alberta and eight days back home in Newfoundland.

biomass   The total quantity or weight of an organism (e.g., codfish) in a given area.

bitumen   A tar-like mixture of sand and oil.

boom-and-bust cycles   A rapid increase in economic activities in a resource-oriented economy, often based on the value of a single commodity, quickly followed by a downturn when the commodity price(s) falls due to a drop in world demand, which is usually associated with a contraction in the business cycle.

BP (before present)   A time scale with a base year of 1950. Past time is measured from this base year, which is considered 'the present' or the beginning of the present era. For example, 18,000 BP means 18,000 years before 1950, or 16,050 BC (before Christ).

British Columbia–Alberta–Saskatchewan Trade, Investment, and Labour Mobility Agreement   This agreement gives businesses and workers in all three provinces seamless access to a larger range of opportunities across all sectors, including energy, transportation, labour mobility, business registration, and government procurement. In 2009, Saskatchewan signed this agreement.

business cycle   The world market (capitalist) economy follows a series of irregular fluctuations in the pace of economic activity. These fluctuations consist of four phases: 'contraction' (a slowdown in the pace of economic activity); 'trough' (the lowest level of economic activity); 'expansion' (a sharp increase in the pace of economic activity); and 'peak' (the maximum level of economic activity).

'Buy America'   In an effort to stimulate the US economy, Washington introduced the 'Buy America' provision, which means that for its economic stimulus package, priority is given to US iron, steel, and other manufactured goods for use in public works and building projects supported with taxpayer-funded recovery money.

Canada Line   Part of Greater Vancouver's rapid transit system. The Canada Line extends from the Waterfront Station in Downtown Vancouver to Richmond with a branch line to Vancouver International Airport.

Canada Pulp and Paper Green Transformation Program   This program paid $0.16 per litre credit for black liquor produced by Canadian mills between 1 January 2009 and 31 December 2009. Firms had to demonstrate such environmental benefits from this program as improvements to their energy efficiency or their capacity to produce alternative energy.

capitalist world-systems theory   Wallerstein's theoretical Marxian framework on the workings of modern capitalism. His division of the world economy into three spatial units and their economic relationships forms a key aspect of the capitalist world system. Industrial nations represent the core; developing nations form the periphery; and semi-peripheries are those countries that are partly industrialized.

carbon sequestration   Carbon capture and storage technology involving the capturing of CO_2 and other greenhouse gas (GHG) emissions from fossil-fuel power stations or other large carbon emitters and the storing of the CO_2 in deep, stable geological formations.

carbon tax   An emissions pricing policy (cap-and-trade system or emissions tax). Emitters would pay for every tonne by purchasing emission allowances auctioned by government.

Cascadia   The name proposed for an independent sovereign state advocated by a grassroots movement in the Pacific Northwest, which would include British Columbia, Washington, and Oregon; see *Ecotopia*.

census metropolitan area   An urban area with a population of at least 100,000, together with adjacent smaller urban centres and even rural areas that have a high degree of economic and social integration with the larger urban area.

Château Clique   The political elite of Lower Canada, composed of an alliance of officials and merchants who had considerable political influence with the British-appointed governor; similar to the Family Compact in Upper Canada.

chem fallow   Summer fallow land that is treated with herbicides to control weeds.

chernozemic   A soil order identified by a well-drained soil that is often dark brown to black in colour; associated with the grassland and parkland natural vegetation types and located in the Prairies climatic zone.

chinook   A dry, warm, downslope wind in the lee of a mountain range. Also called a rain shadow wind because it has dropped most of its moisture on windward slopes.

circumpolar countries   The eight nations associated with the circumpolar area are Canada, Denmark (including Greenland

and the Faeroe Islands), Finland, Iceland, Norway, Russian Federation, Sweden, and the United States of America. Five of these—Canada, Denmark, Norway, Russia, and the US—have a territorial claim to portions of the Arctic seabed.

cirques Large, shallow depressions found in mountains at the head of glacial valleys that are caused by the plucking action of alpine glaciers.

Clean Energy Fund Federal $1 billion fund for research, development, and demonstration projects to advance carbon capture and storage technology and other clean energy research.

climate An average condition of weather in a particular area over a very long period of time.

climategate The release, by a hacker in November 2009, of e-mails and other documents from the Climatic Research Unit (CRU) of the University of East Anglia. The subsequent dissemination of this material caused a controversy, throwing doubt on the accuracy of their predictions of global warming.

climatic zone A geographic area where similar types of weather occur.

commercial forests Forest lands able to grow commercial coniferous (softwoods), deciduous (hardwoods), and mixed woods timber within an acceptable time frame.

comprehensive land-claim agreement Agreement based on territory claimed by Aboriginal peoples that was never ceded or surrendered by treaty. Such agreements extinguish the Aboriginal land claim to vast areas in exchange for a relatively small amount of land, capital, and the organizational structure to manage their lands and capital.

container A sealed steel 'box' of standardized dimensions (measured in 20-foot equivalent units or 'TEUS') for transporting cargo.

continental air masses Homogeneous bodies of air that have taken on moisture and temperature characteristics of the land mass of their origin. Continental air masses are normally dry and cold in the winter and dry and hot in the summer.

continental drift The movement of the earth's crust; also known as plate tectonics.

continentalism To the federal government, continentalism refers to policies, like the Free Trade Agreement, that promote Canadian trade and economic ties with the United States. Washington views continentalism along the same lines as Canada. However, earlier in its history, the United States saw itself spreading across all of North America and this ideology was expressed by the concept of Manifest Destiny.

continuous cropping A popular farming practice in grain-growing areas where the stubble left after harvest is not tilled; stubble serves to control weeds and reduce soil erosion by wind.

continuous permafrost Extensive areas of permanently frozen ground in the Arctic, where at least 80 per cent of the ground is permanently frozen.

convectional precipitation An upward movement of moist air that causes the air to cool, resulting in condensation and then precipitation.

Copenhagen Climate Conference Conference held in Copenhagen, Denmark, in 2009 with the aim of replacing the Kyoto Protocol with another climate protocol with enforceable greenhouse gas emissions targets for all nations. No agreement was reached.

core An abstract area or real place where economic power, population, and wealth are concentrated; sometimes described as an industrial core, heartland, or metropolitan centre.

core/periphery model A theoretical concept based on a dual spatial structure of the capitalist world and a mutually beneficial relationship between its two parts, which are known as the core and the periphery. While both parts are dependent on each other, the core (industrial heartland) dominates the economic relationship with its periphery (resource hinterland) and thereby benefits more from this relationship. The core/periphery model can be applied at several geographical levels: international, national, and regional.

corporate and political elite A concept of society popular among sociologists who argue that common folk are powerless while corporate and political leaders hold the reins of power. In doing so, corporate leaders advance their economic interests with the active support of politicians who, in turn, gain from this relationship.

country food Food, primarily game, such as caribou, fish, and sea mammals, obtained by Aboriginal peoples from the land and sea. Although Indians, Inuit, and Métis now live in settlements, they fish and hunt for cultural and economic reasons.

creative class Culture workers, from artists to computer programmers, who, Richard Florida argues, are the key to a flourishing and progressive city and who are attracted to urban centres rich in diversity and culture.

Crow Benefit Signed in 1897, the Crow's Nest Pass Agreement between the Canadian Pacific Railway and the federal government ensured that the rail rates for grain were low, and in this way helped to overcome the disadvantage of a long rail distance to ports. The rail subsidy ended in 1995, placing the full burden on farmers.

crude birth rate The number of births per 1,000 people in a given year.

crude death rate The number of deaths per 1,000 people in a given year.

cryosolic A soil order associated with permafrost and poorly drained land; soil is either lacking or extremely thin; associated with the tundra and polar desert vegetation types and located in the Arctic climatic zone.

culture The sum of attitudes, habits, knowledge, and values shared by members of a society and passed on to their children.

culture areas Regions within which the population has a common set of attitudes, economic and social practices, and values.

demographic transition theory The historical shift of birth and death rates from high to low levels in a population. The decline in mortality precedes the decline in fertility, resulting in a rapid population growth during the transition period.

demography The scientific study of human populations, including their size, composition, distribution, density, growth, and related socio-economic characteristics.

denudation The process of breaking down and removing loose material found at the surface of the earth. In this way, erosion and weathering lead to a reduction of elevation and relief in landforms.

deposition The deposit of material on the earth's surface by various processes such as ice, water, and wind.

discontinuous permafrost Permanently frozen ground mixed with unfrozen ground in the Subarctic. At its northern boundary about 80 per cent of the ground is permanently frozen, while at its southern boundary about 30 per cent of the ground is permanently frozen.

dispute settlement provisions Binding arbitration to resolve trade disputes, as built into the FTA and NAFTA. Each country has a panel of five members. After deliberations, the panel makes a recommendation to the Free Trade Commission, which comprises cabinet-level representatives or their designates, as to what action should be taken. The Commission is then expected to take the appropriate political action.

Dominion Land Survey This survey method divided Western Canada into one-square-mile sections to allow ownership of specific land units by homesteaders and others.

drainage basin Land sloping towards the sea; an area drained by rivers and their tributaries into a large body of water.

drumlins Landforms created by the deposit of glacial till and shaped by the movement of the ice sheet.

Dry Belt An agricultural area in the semi-arid parts of Alberta and Saskatchewan where crop failures due to drought are more common that is primarily devoted to grain farms and cattle ranches.

Dutch disease A theory describing the apparent relationship within a country between its expanding energy resource sector and a subsequent decline in the manufacturing sector. In 1977, the term first appeared in *The Economist* to describe the phenomenon of a declining manufacturing sector in the Netherlands, which at the same time was enjoying increased revenues from the export of its natural gas but also seeing its exchange rate with other countries increasing.

Eastern Townships An area of Québec in the Appalachian Uplands lying south and east of the St Lawrence Lowland and near the US border. Originally occupied by less than 2,000 Loyalists, most English-speaking settlers arrived after 1791 from neighbouring New England and the British Isles. By the 1870s, French-Canadians formed the majority. The region is now referred to as Estrie.

economic leakage The loss of economic benefits from one jurisdiction to another, which dulls the potential impact of large-scale projects in the local area. In the case of the Territorial North, construction of megaprojects sees the demand for goods and labour satisfied largely by businesses and labour pools outside the region and, similarly, the spending of wages taking place outside the region.

economic structure The sectors of a national, regional, or local economy—primary (e.g., resource extraction and harvesting), secondary (e.g., manufacturing, construction), and tertiary (services)—and the extent to which the whole economy is driven by each of these sectors.

economies of scale A reduction in unit costs of production resulting from an increase in output.

Ecotopia A narrow band along the Pacific coast from northern California to Alaska, defined as a vernacular region with distinctive economic and cultural features by Joel Garreau in his 1981 book, *The Nine Nations of North America*. Ecotopia—an ecological utopia—focuses on a 'green' world with an emphasis on 'quality of life' and a sustainable economy.

ecumene The portion of the land that is settled.

Eeyou Istchee Created in 2007 as a political unit equivalent to a Québec regional county municipality and administered by the Cree Regional Authority; it consists of Jamésie (Cree lands defined following the James Bay and Northern Québec Agreement) plus Whapmagoostui and Cree Village, which are north of the fifty-fifth parallel and therefore within Nunavik.

environmental determinism The assumption that human activities are controlled by the physical environment. Now considered far too deterministic, it was a popular philosophical position of geographers in the late nineteenth and early twentieth centuries.

erosion The displacement of loose material by geomorphic processes such as wind, water, and ice by downward movement in response to gravity.

eskers Long, sinuous mounds of sand and gravel deposited on the bottom of a stream flowing under a glacier; these eskers appear on the land surface after the glacier has retreated.

Estrie An administrative region that overlaps most of the area formerly called the Eastern Townships.

ethnic group People who have a shared awareness of a common identity and who identify themselves with a particular culture.

ethnic origin A Statistics Canada definition, which refers to the ethnic or cultural origins of the respondent's ancestors. An ancestor is someone from whom a person is descended and is usually more distant than a grandparent.

ethnoburbs Suburban residential and business areas with a significant ethnic character composed of new Canadians.

These cultural enclaves in urban areas reflect the pluralistic nature of Canadian society and the very malleable Canadian identity, which allows Canadians to accept and relish such ethnic footprints in their urban landscapes.

ethnocentricity   The viewpoint that one's ethnic group is central and superior, providing a standard against which all other groups are judged.

evapotranspiration   Part of the water cycle: the sum of evaporation of water from the soil to the air and the transpiration of water from plants and its subsequent loss as vapour.

Family Compact   A group of officials who dominated senior bureaucratic positions, the executive and legislative councils, and the judiciary in Upper Canada

faulting   The breaking of the earth's crust as a result of differential movement of the earth's crust; often associated with earthquakes.

fault line   A crack or break in the earth's crust. A complex fault line is known as a fault zone; major fault lines exist between two tectonic plates.

faultlines   Application of a geological phenomenon to the economic, social, and political cracks that divide regions and people.

Fertile Belt   Area of long-grass and parkland natural vegetation in Western Canada associated with black and dark-brown chernozemic soils. It supports a mixed farming area where crop failures due to drought are less common.

fertility rate   The number of births per 1,000 people in a given year; also called crude birth rate; not to be confused with the 'general fertility rate', which is the number of live births per 1,000 women who are of child-bearing age—15 to 44 years—in a given year.

final demand linkages   Expenditures of income generated in the production and export of staples.

First Nations people   By Statistics Canada definition, those Aboriginal persons who report a single response of 'North American Indian' to the Aboriginal identity census question.

folding   The bending of the earth's crust.

forward linkages   Investments in staples production and processing, such as a crushing mill for processing canola.

francophones   Those whose mother tongue is French.

Free Trade Agreement (FTA)   Trade agreement between Canada and the United States enacted in 1989.

French Canadians   Canadians with roots to Québec and who likely still speak French.

friction of distance   The effect of distance on spatial interaction; that is, as distance increases, the number of spatial interactions (such as telephone calls or trade in goods) diminishes.

frontal precipitation   When a warm air mass is forced to rise over a colder air mass, condensation and then precipitation occur.

glacial erosion   The scraping and plucking action of moving ice on the surface of the land.

glacial spillways   Deep and wide valleys formed by the flow of massive amounts of water originating from a melting ice sheet or from water escaping from glacial lakes.

glacial striations   Scratches or grooves in the bedrock caused by rocks embedded in the bottom of a moving ice sheet or glacier.

glacial troughs   U-shaped valleys carved by alpine glaciers.

global circulation system   The movement of ocean currents and wind systems that redistribute energy around the world.

globalization   An economic/political/social process driven by international trade and investment that leads to a single world market and wide-ranging impacts on the environment, cultures, political systems, and economic development. While most large businesses position globalization as a positive force for world economic growth, reducing economic and social disparities between nations, and encouraging the spread of democracy, others consider it a negative force causing environmental degradation, exploitation of the developing world, and domination of world politics and culture by a few powerful countries led by the US.

Great American Desert   The treeless Great Plains as described by American explorers in the nineteenth century; in fact, such lands have a semi-arid climate and a grasslands vegetation cover.

Green Energy Act   Ontario legislation, passed in 2009, that makes it easier to bring renewable energy projects into production.

greenhouse effect   The absorption of long-wave radiation from the earth's surface by the atmosphere.

greenhouse gases   Water vapour, carbon dioxide, and other gases that make up less than 1 per cent of the earth's atmosphere but are essential to maintaining the temperature of the earth.

grooves of geography   Physiographic structure that facilitates exchanges between adjacent regions, as with Canada and the United States, where the physiography has a north–south alignment.

gross domestic product (GDP)   An estimate of the total value of all materials, foodstuffs, goods, and services produced by a country or province in a particular year.

groundfish   Fish that live on or near the bottom of the sea. The most valuable groundfish are cod, halibut, and sole.

Gulf Stream   A warm ocean current paralleling the North American coast that flows from the Gulf of Mexico towards Newfoundland.

habitants   French peasants who settled the land in New France under a form of feudal agriculture known as the seigneurial system. After the British Conquest, the seigneurial system continued until the mid-nineteenth century, marking a significant difference between Upper and Lower Canada.

harmonization   Political process that leads to a common set of policies and programs for two or more countries.

heartland   A geographic area in which a nation's industry, population, and political power are concentrated; also known as a core.

'hewers of wood and drawers of water'   Biblical phrase applied by sociologists and others to the labouring classes of capitalism doing the most menial, low-paid work necessary for the operation of capitalist society. Within the context of the core/periphery model, this term refers to periphery regions where primary production prevails; core areas, on the other hand, focus on the processing of those raw products. Its application to a country's economy refers to the exporting of raw materials rather than of finished goods.

hinterland   A geographic area based on resource development that supplies the heartland with many of its primary products; also known as a periphery.

Hobson's choice   A choice of taking what is offered or nothing.

hollowing-out   The relocation of manufacturing plants in one country to another, which leaves the economy of the original country much weakened.

Holocene epoch   The current geological division of the Geological Time Chart. It began some 11,000 years ago and is associated with the warm climate following the last ice age.

homeland   A land or region where a relatively homogeneous people and their ancestors have been born and raised, and thus have developed a strong attachment to that place; a sense of place.

homesteader   A settler who obtained land. In Western Canada, quarter sections were available as homesteads under the federal government's plan known as the Dominion Lands Act where a settler paid a $10 fee for a quarter section.

horizontal drilling   Recently developed technology used in drilling for oil and gas, as opposed to vertical oil and gas drilling, which has existed for a long time.

hydraulic fracturing   A method used to fracture rock formations in order to allow oil or natural gas to flow from impervious geological strata.

hydrometallurgy   A process that produces nickel, copper, and cobalt directly from ore, thus avoiding the smelting process and eliminating environmentally unfriendly sulphur dioxide and dust emissions.

ice age   A geological period of severe cold accompanied by the formation of continental ice sheets. The most recent ice age, the Pleistocene Ice Age, began some 2 million years ago and ended with the beginning of the Holocene Epoch some 11,000 years ago.

igneous rocks   Rock formed when the earth's surface first cooled or when magma or lava that has reached the earth's surface cools.

industrial phase   Each major change in the evolution of the capitalist economic system is called a stage or phase. The industrial stage marks the shift from a predominantly agricultural economy to an industrial one.

Inland Passage   The Inland Passage refers to the protected waterway of the Pacific Ocean lying between the BC mainland and Vancouver Island and Haida Gwaii.

inland terminals   Rail hubs linked to a port by regular rail services.

Inuit   People descended from the Thule, who migrated into Canada's Arctic from Alaska about 1,000 years ago. The Inuit do not fall under the Indian Act, but are identified as an Aboriginal people under the Constitution Act, 1982.

Irish famine   The great famine in Ireland took place between 1845 and 1852 when the principal crop and source of food, the potato, was devastated by blight, causing widespread crop failures. Many Irish immigrated to Atlantic Canada, especially to Saint John, New Brunswick.

isostatic rebound   The gradual uplifting of the earth's crust following the retreat of an ice sheet that, because of its weight, depressed the earth's crust. Also known as postglacial uplift.

James Bay and Northern Québec Agreement   The 1971 announcement of the James Bay Project triggered a series of events that quickly led to a negotiated settlement and, in 1975, an agreement between the Cree and Inuit of northern Québec and the federal and Québec governments. In this modern treaty, Aboriginal title was surrendered by the Inuit and Cree in exchange for specific rights, including self-government and benefits (cash and financial support for the hunting economy).

jihadists   Within Islam, 'jihad' connotes a wide range of meanings, from an inward spiritual struggle to a political struggle to further the Islamic cause. 'Jihad by the sword' refers to a holy war, and those Muslims who pursue acts of terror and guerrilla warfare for a particular political-religious cause sometimes are referred to as 'jihadists'.

just-in-time principle   A system of manufacturing in which component parts are delivered from suppliers at the time required by the manufacturer, so that manufacturers do not bear the cost burden of building and maintaining large inventories; air pollution from heavier vehicular traffic is an ancillary consequence.

Kativik Regional Government   Administrative organization for Inuit in Nunavik. Formed in 1978 after the James Bay and Northern Québec Agreement.

kimberlite pipes   Intrusions of igneous rocks in the earth's crust that take a funnel-like shape. Diamonds are sometimes found in these rocks.

knowledge-based economy   Sector of post-industrial economy based on the use of inventions and scientific knowledge to produce new products and/or services, often in engineering, management, and computer technology fields. Some consider it as part of the information society.

Labrador Current   Cold ocean current flowing south in the North Atlantic from Greenland and Labrador.

Lake Agassiz   Largest glacial lake in North America that covered much of Manitoba, northwestern Ontario, and eastern Saskatchewan.

latitude   An imaginary line parallel to the equator that encircles the globe.

light sweet crude   The most highly valued crude oil, which because of its low level of sulphur has a pleasant smell and, more importantly, requires little processing to become gasoline, kerosene, and diesel fuel.

liquefied natural gas (LNG)   A liquid form of natural gas chilled to −162°C. The cooling process, called liquefaction, reduces the volume to one six-hundredth of its original volume. As a liquid, it can be loaded onto special tankers and transported by sea around the world, making a once regional commodity a global one. At regasification terminals, the LNG is warmed until it returns to a gaseous state.

Lomonosov Ridge   An 1,800-km-long underwater ridge with a height of 3,500 m above the seabed stretching from the New Siberian Islands of Russia over the central part of the Arctic Ocean to Ellesmere Island of Canada. Its width varies from 60 to 200 km. The Lomonosov Ridge was discovered in 1948 by a Soviet scientific expedition.

longitude   An imaginary line that runs through both the North and South poles.

Lower Mainland   A local term describing Vancouver and the surrounding area extending from the North Shore Mountains to the border with the United States and eastward to include the Fraser Valley and the town of Hope.

Loyalists   Colonists who supported the British during the American Revolution. About 40,000 American colonists who were loyal to Britain resettled in Canada, especially in Nova Scotia and Québec.

luvisolic   A soil order identified by a well-drained soil that is often grey-brown in colour; associated with the broadleaf and mixed forest natural vegetation types in the Great Lakes–St Lawrence climatic zone.

Makivik Corporation   A non-profit organization owned by the Inuit of Nunavik and created in 1978 pursuant to the JBNQA. Its central mandate is the protection of the integrity of the JBNQA, and Makivik focuses on the political, social, and economic development of the Nunavik region.

marine air masses   Large homogeneous bodies of air with moisture and temperature characteristics similar to the ocean where they originated. Marine air masses normally are moist and relatively mild in both winter and summer.

megaprojects   Large-scale construction projects, often related to resource extraction, that exceed $1 billion and take more than two years to complete.

metamorphic rocks   Rocks formed from igneous and sedimentary rocks by means of heat and pressure.

Métis   People of mixed biological and cultural heritage, usually either French–Indian or English– or Scottish–Indian. The joining of blood lines between Indians and Europeans took place during the fur trade and continues today. Originally, the term was more narrowly applied to French–Indian people who settled in the Red River area and who developed a distinct hunting economy and society based on the French language and the Roman Catholic religion.

mortality rate   The number of deaths per 1,000 people in a given year; also called crude death rate.

muskeg   A wet, marshy area found in areas of poor drainage, such as the Hudson Bay Lowland. Muskeg contains peat deposits.

nation   A territory that is politically independent; a group of people with similar cultural characteristics and a shared historical experience that make them self-consciously aware of their uniqueness as a group.

National Research Council of Canada   The government of Canada's leading agency for research, development, and technology-based innovation. NRC conducts research as well as provides research funding to universities and research parks.

Native settlements   Small Aboriginal centres, often found on reserves or in remote, northern locations.

net interprovincial migration   Annual estimates of net migration by provinces and territories determined by the number of people arriving and leaving each province and territory as permanent residents; based on Child Tax Benefit data and income tax records.

net migration   The net effect of immigration and emigration on a country's population in a given period.

newsprint   A general term used to describe very thin paper used primarily in the publication of newspapers.

non-status Indians   Those of Amerindian ancestry who are not registered as Indians under the Indian Act.

NORAD   North American Air Defence Command, created in September 1957 by Canada and the US and headquartered in Colorado Springs, Colorado, as a binational command, centralizing operational control of continental air defences against the threat of Soviet bombers. In March 1981, the name was changed to North American Aerospace Defence Command. Since 11 September 2001, NORAD is responsible for protecting North America from domestic as well as from foreign air attacks.

nor'easters   Strong winds off the North Atlantic from the northeast that bring stormy weather.

North American Free Trade Agreement (NAFTA)   North American Free Trade Agreement between Canada, the United States, and Mexico that came into effect in January 1994, forming the world's largest free trade area. The Agreement has increased trade among the three countries and rearranged the location of labour-intensive manufacturing firms to Mexico, where wages are much lower than in either the United States or Canada. The agreement was not extended to include state, provincial, and local purchasing markets because Canadian provinces did not wish to participate.

North American Security Perimeter   An array of strategies intended to protect America from terrorist attacks. As the US continues to search for a balance between security and trade, cross-border commerce and tourism have been hampered. For example, Americans and Canadians must now have a valid passport to cross the Canada–US border.

Northeast Energy Corridor   A proposed electric transmission system from Saint John, New Brunswick, to Maine in the US. The plan includes transmission lines capable of carrying between 1,200 and 1,500 megawatts of electricity from wind generators, as well as a natural gas co-generator that would supply base-load power for the line.

northern frontier   View of Canada's North as a place of resource wealth to be exploited.

northern lights (aurora borealis)   The visible portions of the dissipation of solar energy carried to the earth's magnetosphere by solar winds. The energy is visible as rapidly moving light that appears as white or green or red flashes or 'curtains' of light across the sky.

Northwest Passage   Sea route(s) through the Arctic Ocean connecting the Atlantic and Pacific Oceans that can be traversed only in the summer.

North West Transportation Corridor   Stretching from Prince Rupert across northern BC and into Western Canada, the North West Transportation Corridor centres on the CN rail route and the Yellowhead Highway.

nunatak   An unglaciated area of a mountain that stood above the surrounding ice sheet.

Nunavik   Homeland of the Inuit of northern Québec and a semi-autonomous political region within that province.

Oregon Territory   Territory in the Pacific Northwest stretching from 42° N to 54° 40′ N, the possession of which was disputed between the US and Great Britain (now composed of British Columbia, Washington, and Oregon). The Treaty of 1846 between the United States and Great Britain determined the boundary at the forty-ninth parallel but with Vancouver Island remaining within British North America.

Organization of Petroleum Exporting Countries (OPEC)   Organization founded in 1960 to control world prices and thus obtain a 'fair' price for its members and other petroleum producers. Currently, there are 12 members: Iran, Iraq, Kuwait, Saudi Arabia, and Venezuela are founding members; Algeria, Angola, Ecuador, Libya, Nigeria, Qatar, and United Arab Emirates are 'full members'.

orogeny   Mountain-building, a geologic process that takes place as a result of plate tectonics (movement of huge pieces of the earth's crust). The result is distinctive structural change to the earth's crust where mountains often are formed.

orographic precipitation   Rain or snow created when air is forced up the side of a mountain, thereby cooling the air and causing condensation followed by precipitation.

orographic uplift   Air forced to rise and cool over mountains. If the cooling is sufficient, water vapour condenses into clouds and rain or snow occurs.

outsourcing   Arrangement by a firm to obtain some parts or services from other firms.

Palliser's Triangle   Area of short-grass natural vegetation in southern Alberta and Saskatchewan determined by Captain John Palliser, who led an expedition organized by the British Colonial Office and the Royal Geographical Society to survey the Canadian West in 1857–60, to be a northern extension of the Great American Desert and therefore unsuitable for agricultural settlement.

pass laws   The pass system, never part of the Indian Act, instituted following the 1885 Northwest Rebellion whereby government officials tried to prevent Indians from leaving their reserves. Officials feared that another rebellion could occur if the Indian tribes united. Under the system, Indians were permitted to leave their reserves only if they had a written pass from the local Indian agent.

patriation   The act of bringing legislation, especially a Constitution, under the authority of the autonomous country to which it applies.

patterned ground   The natural arrangement of stones and pebbles in polygonal shapes found in the Arctic where continuous permafrost is subjected to frost-shattering as the principal erosion process.

peneplain   A more or less level land surface caused by the wearing down of ancient mountains; represents an advanced stage of erosion.

periphery   The weakly developed area surrounding an industrial core; also known as a hinterland.

permafrost   Permanently frozen ground.

physiographic region   A large geographic area characterized by a single landform; for example, the Interior Plains.

physiography   A study of landforms, their underlying geology, and the processes that shape these landforms; geomorphology.

Pineapple Express   A strong and persistent flow of warm air associated with heavy rainfall that originates in the waters adjacent to the Hawaiian Islands.

pingos   Hills or mounds that maintain an ice core and that are found in areas of permafrost.

placelessness   The reverse of 'sense of place'. Placeless landscapes have no distinguishing features and could be found in very many places. Examples are strip malls, cookie-cutter theme parks, and service stations.

Pleistocene epoch   A minor division of the Geological Time Chart beginning nearly 2 million years ago. The glaciation of the Pleistocene Epoch was not continuous but consisted of several (likely, four) glacial advances interrupted by interglacial stages during which the ice retreated and a comparatively mild climate prevailed. The last advance, called the Wisconsin, ended about 11,000 years ago.

pluralism   The social acceptance of a diversity of views in preference to a single approach or method of interpretation. Under pluralism, conflicting values may be considered equally important.

pluralistic society   Societies where small groups within the larger society are permitted to maintain their unique cultural identities; multiculturalism.

podzolic   A soil order, often grey in colour, identified by poor drainage; associated with the boreal forest and the coastal rain forest and with climates that have large amounts of precipitation, such as the Pacific, Atlantic, and Subarctic climatic zones.

polynyas   Areas of open water surrounded by sea ice.

population density   The total number of people in a geographic area divided by the land area; population per unit of land area.

population distribution   The dispersal of a population within a geographic area.

population growth   The rate at which a population is increasing or decreasing in a given period due to natural increase and net migration; often expressed as a percentage of the original or base population.

population increase   The total population increase resulting from the interaction of births, deaths, and migration in a population in a given period of time.

population strength   The equating of population size with economic and political power.

Port Royal   The establishment founded in the summer of 1605 on the north shore of the Annapolis Basin near the mouth of the Annapolis River by a French colonizing expedition led by Pierre du Gua de Monts and Samuel de Champlain.

postglacial uplift   The gradual rising of the earth's crust following the retreat of an ice sheet that, because of its weight, depressed the earth's crust. Also known as isostatic rebound.

post-industrial phase   A stage in capitalism marked by the shift from an industrial economy based on manufacturing to an economy in which service industries, particularly high-technology industries, become the dominant economic activities.

potash   A general term for potassium salts. The most important potassium salt is sylvite (potassium chloride). Potassium (K) is a nutrient essential for plant growth.

Poundmaker   An outstanding political leader of the Plains Cree. During the tumultuous years surrounding treaty-making, he played a key role in setting the terms for Treaty No. 6. Later, Poundmaker sought a peaceful solution for the desperate plight of his people. Unable to control his warriors, he took a conciliatory position in the Cree uprising and arranged for a surrender of his people in 1885. Poundmaker was charged with treason and sentenced to three years in prison. He died shortly after his early release, a man broken by the losses his people had endured and by his time in prison.

power of place   Economic, political, and cultural power of a particular place—city, region, country—derived from various geographic attributes related to location, as well as the extent to which this power is used by those who reside there.

pre-industrial phase   A stage in the evolution of the capitalist economic system that predates capitalism, characterized by mercantile trade but without major manufacturing or industrialization based on the large-scale exploitation of energy, such as coal, steam, and hydroelectricity.

primary prices   The prices for commodities such as foodstuffs, raw materials, and other primary products.

primary products   Goods derived from agriculture, fishing, logging, mining, and trapping; products of nature with no or little processing.

primary sector   Economic sector involving the direct extraction/production of natural resources that includes agriculture, fishing, logging, mining, and trapping.

producer services   Services that have enabled firms and regions to maintain their specialized roles in marketing, advertising, administration, finance, and insurance industries. Producer services are one of several parts of the growing service sector of the economy.

Provincial Agricultural Land Commission   An independent British Columbia agency responsible for administering the province's land-use zone in favour of agriculture. This agency also manages the Agricultural Land Reserve (ALR), which extends over 4.7 million hectares.

push-pull model   One of the laws of migration, as devised by Ravenstein, whereby certain negative factors (e.g., lack of employment opportunities or human security) in the migrant's present location push him/her to migrate, just as certain positive factors (e.g., economic opportunities, more amenable climate, or greater human security) in the location of destination pull the migrant to relocate.

Quaternary Period   The geological period consisting of the Pleistocene and Holocene epochs.

Québecers   English translation of the French word 'Québécois'.

Québec Inc.   Originally, a group of powerful companies created in the wake of the Quiet Revolution; with the help of state subsidies that bolstered Québec's economic autonomy the term broadened to refer to all francophone business people.

Québécois   A term that has evolved from referring to French-speaking residents of Québec to meaning all residents of Québec.

Quiet Revolution   A period in Québec during the Liberal government of Jean Lesage (1960–6) characterized by social, economic, and educational reforms and by the rebirth of pride and self-confidence among the French-speaking members of Québec society, which led to a resurgence of francophone ethnic nationalism. During this time, the secular nationalist movement gained strength.

rain shadow effect   A dry area on the lee side of mountains where air masses descend, causing them to become warmer and drier.

rate of natural increase   The surplus (or deficit) of births over deaths in a population per 1,000 people in a given time period.

recent immigrants   Statistics Canada term that refers to landed immigrants who arrived in Canada within five years prior to a given census.

Red River Migration   The Hudson's Bay Company had perhaps as many as 1,000 settlers from Fort Garry travel by horse-drawn wagons to Fort Vancouver in 1841 to shore up the British claim to the Oregon Territory.

region   An area of the earth's surface defined by its distinctive human and/or natural characteristics. Boundaries between regions often are transition zones where the main characteristics of one region merge into those of a neighbouring region. Geographers use the concept of regions to study parts of the world.

regional consciousness   Identification with a place or region, including the strong feeling of belonging to that space and the willingness to advocate for regional interests.

regional core   Within the core/periphery model, cores can occur at different geographic levels. A regional core is an area (often a large city) that dominates trade and stimulates economic growth in the region.

regional geography   The study of the geography of regions and the interplay between physical and human geography, which results in an understanding of human society, its physical geographical underpinnings, and a sense of place.

regional identity   Persons' association with a place or region and their sense of belonging to a collectivity.

regionalism   The division of countries or areas of the earth into different natural/political/cultural parts.

regional self-interest   The aspirations, concerns, and interests of people living in a region and acted on by local politicians. Sometimes such efforts are designed to improve the prospects of their region at the expense of other regions or of the federal government.

regional service centres   Urban places where economic functions are provided to residents living within the surrounding area.

reserve   Under the Indian Act, lands 'held by her Majesty for the use and benefit of the bands for which they were set apart; and subject to this Act and to the terms of any treaty or surrender'.

residual uplift   The final stages of isostatic rebound.

resource frontier   The perception of the Territorial North as a place of great mineral wealth that awaits development by outsiders.

resource town   An urban place where a single economic activity focused on resource extraction (e.g., mining, logging, oil drilling) dominates the local economy; single-industry town. Also, a company town built near an isolated mine site to house the mine workers and their families.

restrained rebound   The first stage of isostatic rebound.

restructuring   Economic adjustments made necessary by fierce competition, whereby companies are driven to reduce costs by reducing the number of workers at their plants.

Robinson treaties   Two 1850 treaties signed between the Crown (Canada West) and the Ojibwa Indians of Lake Superior and the Ojibwa Indians of Lake Huron. Under the terms of both agreements, the Crown secured an area of 52,400 square miles mostly in central and northern Ontario. For the first time, reserves were part of the agreement. The Crown paid the Ojibwa Indians a lump sum as well as an annual payment in perpetuity.

Scottish Highland clearances   Forced displacements of poor tenant farmers in the Scottish Highlands during the eighteenth and nineteenth centuries. Migration ensued and, in 1812, Scottish settlers arrived at Fort Garry to found Lord Selkirk's experimental colony. Most, perhaps 100,000, settled in Nova Scotia. The clearances were part of a process of change of estate land use from small farms to large-scale sheep herding.

sea ice   Ice formed from ocean water. Types of sea ice include: (1) 'fast ice' frozen along coasts and that extends out from land; (2) 'pack ice', which is floating, consolidated sea ice detached from land; (3) the 'ice floe', a floating chunk of sea ice that is less than 10 km (six miles) in diameter; and (4) the 'ice field', a chunk of sea ice more than 10 km (six miles) in diameter.

Sea-to-Sky Highway   Highway that winds through the spectacular Coast Mountains, linking communities from West Vancouver to Whistler. Since rock slides occur frequently, the highway was widened and straightened to reduce the chances of rock slides and to improve safety and reliability for the 2010 Winter Olympic Games.

secondary sector   The sector of the economy involved in processing and transforming raw materials into finished goods; the manufacturing sector of an economy.

sedimentary rocks   Rocks formed from the layered accumulation in sequence of sediment deposited in the bottom of an ocean.

seigneurs   Members of the French elite—high-ranking officials, military officers, the nascent aristocracy—who were awarded land in New France by the French king. A seigneur was an estate owner who had peasants (habitants) to work his land.

sense of place   The special and often intense feelings that people have for the area where they live. Considered a social product, sense of place can be applied to different geographic levels, i.e., local, regional, and even national levels. These feelings are derived from a combination of experiences: some from natural factors such as climate, others from cultural factors such as language. A sense of place is a powerful bond between people and their region.

settlement area The geographic extent of a comprehensive land claim. While less than 25 per cent of this land is allocated to the Aboriginal beneficiaries as a collective (not individual) landholding, the entire area is subject to the environmental and wildlife regulations exercised by the settlement area's co-management boards.

sex ratio The ratio of males to females in a given population; usually expressed as the number of males for every 100 females.

***shariah* law** Islamic religious law based on the Koran, the Muslim holy book; under most interpretations, Islamic law gives men more rights than women in matters of inheritance, divorce, and child custody.

softwood forest The predominant forest in Canada. Softwood forests consist mainly of coniferous trees, characterized by needle-like foliage.

soil orders Classes of soil based on observable soil properties and soil-forming processes. In Canada there are nine soil orders, including chernozemic, cryosolic, and podzolic.

sovereignty-association A concept designed by the Parti Québécois under the Lévesque government and employed in the 1980 referendum. This was based on the vision of Canada as consisting of two 'equal' peoples. Sovereignty-association called for Québec sovereignty within a partnership with Canada based on an economic association.

spawning biomass The total quantity or weight of a species at sexual maturity in a given area that can reproduce. Cod, for example, reach sexual maturity around the age of seven.

specific land claims Claims made by treaty Indians to rectify shortcomings in the original treaty agreement with a band or that seek to redress failure on the part of the federal government to meet the terms of the treaty. Many of these have involved the unilateral alienation of reserve land by the government.

sporadic permafrost Pockets of permanently frozen ground mixed with large areas of unfrozen ground. Sporadic permafrost ranges from a trace of permanently frozen ground to an area having up to 30 per cent of its ground permanently frozen.

staples thesis Harold Innis's idea that the history of Canada, especially its regional economic and institutional development, was linked to the discovery, utilization, and export of particular staple resources in Canada's vast frontier. It was expected that economic diversification would take place, making the region less reliant on primary resources. Innis proposed this thesis in the early 1930s, and his ideas continue to influence Canadian scholars.

staple trap The economic and social consequences on a region and its population following the exhaustion of its resources; the opposite outcome of the positive outcome of economic diversification anticipated in the staples thesis.

status (registered) Indians Aboriginal peoples who are registered as Indians under the Indian Act.

strata Layers of sedimentary rock.

Stryker Slang term for a prospective gang member and the title of a 2004 film by Winnipeg filmmaker Noam Gonick.

subsidence A downward movement of the ground. Subsidence occurs in areas of permafrost when large blocks of ice within the ground melt, causing the material above to sink or collapse.

summer fallow The farming practice of leaving land idle for a year or more to accumulate sufficient soil moisture to produce a crop or to restore soil fertility; summer fallowing is being replaced by continuous cropping.

Sunshine Coast As part of the Georgia Basin, the Sunshine Coast lies in the rain shadow of Vancouver Island. It extends from Gibsons to Powell River and this area is noted for more sunshine and less rainfall than Vancouver.

super cycle theory Theory based on two premises: (1) that demand will tend to outstrip supply and thus keep prices high; and (2) that in a global economic downturn, demand from industrializing countries will keep price declines to a minimum.

Sustainable Development Technology Canada A foundation created by the Canadian government to support the development and demonstration of clean technologies—solutions that address issues of clean air, greenhouse gases, clean water, and clean soil to deliver environmental, economic, and health benefits to Canadians.

sustainable resource use Use of renewable resources (e.g., forests, fish stocks) when the rate of consumption equals (or is less than) the resource's natural rate of replenishment.

tailings Waste or residue from a mining operation, for example, the non-recyclable wastewater left after bitumen is turned into oil, which consists of sand, clay, and a number of chemicals, including naphthenic acids (NAs), polycyclic aromatic hydrocarbons (PAHs), arsenic, and mercury.

terraces Old sea beaches left after the sea has receded; old flood plains created when streams or rivers cut downward to form new and lower flood plains. The old flood plain (now a terrace) is found along the sides of the stream or river.

terra nullius The doctrine according to which European countries claimed legal right to ownership of the land occupied by Indians and Inuit because the land was not cultivated and lacked permanent settlements.

tertiary/quaternary sector The economic sector engaged in services such as retailing, wholesaling, education, and financial and professional services; the quaternary sector, for which at present statistical data are not compiled, involves the collection, processing, and manipulation of information.

till Unsorted glacial deposits.

traditional ecological knowledge Familiarity with and knowledge of the natural surroundings. Such knowledge of the environment and wildlife in a particular area or ecosystem accumulates over centuries among a group of people who live close to the land/sea; also called local ecological knowledge.

tragedy of the commons The destruction of renewable resources that are not privately owned, such as fisheries and forests. Historically, common pasture was available for the livestock of all people within a community, but in the absence of some form of collective control some individuals may maximize the use of the resource for personal gain; such use, in total, overwhelms the capacity of the resource to maintain and regenerate itself.

transmission technology High-voltage transmission lines that reduced power loss and thus made shipping electricity over long distances viable.

treaty Indians Aboriginal peoples who are descendants of Indians who signed a numbered treaty and who benefit from the rights described in the treaty. All treaty Indians are status Indians, but not all status Indians are treaty Indians.

Tyrrell Sea Prehistoric Hudson Bay as the Laurentide Ice Sheet receded. Its extent was considerably greater than that of present-day Hudson Bay because the land had been depressed by the weight of the ice sheet.

undercounting Population counts in the census enumeration process that have 'missed' households, producing for that particular population a figure that is unrealistically low. When challenged by local governments, Statistics Canada will conduct a reassessment and may readjust the population count. Local governments seek the highest possible population count because their federally derived revenues are often calculated on a per capita basis.

underemployment Several definitions may be used; a classical one refers to workers who are employed, but not in the desired capacity, whether in terms of compensation, hours, or level of education, skill, and experience. In this text, 'underemployment' refers to persons in small communities where the very few jobs are already filled and, because these potential workers are aware that no job opportunities exist, they do not seek jobs elsewhere.

unemployment Lack of paid work, but this term and statistics based on it measure only those who are seeking paid employment.

upgrader A processing plant that breaks large hydrocarbon molecules (such as bitumen) into smaller ones by increasing the hydrogen to carbon ratio. The product is supplied to refineries, which will process it into gasoline, jet fuel, diesel, propane, and butane.

urban areas Communities with economic and social functions that differentiate them from rural places; the common practice of defining urban population is by a specified size that assumes the presence of urban economic and social functions. Statistics Canada considers all places with a combination of a population of 1,000 or more and a population density of at least 400 per km^2 to be urban areas. People living in urban areas make up the urban population. People living outside of urban areas are considered rural residents and, by definition, constitute the rural population.

value added The difference between a firm's sales revenues and the cost of its materials.

value-added production Manufacturing that increases the value of primary (staple) goods.

visible minority A term used in Canada to describe non-Caucasian persons. Statistics Canada's definition of this group is based on the Employment Equity Act, which defines visible minorities as 'persons, other than Aboriginal peoples, who are non-Caucasian in race or non-white in colour'.

weathering The decomposition of rock and particles in situ.

western alienation Feeling on the part of those in Western Canada and BC—derived from past government actions and a natural periphery response to the core—that they have little influence on federal policy and that Central Canada controls the government in Ottawa.

Western Sedimentary Basin Within the geological structure of the Interior Plains, normally flat sedimentary strata that are bent into a basin-like shape. These basins often contain petroleum deposits.

winter roads Temporary ice roads over muskeg, lakes, and rivers built during the winter to provide ground transportation for freight and travel to remote communities.

zebra mussel A small freshwater mollusc (*Dreissena polymorpha*). In 1988 the zebra mussel gained a foothold in the Great Lakes, likely from having attached itself to ships coming from Europe. It has since spread and only the higher salinity in the lower St Lawerence River has halted its expansion downstream.

WEBSITES

Chapter 1

atlas.nrcan.gc.ca/site/english/maps/reference/
provincesterritories
The *Atlas of Canada* has prepared a set of provincial, territorial, and regional maps, which appear in Chapters 5–10 of this book.

geography.about.com/od/politicalgeography/a/
coreperiphery.htm
A summary of the core/periphery theory is presented here.

www.allacademic.com//meta/p_mla_apa_research_citation/
0/7/3/8/7/pages73870/p73870-1.php
The North American Security Perimeter and its implications for North American community is discussed here.

Chapter 2

www.ec.gc.ca/cc/default.asp?Lang=En
Environment Canada's website for climate change.

atlas.nrcan.gc.ca
The *Atlas of Canada* site contains a variety of maps describing the physical nature of Canada.

nrcan-rncan.gc.ca
Natural Resources Canada provides information on many resource topics.

www.nasa.gov/topics/earth/features/arctic_thinice.html
NASA's website has satellite images of the Arctic Ocean, including late summer when the Northwest Passage is ice-free.

Chapter 3

www.american-indians.net
North American Aboriginal Culture Regions are found at this site.

atlas.nrcan.gc.ca/site/english/maps/historical/
aboriginalpeoples/circa1823/1
The Hudson's Bay Company undertook a census of Aboriginal peoples in 1822. The *Atlas of Canada* has produced a map and accompanying text describing this very detailed census.

atlas.nrcan.gc.ca/site/english/maps/historical/
territorialevolution
atlas.nrcan.gc.ca/sitefrancais/english/maps/historical/
territorialevolution/territorial_animation.gif/image_view
A set of historic maps that illustrate the territorial evolution of Canada and an animated version of Canada's territorial development are found here.

faculty.marianopolis.edu/c.belanger/quebechistory/events/
quiet.htm
An account of the Quiet Revolution in Québec.

archives.cbc.ca/politics/federal_politics/topics/1891/
CBC Digital Archives provides a close look at 'separation anxiety' in Canada associated with the 1995 Referendum.

www.mhs.mb.ca/docs/pageant/13/selkirksettlement1.shtml
The Manitoba Historical Society has documented Lord Selkirk's land grant of 1811.

wiki.xiaoyaozi.com/en/Haldimand_Proclamation.htm
The Haldimand Proclamation.

Chapter 4

www.bank-banque-canada.ca/en/mpr/pdf/2010/
mprsumjan10.pdf
Bank of Canada Monetary Policy Report Summary, January 2010.

www.statcan.gc.ca/daily-quotidien/090929/
dq090929b-eng.htm
Statistics Canada's 2009 population estimates.

www12.statcan.ca/census-recensement/2006/as-sa/97-558/
index-eng.cfm?CFID=2953573&CFTOKEN=43245102
Statistics Canada provides a detailed account of the demography and social characteristics of Aboriginal peoples from the 2006 census.

www.statcan.gc.ca/pub/11-008-x/2009001/article/
10864-eng.htm
Linda Gionet's article, 'First Nations People: Selected Findings of the 2006 Census', *Canadian Social Trends*, provides another perspective.

www.statcan.gc.ca/pub/11-621-m/11-621-m2009077-eng.htm
Statistics Canada's account of manufacturing in 2008.

Chapter 5

www.statcan.gc.ca/pub/75-001-x/2009102/pdf/
10788-eng.pdf
Statistics Canada article (2009) on the factors behind the decline of manufacturing in Ontario, Canada, and other industrial countries.

www.ic.gc.ca/eic/site/auto-auto.nsf/eng/h_am01302.html
Industry Canada's website for statistics, analysis, and industrial profiles of the automobile industry.

www.scotiacapital.com/English/bns_econ/bns_auto.pdf
An analysis of North American automobile industry production and sales.

www.greenenergyact.ca/
Ontario's Green Energy Act.

www.ainc-inac.gc.ca/ai/mr/is/eac-eng.asp
A chronology of the Six Nations land dispute is maintained by Indian and Northern Affairs Canada.

www.sdtc.ca/en/about/index.htm
Sustainable Development Technology Canada provides start-up funding for 'clean' technologies.

Chapter 6

www.greatlakes-seaway.com/en/
Official site of the Great Lakes–St Lawrence Seaway system.

www.canadiangeographic.ca
History of the St Lawrence Seaway plus archival photographs and a tour of the seaway.

www.qc.ec.gc.ca/csl/pub/pub004_e.html
A scientific account of the environmental impact of the zebra mussel in the St Lawrence River by the St Lawrence Centre of Environment Canada.

www.mrnf.gouv.qc.ca/mines/strategie/index.jsp
Details of the Québec government's mineral strategy are presented at this site.

www.mrnf.gouv.qc.ca/presse/communiques-detail.jsp?id=7699
Québec's wind-generated electric program is described at this site.

www.mrnf.gouv.qc.ca/energie/hydroelectricite/developpement.jsp
Québec government announcement of its hydroelectric projects to 2010.

Chapter 7

earthquakescanada.nrcan.gc.ca/histor/20th-eme/1949-eng.php
Details on the 1949 earthquake off Haida Gwaii.

www.th.gov.bc.ca/PacificGateway/documents/PGS_Action_Plan_043006.pdf
Pacific Gateway Strategy Action Plan.

forum.skyscraperpage.com/showthread.php?t=151987&page=2
Maps and diagrams of highway construction projects designed to create a super east–west highway corridor.

www.llbc.leg.bc.ca/public/PubDocs/bcdocs/322107/north_east_coal_facts.pdf
In 1982, the BC government described its high hopes for the Northeast Coal Project.

mmsd.mms.nrcan.gc.ca/stat-stat/prod-prod/PDF/ib2010_e.pdf
Natural Resources Canada report on its estimates for Canadian mineral production in 2009, with several tables.

Chapter 8

www.oilsands.alberta.ca/519.cfm
Alberta's oil sands: facts, statistics, and video produced by the Alberta government.

www.treasuryboard.alberta.ca/ResponsibleActions.cfm
Alberta government report: *Responsible Actions: A Plan for Alberta's Oil Sands.*

www.cbc.ca/edmonton/features/dirtyoil/
CBC Edmonton: 'Alberta Oil Sands: Black Gold or Black Eye?'

pubs.pembina.org/reports/climate-leadership-report-en.pdf
Pembina Institute and David Suzuki Foundation report on greenhouse-gas emissions and carbon tax.

www.defenseindustrydaily.com/Boeings-Skyhook-Shot-Redefining-the-Aerial-Heavy-Lifting-Market-04970/
Information on Skyhook.

Chapter 9

www.parl.gc.ca/information/library/PRBpubs/prb0820-e.pdf
Canada's new equalization formula.

www.eleanorbeaton.com/userfiles/file/Chatelaine%20-%20Oil%20Patch%20Widows.pdf
'No Man's Land'—story of Newfoundland and commuting.

www.geography.ryerson.ca/jmaurer/702art/702Clapp1999.pdf
The resource cycle in forestry and fishing.

nbwoodlotowners.ca///uploads//Website_Assets/APEC%27s_Atlantic_Lumber.pdf
Atlantic Canada's forest industry.

www.offshore-technology.com/projects/hibernia/
Hibernia, Jeanne d'Arc Basin, and its technology.

www.sysco.ns.ca/history.htm
The history of the steel plant at Cape Breton.

Chapter 10

apecs.arcticportal.org/index.php?option=com_
jreviews&Itemid=131
The Association of Polar Early Career Scientists (APECS) has launched a discussion forum called 'Polar Literature Discussion Webpage' where recent literature is cited.

arctic-council.org/article/about
Formation and mandate of the Arctic Council, as well as Canada's role in the Arctic Council.

www.scientificamerican.com/article.cfm?id=drawing-lines-
in-the-sea&page=2
In this issue of *Scientific American*, circumpolar countries are described as drawing lines in the Arctic Ocean to define their territorial limits. These lines represent international accepted boundaries but also those that are in dispute.

www.uphere.ca/node/175
In December 2007, the northern magazine, *Up Here*, featured an article by Jack Danylchuk on the unwanted legacy of arsenic.

www.stats.gov.nt.ca
The Statistical Bureau of the Northwest Territories provides not only a wide variety of statistical data but also links to the Yukon and Nunavut.

www.ipycanada.ca
Canada's website for International Polar Year scientific activities.

Chapter 11

www.canadiangeographic.ca/blog/posting.asp?ID=30
What's at stake for Canada in the Arctic?

www.nationalpost.com/m/story.html?id=1100168
Good public policy? Ottawa and Ontario dole out $4 billion to Chrysler and General Motors.

www.capp.ca/GetDoc.aspx?DocID=141879
Oil sands economic impacts across Canada.

www.scotiacapital.com/English/bns_econ/bnscomod.pdf
Scotiabank Commodity Price Index, 21 Dec. 2009.

NOTES

Chapter 1

1. In the late nineteenth century, geographers believed that the physical environment determined human affairs. That position was rejected, although geographers recognize that the environment does exert a strong influence on the nature of human activities in various regions of the world. Students can find a more complete discussion of environmental determinism and other philosophical options in geography, including possibilism, positivism, humanism, and Marxism, in William Norton, *Human Geography*, 7th edn (Toronto: Oxford University Press, 2010). The writings of most geographers reflect either one of these philosophical positions or some combination. The core/periphery model, for example, sprang from Marxist scholarship. Because of the model's powerful spatial implications, other scholars holding different philosophical positions have modified this theory by removing its economically deterministic character. Instead, they accept that external forces (such as global institutions like the WTO) and internal forces (such as federal–provincial agreements like equalization payments) can modify the impact of the physical environment on regional development. Hinterlands, therefore, are not locked into a single outcome because of their physical geography.

2. While Canadians and Americans occupy the same continent, historic events and geographic differences laid the foundation for the emergence of two different societies within one continent. These differences are found in many aspects of the two societies. Each country has a different approach to gun control legislation, multiculturalism, and a national health-care system. As a result of these and other differences, some scholars believe that Canadians are more trusting of their governments and more tolerant of social diversity than are Americans (Hartz, 1995; Lipset, 1990; Lemon, 1996; Saul, 1997).

3. With the federal government's recognition of Aboriginal title to ancestral lands in 1973 following the judgement by the Supreme Court of Canada in *Calder v. Attorney General*, comprehensive land-claim settlements have followed this watershed decision, and these agreements have transformed the cultural, economic, and political landscape of the Territorial North and the northern areas of many provinces. Already the James Bay Cree, Inuit, and Innu of Québec, the Inuvialuit, Gwich'in, Sahtu Dene, and Dogrib of the Northwest Territories, the First Nations in Yukon, the Inuit of Nunavut, the Nisga'a of BC, and the Inuit of Labrador have obtained land-claim settlements and received land and cash in exchange for surrendering their Aboriginal rights to vast areas of land. In 1975, the James Bay Cree and Inuit achieved a regional form of self-government in northern Québec. Following 1995, self-government became a reality in land-claim negotiations and those groups who settled before 1995 can return to the negotiation table on the issue of self-government. While the North remains a resource hinterland, land-claim settlements ensure that those Aboriginal peoples will have more control over resource development and will therefore be better able to protect the wildlife and their hunting lifestyle.

4. A group of Muslim women visited the town council to present a fuller picture of Islam and its customs. 'There was a real exchange', Najat Boughaba told Radio-Canada. 'There were people who reached out to us. I really think our visit to Hérouxville benefited both sides' (CBC News, 2007). Najat Boughaba was struck by how little the townspeople knew about Islam and its customs. Following this exchange, the town council in Hérouxville amended its immigrant code of conduct to remove references to 'no stoning of women in public' and 'no female circumcision'. On 8 February 2007, Québec Premier Jean Charest announced the establishment of the Bouchard-Taylor Commission, headed by sociologist Gérard Bouchard and philosopher Charles Taylor, formally known as the Consultation Commission on Accommodation Practices Related to Cultural Differences, in response to public discontent concerning reasonable accommodation.

5. Trade disputes remain troublesome. In April 2009, for example, the United States rekindled the softwood lumber dispute by imposing a 10 per cent tariff on lumber from four provinces that, according to Washington, exported more to the US than the 2006 lumber agreement specified. The Free Trade Agreement, which was replaced by the North American Free Trade Agreement, really means freer trade rather than free trade. As we have seen, three trade agreements—the Auto Pact (1965; nullified in 2001), the FTA (1989), and NAFTA (1994)—led to a realignment of the Canadian economy so that it was more thoroughly integrated with the North American economy. These trade agreements saw more and more manufactured goods exported to the United States, thus breaking the old pattern of exporting primarily low value unprocessed or semiprocessed resource products, and the volume of exports to the United States grew dramatically. In spite of these trade agreements, however, Washington is prepared to defend US business interests by imposing trade barriers, as has been the case with duties on softwood lumber and grain, to restrict the natural flow of certain Canadian goods into

the United States. The purpose of these duties is twofold: (1) to protect US farmers and forest companies in the short run by imposing duties on targeted Canadian exports, and (2) to force Canada to accept a long-term agreement that will limit its grain and lumber exports.

Chapter 2

1. Canadians have various visions of themselves, their region, and their country. For the most part, these visions are rooted in the physical nature and historical experiences that have affected Canada and its regions. For example, people see Canada as a northern country because of its location in North America and because of its climates, which are often noted for long, cold winters. Professor Louis-Edmond Hamelin's concept of nordicity exemplifies the impact of 'northernness' from a geographer's perspective. Hamelin (1979) provides a measure of 'northernness' according to five geographic zones—Extreme North, Far North, Middle North, Near North, and Ecumene (southern Canada). Songwriters, too, have been intrigued by Canada's northern nature, and well-known writers, from Jack London and Robert Service to Margaret Atwood, Pierre Berton, and Farley Mowat, have written about the North and, in doing so, have etched out another parameter of Canadian identity.

2. Tectonic forces press, push, and drag portions of the earth's crust in a slow but steady movement. This process is the basis for the theory of continental drift. In 1912, Alfred Wegener suggested that long ago all the earth's continents formed one huge land mass. Tectonic action caused the breakup of this land mass into huge slabs. These slabs of the earth's crust drifted slowly on the molten mass (magma) beneath the earth's crust.

3. The Champlain Sea covered Anticosti Island and the northern tip of the island of Newfoundland. For the purposes of this text, the eastern extent of this physiographic region ends just east of Québec City.

4. The mean annual temperature of a location on the earth's surface is a measure of the energy balance at that point. Solar energy is the source of heat for the earth, and this energy is returned to the atmosphere in a variety of ways. Therefore, a global energy balance exists. However, there are regional energy surpluses and deficits in different parts of the world. For example, the Arctic has an energy deficit, while the tropics have a surplus. These energy differences drive the global atmospheric and oceanic circulation systems. When the mean annual temperature is below zero Celsius, it indicates that an energy deficit exists.

5. French and Slaymaker (1993) discuss the effects of global warming and climate change on Canada's North. They maintain that climate change caused by the greenhouse effect would alter Canada's cold environments more dramatically than the country's other environments. The authors describe possible changes to sea ice, permafrost, snow cover, sea level, and natural vegetation. There are four key factors for such a remarkable climatic change in northern Canada:

- The percentage of carbon dioxide in the atmosphere over the Territorial North is much greater than that over southern Canada.

- The thinning of the ozone layer over the Territorial North permits more solar radiation to enter the atmosphere.

- As permafrost melts, the release of methane gases from the muskeg in northern Canada will add more greenhouse gases to the northern atmosphere.

- The reduction in the duration of snow cover will expose the northern lands to solar radiation for a longer time.

Chapter 3

1. Under the terms of the British North America Act, the Dominion of Canada was composed of four provinces (Ontario, Québec, New Brunswick, and Nova Scotia). Modelled after the British parliamentary and monarchical system of government, the newly formed country had a Parliament made up of three elements: the head of government (a governor general who represented the monarch), an upper house (the Senate), and a lower house (the House of Commons). This Act was modified several times to accommodate Canada's evolving political needs and its gradual movement to independent nationhood. The patriation of Canada's Constitution in 1982 removed the last vestige of Canada's political dependence on the United Kingdom, although Canada still recognizes the British monarch as its symbolic head.

The British North America Act was based on the highly centralized government of the United Kingdom in the 1860s. However, this Act assigned specific powers to the provinces in order to satisfy Québec's demand for control over its culture. The Canadian political system that emerged, therefore, allowed for regionalized politics. For example, political parties in the House of Commons sometimes serve regional interests. In the 1920s, the Progressive Party represented the concerns of farmers in Western Canada, while the pro-independence Bloc Québécois, which was formed in 1990, not only serves the interests of Québec but is also active in the separatist movement. Furthermore, while the House of Commons is based on the principle of representation by population, Senate membership is based on the principle of equal regional representation. However, because senators are appointed by the Prime Minister and not elected by the people in the different regions of the country, the Senate fails to provide a regional counterweight to the House of Commons.

2. A group of Irish Americans, known as Fenians, was struggling for Irish independence. They believed that attacking British possessions in North America would advance the cause of a free Ireland. Between 1866 and 1870, the Fenians launched several raids across the border into Canada. The United States did not encourage these raids and eventually forced the Fenians to disband. By the end of the American Civil War, Anglo-American relations again were strained because of Britain's tacit support for the Confederacy in the American Civil War. For that reason, the United States withdrew from the Reciprocity Treaty in 1866. This treaty, a free trade agreement between British North America and the United States, began in 1854; the subsequent years were prosperous ones for British North America. Its loss forced the Province of Canada to seek an alternative economic union with the other British colonies in North America.

3. Representation in the House of Commons is readjusted after each decennial (10-year) census in accordance with the Constitution Act, 1867 (formerly the BNA Act) and the Electoral Boundaries Readjustment Act (1985, as amended). On 13 June 1992, following the release of the population figures from the 1991 census, the Chief Electoral Officer of Canada published in the *Canada Gazette* the results of the calculations required by the Constitution Act, which meant an increase in the number of seats in the House of Commons from 295 to 301. In 2004, a readjustment took place when seven new ridings were created for a total of 308 members of the House of Commons. The formula for determining the number of seats for each province and territory is available at: http://www.elections.ca/scripts/fedrep/federal_e/repform_e.htm. In legislation introduced in 2010—on April Fool's Day!—the federal government announced that Ontario is to receive 18 additional seats in the House of Commons, while British Columbia and Alberta will receive an additional seven and five, respectively. Representation from other provinces would remain the same, so that the House of Commons, possibly following the first election after the 2011 census, would have 338 members (Delacourt, 2010).

4. The Manitoba Act of 1870 recognized the legal status of farms and other lands occupied by the Métis as 'fee simple' private property. As well, the Act provided that 1.4 million acres (566,580 ha) be reserved for the children of Métis. The land was allocated to these Métis in 240-acre (97-ha) parcels, plus 160 acres (65 ha) in 'scrip' for each adult head of a family. These lands were distributed after 1875, but much of the 'scrip' land was sold and then occupied by non-Métis. For more on this subject, see Tough (1996: ch. 6).

5. Pontiac, the Odawa chief in the Ohio Valley, led a successful uprising against the British in 1763. By capturing the forts in the Ohio Territory, he exposed Britain's precarious hold on this region, which the British had just obtained from the French. However, Pontiac and his followers could not hold these forts against the British because they could no longer obtain ammunition and muskets from the French. Pontiac concluded that his best move would be to make peace with Britain. The British came to the same conclusion, though for other reasons. Without the help of Pontiac and the other chiefs in this region, Britain would lose control over these lands. Britain therefore had to form an alliance with them. With that objective in mind, George III announced an important concession to these Indians in the Royal Proclamation of 1763, namely, that the King recognized them as valued allies and that the land they used to hunt and trap was 'Indian land' within the British Empire.

6. In 1215, King John of England was forced to sign the Magna Carta. In this charter, he promised to stop interfering with the Church and the law and to consult regularly with the country's leaders before collecting new taxes.

7. While the French language was not recognized by the Québec Act, 1774, the Governor made use of the French language in conducting his business with local officials. For example, the judges appointed by the Governor had to know both languages in order to facilitate the business of the court. In short, while English was the official language of British North America, the British colony of Québec functioned in both the French and English languages.

8. Louis Riel was the Métis political and spiritual leader in the late nineteenth century. This controversial figure is considered both a Father of Confederation and a traitor to the country. Riel, who was born in the Red River Colony in 1844, studied for the priesthood at the Collège de Montréal. The founder of Manitoba and the central figure in both the Red River Rebellion (1869–70) and the North-West Rebellion (1885), he was captured shortly after the Battle of Batoche, where the Métis forces were defeated. After a trial in Regina, the jury found Riel guilty of treason but recommended clemency. Appeals were made to Manitoba's Court of Queen's Bench and to the Judicial Committee of the Privy Council. Both appeals were dismissed. A final appeal went to the federal cabinet, but the government of John A. Macdonald wanted Riel executed. Riel was hanged in Regina on 16 November 1885. His body was interred in the cemetery at the Cathedral of St Boniface in Manitoba.

Riel's execution has had a lasting effect on Canada. In Québec, French Canadians felt betrayed by the Conservative government and federalism. Riel's execution was proof for Canada's French-speaking population that they could not count on the federal government to look after French-Canadian interests. It was also a blow against a francophone presence in the West. In Ontario the hanging of Louis Riel satisfied the anti-Catholic and anti-French majority. For the Orange Order (the Protestant fraternal society that blamed Riel for the death of one of their members, Thomas Scott, who was executed by a Métis firing squad during the Red River Rebellion), Riel's execution was long overdue. In the West, Riel's hanging resulted in the marginalization of both the Métis and Indian tribes, especially those who participated in the uprising. Today, historians and many Canadians accept the view that these Rebellions were, from the point of view of the Metis, resistances to a threat to their way of life.

9. A recent example is the political fallout from the Québec referendum. Prime Minister Chrétien sought to fulfill his verbal promises made in the closing days before the vote on the 1995 referendum. In a House of Commons resolution, the federal government proposed three concessions to Québec: (1) a veto over constitutional changes; (2) recognition of Québec's distinct society status; and (3) devolution of federal powers to Québec. In the case of the veto, Ottawa was prepared to 'lend' its constitutional veto to Québec, Ontario, Atlantic Canada, and the four western provinces. Not only was the federal government committing itself to seeking permission from these four regions before putting its stamp of approval on any constitutional change, it was also recognizing that Canada consisted of four major regions. The premiers of Alberta and British Columbia reacted negatively to that concept of regionalism. British Columbians in particular saw this arrangement as another example of Ottawa's failure to recognize the west coast as a 'distinct and powerful' part of Canada. The federal government retreated from this issue and quickly amended its resolution to extend the veto to British Columbia. In December 1995, this resolution passed in both the House of Commons and the Senate. It then became the law of the land that Canada consists of five major regions!

10. In 1974, the Québec Liberal government passed Bill 22 (Loi sur la langue officielle), which made French the language of government and the workplace. English was no longer an official language in Québec. In 1977, the Parti Québécois government introduced a much stronger language measure in the form of Bill 101 (Charte de la langue française). This legislation eliminated English as one of the official languages of Québec and required the children of all newcomers to Québec to be educated in French. Four years later, Bill 178 required all commercial signs to use only French. The French language has made modest gains outside of Québec. In 1969, New Brunswick passed an Official Languages Act, which gave equal status, rights, and privileges to English and French, and the federal Parliament passed the Official Languages Act, which declared the equal status of English and French in Parliament and in the Canadian public service.

11. Separatists argue that Québec does not have enough powers, that is, Québec is subordinate to Ottawa. For separatists, the solution lies in independence, whether achieved through Parizeau's 'chicken-plucking' strategy or Bouchard's 'winning' referendum strategy. A federalist counter-argument is that, as a member of a federation, Québec has many 'exclusive' powers, such as power over language and education. However, circumstances may force Ottawa to make a decision that adversely affects some provinces while favouring others. In 1982, the patriation of the British North America Act, renamed the Constitution Act, 1867, was such a decision. The Constitution Act, 1982, which was entrenched at the same time, added to the British North America Act in several ways, but without a doubt the most important addition has been the Charter of Rights and Freedoms. These rights and freedoms strengthen the rights of individuals and weaken collective rights. Prime Minister Trudeau, who conceived of society as an agglomeration of individuals (not collectivities), whose rights accrued to them as individuals, saw the Charter as protecting individuals from governments that try to suppress individual rights. Then, too, there is the Supreme Court of Canada's changed role, which has become more proactive with the adjudication of Charter cases.

Chapter 4

1. The American railway industry has been going through a major restructuring process since the 1970s. Canada joined this process as the flow of goods between Canada and the United States increased in the 1990s. This pressure came from the increased trade stemming from the Canada–US Free Trade Agreement in 1989 and later with the North American Free Trade Agreement in 1994. While the vast majority of the trade is between Canada and the US, trade with Mexico is anticipated to increase in the twenty-first century. After CN purchased Illinois Central it then merged with the Burlington Northern Santa Fe Corporation in 1999. The new CN offers its customers shorter transit times and access to key ports in North America. Canadian exports to the US make up most of CN's traffic. These exports include petroleum and chemicals, forest products, automobiles and automobile parts, grain, coal, minerals, and fertilizer.

   Canadian National Railway had its origins in the amalgamation of five financially failing railways: from 1917 to 1923, the Grand Trunk, the Grand Trunk Pacific, the Intercolonial, the Canadian Northern, and the National Transcontinental were combined to form the publicly operated Canadian National Railway. In 1993, the Canadian government privatized this Crown corporation.

2. Race, unlike ethnicity, is based on physical characteristics. Racial types are frequently assigned a set of social characteristics, which is known as 'stereotyping'. Sociologists define 'race' as the socially constructed classification of persons into categories on the basis of real or imagined physical characteristics such as skin colour. Others consider 'race' a means of creating major divisions of humankind on the basis of distinct physical characteristics.

Chapter 5

1. As a general rule, wages in the primary and secondary sectors are often higher than the national average, while wages in certain tertiary jobs are below the national average. In fact, many jobs in the service sector are associated with low hourly wages and part-time employment. These are often disparagingly referred to as 'hamburger-flipping' jobs or McJobs.

2. At the beginning of the twentieth century, the Ontario government sought to develop the northeastern section of the province, especially agricultural lands in the Clay Belt, forest stands, and mineral deposits in the Canadian Shield. Between 1903 and 1909, the government-financed Temiskaming and Northern Ontario Railway was built from the Canadian Pacific Railway line at North Bay northward to the small town of Cochrane on the National Transcontinental Railway (now the CN). Over the next decade, branch lines were extended to Cobalt, Timmins, and Iroquois Falls. Overall, the Ontario government was pleased with the railway's impact on resource development. Plans were made to build the railway farther north—into the Hudson Bay Lowland. By 1932, the Temiskaming and Northern Ontario Railway stretched from Cochrane to Moosonee at the southern tip of James Bay, but further developments did not occur in the resource-scarce Hudson Bay Lowland.

3. In 1970, mercury was discovered in the fish near the Grassy Narrows Reserve, which is about 500 km downstream from the pulp mill. Levels of methyl mercury in the aquatic food chain were 10 to 50 times higher than those in the surrounding waterways (Shkilnyk, 1985: 189). These levels were similar to those found in the fish of Minamata Bay, Japan. Over 100 residents of this Japanese village died from mercury poisoning in the 1960s, and over 1,000 people suffered irreversible neurological damage. Because they depend on fish and game, the Ojibwa at the Grassy Narrows Reserve ate fish on a daily basis and many complained of mercury-related illnesses. While, unlike the Minamata incident, no one died, the economic and social impact on the Ojibwa was nevertheless an industrial tragedy of immense proportions.

4. Toronto has had various forms of regional government. About 40 years ago, the Ontario government established a form of regional government by combining the City of Toronto with the surrounding centres of Etobicoke, North York, Scarborough, York, and East York into the municipality of Metropolitan Toronto (Metro Toronto). These six jurisdictions became one large urban area for regional planning purposes, but each retained its city government. As Metro Toronto continued to grow, it spread into adjacent jurisdictions. In 1988, the Ontario government created the Greater Toronto Area (GTA). Often called the Metro Toronto Region, it consists of Metro Toronto and the regional municipalities of Halton, Peel, York, and Durham. In 1997, the provincial government passed a bill to change the municipal government structure to ensure more efficient, cost-effective services. The legislation came into effect in 1998, amalgamating six former municipalities (Etobicoke, York, Toronto, East York, North York, and Scarborough) and the regional municipalities into the new City of Toronto.

5. Toronto might solve its traffic problem by imposing a tax on automobiles entering the city's downtown. The first world-class city to experiment with a road toll was London, England. In 2002, London began charging drivers of automobiles £5 a day (about $12) to enter and leave the centre of London between 7 a.m. and 6:30 p.m. on weekdays, and this has resulted in a sharp decline in the volume of traffic, a great improvement in urban mobility, and a modest reduction in air pollution.

Chapter 6

1. By responsible government, Lord Durham meant a 'political system in which the Executive is directly and immediately responsible to the Legislature, in which the ministers are members of the Legislature, chosen from the party which includes the majority of the elected representatives of the people' (Lucas, 1912: I, 138).

2. In 1912, Québec gained northern territories inhabited by the Inuit and Cree. Ottawa ceded these lands to Québec with the understanding that the Québec government would be responsible for settling land claims with the Aboriginal peoples in these territories. At the time of the 1995 referendum, the Cree in northern Québec, in response to the separatist claim to territorial independence, declared that they have the right to secede from Québec. They argued that if Québec has the right to secede from Canada, then the Cree have the right to secede from Québec. From a geopolitical perspective, the partitioning of Canada or Québec makes sense only to those supporting ethnic nationalism.

3. For the most recent federal election the House of Commons had 308 seats. Québec kept its 75 ridings because of a law that guarantees each province no fewer seats than it had in 1976. Reapportionment takes place every 10 years based on population figures from the census. The last reapportionment, based on the 2001 census, added eight new seats to the Commons for Ontario, BC, and Alberta. For more on this, see above, Chapter 3, note 3.

4. Camille Laurin, the father of Bill 101, declared that French was the province's only official language. Bill 101 required the children of immigrants to go to French schools and made the presence of French compulsory in the workplace and on commercial signs.

5. The case of Churchill Falls is an interesting one. The divide between the Atlantic Ocean and Hudson Bay marks the Labrador–Québec boundary. For historical reasons, the Québec government does not formally recognize this boundary, but it does treat the area as part of Newfoundland. Newfoundland owns the large Churchill Falls hydroelectric project, but virtually all this power is purchased by Hydro-Québec. Hydro-Québec then transmits it across Québec to markets in the Great Lakes–St Lawrence Lowlands and the United States. In 2009, for the first time, Newfoundland and Labrador is selling its small share of electricity directly to the United States through Hydro-Québec transmission lines. Under a five-year arrangement, Hydro-Québec will receive $20 million a year to transport approximately 130 megawatts of electricity from the Churchill Falls power site.

6. While the same low electricity rates were provided to aluminum producers in Québec, US aluminum firms did not challenge the rates because they own facilities in both countries. The 13 companies with risk-sharing contracts are: Norsk Hydro Canada Inc., Aluminerie Alouette Inc., Quebec-Cartier Mining Co., Cafco Industries Ltd, Timminco Ltd, QIT-Fer et Titane Inc., PPB Canada Inc., Reynolds Metals Co., Argonal, Hydrogenal, SKW Canada Inc., ABI Inc., and Aluminerie Lauralco Inc.

Chapter 7

1. In December 2009 the British Columbia government, as part of a reconciliation agreement with the Haida, officially renamed the Queen Charlotte Islands as Haida Gwaii, which, in Haida, means 'islands of the people'. Since the 1980s the archipelago of more than 150 islands had commonly been referred to both as Haida Gwaii and as the Queen Charlottes (CBC News, 2009).

2. Besides the issue of luring British Columbia into Confederation, there were other political reasons for constructing a transcontinental railway. First, there was the urgent need to exert political control over the newly acquired but sparsely settled lands in Western Canada. As in British Columbia, the perceived threat to this territory was from the United States. Second, there was the need to create a larger market for manufactured goods produced by the firms in southern Ontario and Québec.

3. The exact number of people in British Columbia in 1871 is not known. How many Aboriginal peoples is a guess because their numbers declined sharply as they came into contact with European diseases. Similarly, the number of Americans and people from other countries who remained in the country after the gold rush is unknown. Certainly most moved on to the next gold rush, but some stayed. The gold rush of 1858 may have attracted about 25,000 Americans who sailed from San Francisco to New Westminster at the mouth of the Fraser River.

4. The provincial government created the Technical University of British Columbia in 1997 and this virtual university began to provide on-line courses in 2000. The objective is to prepare students for the high-tech industry. The university offers certificate programs in electronic commerce and software development. High-technology companies are encouraged to locate offices and research laboratories near its Surrey campus. Students undertake internships and co-operative work sessions with high-technology firms. Its theoretical basis lies in the concept of high-tech clusters around a university. In 2002, the BC government placed the Technical University under the aegis of Simon Fraser University and it was renamed SFU Surrey.

5. In 1950, Alcan secured the rights to the water in the Nechako River system until 1999. Further hydroelectric development would extend this right to the Nechako waters. Hence, Alcan commenced construction of the Kemano Completion Project. When the British Columbia government signed this long-term agreement, it saw the industrial development as the key to opening up the province's northwest. At that time, Victoria imagined that Kitimat would become an industrial complex with a population quickly reaching 20,000.

Chapter 8

1. Two novels illustrate the powerful impact of settlement on the Indians and Métis. Rudy Wiebe's *The Temptations of Big Bear* (1973) focuses on the Cree; Guy Vanderhaegh's *The Englishman's Boy* (1996) looks at the Cypress Hills Massacre.

2. The Cree attacked the outpost at Frog Lake, laid siege to Fort Battleford, and defeated the North West Mounted Police at Fort Pitt and Cut Knife Hill. With the arrival of the Canadian militia from eastern Canada, the Cree were defeated.

3. It was only in 1960 that the federal government extended the rights and privileges accorded to Canadian citizens to its Aboriginal peoples. Until then, Indians could not vote in provincial and federal elections without losing their status. The right to vote marked the start of a long journey to find a place in Canadian society. This journey is far from over.

4. Summer fallowing accomplished two goals: it conserved moisture and controlled weeds. In the 1990s, farmers began to abandon this technique because advances in technology enabled them to accomplish these goals without having to resort to summer fallowing.

5. The search for a quicker-maturing wheat began in 1892 when a cross was made between Red Fife, a popular wheat grown on the Prairies, and an earlier maturing wheat. After a decade of trials, Marquis wheat was tested at the Dominion Experimental Farms at Indian Head, Saskatchewan. By 1910, this variety of wheat made Canada famous for producing an exceptionally high-quality, hardy spring wheat.

6. Western alienation is based on past experience of real or perceived 'abuse' by big business, such as the CPR, and big government, meaning government controlled by eastern interests. Such feelings were deeply felt by homesteaders who settled the West. By early in the twentieth century, western alienation was particularly strong and the subsequent resentment was often aimed at the Canadian Pacific Railway. In fact, farmers regarded this railway company as the most rapacious agent of eastern Canadian interests. Not only had the company obtained millions of acres of fertile land, but its real estate offices often manipulated station sites to ensure their location on Canadian Pacific Railway property. Since farmland increased in value with proximity to rail-loading sites, such Canadian Pacific Railway land increased in value and could be sold to settlers at a higher price. Even more galling to westerners was the fact that CPR landholdings could not be taxed for 20 years. To extend this tax-free period, the CPR sometimes delayed selecting land, which meant that large areas were not available for homesteading because the railway could opt to select such lands as part of its grant.

7. Alberta, Manitoba, and Saskatchewan did not gain full control over Crown lands, and therefore over natural resources, until 1930. At that time, Ottawa transferred federal property to these provinces, giving them access to lucrative sources of taxation associated with natural resource developments. Until then, taxes from these lands went to Ottawa, supposedly to pay for railway building in the West. At first, the revenue from natural resources was small, but with the discovery of oil at Leduc, Alberta, in 1947, royalties from the production of petroleum provided most of the revenue for Alberta. Equally important, the exploitation of the oil and gas deposits in Alberta grew rapidly and eventually resulted in a variety of petroleum-processing plants and pipeline-construction firms in Alberta. Later, more modest energy and mineral developments were found in Saskatchewan, British Columbia, and Manitoba.

8. For example, in June 1999, the Organization for Economic Co-operation and Development announced that 1998 wheat subsidies in American dollars were $141 a tonne in the European Union, $61 a tonne in the US, and $8 a tonne in Canada (Hursh, 1999: C9).

9. Changes in the structure of farming are also anticipated. With expansion of the livestock and meat-processing industries, the number of farm workers may also increase. Large-scale hog farms, for instance, require farm workers. Also, the increase in specialty crops is causing more farmers to undertake contract farming; for example, they grow a certain quantity of canary seed for a set price. Some of these crops are now processed locally, a trend that is expected to intensify.

10. Farmers are becoming less sheltered from economic forces. Through contract farming, agribusiness—the sector of the economy that provides inputs to farms, such as chemical fertilizers, and procures agricultural products from farms for processing and distribution to consumers—is increasingly affecting the daily lives of farmers. Some fear that in the next severe economic downturn or drought, family farming operations will be replaced by corporate farms. In fact, the trend towards contract farming by agribusiness may be the first sign of such an ownership shift. While not yet apparent in the grain industry, agribusiness has established itself in specialty crops, poultry, and hog farming.

11. Farmers' decisions are affected by a complex set of variables, but ultimately the cost of doing business is the determining factor. For instance, the increased production of canola may affect the price of livestock feed. A protein by-product derived from the crushing of canola seeds is suitable for livestock. Instead of importing a similar protein supplement and paying for the built-in freight costs, using a local canola-based product to feed livestock may reduce costs.

12. Since the early 1990s, sawmills in the interior of British Columbia have exhausted local supplies of timber. They have had to purchase logs from Alberta ranchers who own timber lands in the foothills around Calgary.

Chapter 9

1. On 30 April 1999, the Newfoundland House of Assembly gave unanimous consent to a constitutional amendment that would officially change the name of the province to Newfoundland and Labrador.

2. The Maritime region was a focal point for the struggle between France and Britain. Indian allies were critical for both European powers. 'Involvement with the French brought benefits—and immediate consequences. Before they realized the full scope of the French–British rivalry, the Mi'kmaq and the Maliseet discovered they had already chosen sides' (Coates, 2000: 31).

3. As with Fort McMurray (Wood Buffalo), Alberta, Statistics Canada has created a geographically sprawling census metropolitan area—Cape Breton—out of Sydney and other Cape Breton Island municipalities.

4. Coal, a combustible sedimentary rock formed from the remains of plant life during the Carboniferous Age (a geological period in the Paleozoic era), is classified into four types: anthracite, bituminous, sub-bituminous, and lignite. Anthracite is the highest grade of coal, while bituminous is used in the iron and steel industry and for generating thermoelectric power. Most coal mined in Nova Scotia has been bituminous coal.

5. In 2006 the Brazilian mining company CVRD (now named Vale) purchased the Canadian nickel-mining company Inco, including its Sudbury operations and the Voisey's Bay mine site, for approximately $17 billion. Vale Inco is now a Canadian-based subsidiary of its Brazilian parent company.

6. On 26 May 2004 the Labrador Inuit voted 76 per cent in support of the agreement, with an 86.5 per cent voter turnout. The provincial and federal governments passed legislation in 2004 and 2005, respectively, giving legal effect to the Labrador Inuit Land Claims Agreement Act. This agreement contains the provision that the Labrador Inuit will receive 25 per cent of the revenue from mining and petroleum production on their settlement land, as well as 5 per cent of provincial royalties from the Voisey's Bay project. The federal environmental review panel examining the possible environmental and social impacts of the Voisey's Bay Nickel Company's mine proposal expressed concern about the disposal of the 15,000 tonnes of mine tailings to be produced each day. The mining company proposes to deposit the toxic tailings in a pond and prevent this from draining into surrounding streams and rivers by building two dams. While federal and provincial environmental officials are satisfied with the company's solution to the tailings problem, local people, especially the Inuit and Innu, are skeptical and worried about the effects of these toxins on the wildlife they depend on for food.

Chapter 10

1. There is a third climatic zone in the Territorial North—the Cordillera. However, the Cordillera climate, often described as a mountain climate, is affected by elevation (as elevation increases, temperature drops). North of 60° N, the Cordillera climate is also affected by latitude so that boreal natural vegetation is found at lower elevations and tundra natural vegetation at high elevations.

2. Sharing food was essential for small hunting groups to be able to live together harmoniously. By sharing in all hunters' successes, they could adjust for the vagaries of individual luck and reduce the threat of starvation. Today, the sharing of food remains a pivotal component of Aboriginal culture and the Native economy.

3. Sometime around the year 1000 the Vikings made contact with the ancestors of the Inuit, the Thule (c. AD 1000 to 1600). The Thule originated in Alaska where they hunted bowhead whales and other large sea mammals. They quickly spread their whaling technology across the Arctic, travelling in skin boats and dogsleds. With the onset of the Little Ice Age in the fifteenth century, climate conditions affected the distribution of animals and the Thule who were dependent on them. An increased amount of sea ice blocked the large whales from their former feeding grounds, resulting in the collapse of the Thule whale hunt. With the loss of their main source of food, the Thule had to rely more and more on locally available foods, usually some combination of seal, caribou, and fish. By the eighteenth century, the Thule culture had disappeared and had been replaced by the Inuit hunting culture.

4. Today, capital and operating costs are still serious problems for hunters. Transportation from a settlement to hunting/trapping areas is a major expenditure. Snowmobiles, for example, may cost as much as $20,000 in a northern retail store (in the 1990s, they cost $5,000 to $10,000), and the price of gasoline to fuel them is significantly higher in the North. With a continuing rise in the prices of manufactured goods, the cost of a snowmobile 10 years from now may double again. Since a snowmobile used for long-distance travel has a short lifespan, the hunter/trapper must replace it about every three or four years.

5. The first comprehensive land-claim submission came from the Dene/Métis in 1974. The land claimed by the Dene/Métis extended over most of the Mackenzie Basin north of 60° N. This land was called Denendeh. In 1988, an Agreement-in-Principle was signed by the two negotiating parties (the federal government and Dene Nation). This agreement called for a cash payment of $500 million over 15 years plus title to 181,230 km^2 of land. Nearly 6 per cent of this land (10,000 km^2) was to include subsurface rights for the Dene/Métis. As well, the Dene/Métis would obtain a share of federal resource royalties, including those generated by the Norman Wells oil field. Chiefs and elders from the Great Slave Lake area refused to approve this agreement for two main reasons—it contained no reference to self-government and it called for the surrender of Aboriginal rights. With this rejection, the Gwich'in and Sahtu/Métis subsequently negotiated their own comprehensive land-claim agreements with Ottawa.

6. The Inuvialuit broke away from the other Inuit who were seeking a common land-claim settlement that would have stretched from the Arctic Coast of the Yukon to Baffin Island. The Inuvialuit wanted to advance their own land claim before the huge oil and gas deposits in the Beaufort Sea were developed. The Inuvialuit, with their separate agreement, hoped to obtain land with oil and gas deposits and achieve a taxation arrangement similar to that of the

Arctic Slope Regional Corporation of the Inupiat in Alaska near Prudhoe Bay. This municipality received substantial tax revenues from the oil companies.

7. Under the IFA, Canada agreed to transfer $45 million in 1977 dollars. By 1984, these funds were valued at $152 million. The payments to the Inuvialuit Regional Corporation, the business corporation of the IFA, began on 31 December 1984 with $12 million. By 1997, there had been three annual payments of $1 million beginning 31 December 1985; five annual payments of $5 million beginning 31 December 1988; four annual payments of $20 million beginning 31 December 1993; and a final payment on 31 December 1997 of $32 million (Canada, 1985: 107).

8. This pipeline would be constructed by Canadian Arctic Gas Pipeline Ltd, which represented 27 Canadian and American oil producers, such as Exxon, Gulf, and Shell, and Trans-Canada Pipelines. Later, a rival bid was made by Foothills Pipeline Ltd, which consisted of Alberta Gas Trunk Line and Westcoast Transmission. Prudhoe Bay is located along the North Slope of Alaska. Alaskan oil accounts for 20 per cent of US oil production. An oil pipeline, known as the Trans-Alaska Pipeline System (TAPS), transports Alaskan oil from Prudhoe Bay to Valdez and then by tanker to markets along the west coast of the United States.

Chapter 11

1. The potential outcomes of Québec separating are sobering to consider. Many argue that, once split, the durability of 'the rest of Canada' is uncertain. In today's world, discontiguous states are rare. Of the nearly 200 nations in the world, fewer than 10 states are divided. Most of these separated enclaves are small, such as Kaliningrad, which is separated from Russia by Lithuania and Belarus. Others have evolved into separate states, as did Pakistan, which existed as two parts for 25 years until East Pakistan separated to form a new state, Bangladesh, in 1972. If Québec separated, Atlantic Canada would be cut off from the rest of the country. Under those circumstances, would Atlantic Canada join the United States?

BIBLIOGRAPHY

Chapter 1

Adams, Michael. 2003. *Fire and Ice: The United States, Canada and the Myth of Converging Values*. Toronto: Penguin.

Agnew, John. 2002. *American Space/American Place: Geographies of the Contemporary United States*. New York: Routledge.

Azzi, Stephen. 2006. 'Debating Free Trade', *National Post*, 14 Jan., A18.

Barnes, Trevor, ed. 1993. 'Focus: A Geographical Appreciation of Harold A. Innis', *Canadian Geographer* 37, 4: 352–64.

Blackwell, Tom. 2006. 'Be "Less Religious" about Sovereignty, Manley Urges', *National Post*, 7 Feb., A7.

Burney, Derek. 2007. 'From the FTA, Lessons in Leadership: Free Trade Turns 20', *Globe and Mail*, 9 Oct., A19.

CBC News. 2007. 'Hérouxville Drops Some Rules from Controversial Code', 13 Feb. At: <www.cbc.ca/canada/montreal/story/2007/02/13/qc-herouxville20070213.html>.

Cresswell, Tim. 2005. *Place: A Short Introduction*. London: Blackwell.

Columbo, John Robert, ed. 1987. *New Canadian Quotations*. Edmonton: Hurtig.

De Blij, H.J., and Alexander B. Murphy. 2006. *Human Geography: Culture, Society, and Space*, 8th edn. Toronto: John Wiley.

Dicken, Peter. 2007. *Global Shift: Mapping the Changing Contours of the World Economy*, 5th edn. New York: Guilford.

Frank, A.G. 1969. *Capitalism and Underdevelopment in Latin America*. New York: Monthly Review Press.

Friedmann, John. 1966. *Regional Development Policy: A Case Study of Venezuela*. Cambridge, Mass.: MIT Press.

Garreau, Joel. 1981. *The Nine Nations of North America*. Boston: Houghton Mifflin.

Hamilton, Graeme. 2007. 'Welcome! Leave Your Customs at the Door', *National Post*, 30 Jan., A1, A7.

Hare, Kenneth F. 1968. 'Canada', in John Warkinton, ed., *Canada: A Geographical Interpretation*. Toronto: Methuen, 3–12.

Hart, Michael. 1991. 'A Lower Temperature: The Dispute Settlement Experience under the Canada–United States Free Trade Agreement', *American Review of Canadian Studies* (Summer/Autumn): 193–205.

Hartz, Louis. 1995. *The Liberal Tradition in America*. New York: Harcourt Brace Jovanovich.

Hayter, Roger, and Trevor J. Barnes. 2001. 'Canada's Resource Economy', *Canadian Geographer* 45, 1: 36–41.

Hiller, Harry. 2000. 'Region as a Social Construction', in Keith Archer and Lisa Young, eds, *Regionalism and Party Politics in Canada*. Toronto: Oxford University Press.

Hutchison, Bruce. 1942. *The Unknown Country: Canada and Her People*. Toronto: Longmans, Green and Company.

Innis, Harold. 1930. *The Fur Trade in Canada: An Introduction to Canadian Economic History*. New Haven: Yale University Press.

Khondaker, Jafar. 2007. 'Canada's Trade with China: 1997 to 2006', *Canada Trade Highlight Series*, No. 1 Statistics Canada Catalogue no. 65–508–XWE. At: <www.statcan.gc.ca/pub/65-508-x/65-508-x2007001-eng.htm>.

Konrad, Victor, and Heather Nicol. 2008. *Beyond Walls: Reinventing the Canada–US Borderlands*. Aldershot: Ashgate.

Lemon, James T. 1996. *Liberal Dreams and Nature's Limits: Great Cities of North America Since 1600*. Toronto: Oxford University Press.

Lewington, Jennifer. 2007. 'Immigrants and Integration—Is the City Ready to Listen?', *Globe and Mail*, 29 Jan. At: <www.theglobeandmail.com/servlet/story/RTGAM.20070116.wimmig16home/BNStory/National/home>.

Lipset, Seymour M. 1990. *Continental Divide: The Values and Institutions of the United States and Canada*. New York: Routledge.

Macdonald, Neil. 2009. 'Interview with US Homeland Security Secretary Janet Napolitano', CBC News, 20 Apr. At: <www.cbc.ca/canada/story/2009/04/20/f-transcript-napolitano-macdonald-interview.html>.

Paasi, Anssi. 2003. 'Region and Place: Regional Identity in Question', *Progress in Human Geography* 27, 4: 475–85.

Parkinson, David. 2009. 'Despite Some Price Woes, Commodities Bull Market Is Thriving', *Globe and Mail*, 17 July, B8.

Peet, Richard, and Elaine Hartwick. 2009. *Theories of Development: Contentions, Arguments, Alternatives*, 2nd edn. New York: Guilford.

Norcliffe, Glen. 2001. 'Canada in a Global Economy', *Canadian Geographer* 45: 14–30.

Northwest Territories, Bureau of Statistics. 2006. *Northwest Territories—2005 . . . by the Numbers*. At: <www.stats.gov.nt.ca/Statinfo/Generalstats/bythenumbers/BTNhome(dvo).html>.

Norton, William. 2010. *Human Geography*, 7th edn. Toronto: Oxford University Press.

Potter, Mitch. 2010. 'Secretive Family and a Vital U.S. Link Create International Span of Mystery', *Toronto Star*, 16 Feb., A1, A17.

Relph, E.C. 1976. *Place and Placelessness*. London: Pion.

Resnick, Philip. 2000. *The Politics of Resentment: British Columbia Regionalism and Canadian Unity*. Vancouver: University of British Columbia Press.

Rumney, Thomas A. 2010. *Canadian Geography: A Scholarly Bibliography*. Lanham, Md: Scarecrow Press.

Samuelson, Paul. 1976. *Economics*. New York: McGraw-Hill.

Saul, John Ralston. 1997. *Reflections of a Siamese Twin: Canada at the End of the Twentieth Century*. Toronto: Viking.

Simco, Luke. 2009. 'Wallin Addresses SARM Convention', *StarPhoenix* (Saskatoon), 13 Mar., A5.

Simpson, Jeffrey. 1993. *Faultlines: Struggling for a Canadian Vision*. Toronto: HarperCollins.

Statistics Canada. 1997. *A National Overview: Population and Dwelling Counts*, 1996 Census of Canada (Catalogue no. 93–357–XPB). Ottawa: Industry Canada.

———. 2006a. 'Land and Freshwater Area, by Province and Territory, 2006'. At: <www40.statcan.ca/l01/cst01/phys01.htm>.

——. 2006b. 'Trade'. At: <www41.statcan.ca/1130/ceb1130_000_e.htm>.

——. 2007a. 'Population and Dwelling Counts for Canada, Provinces and Territories, 2006 and 2001 Censuses—100% data'. At: <www12.statcan.ca/english/census06/data/popdwell/Table.cfm?T=101>.

——. 2007b. 'Gross Domestic Product, Expenditure-based, by Province and Territory'. At: <www40.statcan.ca/l01/cst01/econ15.htm>.

——. 2007c. *The Evolving Linguistic Portrait, 2006 Census: The Proportion of Francophones and of French Continue to Decline.* 2006 Census: Analytic Series. At: <www12.statcan.ca/english/census06/analysis/language/continue_decline.cfm>.

——. 2007d. 'Canadian Statistics: Distribution of Employed People, by Industry, by Province'. At: <www40.statcan.ca/l01/cst01/labor21c.htm>.

——. 2007e. 'Canada's International Merchandise Trade', *The Daily*, 13 Feb. At: <www.statcan.ca/Daily/English/070213/td070213.htm>.

——. 2007f. 'Export of Goods on a Balance-of-Payments Basis, by Product'. At: <www40.statcan.ca/l01/cst01/gblec04.htm>.

——. 2008. 'Aboriginal Identity Population by Age Groups, Median Age and Sex, 2006 Counts, for Canada, Provinces and Territories—20% Sample Data'. At: <www12.statcan.ca/english/census06/data/highlights/Aboriginal/pages/Page.cfm?Lang=E&Geo=PR&Code=01&Table=1&Data=Count&Sex=1&Age=1&StartRec=1&Sort=2&Display=Page>.

Tuan, Yi-Fu. 1977. *Space and Place: The Perspective of Experience.* Minneapolis: University of Minnesota Press.

——. 1980. *Landscapes of Fear.* Oxford: Blackwell.

VanderKlippe, Nathan. 2009. 'T. Boone Pickens Bets on Natural Gas', *Globe and Mail*, 19 June. At: <www.theglobeandmail.com/globe-investor/t-boone-pickens-bets-on-natural-gas/article1186003/>.

Wallerstein, Immanuel. 1974. *The Modern World System: Capitalist Agriculture and the Origins of the European World Economy in the Sixteenth Century.* New York: Academic Press.

——. 1979. *The Capitalist World Economy.* Cambridge: Cambridge University Press.

Watkins, M.H. 1963. 'A Staple Theory of Economic Growth', *Canadian Journal of Economics and Political Science* 29, 2: 141–58.

——. 1977. 'The Staple Theory Revisited', *Journal of Canadian Studies* 12, 5: 83–95.

Yukon. 2006. *Yukon Labour Force Survey Review, 2005.* At: <www.eco.gov.yk.ca/stats/onetime/lforcerev05.pdf>.

Chapter 2

BBC News. 2010. 'Q & A: Professor Phil Jones', 13 Feb. At: <news.bbc.co.uk/2/hi/8511670.stm>.

Bone, Robert M., Shane Long, and Peter McPherson. 1997. 'Settlements in the Mackenzie Basin: Now and in the Future 2050', in Cohen (1997: 265–74).

Bouchard, Mireillle. 2001. 'Un défi environnemental complexe du XXIe siècle au Canada: L'identification et la compréhension de la réponse des environnements face aux changements climatiques globaux', *Canadian Geographer* 45, 1: 54–70.

CBC News. 2004. 'Inside Walkerton: Canada's Worst-ever E. Coli Contamination', 20 Dec. At: <canadaonline.about.com/gi/dynamic/offsite.htm?site=http://www.cbc.ca/news/background/walkerton/>.

Christopherson, Robert W. 1998. *Geosystems: An Introduction to Physical Geography*, 3rd edn. Upper Saddle River, NJ: Prentice-Hall.

Cohen, Stewart J., ed. 1997. *The Final Report of the Mackenzie Basin Impact Study.* Downsview, Ont.: Environment Canada.

Conrad, Cathy T. 2009. *Severe and Hazardous Weather in Canada: The Geography of Extreme Events.* Toronto: Oxford University Press.

Dearden, Philip, and Bruce Mitchell. 2005, 2009. *Environmental Change and Challenge: A Canadian Perspective*, 2nd and 3rd edns. Toronto: Oxford University Press.

de Loë, R. 2000. 'Floodplain Management in Canada: Overview and Prospects', *Canadian Geographer* 44, 4: 354–68.

Fisheries and Oceans Canada. 2003. 'Fast Facts'. At: <www.dfo-mpo.gc.ca/communic/facts-info/facts-info_c.htm>.

French, H.M., and O. Slaymaker, eds. 1993. *Canada's Cold Environments.* Montréal and Kingston: McGill-Queen's University Press.

Hamelin, Louis-Edmond. 1979. *Canadian Nordicity: It's Your North, Too*, trans. William Barr. Montreal: Harvest House.

Hare, F. Kenneth, and Morley K. Thomas. 1974. *Climate Canada.* Toronto: Wiley.

Intergovernmental Panel on Climate Change (IPCC). 2010. *The IPCC Assessment Reports.* At: <www.ipcc.ch/>.

Laycock, A.H. 1987. 'The Amount of Canadian Water and Its Distribution', in M.C. Healey and R.R. Wallace, eds, *Canadian Arctic Resources.* Ottawa: Department of Fisheries and Oceans, 13–42.

McKibben, Bill. 2010. *Earth: Making a Life on a Tough Planet.* New York: Times Books.

NASA. 2008. 'Arctic Sea Ice Reaches Lowest Coverage for 2008'. At: <www.nasa.gov/topics/earth/sea_ice_nsidc.html>.

Norton, William. 2010. *Human Geography*, 7th edn. Toronto: Oxford University Press.

Pachauri, R.K., and A. Reisinger, eds. 2007. *Climate Change 2007: Synthesis Report.* Geneva: IPCC Secretariat.

Rasid, Harun, Wolfgang Haider, and Len Hunt. 2000. 'Post-flood Assessment of Emergency Evacuation Policies in the Red River Basin, Southern Manitoba', *Canadian Geographer* 44, 4: 369–86.

Shabbar, Amir, Barrie Bonsal, and Madhav Khandekar. 1997. 'Canadian Precipitation Patterns Associated with Southern Oscillation', *Journal of Climate* 10: 3016–27.

Slocombe, D. Scott, and Phillip Dearden. 2002. 'Protected Areas and Ecosystem-Based Management', in Dearden and Rick Rollins, eds, *Parks and Protected Areas in Canada: Planning and Management*, 2nd edn. Toronto: Oxford University Press, 295–320.

Chapter 3

Bone, Robert. 2009. *The Canadian North: Issues and Challenges*, 3rd edn. Toronto: Oxford University Press.

Brownlie, Robin Jarvis. 2003. *A Fatherly Eye: Indian Agents, Government Power, and Aboriginal Resistance in Ontario, 1918–1939*. Toronto: Oxford University Press.

Bumsted, J.M. 2007. *A History of the Canadian Peoples*, 3rd edn. Toronto: Oxford University Press.

Canada. 1882. *Census of Canada 1880–81*, vol. 1. Ottawa: MacLean, Rogers & Company.

———. 1892. *Census of Canada 1890–91*, vol. 1. Ottawa.

Canada, Department of Finance. 2009a. 'Equalization Program'. At: <www.fin.gc.ca/fedprov/eqp-eng.asp>.

———. 2009b. 'Federal Transfers to Provinces and Territories'. At: <http://www.fin.gc.ca/access/fedprov-eng.asp#Major>.

Cardinal, Harold. 1969. *The Unjust Society: The Tragedy of Canada's Indians*. Edmonton: Hurtig.

Carter, Sarah. 2004. '"We Must Farm To Enable Us To Live": The Plains Cree and Agriculture to 1900', in R. Bruce Morrison and C. Roderick Wilson, eds, *Native Peoples: The Canadian Experience*. Toronto: Oxford University Press, 320–40.

Cook, Ramsay. 1993. *The Voyages of Jacques Cartier*. Toronto: University of Toronto Press.

Delacourt, Susan. 2010. 'Ontario to Get 18 New MPs', *Toronto Star*, 2 Apr., A1, A19.

Dickason, Olive Patricia, with David T. McNab. 2009. *Canada's First Nations: A History of Founding Peoples from Earliest Times*, 4th edn. Toronto: Oxford University Press.

Garreau, Joel. 1981. *The Nine Nations of North America*. Boston: Houghton Mifflin.

Globe and Mail. 1995. 'No—by a Whisker', 31 Oct., A1.

Harris, R. Cole. 1997. *The Resettlement of British Columbia: Essays on Colonialism and Geographical Change*. Vancouver: University of British Columbia Press.

———. 2003. *Making Native Space: Colonialism, Resistance, and Reserves in British Columbia*. Vancouver: University of British Columbia Press.

———. 2009. *The Reluctant Land: Society, Space, and Environment in Canada before Confederation*. Vancouver: University of British Columbia Press.

——— and John Warkentin. 1991. *Canada before Confederation: A Study in Historical Geography*. Ottawa: Carleton University Press.

Indian and Northern Affairs Canada (INAC). 2009. 'Registered Indian Population by Sex and Residence 2007'. At:<www.ainc-inac.gc.ca/ai/rs/pubs/sts/ni/rip/rip07/rip07-eng.asp#Sum>.

Kerr, Donald, and Deryck W. Holdsworth, eds. 1990. *Historical Atlas of Canada, Volume III: Addressing the Twentieth Century 1891–1961*. Toronto: University of Toronto Press.

Lower, J. Arthur. 1983. *Western Canada: An Outline History*. Vancouver: Douglas & McIntyre.

McVey, Wayne W., and W.E. Kalbach. 1995. *Canadian Population*. Toronto: Nelson Canada.

Miller, J.R. 2000. *Skyscrapers Hide the Heavens: A History of Indian–White Relations in Canada*, 3rd edn. Toronto: University of Toronto Press.

Milloy, John S. 1999. *A National Crime: The Canadian Government and the Residential School System*. Winnipeg: University of Manitoba Press.

Moffat, Ben. 2002. 'Geographic Antecedents of Discontent: Power and Western Canadian Regions 1870 to 1935', *Prairie Perspectives* 5: 202–28.

Nemni, Max. 1994. 'The Case against Quebec Nationalism', *American Review of Canadian Studies* 24, 2: 171–96.

Romaniuc, Anatole. 2000. 'Aboriginal Population of Canada: Growth Dynamics under Conditions of Encounter of Civilisations', *Canadian Journal of Native Studies* 20: 95–137.

Saul, John Ralston. 1997. *Reflections of a Siamese Twin: Canada at the End of the Twentieth Century*. Toronto: Viking.

Sifton, Clifford. 1922. 'The Immigrants Canada Wants', *Maclean's* 35, 7: 16.

Simpson, Jeffrey. 1993. *Faultlines: Struggling for a Canadian Vision*. Toronto: HarperCollins.

Smith, P.J. 1982. 'Alberta Since 1945: The Maturing Settlement System', in L.D. McCann, ed., *Heartland and Hinterland: A Regional Geography of Canada*. Scarborough, Ont.: Prentice-Hall.

Statistics Canada. 2003. *Historical Statistics of Canada*. Ottawa: Statistics Canada Catalogue no. 11–516–XIE. At: <www.statcan.ca/english/freepub/11-516-XIE/sectiona/sectiona.htm>.

———. 2008. 'Aboriginal Identity Population by Age Groups, Median Age and Sex, 2006 Counts, for Canada, Provinces and Territories'. At: <www12.statcan.gc.ca/english/census06/data/highlights/Aboriginal/pages/Page.cfm?Lang=E&Geo=PR&Code=01&Table=1&Data=Count&Sex=1&Age=1&StartRec=1&Sort=2&Display=Page>.

Taylor, Charles. 1993. *Reconciling the Solitudes*. Montréal and Kingston: McGill-Queen's University Press.

Thomas, David Hurst. 1999. *Exploring Ancient Native America: An Archaeological Guide*. New York: Routledge.

Tough, Frank. 1996. *As Their Natural Resources Fail: Native Peoples and the Economic History of Northern Manitoba, 1870–1930*. Vancouver: University of British Columbia Press.

Tracie, Carl J. 1996. *Toil and Peaceful Life: Doukhobor Village Settlement in Saskatchewan, 1899–1918*. Regina: Canadian Plains Research Centre, University of Regina.

Villeneuve, Paul. 1993. 'Allocution présidentielle: L'invention de l'avenir au nord de l'amérique', *Le géographe canadien* 37, 2: 98–104.

Williams, Glyndwr. 1983. 'The Hudson's Bay Company and the Fur Trade, 1670–1870', *The Beaver* (Autumn): 4–81.

Chapter 4

Association of University Research Parks. 2009. 'AURP Names 2009 Awards of Excellence Recipients'. At: <www.aurp.net/more/awardrecipients2009.cfm>.

Bank of Canada. 2010. *Monetary Policy Report Summary*, Jan. At: <www.bank-banque-canada.ca/en/mpr/pdf/2010/mprsumjan10.pdf>.

Beaujot, Roderic. 1991. *Population Change in Canada: The Challenges of Policy Adaptation*. Toronto: McClelland & Stewart.

Bell, Daniel. 1976. *The Coming of the Post Industrial Society*. New York: Basic Books.

Bernard, André. 2009. 'Trends in Manufacturing Employment', *Perspective*. Statistics Canada Catalogue no. 75-001xs. At: <www.statcan.gc.ca/pub/75-001-x/2009102/article/../pdf/10788-eng.pdf>.

Bernier, Jacques. 1991. 'Social Cohesion and Conflicts in Quebec', in Guy M. Robinson, ed., *A Social Geography of Canada*. Toronto: Dundurn Press.

Bone, Robert M. 2009. *The Canadian North: Issues and Challenges*, 3rd edn. Toronto: Oxford University Press.

Bourne, Larry S., and Damaris Rose. 2001. 'The Changing Face of Canada: The Uneven Geographies of Population and Social Change', *Canadian Geographer* 45, 1: 105–19.

Britton, John N.H., ed. 1996. *Canada and the Global Economy: The Geography of Structural and Technological Change*. Montréal and Kingston: McGill-Queen's University Press.

Canada. 1884. *Census of Canada, 1880–1881*. Ottawa: Department of Agriculture.

Cardinal, Harold. 1969. *The Unjust Society: The Tragedy of Canada's Indians*. Edmonton: Hurtig.

Carney, Mark. 2009. 'Canada's Economic Outlook and Framework for Unconventional Monetary Policy', opening statement to the Standing Senate Committee on Banking, Trade and Commerce, Ottawa, 6 May. At: <www.bis.org/review/r090507d.pdf>.

Chui, Tina, Kelly Tran, and Hélène Maheux. 2008. 'Immigration in Canada: A Portrait of the Foreign-born Population, 2006 Census: Immigration: Driver of Population Growth'. At: <www12.statcan.ca/census-recensement/2006/as-sa/97-557/figures/c1-eng.cfm>.

CIA. 2007. The World Fact Book: Canada. Washington. At: <www.cia.gov/library/publications/the-world-factbook/print/ca.html>.

Denevan, William M. 1992. 'The Pristine Myth: The Landscape of the Americas in 1492', *Annals, Association of American Geographers* 82, 3: 369–85.

Dickason, Olive Patricia, with David T. McNab. 2009. *Canada's First Nations: A History of Founding Peoples from Earliest Times*, 4th edn. Toronto: Oxford University Press.

Drucker, Peter. 1969. *The Age of Discontinuity*. London: Heinemann.

Fife, Robert. 2002. 'Migrants Must Spread Out: Ottawa', *National Post*, 22 June, A1, A9.

Florida, Richard. 2002a. 'The Economic Geography of Talent', *Annals, Association of American Geographers* 92: 743–55.

———. 2002b. *The Rise of the Creative Class: And How It's Transforming Work, Leisure, Community and Everyday Life*. New York: Basic Books.

———. 2005. *Cities and the Creative Class*. London: Routledge.

———. 2008. *Who's Your City? How the Creative Economy Is Making Where to Live the Most Important Decision of Your Life*. New York: Basic Books.

Foot, David, with D. Stoffman. 1996. *Boom, Bust and Echo: How to Profit from the Coming Demographic Shift*. Toronto: Macfarlane, Walter & Ross.

Gilbert, Anne. 2001. 'Le français au Canada, entre droits et géographie', *Canadian Geographer* 45, 1: 175–9.

Gionet, Linda. 2009. 'First Nations People: Selected Findings of the 2006 Census', *Canadian Social Trends* No. 87. At: <www.statcan.gc.ca/pub/11-008-x/2009001/article/10864-eng.htm>.

Harrison, Brian, and Louise Marmen. 1994. *Focus on Canada: Languages in Canada*. Catalogue no. 96-313E. Ottawa: Minister of Industry, Science and Technology.

Harvey, David. 1989. *The Urban Experience*. Baltimore: Johns Hopkins University Press.

Herberg, Edward N. 1989. 'Identity, Cultural Production and the Vitality of Francophone Communities outside Quebec', in Leen d'Haenens, ed., *Images of Canadianness: Visions on Canada's Politics, Culture, Economics*. Ottawa: University of Ottawa Press.

Hiebert, Daniel. 2000. 'Immigration and the Changing Canadian City', *Canadian Geographer* 44, 1: 25–43.

——— and David Ley. 2003. 'Assimilation, Cultural Pluralism and Social Exclusion among Ethnocultural Groups in Vancouver', *Urban Geography* 24, 1: 16–44.

Javed, Noor. 2010. '"Visible Minority" Will Mean "White" by 2031', *Toronto Star*, 10 Mar., A3.

Jiminez, Marina, and Kim Lunman. 2004. 'Canada's Biggest Cities See Influx of New Immigrants', *Globe and Mail*, 19 Aug.

Kyle, Cassandra. 2009. 'Solido Receives $1.5M Tax Rebate', *StarPhoenix* (Saskatoon), 17 July, C9. At: <www.thestarphoenix.com/Technology/Solido+receives+rebate/1800053/story.html>.

Ley, David. 1999. 'Myths and Meanings of Immigration and the Metropolis', *Canadian Geographer* 43, 1: 2–18.

——— and Daniel Hiebert. 2001. 'Immigration Policy as Population Policy', *Canadian Geographer* 45, 1: 120–5.

Li, Peter S. 2003. *Destination Canada: Immigration Debates and Issues*. Toronto: Oxford University Press.

McVey, Wayne W., and W.E. Kalbach. 1995. *Canadian Population*. Toronto: Nelson Canada.

Martin, Don. 2005. 'The Paycheque Exiles', *National Post*, 6 Sept., A18.

Mooney, James. 1928. *The Aboriginal Population of America North of Mexico*. Smithsonian Miscellaneous Collections. Washington: Smithsonian Institution.

Newhouse, David R. 2000. 'From the Tribal to the Modern: The Development of Modern Aboriginal Societies', in Ron F. Laliberte et al., eds, *Expressions in Canadian Native Studies*. Saskatoon: University of Saskatchewan Extension Press, 395–409.

———. 2007. 'Aboriginal Languages in Canada: Emerging Trends and Perspectives on Second Language Acquisition', *Canadian Social Trends* no. 83, Statistics Canada, Catalogue no. 11–008.

Ostry, Bernard. 2005. 'Canada's Ethnic Makeup Is Branching Out in New Directions', *Globe and Mail*, 15 Nov., A25.

Peters, Evelyn J. 2010. 'Aboriginal People in Canadian Cities', in Trudi Bunting, Pierre Filion, and Ryan Walker, eds, *Canadian Cities in Transition: New Directions in the Twenty-First Century*, 4th edn. Toronto: Oxford University Press, ch. 22.

PriceWaterhouseCoopers Canada. 2009. *SR&ED Tax Clips: 2009 Provincial and Territorial R&D Tax Credits (May 22, 2009)*. At: <www.pwc.com/ca/en/sred/tax-clips/2009-provincial-territorial-credits.jhtml>.

Ravenstein, Ernest George. 1885. 'The Laws of Migration', *Journal of the Statistical Society of London* 48, 2: 167–235.

———. 1889. 'The Laws of Migration', *Journal of the Royal Statistical Society* 52, 2: 241–305.

Royal Commission on Bilingualism and Biculturalism. 1970. Report. *Book IV: Cultural Contributions of the Other Ethnic Groups*. Ottawa: Queen's Printer.

Saul, John Ralston. 1997. *Reflections of a Siamese Twin: Canada at the End of the Twentieth Century*. Toronto: Viking.

Scoffield, Heather. 2009a. 'Recession Toll Harsh, but Easing', *Globe and Mail*, 1 June. At: <www.globeinvestor.com/servlet/story/RTGAM.20090601.wgdp0601/GIStory/>.

———. 2009b. 'Canada's Recovery in Sight, Says Report', *Globe and Mail*, 30 July. At: <www.theglobeandmail.com/report-on-business/canadas-recovery-in-sight-report-says/article1236130/>.

Statistics Canada. 1997a. '1996 Census: Mother Tongue, Home Language and Knowledge of Languages', *The Daily*, 2 Dec. At: <www.statcan.ca/Daily/English/>.

———. 1997b. *A National Overview: Population and Dwelling Counts*. Catalogue no. 93–357–XPB. Ottawa: Minister of Industry.

———. 1997c. *Mortality—Summary List of Causes, 1995*. Catalogue no. 84–209–XPB. Ottawa: Minister of Industry.

———. 2002a. *Census of Canada 2001—Census Geography. Highlights and Analysis: Canada's 2001 Population*. Ottawa. At: <www12.statcan.ca/English/census01/>.

———. 2002b. 'Census of Population: Language, Mobility and Migration', *The Daily*, 10 Dec. At: <www.statcan.ca/Daily/English/02120/d021210a.htm>.

———. 2002c. *A Profile of the Canadian Population: Where We Live, 2001 Census*. At: <www.statcan.ca/bsolc/english/bsolc?catno=96F0030XIE2001001>.

———. 2003a. *Census of Canada 2001—Canada's Ethnocultural Portrait: The Changing Mosaic*. Analytical Series 96F0030XIE2001008. Ottawa, 21 Jan. At: <www12.statcan.ca/english/census01/products/analytic/companion/etoimm/canada.cfm>.

———. 2003b. 'Components of Population Growth', 28 Feb. Ottawa. At: <www.statcan.ca/English/Pgdb/demo33a.htm>.

———. 2003c. *Census of Canada 2001—Aboriginal Peoples of Canada: A Demographic Profile*. Analysis series 96F0030XIE2001007. Ottawa, 21 Jan. At: <www12.statcan.ca/english/census01/products/analytic/companion/abor/contents.cfm>.

———. 2003d. *Census of Canada 2001—Religions in Canada*. Analysis series 96F0030XIE2001015. At: <www12.statcan.ca/english/census01/products/analytic/companion/rel/contents.cfm>.

———. 2006a. 'Components of Population Growth, by Province and Territory, July 1, 2005 to June 30, 2006'. At: <www40.statcan.ca/l01/cst01/demo33c.htm>.

———. 2006b. *Annual Demographic Statistics 2005*. Catalogue no. 91–213–XIB. At: <www.statcan.ca/english/freepub/91-213-XIB/0000591-213-XIB.pdf>.

———. 2006c. 'Labour Force Survey', *The Daily*, 10 Feb. At: <www.statcan.ca/Daily/English/060210/d060210a.htm>.

———. 2006d. 'Canada's Population by Age and Sex', *The Daily*, 26 Oct. At: <www.statcan.ca/Daily/English/061026/d061026b.htm>.

———. 2007a. 'Population by Year, by Province and Territory'. At: <www40.statcan.ca/l01/cst01/demo02a.htm?sdi=population%20change%20provinces>.

———. 2007b. 'Community Profiles'. At: <www12.statcan.ca/english/census06/data/profiles/community/Search/SearchForm_Results.cfm?Lang=E>.

———. 2007c. 'Portrait of the Canadian Population in 2006: Subprovincial Population Dynamics: Canada's Population Becoming More Urban'. At: <www12.statcan.ca/english/census06/analysis/popdwell/Subprov1.cfm>.

———. 2007d. 'Portrait of the Canadian Population in 2006: Subprovincial Population Dynamics: Vast Majority of Canada's Population Growth Is Concentrated in Large Metropolitan Areas'. At: <www12.statcan.ca/english/census06/analysis/popdwell/Subprov3.cfm>.

———. 2007e. *2006 Census Dictionary*. At: <www12.statcan.ca/english/census06/reference/dictionary/index.cfm>.

———. 2007f. 'Dwelling Counts, for Canada, Provinces and Territories, and Census Subdivisions (Municipalities), 2006 and 2001 Censuses—100% Data'. At: <www12.statcan.ca/english/census06/data/popdwell/Filter.cfm?T=308&S=1&O=A>.

———. 2007g. 'Population and Growth Components (1851–2001 Censuses)'. At: <www40.statcan.ca/l01/cst01/demo03.htm>.

———. 2007h. 'Births and Birth Rate, by Province and Territory'. At: <www40.statcan.ca/l01/cst01/demo04b.htm>.

———. 2007i. 'Deaths and Death Rate, by Province and Territory'. At: <www40.statcan.ca/l01/cst01/demo07b.htm>.

———. 2007j. 'Population and Dwelling Counts, for Canada, Provinces and Territories, 2006 and 2001 Censuses—100% Data'. At: <www.12.statcan.ca/english.census06/data/popdwell/Table.cfm?T=101>.

———. 2008a. 'Aboriginal Peoples in Canada in 2006: Inuit, Métis and First Nations, 2006 Census', *The Daily*, 15 Jan. At: <www.statcan.gc.ca/daily-quotidien/080115/tdq080115-eng.htm>.

———. 2008b. 'Language (including Language of Work)', Releases no. 4 and 6. At: <www12.statcan.gc.ca/census-recensement/2006/rt-td/lng-eng.cfm>.

———. 2008c. 'Aboriginal Identity Population, 2006 Counts, Percentage Distribution, Percentage Change for Both Sexes, for Canada, Provinces and Territories—20% Sample

Data', *Aboriginal Peoples Highlight Tables*. 2006 Census. Statistics Canada Catalogue no. 97–558–XWE2006002. At: <www12.statcan.ca/english/census06/data/highlights/aboriginal/index.cfm?Lang=E>.

———. 2008d. 'The Evolving Linguistic Portrait, 2006 Census: The Proportion of Francophones and of French Continue to Decline'. At: <www12.statcan.ca/census-recensement/2006/as-sa/97-555/p6-eng.cfm>.

———. 2008e. 'Mandate, Responsibilities and Objectives'. At: <www.statcan.gc.ca/about-apercu/mandate-mandat-eng.htm>.

———. 2009a. 'Components of Population Growth, by Province and Territory, 2007–2008'. At: <www40.statcan.gc.ca/l01/cst01/demo33a-eng.htm>.

———. 2009b. 'Ethnocultural Portrait of Canada: Ethnic Origins, 2006 Counts, for Canada, Provinces and Territories'. At: <www12.statcan.ca/english/census06/data/highlights/ethnic/pages/Page.cfm?Lang=E&Geo=PR&Code=01&Data=Count&Table=2&StartRec=1&Sort=3&Display=All&CSDFilter=5000>.

———. 2009c. 'Table 2. Quarterly Demographic Estimates'. At: <www.statcan.gc.ca/daily-quotidien/090326/t090326a2-eng.htm>.

———. 2009d. 'Annual Demographic Estimates: Canada, Provinces and Territories. At: <www.statcan.gc.ca/pub/91-215-x/2008000/5200700-eng.htm>.

———. 2009e. 'Labour Force, Employed and Unemployed, Numbers and Rates, by Province'. At: <www40.statcan.ca/l01/cst01/labor07c-eng.htm>.

———. 2009f. 'Gross Domestic Product, Expenditure-based, by Province and Territory'. At: <www40.statcan.gc.ca/l01/cst01/econ15-eng.htm>.

———. 2009g. 'Population, Age Distribution and Median Age by Province and Territory, as of July 1, 2008'. At: <www.statcan.gc.ca/daily-quotidien/090115/t090115c1-eng.htm>.

———. 2009h. *Aboriginal Peoples in Canada in 2006: Inuit, Métis and First Nations, 2006 Census*. 2006 Analysis Series. At: <www12.statcan.gc.ca/census-recensement/2006/as-sa/97-558/p2-eng.cfm#02>.

———. 2009i. 'Births', *The Daily*, 22 Sept. At: <www.statcan.gc.ca/daily-quotidien/090922/dq090922b-eng.htm>.

———. 2009j. *Immigration in Canada: A Portrait of the Foreign-born Population, 2006 Census: Immigrants Came from Many Countries*. At: <www12.statcan.ca/census-recensement/2006/as-sa/97-557/figures/c2-eng.cfm>.

———. 2009k. *Canadian Demographics at a Glance: Population Growth in Canada*. At: <www.statcan.gc.ca/pub/91-003-x/2007001/figures/4129879-eng.htm>.

———. 2010. 'Ethnocultural Portrait of Canada', Highlight Tables, 2006 Census. At: <www12.statcan.ca/census-recensement/2006/dp-pd/hlt/97-562/sel_geo.cfm?Lang=E&Geo=PR&Table=2>.

Taylor, Charles. 1994. *Multiculturalism: Examining the Politics of Recognition*. Princeton, NJ: Princeton University Press.

Trewartha, G.T., A.H. Robinson, and E.H. Hammond. 1967. *Elements of Geography*, 5th edn. New York: McGraw-Hill.

Walks, R. Alan, and Larry S. Bourne. 2006. 'Ghettos in Canada's Cities? Racial Segregation, Ethnic Enclaves and Poverty Concentration in Canadian Urban Areas', *Canadian Geographer* 50, 3: 273–97.

Woods, Alan. 2006. 'Dual Citizenship Faces Review', *National Post*, 21 Sept. At: <www.canada.com/national-post/news/story.html?id=fb2d75ab-8880-4945-8537-1508186a4964&k=61921>.

Chapter 5

Anderson, Fiona. 2006. 'Dropping Lumber Demand Threatens Canadian Exporters', *StarPhoenix* (Saskatoon), 23 Sept., D7.

Atkin, David. 2000. 'In a Hot-Wired World, Everything's Personal', *National Post*, 15 Nov., C1, C6–7.

Avery, Simon. 2006. 'In Ottawa, It's a Low-Budget Tech Rebound', *Globe and Mail*, 8 Nov. At: <www.theglobeandmail.com/servlet/story/RTGAM.20061108.wottawaa1108/BNStory/Technology/home>.

Bagnall, James. 2003. 'Ottawa Slipping Off the Tech Map', *National Post*, 22 Aug., FP5.

Beatty, Stephen. 2009. 'Speech: Toyota Canada's Managing Director at UofG', 10 Mar. At: <news.guelphmercury.com/printArticle/450846>.

Bernard, André. 2009. 'Trends in Manufacturing Employment', *Perspectives*, Feb. Statistics Canada Catalogue no. 75–001–X. At: <www.statcan.gc.ca/pub/75-001-x/2009102/pdf/10788-eng.pdf>.

Bertin, Oliver, and Greg Keenan. 2003. 'Algoma to Axe 600, Cut Costs in Latest Bid to Stay Profitable', *Globe and Mail*, 15 May, B1.

Bone, Robert M. 2009. *The Canadian North: Issues and Challenges*, 3rd edn. Toronto: Oxford University Press.

Brean, Joseph. 2003. 'Outbreak Predicted to Cost City $1-billion', *National Post*, 3 May, A6.

Brent, Paul. 2002. 'Car Output Still Higher in Canada', *National Post*, 15 Aug., FP5.

———. 2003. 'Daimler Calls Off Windsor Plant Plans', *National Post*, 23 May, FP5.

Britton, John N.H. 1996. 'High-Tech Canada', in Britton, ed., *Canada and the Global Economy: The Geography of Structural and Technological Change*. Montréal and Kingston: McGill-Queen's University Press.

Campion-Smith, Bruce. 2006. 'Mayor Fumes at Kashechewan Move', *Toronto Star*, 9 Nov. At: <www.thestar.com/NASApp/cs/ContentServer?pagename=thestar/Layout/Article_Type1&c=Article&cid=1163026213121&call_pageid=968332188774&col=968350116467>.

CBC News. 2006a. 'Caledonia Land Claim: Historical Timeline', 1 Nov. At: <www.cbc.ca/news/background/caledonia-land-claim/historical-timeline.html>.

———. 2006b. 'Ford to Close Ontario Plant, Cut 10,000 More Salaried Jobs', 15 Sept. At: <www.cbc.ca/money/story/2006/09/15/ford-cuts.html>.

———. 2009. 'Ontario Pushes Electric Cars as Auto-Sector Boost', 15 July. At: <www.cbc.ca/canada/story/2009/07/15/ont-electric-cars511.html>.

———. 2010. 'New Canada–U.S. Border Route Dispute Ends', 9 Apr. At: <www.cbc.ca/canada/windsor/story/2010/04/09/wdr-border-access-route-100409.html?ref=rss>.

Chevralier, Patrick. 2009. 'Gold', *Canadian Mineral Yearbook 2005*. At: <www.nrcan-rncan.gc.ca/mms-smm/busi-indu/cmy-amc/2005revu/gol-eng.htm>.

Cole, Trevor. 2009. 'Hamilton's Dead. Or Is It?', *Globe and Mail*, 28 Aug. At: <www.theglobeandmail.com/report-on-business/rob-magazine/hamiltons-dead-or-is-it/article1264739/>.

Courchene, Thomas J., with Colin R. Telmer. 1998. *From Heartland to North American Region State: The Social, Fiscal and Federal Evolution of Ontario*. Monograph Series on Public Policy, Centre for Public Management. Toronto: Faculty of Management, University of Toronto.

Cross, Philip. 2009. '2008 in Review', *Canadian Economic Observer*. At: <www.statcan.gc.ca/pub/11-010-x/2009004/part-partie3-eng.htm >.

CTV News. 2006. 'Canada, U.S. Agree on Softwood Framework', 26 Apr. At: <www.ctv.ca/servlet/ArticleNews/story/CTVNews/ 20060426/softwood_deal_060426/20060426?hub=Canada>.

Darling, Graham. 2007. 'Land Claims and the Six Nations in Caledonia Ontario', Centre for Constitutional Studies. At: <www.law.ualberta.ca/centres/ccs/Current-Constitutional-Issues/Land-Claims-and-the-Six-Nations-in-Caledonia-Ontario.php>.

Detroit River International Crossing. 2009. *Forum for the Private Sector—April 23, 2009*. At: <www.partnershipborderstudy.com/pdf/2009-04-23DRICForumBooklet.pdf>.

Dickason, Olive Patricia, with David T. McNab. 2009. *Canada's First Nations: A History of Founding Peoples from Earliest Times*, 4th edn. Toronto: Oxford University Press.

Dicken, Peter. 1992. *Global Shift: The Internationalization of Economic Activity*, 2nd edn. New York: Guilford Press.

Drummond, Don, and Derek Burleton. 2008. 'Time for a Vision of Ontario's Economy'. At: <www.td.com/economics/special/db0908_ont.pdf>.

Environment Canada. 2006. 'Information on Greenhouse Gas Sources and Sinks'. At: <www.ec.gc.ca/pdb/ghg/onlineData/dataAndReports_e.cfm>.

Evans, Mark. 2006. 'The "Quiet Boom": High-Tech Turnaround Drives Ottawa, Waterloo Growth', *National Post*, 31 Jan. At: <www.canada.com/nationalpost/financialpost/story.html?id=2a0b745b-eded-4269-af0f-f454d25a8ca6>.

Export Development Canada. 2007. 'Commodity Price Update', 28 May. At: <www.edc.ca/english/docs/commpric_e.pdf>.

Ferley, Paul, and Robert Hogue. 2009. 'Has the Storm Passed?', *Provincial Outlook*, June. Royal Bank of Canada. At: <www.rbc.com/economics/market/pdf/provfcst.pdf>.

Florida, Richard. 2002. *The Rise of the Creative Class: And How It's Transforming Work, Leisure, Community and Everyday Life*. New York: Basic Books.

French, Cameron. 2007. 'U.S. Security-related Delays at Border Cost Billions, Canadian Producers Say', *International Herald Tribune*. At: <www.iht.com/articles/2007/03/05/business/canada.php>.

Gomes, Carlos. 2009. 'Auto Lending Conditions Improve in North America', *Global Auto Report*, 25 June. Scotia Economics. At: <www.scotiacapital.com/English/bns_econ/bns_auto.pdf>.

———. 2010. 'Demographic Boom Turning To Bust for North American Auto Sales', *Global Auto Report*, 26 Feb. Scotiabank Group. At: <www.scotiacapital.com/English/bns_econ/bns_auto.pdf>.

Hoffman, Andy. 2009. 'At a Secret Facility in Sudbury, 27 Highly Skilled Workers Cut and Polish Diamonds as Part of Ontario's Ambition to Become a Key Line in the Global Diamond Trade: All Are from Vietnam', *Globe and Mail*, 15 Oct., B1, B6.

Indian and Northern Affairs Canada (INAC) 2009. 'Chronology of Events at Caledonia'. At: <www.ainc-inac.gc.ca/ai/mr/is/eac-eng.asp>.

JAMA Canada. 2010. 'Monthly and Annual Statistics for JAMA Canada Member Companies', 18 Mar. At: <www.jama.ca/jamastats/>.

Keenan, Greg. 2003. 'Layoffs Mount among Auto Makers', *Globe and Mail*, 26 Mar., B4.

———. 2007. 'Chrysler to Axe 20% of Canadian Work Force', *Globe and Mail*, 8 Feb. At: <www.theglobeandmail.com/servlet/story/RTGAM.20070208.wchrysler08/BNStory/Business/home>.

———. 2009. 'Windsor Gets Bailout Dividend: New Chrysler Minivan', *Globe and Mail*, 20 Aug. At: <www.theglobeandmail.com/report-on-business/windsor-gets-bailout-dividend-new-chrysler-minivan/article1257692/>.

Kowaluk, Russell, and Rob Larmour. 2009. 'Manufacturing: The Year 2008 in Review', Statistics Canada. At: <www.statcan.gc.ca/bsolc/olc-cel/olc-cel?lang=eng&catno=11-621-M2009077>.

McCarthy, Shawn. 1997. 'Ottawa Trades Bureaucrats for Bytes', *Globe and Mail*, 30 Aug., B1, B4.

Marr, Lisa Grace. 2010. 'Hamilton Losing Siemens Turbine Jobs', *Toronto Star*, 12 Mar. At: <www.thestar.com/business/article/778745-hamilton-losing-siemens-turbine-jobs>.

Matteo, Livio Di. 2006. 'Breakaway Country', *National Post*, 6 Sept., FP19.

National Post. 2007. 'Ipperwash Report', 1 June, A14.

Natural Resources Canada. 2006. *Mill Closures and Mill Investments in the Canadian Forest Sector: The State of Canada's Forests, 2005–2006*. At: <www.nrcan.gc.ca/cfs-scf/national/what-quoi/sof/sof06/mergers_e.html>.

———. 2008. 'Mineral Production of Canada, by Province and Territory'. At: <mmsd.mms.nrcan.gc.ca/stat-stat/prod-prod/PDF/2008p.pdf>.

———. 2010a. *Price Monitor*, 17 Mar. At: <canadaforests.nrcan.gc.ca/article/pricemonitor>.

———. 2010b. 'The Global Recession Reduced Canada's Mineral Production in 2009', *Information Bulletin*, Mar. At: <mmsd.mms.nrcan.gc.ca/stat-stat/prod-prod/PDF/ib2010_e.pdf>.

Norcliffe, Glen. 1996. 'Mapping Deindustrialization: Brian Kipping's Landscape of Toronto', *Canadian Geographer* 40, 3: 266–73.

OICA (International Organization of Motor Vehicle Manufacturers). 2009. 'Production Statistics'. At: <oica.net/category/production-statistics/>.

Ontario Medical Association (OMA). 2005. *The Illness Costs of Air Pollution in Ontario: A Summary of Findings*. Toronto. At: <www.oma.org/Health/smog/icap.asp>.

Plasco Energy Inc. 2009. At <cleantechnologyreport.ca/comp_profiles/CP-RW-PlascoEnergyGroupInc.pdf>.

Potter, Mitch. 2010. 'Secretive Family and a Vital U.S. Link Create International Span of Mystery', *Toronto Star*, 16 Feb, A1, A17.

Preston, Valerie, and Lucia Lo. 2000. 'Canadian Urban Landscape Examples—21: "Asian Theme" Malls in Suburban Toronto: Land Use Conflicts in Richmond Hill', *Canadian Geographer* 44, 2: 182–90.

Shkilnyk, A.M. 1985. *A Poison Stronger Than Love: The Destruction of an Ojibwa Community*. New Haven: Yale University Press.

Statistics Canada. 2002. '2001 Census: Population and Dwelling Counts, for Census Metropolitan Areas and Census Agglomerations, 2001 and 1996 Censuses', 16 July. At: <www.statcan.ca/English/IPS/Data/93F0050XCB2001013.htm>.

———. 2003. 'Exports of Goods on a Balance-of-Payments Basis', *Canadian Statistics, International Trade*. At: <www.statcan.ca/english/Pgdb/gblec04.htm>.

———. 2006a. 'Exports of Goods on a Balance-of-Payments Basis', *Canadian Statistics, International Trade*. At: <www40.statcan.ca/l01/cst01/gblec04.htm>.

———. 2006b. 'Canadian Statistics: Distribution of Employed People, by Industry, by Province'. At: <www40.statcan.ca/l01/cst01/labor21c.htm>.

———. 2007a. 'Population Urban and Rural, by Province and Territory: 1851 to 2001'. At: <www40.statcan.ca/l01/cst01/demo62g.htm>.

———. 2007b. 'Census of Agriculture Counts 57,211 Farms in Ontario', *2006 Census of Agriculture*. At: <www.statcan.ca/english/agcensus2006/media_release/on.htm>.

———. 2007c. 'Population and Dwelling Counts, for Census Metropolitan Areas and Census Agglomerations, 2006 and 2001 Censuses—100% Data'. At: <www12.statcan.ca/english/census06/data/popdwell/Table.cfm?T=201&S=3&O=D&RPP=150>.

———. 2007d. 'Study: Canada's Changing Auto Industry', *The Daily*, 17 May. At: <www.statcan.ca/Daily/English/070517/d070517c.htm>.

———. 2009a. 'Provincial and Territorial Economic Accounts', *The Daily*, 27 Apr. At: <www.statcan.gc.ca/daily-quotidien/090427/dq090427a-eng.htm>.

———. 2009b. 'Canada's Population Estimates: Quarterly Population Estimate', Table 2, 23 Dec. At: <www.statcan.gc.ca/daily-quotidien/091223/t091223b2-eng.htm>.

———. 2009c. 'Chart 21.1 (data) Unemployment Rate', 16 Jan. At: <www41.statcan.gc.ca/2008/2621/grafx/htm/ceb2621_000_1-eng.htm#table>.

———. 2009d. 'Distribution of Employed People, by Industry, by Province'. At: <www40.statcan.gc.ca/l01/cst01/labor21c-eng.htm>.

———. 2009e. 'Net Interprovincial Migration', *Quarterly Demographic Estimates*, Table 2-12. At: <www.statcan.gc.ca/pub/91-002-x/2009001/t032-eng.htm>.

———. 2009f. 'Toronto: Place of Birth for the Immigrant Population', *Immigration and Citizenship Highlight Tables, 2006 Census*. At: <census2006.ca/census-recensement/2006/dp-pd/hlt/97-557/T404-eng.cfm?SR=1>.

———. 2009g. 'Population and Demography'. At:<www41.statcan.gc.ca/2008/3867/ceb3867_000-eng.htm>.

———. 2010. 'Labour Force, Employed and Unemployed, Numbers and Rates, by Province', 29 Jan. At: <www40.statcan.gc.ca/l01/cst01/labor07b-eng.htm>.

Steinhart, David. 1999. 'Ste-Thérèse Truck Plant Reopens under Paccar Label', *National Post*, 3 Aug., C7.

Transport Canada. 2009. 'Detroit River International Crossing Project Receives Environmental Approval', 3 Dec. At: <www.tc.gc.ca/eng/mediaroom/releases-2009-h166e-5754.htm>.

Chapter 6

Blackwell, Richard. 2008. 'Quebec Snags New Silicon Plant', *Globe and Mail*, 25 Aug., B1.

Bombardier. 2010. Media Centre. At: <www.bombardier.com/en/corporate/media-centre/press-releases>.

Bone, Robert M. 2009. *The Canadian North: Issues and Challenges*, 3rd edn. Toronto: Oxford University Press.

Bouchard, Gérard, and Charles Taylor. 2008. *Building the Future: A Time for Reconciliation (Abridged Report)*. Québec: Commission de consultation sur les pratiques d'accommodement reliées aux differences culturelles. At: <www.accommodements.qc.ca/index-en.html>.

Bouchard, Lucien, et al. 2006. *Manifest—Pour un Québec lucide*. At: <www.pourunquebeclucide.com/cgi-cs/cs.waframe.content?topic=28226&lang=1>.

Bumsted, J.M. 2007. *A History of the Canadian Peoples*, 3rd edn. Toronto: Oxford University Press.

Canada. 2006.' Net Merchantable Volume of Roundwood Harvested by Category and Province/Territory, 1940–2005', National Forestry Database Program, Table 5.1. At: <www.nfdp.ccfm.org/compendium/products/tables_index_e.php>.

Canada's Innovation Leaders. 2006. 'Canada's Top 20 Research Communities 2006', *National Post*, Research infosource supplement, 3 Nov., 7.

Canadian Hydropower Association. 2009a. *Hydropower in Canada: Past, Present and Future*. At: <www.canhydropower.org/hydro_e/pdf/hydropower_past_present_future_en.pdf>.

———.2009b. 'The Canadian Hydropower Association Applauds the Launch of the Romaine Hydroelectric Project'. At: <www.canhydropower.org/hydro_e/pdf/2009-05-14_La_Romaine_news_release.pdf>.

CBC News. 2010. 'Hydro-Québec Signs Vermont Power Deal', 11 Mar. At: <www.cbc.ca/money/story/2010/03/11/mtl-qc-vermont-power-deal.html?ref=rss>.

Cross, Philip. 2009. '2008 in Review', *Canadian Economic Observer*. At: <www.statcan.gc.ca/pub/11-010-x/2009004/part-partie3-eng.htm>.

Dougherty, Kevin. 2007. 'Quebec the First to Announce Carbon Tax', *National Post*, 7 June, A1, A10.

Duhaime, Gérard, Nick Bernard, and Robert Comtois. 2005. 'An Inventory of Abandoned Mining Exploration Sites in Nunavik, Canada', *Canadian Geographer* 49, 3: 260–71.

Ferrao, Vincent. 2006. 'Recent Changes in Employment by Industry', *Perspectives on Labour and Income* (Statistics Canada) 7, 1. At: <www.statcan.ca/english/freepub/75-001-XIE/10106/art-1.htm>.

Fischer, David Hackett. 2008. *Champlain's Dream*. New York: Simon & Schuster.

George, Jane. 2009. 'Makivik Nickel Share Dwindles to $6.8 Million', *Nunatsiaq News*, 19 June. At: <www.nunatsiaqnews.com/archives/2009/906/90619/news/nunavik/90619_2260.html>.

Hodgins, Bruce W., Richard P. Bowles, James L. Hanley, and George A. Rawlyk. 1974. *Canadiens, Canadians and Québécois*. Scarborough, Ont.: Prentice-Hall.

Hornig, James F., ed. 1999. *Social and Environmental Impacts of the James Bay Hydroelectric Project*. Montréal and Kingston: McGill-Queen's University Press.

Hydro-Québec. 2006a. 'Eastmain-1 Hydroelectric Development'. At: <www.hydroquebec.com/eastmain1/en/batir/fiche_7.html>.

———. 2006b. 'Eastmain-1 Hydroelectric Development'. At: <www.hydroquebec.com/eastmain1/en/batir/etapes_photos.html?list=4>.

———. 2009a. 'Eastmain-1-A Powerhouse'. At: <www.hydroquebec.com/rupert/en/batir/etapes_photos.html?list=7>.

———. 2009b. 'Projet de la Romaine'. At: <www.hydroquebec.com/romaine/projet/galerie.html>.

———. 2009c. *Romaine Complex: Project Information*, Feb. At: <www.hydroquebec.com/romaine/pdf/doc_info_2009_en.pdf>.

———. 2010. 'Eastmain-1-A/Sarcelle/Rupert Project: Schedule', 25 Jan. At: <www.hydroquebec.com/rupert/en/echeancier.html>.

Jenish, D'Arcy. 2009. 'Inland Superhighway', *Canadian Geographic* 129, 4: 34–48.

Kativik Regional Government. 2009. 'Proposed Timeline for the Creation of the Nunavik Regional Government (NRG)', *Towards a Government of Nunavik*. At: <www.nunavikgovernment.ca/en/archives/nunavik_regional_government/towards_a_government_of_nunavik.html>.

Koplow, Douglas. 1994. 'Federal Energy Subsidies and Recycling: A Case Study', *Resource Recycling*, Nov. At: <www.p2pays.org/ref/06/05940.pdf>.

Legault, Josée. 2009. 'Sabia Meeting Shows How Much Quebec Inc. Has Changed', *The Gazette* (Montréal), Apr. At: <www.vigile.net/Sabia-meeting-shows-how-much>.

Leger, Kathryn. 2000. 'Biochem Sold for $5.9-billion', *National Post*, 12 Dec., C1, C6.

Lucas, C.P. 1912. *Lord Durham's Report of the Affairs of British North America*, vol. 1. Oxford: Clarendon Press.

McCutcheon, Sean. 1991. *Electric Rivers: The Story of the James Bay Project*. Montréal: Black Rose Books.

Marotte, Bertrand. 2009a. 'With Glowing Hearts, an Algerian City Rises', *Globe and Mail*, 9 July. At: <www.theglobeandmail.com/globe-investor/with-glowing-hearts-an-algerian-city-rises/article1212917/>.

———. 2009b. 'Bastion of Quebec Inc. Cuts Its Final Deal', *Globe and Mail*, 17 July, B1.

Marowits, Ross. 2009. 'Bombardier Cancels Learjet Order', *Globe and Mail*, 20 Aug. At: <www.theglobeandmail.com/globe-investor/bombardier-cancels-learjet-order/article1258306/>.)

———. 2010. 'SNC-Lavalin Focuses on Global Growth', *Globe and Mail*, 17 Mar., B3.

Martin, Thibault, and Steven M. Hoffman, eds. 2008. *Power Struggles: Hydro Development and First Nations in Manitoba and Quebec*. Winnipeg: University of Manitoba Press.

Moreault, Éric. 2009a 'Le Saint-Laurent encore «vulnérable» à la pollution', *Le Soleil*, 30 juin. At: <www.cyberpresse.ca/le-soleil/actualites/environnement/200906/30/01-880288-le-saint-laurent-encore-vulnerable-a-la-pollution.php>.

———. 2009b. 'Une autre "erreur boréale"?', *Le Soleil*, 6 julliet. At: <www.cyberpresse.ca/le-soleil/actualites/environnement/200907/05/01-881433-une-autre-erreur-boreale.php>.

Mills, David. 1988. 'Durham Report', in James H. Marsh, ed., *The Canadian Encyclopedia*, 2nd edn. Edmonton: Hurtig, 637–8.

Murphy, Michael. 2007. *Canada: The State of the Federation, 2005. Quebec and Canada in the New Century: New Dynamics, New Opportunities*. Montréal and Kingston: McGill-Queen's University Press.

Natural Resources Canada. 2009a. 'High Nonmetal Prices Help Push Canadian Mineral Production to Over $45 Billion in 2008', *Information Bulletin*, Mar. At: <mmsd.mms.nrcan.gc.ca/stat-stat/prod-prod/PDF/ib2009_e.pdf>.

———. 2009b. 'Key Facts: Quebec', 17 July. At: <canada-forests.nrcan.gc.ca/keyfacts/qc>.

———. 2010. 'The Global Recession Reduced Canada's Mineral Production in 2009', *Information Bulletin*, Mar. At: <mmsd.mms.nrcan.gc.ca/stat-stat/prod-prod/PDF/ib2010_e.pdf>.

Paquin, Stephane. 1999. *Invention d'un mythe: Le pacte entre deux peuples fondateurs*. Montréal: VLB.

Padova, Allison. 2004. *Trends in Containerization at Canadian Ports*. Ottawa: Library of Parliament. At: <www.parl.gc.ca/information/library/PRBpubs/prb0575-e.htm>.

Québec. 2003. The Québec Forest Industry, Québec Forest Industries Association. At: <www.aifq.qc.ca/english/stats/papers01.html>.

———. 2006a. The Québec Forest Industry, Québec Forest Industries Association. At: <www.cifq.qc.ca/html/english/sciage/statistiques.php>.

————. 2006b. 'Québec Statistiques Touristiques: Le tourisme au Québec en bref—2004'. At: <www.google.ca/search?hl=en&client=firefox-a&rls=org.mozilla%3Aen-US%3Aofficial_s&q=le+tourisme+au+quebec+en+bref&btnG=Search&meta=>.

Reuters. 2009. 'Lavalin Buys Stake in Russian Firm', *Globe and Mail*, 16 Aug. At: <www.theglobeandmail.com/report-on-business/snc-lavalin-buys-stake-in-russian-firm/article1252431/>.

Richardson, Boyce. 1975. *Strangers Devour the Land: The Cree Hunters of the James Bay Area versus Premier Bourassa and the James Bay Development Corporation*. Toronto: Macmillan.

Robitaille, Antoine. 2008. 'Charest mise sur le Nord Charest', *Le Devoir.Com*, 29 Apr. At: <translate.google.com/translate?hl=en&sl=fr&u=http://www.ledevoir.com/2008/09/29/208131.html&ei=b6JnSpWOHYfQlAfSpLnACQ&sa=X&oi=translate&resnum=6&ct=result&prev=/search%3Fq%3Dplan%2Bnord%2BPremier%2BChare st%26hl%3Den>.

Rodrigue, Jean-Paul, Claude Comtois, and Brian Slack. 2009. *The Geography of Transportation Systems*, 2nd edn. At: <www.people.hofstra.edu/geotrans/eng/ch4en/appl4en/ch4a3en.html>.

Roslin, Alex. 2001. 'Cree Deal a Model or Betrayal?', *National Post*, 10 Nov., FP7.

Royal Bank of Canada. 2010. *Provincial Outlook: March 2010*. At: <www.rbc.com/economics/market/pdf/provfcst.pdf>.

Salisbury, Richard Frank. 1986. *A Homeland for the Cree: Regional Development in James Bay, 1971–1981*. Montréal and Kingston: McGill-Queen's University Press.

Scoffield, Heather. 2009. 'Few Bumps in La Belle Province's Recession Ride', *Globe and Mail*, 4 Aug., B3.

Séguin, Rhéal. 2009. 'Tiny Quebec Town Is Sitting on a Gold Mine', *Globe and Mail*, 14 July. At: <www.theglobeandmail.com/news/national/tiny-quebec-town-is-sitting-on-a-gold-mine/article1217078>.

Service Canada. 2006. 'Sectorial Outlook: Province of Québec'. At: <www150.hrdc-drhc.gc.ca/imt/regional/english/sec_out/2006-2008/introduction/index.html>.

Simard, Jean-Jacques, et al. 1996. *Tendances Nordiques: Les changements sociaux 1970–1990 chez les Cris et les Inuit du Québec: Un enquête statistique exploratoire*. Québec City: GETIC, Université Laval.

Simpson, Jeffrey. 2009. 'Champlain's Dream Lives on in North America', *Globe and Mail*, 23 Oct. At: <www.theglobeandmail.com/news/opinions/champlains-dream-lives-on-in-north-america/article1336484/>.

South Africa Info. 2008. 'Major Milestone for Gautrain', 11 July. At: <www.southafrica.info/business/economy/infrastructure/gautrain-trains.htm>.

Statistics Canada. 2002. '2001 Census: Population and Dwelling Counts, for Census Metropolitan Areas and Census Agglomerations, 2001 and 1996 Censuses', 16 July. At: <www.statcan.ca/english/IPS/Data/93F0050XCB2001013.htm>.

————. 2006a. 'Distribution of Employed People, by Industry, by Province'. At: <www40.statcan.ca/101/cst01/labor21c.htm>.

————. 2006b. 'Canadian Statistics—Net Merchantable Volume of Roundwood Harvested'. At: <www.statcan.ca/english/Pgdb/prim41.htm>.

————. 2007. 'Population and Dwelling Counts, for Census Metropolitan Areas and Census Agglomerations, 2006 and 2001 Censuses—100% Data'. At: <www12.statcan.ca/english/census06/data/popdwell/Table.cfm?T=201&S=3&O=D&RPP=150>.

————. 2008. 'Total Farm Area, Land Tenure, and Land in Crops, by Provinces, 1986 to 2006', *Census of Agriculture*. At: <www40.statcan.gc.ca/l01/cst01/agrc25f-eng.htm>.

————. 2009a. 'Individuals by Total Income Level by Province and Territory'. At: <www40.statcan.ca/l01/cst01/famil105a-eng.htm>.

————. 2009b. 'Gross Domestic Product, Expenditure-based, by Province and Territory'. At: <www40.statcan.gc.ca/cbin/fl/cstprintflag.cgi>.

————. 2009c. 'Canada's Population Estimates', *The Daily*, 23 June. At: <www.statcan.gc.ca/daily-quotidien/090623/dq090623a-eng.htm>.

————. 2009d. 'Distribution of Employed People, by Industry, by Province'. At: <www40.statcan.gc.ca/l01/cst01/labor21c-eng.htm>.

————. 2009e. *Using Languages at Work in Canada, 2006 Census: Provinces and Territories*, 2006 Analysis Series. At: <www12.statcan.gc.ca/census-recensement/2006/as-sa/97-555x/p4-eng.cfm>.

————. 2009f. 'Birth Rate, by Province and Territory'. At: <www40.statcan.gc.ca/l01/cst01/demo04b-eng.htm>.

Tower, Courtney. 2010. 'Port Montreal Sees a Gradual Recovery in 2010', *Journal of Commerce*, 4 Jan. At: <www.joc.com/maritime/port-montreal-sees-gradual-recovery-2010>.

Turgeon, Pierre. 1992. *La Radissonie: le pays de la Baie James*. Montréal: Libre Expressions.

Vanderklippe, Nathan. 2010. 'Shell's 76-year-old Gasoline Facility Becomes Second Major Canadian Plant to Close in Recent Years', *Globe and Mail*, 8 Jan., B1.

Wyman, Diana. 2006. 'Trade Liberalization and the Canadian Clothing Market', *Canadian Economic Observer*. Catalogue no. 11–010. Ottawa. At: <www.statcan.ca/english/ads/11-010-XPB/pdf/dec06.pdf>.

Yakabuski, Konrad. 2008. 'Environment? Economics? Hydro-Québec Project Fails Grade', *Globe and Mail*, 14 Aug., B1.

Chapter 7

Barnes, Trevor, and Roger Hayter. 1997. *Trouble in the Rainforest: British Columbia's Forest Economy in Transition*. Canadian Western Geographical Series, vol. 33. Victoria: Western Geographical Press.

BC Facts. 2009. 'Hollywood North'. At: <www.gov.bc.ca/bcfacts/>.

BC Hydro. 2009. 'Bioenergy Call for Power', 18 Nov. At: <www.bchydro.com/planning_regulatory/acquiring_power/bioenergy_call_for_power.html>.

BC Ministry of Transportation. 2007a. *BC Transportation Financing Authority—Statement of Earnings*. At: <www.bc-budget.gov.bc.ca/2007/sp/trans/default.aspx?hash=8>.

———. 2007b. 'Pacific Gateway'. At: <www.th.gov.bc.ca/PacificGateway/index.htm>.

BC Stats. 2006. *Quick Facts about British Columbia 2006 Edition*. Victoria. At: <www.bcstats.gov.bc.ca/data/bcfacts.asp>.

———. 2009a. *Quick Facts about British Columbia 2009 Edition*. Victoria. At: <www.bcstats.gov.bc.ca/data/bcfacts.asp>.

———. 2009b. 'Exports 1999–2008'. At: <www.bcstats.gov.bc.ca/pubs/exp/exp_ann.pdf>.

BC Treaty Commission. 2007. 'Lheidli T'enneh First Nation Vote to Reject the Treaty', Apr. At: <www.bctreaty.net/files/pdf_documents/April07update.pdf>.

Boei, Bill, and Amy O'Brian. 2003. '10,000 Flee Kelowna, B.C., Blaze', *National Post*, 23 Aug., A1.

Bone, Robert M. 2009. *The Canadian North: Issues and Challenges*, 3rd edn. Toronto: Oxford University Press.

CBC News. 2009. 'Queen Charlotte Islands Renamed Haida Gwaii in Historic Deal', 11 Dec. At: <www.cbc.ca/canada/british-columbia/story/2009/12/11/bc-queen-charlotte-islands-renamed-haida=gwaii.html>.

Christensen, Bev. 1995. *Too Good to Be True: Alcan's Kemano Completion Project*. Vancouver: Talonbooks.

Clapp, R.A. 2008. 'The Resource Cycle in Forestry and Fishing', *Canadian Geographer* 42, 2: 129–44.

CN. 2007. 'Port of Prince Rupert Container Terminal'. At: <www.cn.ca/specialized/ports_docks/prince_rupert/fairview/en_KFPortsPrinceRupert_fairview.shtml>.

CTV News. 2009. 'Huge Fire Impacts 10,000 in Kelowna, BC', 18 July. At: <www.ctvbc.ctv.ca/servlet/an/local/CTVNews/20090718/BC_west_kelowna_fire_090718/?hub=BritishColumbiaHome>.

David Suzuki Foundation. 2005. *Clearcutting Canada's Rainforest, Status Report 2005: Canada's Rainforests under Threat*. At: <www.davidsuzuki.org/files/Forests/DSF-rainforests-2005-5.pdf>.

———. 2009. 'Sea Lice'. At: <www.davidsuzuki.org/Oceans/Aquaculture/Salmon/Sea_Lice.asp>.

Ebner, David. 2009. 'Waiting for His Ships to Come In', *Globe and Mail*, 24 Sept., B3.

Export Development Canada. 2009. 'British Columbia's Exports to Drop Significantly in 2009 amid Forest and Energy Woes, Says EDC', 6 May. At: <www.edc.ca/english/docs/news/2009/mediaroom_16357.htm>.

Fisheries and Oceans Canada. 2009. 'Pacific Salmon Treaty Renewal', 5 Jan. At: <www.dfo-mpo.gc.ca/media/backfiche/2009/pr01-eng.htm>.

Harris, Cole. 1997. *The Resettlement of British Columbia: Essays on Colonialism and Geographical Change*. Vancouver: University of British Columbia Press.

Hayter, Roger. 1992. 'The Little Town That Did: Flexible Accumulation and Community Response in Chemainus, British Columbia', *Regional Studies* 26: 647–63.

Hume, Mark. 2009. 'Millions of Missing Fish Signal Crisis on the Fraser River', *Globe and Mail*, 12 Aug., B1.

Hume, Stephen. 2000. 'Did Francis Drake Discover B.C.?', *National Post*, 5 Aug., B1–B2.

Hutchison, Bruce. 1942. *The Unknown Country: Canada and Her People*. Toronto: Longmans, Green and Company.

Indian and Northern Affairs Canada. 2007. 'Backgrounder—Wei Wai Kum Cruise Ship Port'. At: <www.ainc-inac.gc.ca/nr/prs/m-a2007/2-2892-bk-eng.asp>.

Industry Advisory Group. 2006. *Pacific Gateway Strategy Action Plan 2006–2020*. Vancouver. At: <www.th.gov.bc.ca/PacificGateway/documents/PGS_Action_Plan_043006.pdf>.

Katz, Diane. 2010. 'The Agricultural Land Reserve Doesn't Work—So Let's Get Rid of It', *BCBusiness Online*, 6 Jan. At: <www.bcbusinessonline.ca/bcb/business-sense/2010/01/06/alr-tear-down-wall>.

McCullough, Michael. 1998. *Granville Island: An Urban Oasis*. Vancouver: CMHC.

McKinnell, Skip, Robert Emmett, and Joseph Orsi. 2009. '2009 Salmon Forecasting Forum', North Pacific Marine Science Organization, *PICES Press* 17, 2: 32–3.

McVey, Wayne W., and W.E. Kalbach. 1995. *Canadian Population*. Toronto: Nelson Canada.

Mineral Resources Education Program of BC. 2009. At: <www.bcminerals.ca/files/bc_mine_information.php>.

Mittelstaedt, Martin. 2004. 'Farmed Salmon Laced with Toxins, Study Finds', *Globe and Mail*, 9 Jan., A1.

Morton, Alexandra, Richard Toutledge, Corey Peet, and Aleria Ladwig. 2004. 'Sea Lice (*Lepeophtheirus salmonis*) Infection Rates on Juvenile Pink (*Oncorhynchus gorbuscha*) and Chum (*Oncorhynchus keta*) Salmon in the Nearshore Marine Environment of British Columbia, Canada', *Canadian Journal of Fisheries and Aquatic Sciences* 61, 2: 147–57.

Natural Resources Canada. 2006. 'About Hydroelectric Energy'. At: <www.canren.gc.ca/tech_appl/index.asp?CaId=4&PgId=26>.

———. 2010. 'The Global Recession Reduced Canada's Mineral Production in 2009', *Information Bulletin*, Mar. At: <mmsd.mms.nrcan.gc.ca/stat-stat/prod-prod/PDF/ib2010_e.pdf>.

Nemetz, Peter N. 1990. *The Pacific Rim Investment, Development and Trade*, 2nd rev. edn. Vancouver: University of British Columbia Press.

Nisga'a. n.d. 'Nisga'a History'. At: <www.schoolnet.ca/aboriginal/nisga1/hist-e.html>.

Padova, Allison. 2004. *Trends in Containerization at Canadian Ports*. Ottawa: Library of Parliament. At: <www.parl.gc.ca/information/library/PRBpubs/prb0575-e.htm>.

Prince Rupert Port Authority. 2007. 'The New World Port'. At: <www.rupertport.com/about.htm>.

———. 2010. 'Prince Rupert Records 12-Year-High Cargo Volumes in 2009', 25 Jan. At: <www.rupertport.com/pdf/newsreleases/ppr_records_12yearhigh2009cargovolumes.pdf>.

Provincial Agricultural Land Commission. 2009. 'The Commission'. At: <www.alc.gov.bc.ca/commission/alc_main.htm>.

Rio Tinto Alcan. 2010. 'Kitimat Works Modernization'. At: <www.kitimatworksmodernization.com/pages/modernization-project/about-kitimat-modernization.php>.

Schrier, Dan. 2010. 'B.C. Origin Exports to Selected Destinations', *Exports (BC Origin) 2000–2009*, Mar. At: <www.bcstats.gov.bc.ca/pubs/exp/exp_ann.pdf>.

Statistics Canada. 2006. 'Distribution of Employed People by Industry, by Province, 2005'. At: <www40.statcan.ca/l01/cst01/labor21c.htm>.

———. 2007. 'Population and Dwelling Counts, for Census Metropolitan Areas and Census Agglomerations, 2006 and 2001 Censuses—100% Data'. At: <www12.statcan.ca/english/census06/data/popdwell/Table.cfm?T=201&S=3&O=D&RPP=150>.

———. 2008. *Oil and Gas Extractions—2007*, Table 5-1, 'Commodity Data, Oil and Gas Extractions—Net Production Withdrawals'. Catalogue no. 26–213–X. At: <www.statcan.gc.ca/pub/26-213-x/2007000/t018-eng.pdf>.

———. 2009. 'Distribution of Employed People by Industry, by Province, 2008'. At: <www40.statcan.gc.ca/l01/cst01/labor21c-eng.htm>.

VanderKlippe, Nathan. 2007a. 'Lumber Waste a Hot Commodity', *National Post*, 6 June, FP1, FP6.

———. 2007b. 'A Town's Ship Comes In', *National Post*, 9 June, FP4.

White, P., M. Michalowski, and P. Cross. 2006. 'The West Coast Boom', *Canadian Economic Observer*, May. Ottawa: Statistics Canada Catalogue no. 11–010.

Wynn, Graeme, and Timothy Oke, eds. 1992. *Vancouver and Its Region*. Vancouver: University of British Columbia Press.

Chapter 8

Agriculture and Agri-Food Canada. 2009. *Overview of the Canadian Agriculture and Agri-Food System 2008*. At: <www4.agr.gc.ca/AAFC-AAC/display-afficher.do?id=1228246364385&lang=eng#a2>.

Akinremi, O.O., S.M. McGinn, and H.W. Cutforth. 2001. 'Seasonal and Spatial Patterns of Rainfall on the Canadian Prairies', *Journal of Climate* 14, 9. 2177–82.

Alberta. 2007. 'Oil Reserves'. At: <www.energy.gov.ab.ca/docs/oil/pdfs/AB_OilReserves.pdf>.

Bell, Ian. 1999. 'Dancing with Elephants', *Western Producer*, 2 Sept., 60–1.

Bonsal, B.R., X. Zhang, and W.D. Hogg. 1999. 'Canadian Prairie Growing Season Precipitation Variability and Associated Atmospheric Circulation', *Climate Research* 11: 191–208.

Bradsher, Keith. 2007. 'Rise in China's Pork Prices Signals End to Cheap Output', *New York Times*, 8 June. At: <www.nytimes.com/2007/06/08/business/worldbusiness/08prices.html>.

Bramley, Matthew, Pierre Sadik, and Dale Marshall. 2008. *Climate Leadership, Economic Prosperity: Final Report on an Economic Study of Greenhouse Gas Targets and Policies for Canada*. Pembina Institute and David Suzuki Foundation. At: <pubs.pembina.org/reports/climate-leadership-report-en.pdf>.

Gonick, Noam. 2004. *Stryker*. Winnipeg: Telefilm Canada. At: <www.strykerthemovie.com>.

Grauman, Meny. 2006. 'Alberta Profile—Economic View', *Provincial Pulse* (Scotiabank), 3 Feb., 1.

Greenwood, John. 2007. 'Surging Prices Separate Wheat from the Chaff', *National Post*, 15 June, FP1.

Hall, Angela. 2003. 'Hogs Touchy Issue in Rural Sask.', *StarPhoenix* (Saskatoon), 25 Feb., C8.

Hugenholtz, D.H., and S.A. Wolfe. 2005. 'Biogeomorphic Model of Dune Activation and Stabilization on the Northern Great Plains', *Geomorphology* 70: 53–70.

Hursh, Kevin. 1999. 'Subsidy Complaints Don't Hold Water', *StarPhoenix* (Saskatoon), 28 July, C9.

Jones, David C. 1987. *Empire of Dust: Settling and Abandoning the Prairie Dry Belt*. Edmonton: University of Alberta Press.

Jubinville, Mike. 2009. 'Market Focus—CWB PRO Review', *Farm Credit Canada*, 2 Oct. At: <www.fcc-fac.ca/newsletters/en/express/articles/20091002_e.asp>.

Kroeger, Authur. 2007. *Hard Passage: A Mennonite Family's Long Journey from Russia to Canada*. Edmonton: University of Alberta Press.

———. 2009. *Retiring the Crow Rate: A Narrative of Political Management*. Edmonton: University of Alberta Press.

MacLachlan, Ian. 2001. *Kill and Chill: Restructuring Canada's Beef Commodity Chain*. Toronto: University of Toronto Press.

National Energy Board. 2006. *Canada's Oilsands: Opportunities and Challenges to 2015—An Update*. Calgary: Publication Office, National Energy Board. At: <www.neb.gc.ca/clf-nsi/rnrgynfmtn/nrgyrprt/lsnd/pprtntsndchllngs20152006/pprtntsndchllngs20152006-eng.pdf>.

Natural Resources Canada. 2009. *The State of Canada's Forests 2009*. At: <canadaforests.nrcan.gc.ca/rpt>.

———. 2010. 'The Global Recession Reduced Canada's Mineral Production in 2009', *Information Bulletin*, Mar. At: <mmsd.mms.nrcan.gc.ca/stat-stat/prod-prod/PDF/ib2010_e.pdf>.

Oilsands Developers Group. 2009. 'Oilsands Projects', 30 June. At: <www.oilsandsdevelopers.ca/index.php/test-project-table/>.

Paul, Alec H. 1997. 'Shortlines, Mainlines, Branchlines, Dead Lines: Rural Railways in Southwestern Saskatchewan in the 1990s', in John Welsted and John Everitt, eds, *The Yorkton Papers: Research by Prairie Geographers*. Brandon Geographical Studies No. 2. Brandon, Man.: University of Brandon.

Price, Jacqueline D. 2003. 'Are Factory Farms Fouling Our Water?', *Alberta Views* (May–June): 34–9.

Richards, J. Howard. 1968. 'The Prairie Region', in John Warkentin, ed., *Canada: A Geographical Interpretation*. Toronto: Methuen, ch. 12.

Rodrigue, Jean-Paul, Claude Comtois, and Brian Slack. 2009. *The Geography of Transport Systems*. Toronto: Routledge.

Saskatchewan Ministry of Agriculture. 2009a. 'September Estimate of 2009 Crop Production'. At: <www.agriculture.gov.sk.ca/Default.aspx?DN=d8fe5165-80e6-46b4-ba5f-217e2a0184a2>.

———. 2009b. 'Crop Statistics'. At: <www.agriculture.gov.sk.ca/agriculture_statistics/HBv5_P2.asp>.

———. 2009c. *Agricultural Statistics Fact Sheet*. At: <www.agriculture.gov.sk.ca/Saskatchewan_Agricultural_Statistics_Fact_Sheet>.

Schwartz, Jennifer. 2009. 'US Oil Imports from Canada Hit New High in July', 7 Oct. At: <www.heatingoil.com/blog/us-oil-imports-from-canada-hit-new-high-in-july-1007/>.

Spry, Irene M. 1963. *The Palliser Expedition: An Account of John Palliser's British North American Expedition 1857–1860*. Toronto: Macmillan.

Statistics Canada. 1992. *Census Overview of Canadian Agriculture 1971–1991*. Catalogue no. 93–348. Ottawa: Ministry of Industry, Science and Technology.

———. 2003. 'Agriculture 2001 Census Farm Operations: Provincial/Regional Trends'. At: <www.statcan.ca/english/agcensus2001/first/regions/contents.htm>.

———. 2006. 'Exports of Goods on a Balance-of-Payments Basis, by Product'. At: <www40.statcan.ca/l01/cst01/gblec05.htm>.

———. 2007a. 'Total Farm Area, Land Tenure and Land in Crops, by Province', *2006 Census of Agriculture*. At: <www40.statcan.ca/l01/cst01/agrc25j.htm>.

———. 2007b. 'Population and Dwelling Counts, for Canada, Provinces and Territories, 2006 and 2001 Censuses—100% Data'. At: <www12.statcan.ca/english/census06/data/popdwell/Table.cfm?T=101>.

———. 2008. 'Aboriginal Identity Population, 2006 Counts, Percentage Distribution, Percentage Change for Both Sexes, for Canada, Provinces and Territories—20% Sample Data'. At: <www12.statcan.gc.ca/english/census06/data/highlights/Aboriginal/pages/Page.cfm?Lang=E&Geo=PR&Code=48&Table=>.

———. 2009a. *Energy Statistics Handbook*. Catalogue no. 57–601–X. At: <www.statcan.gc.ca/pub/57-601-x/57-601-x2009002-eng.pdf>.

———. 2009b. 'Distribution of Employed People by Industry, by Province, 2008'. At: <www40.statcan.gc.ca/l01/cst01/labor21c-eng.htm>.

———. 2009c. 'Population, Age Distribution and Median Age by Province and Territory, as of July 1, 2008', Table 1. At: <www.statcan.gc.ca/daily-quotidien/090115/t090115c1-eng.htm>.

———. 2009d. *Aboriginal Peoples in Canada in 2006: Inuit, Métis and First Nations*, 2006 Census. At: <www12.statcan.gc.ca/census-recensement/2006/as-sa/97-558/p3-eng.cfm#01>.

———. 2010. 'Exports of Goods on a Balance-of-Payments Basis, by Product', *Merchandise Exports*, 11 Mar. At: <www40.statcan.gc.ca/l01/cst01/gblec04-eng.htm>.

Vanderhaeghe, Guy. 1996. *The Englishman's Boy*. Toronto: McClelland & Stewart.

Wiebe, Rudy. 1973. *The Temptations of Big Bear*. Toronto: McClelland & Stewart.

Wolfe, S.A., C.H. Hugenholtz, C.P. Evans, D.J. Huntley, and J. Ollerhead. 2007. 'Potential Aboriginal-Occupation-Induced Dune Activity, Elbow Sand Hills, Northern Great Plains, Canada', *Great Plains Research* 17: 173–92.

WTRG Economics. 2009. 'Oil Price History and Analysis'. At: <www.wtrg.com/prices.htm>.

Chapter 9

Bennett, Margaret. 1989. *The Last Stronghold: Scottish Gaelic Traditions in Newfoundland*. Edinburgh: Canongate.

Bissett, Kevin. 2010. 'N.B. Premier Pulls Plug on Power Sale to Quebec', *Toronto Star*, 25 Mar., A8.

Blades, Kent. 1995. *Net Destruction: The Death of Atlantic Canada's Fishery*. Halifax: Nimbus.

Bradfield, Michael. 1991. *Maritime Economic Union: Sounding Brass and Tinkling Symbolism*. Halifax: Canadian Centre for Policy Alternatives.

Canadian Assocation of Petroleum Producers. 2009. 'Expenditures and Revenues', *Statistical Handbook*. At: <www.capp.ca/library/statistics/handbook/Pages/default.aspx#ppkIZWFa77H8>.

Canada, Department of Finance. 2009. 'Equalization Program'. At: <www.fin.gc.ca/fedprov/eqp-eng.asp>.

Canadian Human Rights Commission. 2003. *Report to the Canadian Human Rights Commission on the Treatment of the Innu of Canada by the Government of Canada*, by Professors Constance Backhouse and Donald M. McRae. At: <www.chrc-ccdp.ca/publications/Rapport_Innu_Report/RapportInnuReport_Page3.asp?l=e>.

Cashin, Richard. 1993. *Charting a New Course: Towards the Fishery of the Future*. Ottawa: Department of Fisheries and Oceans.

CBC News. 2007. 'Long Commute, Huge Rewards', 29 Oct. At: <www.cbc.ca/canada/newfoundland-labrador/story/2007/10/29/big-commute.html>.

———.2009. 'The Big Commute', 29 Oct. At: <www.cbc.ca/nl/features/bigcommute/>.

———. 2010. 'NB Power Sale Cancelled', 24 Mar. At: <www.cbc.ca/canada/new-brunswick/story/2010/03/24/nb-nbpower-graham-1027.html>.

Clapp, R.A. 1998. 'The Resource Cycle in Forestry and Fishing', *Canadian Geographer* 42, 2: 129–44.

Coates, Ken S. 2000. *The Marshall Decision and Native Rights*. Montréal and Kingston: McGill-Queen's University Press.

Conrad, Cathy T. 2009. *Severe and Hazardous Weather in Canada: The Geography of Extreme Events*. Toronto: Oxford University Press.

Cox, Kevin. 1994. 'How Hibernia Will Cast Off', *Globe and Mail*, 12 Nov., D8.

Department of Labrador and Aboriginal Affairs. 2009. 'Land Claims: Innu Nation of Labrador'. At: <www.laa.gov.nl.ca/laa/land_claims/index.html>.

Erskine, David. 1968. 'The Atlantic Region', in John Warkentin, ed., *Canada: A Geographical Interpretation*. Toronto: Methuen, 231–80.

Faragher, John Mack. 2005. *A Great and Noble Scheme: The Tragic Story of the Expulsion of the French Acadians from Their American Homeland*. New York: Norton.

Feehan, James P., and Melvin Baker. 2005. *The Renewal Clause in the Churchill Falls Contract: The Origins of a Coming Crisis*, Papers in Political Economy No. 96. London: University of Western Ontario, Political Economy Research Group.

Fisheries and Oceans Canada. 2009. 'Landings: Seafisheries'. At: <www.dfo-mpo.gc.ca/stats/commercial/sea-maritimes-eng.htm>.

Griffiths, N.E.S. 2005. *From Migrant to Acadian: A North American Border People 1604–1755*. Montréal and Kingston: McGill-Queen's University Press.

Hardin, Garrett. 1968. 'The Tragedy of the Commons', *Science* 162: 1243–8.

Holden, Michael. 2008. *Canada's New Equalization Formula*. Ottawa: Library of Parliament, PRB 08–20E.

Llewellyn, Stephen. 2008. 'Nackawic Mill May Shut Down for 4–6 Weeks', *Daily Gleaner* (Fredericton), 11 Nov. At: <dailygleaner.canadaeast.com/rss/article/477340>.

McCarthy, Shawn. 2007. 'Nfld. Grabs a Seat at Oil Table', *Globe and Mail*, 23 Aug. At: <www.reportonbusiness.com/servlet/story/RTGAM.20070823.whebron0823/BNStory/robNews/home>.

MacKenzie, A.A. 1979. *The Irish in Cape Breton*. Antigonish, NS: Formac.

Macpherson, Alan G., ed. 1972. *The Atlantic Provinces: Studies in Canadian Geography*. Toronto: University of Toronto Press.

Matthews, Ralph. 1983. *The Creation of Regional Dependency*. Toronto: University of Toronto Press.

Moore, Oliver. 2009. 'Innu Reach Deal on Lower Churchill Project', *Globe and Mail*, 26 Sept., B1.

Natural Resources Canada. 2009a. 'Key Facts in the Forest Industry, New Brunswick'. At: <canadaforests.nrcan.gc.ca/keyfacts/nb>.

———. 2009b. 'Canada's Forests: Statistical Data', *Forest Management*. At: <canadaforests.nrcan.gc.ca/statsprofile/forest/pe>.

———. 2009c. 'Mineral Production of Canada, by Provinces and Territory'. At: <mmsd.mms.nrcan.gc.ca/stat-stat/prod-prod/2008p-eng.aspx>.

Phillips, David. 1993. *The Day Niagara Falls Ran Dry!* Toronto: Canadian Geographic and Key Porter Books.

Physorg.com. 2006. 'Ocean Study Predicts the Collapse of All Seafood Fisheries by 2050'. At: <www.physorg.com/news81778444.html>.

Power, Thomas P., ed. 1991. *The Irish in Atlantic Canada, 1780–1900*. Fredericton: New Ireland Press.

Roberts, Terry. 2009. 'Construction Begins on $22B Nickel Plant in Long Harbour', 17 Apr. At: <www.dailybusiness-buzz.ca/2009/04/17/construction-begins-on-22b-nickel-plant-in-long-harbour/>.

Samson, Colin. 2003. *A Way of Life That Does Not Exist: Canada and the Extinguishment of the Innu*. St John's: ISER Books.

Statistics Canada. 2002. *Census of Canada 2001—Census Geography. Highlights and Analysis: Canada's 2001 Population*. At: <www12.statcan.ca/English/census01>.

———. 2007a. 'Population and Dwelling Counts, for Census Metropolitan Areas and Census Agglomerations, 2006 and 2001 Censuses—100% Data'. At: <www12.statcan.ca/english/census06/data/popdwell/Table.cfm?T=201&S=3&O=D&RPP=150>.

———. 2007b. 'Labour Force, Employed and Unemployed, Numbers and Rates'. At: <www40.statcan.ca/l01/cst01/labor07c.htm>.

———. 2008. 'Farm Population and Total Population by Rural and Urban Population, by Province', *2001 and 2006 Census of Agriculture and Census of Population*. At: <www40.statcan.gc.ca/l01/cst01/agrc42a-eng.htm>.

———. 2009a. 'Chart 21.1 (data) Unemployment Rate', 16 Jan. At: <www41.statcan.gc.ca/2008/2621/grafx/htm/ceb2621_000_1-eng.htm#table>.

———. 2009b. 'Distribution of Employed People, by Industry, by Province'. At: <www40.statcan.gc.ca/l01/cst01/labor21c-eng.htm>.

———. 2009c. 'Urban Area (UA)', *2006 Census Dictionary*. At: <www12.statcan.ca/census-recensement/2006/ref/dict/geo049-eng.cfm>.

———. 2009d. 'Population, Age Distribution and Median Age by Province and Territory, as of July 1, 2008', Table 1. At: <www.statcan.gc.ca/daily-quotidien/090115/t090115c1-eng.htm>.

———. 2010. 'Labour Force, Employed and Unemployed, Numbers and Rates, by Province'. 29 Jan. At:<www40.statcan.gc.ca/l01/cst01/labor07b-eng.htm>.

Stavely, Michael. 1987. 'Newfoundland Economy and Society at the Margin', in L.D. McCann, ed., *Heartland and Hinterland: A Geography of Canada*, 2nd edn. Scarborough, Ont.: Prentice-Hall, 247–85.

Storey, Keith. 2009. 'Help Wanted: Demographics, Labor Supply and Economic Change in Newfoundland and Labrador', *Challenged by Demography: A NORA Conference on the Demographic Challenges of the North Atlantic Region*, Alta, Norway, 20 Oct.

Chapter 10

Abele, Frances, Thomas J. Courchene, F. Leslie Seidle, and France St-Hilaire, eds. 2009. *Northern Exposure: Peoples, Powers and Prospects in Canada's North*. Montréal: Institute for Research on Public Policy.

Agnew, John. 2005. 'Sovereignty Regimes: Territoriality and Authority in Contemporary World Politics', *Annals, Association of American Geographers* 95, 2: 437–61.

BBC News. 2007. 'Canada to Strengthen Arctic Claim', 10 Aug. At: <news.bbc.co.uk/2/hi/Americas/6941426.stm>.

Berger, Thomas R. 1977. *Northern Frontier, Northern Homeland: The Report of the Mackenzie Valley Pipeline Inquiry*, 2 vols. Ottawa: Minister of Supply and Services.

Bone, Robert M. 2009. *The Canadian North: Issues and Challenges*, 3rd edn. Toronto: Oxford University Press.

Boswell, Randy, and Juliet O'Neill. 2010. 'Clinton Blasts Canada for Exclusivity of Arctic Talks', *Montreal Gazette*, 29 Mar. At: <www.montrealgazette.com/news/Clinton+blasts+Canada+exclusive+Arctic+talks/2740399/story.html>.

Bumsted, J.M. 2010. *The Peoples of Canada: A Pre-Confederation History*, 3rd edn. Toronto: Oxford University Press.

Byers, Michael. 2009. *Who Owns the Arctic? Understanding Sovereignty Disputes in Canada's North*. Vancouver: Douglas & McIntyre.

Canada. 1985. *The Western Arctic Claim: The Inuvialuit Final Agreement*. Ottawa: Department of Indian Affairs and Northern Development.

———. 1991. 'Comprehensive Land Claim Agreement Initialled with Gwich'in of the Mackenzie Delta in the Northwest Territories', Communiqué 1–9171. Ottawa: Department of Indian Affairs and Northern Development.

———. 1993a. 'Formal Signing of Tungavik Federation of Nunavut Final Agreement', Communiqué 1–9324. Ottawa: Department of Indian Affairs and Northern Development.

———. 1993b. *Umbrella Final Agreement between the Government of Canada, Council for Yukon Indians and the Government of the Yukon*. Ottawa: Department of Indian Affairs and Northern Development.

———. 2004. 'Agreements'. Ottawa: Department of Indian and Northern Affairs. At: <www.ainc-inac.gc.ca/pr/agr/index_e.html#Comprehensive%20Claims%20Agreements>.

CBC News. 2007. 'Mackenzie Gas Line Still "Leading Case" Despite Bloating $16.2B Cost Outlook', 12 Mar. At: <www.cbc.ca/cp/business/070312/b031292A.html#skip300x250>.

———. 2009. 'Yellowknife Diamond-cutting Plant in Limbo', 11 June. At: <www.cbc.ca/canada/north/story/2009/06/11/nwt-diamonds.html?ref=rss>.

Chan, Laurie H.M. 2006. 'Food Safety and Food Security in the Canadian Arctic', *Meridian* (Publication of the Canadian Polar Commission): 1–3.

Coates, Ken, P. Whitney Lackenbauer, William Morrison, and Greg Poelzer. 2008. *Arctic Front: Defending Canada in the Far North*. Toronto: Thomas Allen.

Crowe, Keith J. 1991. *A History of the Original Peoples of Northern Canada*, 2nd edn. Montréal and Kingston: McGill-Queen's University Press.

Danylchuk, Jack. 2007 'Giant, Glittering and Tarnished', *Up Here*, 18 Dec. At: <www.uphere.ca/node/175>.

Dickason, Olive Patricia. 2002. *Canada's First Nations: A History of Founding Peoples from Earliest Times*, 3rd edn. Toronto: Oxford University Press.

Drummond, K.J. 2009. *Northern Canada Distribution of Ultimate Oil and Gas Resources*. At: <drummondconsulting.com/NCAN09Report.pdf>.

Ebner, David. 2009. 'Exxon Backs Alaska Gas Pipeline', *Globe and Mail*, 12 June, B14.

Elias, Peter Douglas. 1995. *Northern Aboriginal Communities: Economies and Development*. North York, Ont.: Captus Press.

Globe and Mail. 1993. 'Mackenzie Land Claim Settled', 7 Sept., A1.

Griffiths, Franklyn. 2009. 'Canadian Arctic Sovereignty: Time to Take Yes for an Answer on the Northwest Passage', in Abele et al. (2009: 107–36).

Hamelin, Louis-Edmond. 1978. *Canadian Nordicity: It's Your North, Too*, trans. William Barr. Montréal: Harvest House.

Indian and Northern Affairs Canada (INAC). 2007. *Northern Oil and Gas Annual Report 2006*. At: <www.ainc-inac.gc.ca/oil/ann/ann2006/dev_e.html>.

———. 2009a. *Northern Oil and Gas Annual Report 2008*. At: <www.ainc-inac.gc.ca/nth/og/pubs/ann/ann2008/ann2008-eng.pdf>.

———. 2009b. *Framework Agreement*, 14 Jan. At: <www.ainc-inac.gc.ca/al/ldc/ccl/agr/dcf/dcf-eng.asp>.

Légaré, André. 2008a. 'Canada's Experiment with Aboriginal Self-determination in Nunavut: From Vision to Illusion', *International Journal on Minority and Group Rights* 15: 335–67.

———. 2008b. 'Inuit Identity and Regionalization in the Canadian Central and Eastern Arctic: A Survey of Writings about Nunavut', *Polar Geography* 31, 3 and 4: 99–118.

McGhee, Robert. 1996. *Ancient People of the Arctic*. Vancouver: University of British Columbia Press.

MacLachlan, Letha. 1996. *NWT Diamonds Project: Report of the Environmental Assessment Panel*. Ottawa: Canadian Environmental Assessment Agency.

McRae, Donald. 2008. 'An Arctic Agenda for Canada and the United States', in *From Correct to Inspired: A Blueprint for Canada–US Engagement*. At: <www.carleton.ca/ctpl/conferences/documents/BackgroundPapers-Final.pdf>.

Marcus, Alan R. 1995. *Relocating Eden: The Image and Politics of Inuit Exile in the Canadian Arctic*. Hanover, NH: University Press of New England.

Natural Resources Canada. 2007. 'High Metal Prices Spur Mineral Production to a Record $34 billion in 2006', *Information Bulletin*, Mar. At: <www.nrcan.gc.ca/mms/pdf/minprod-07_e.pdf>.

———. 2010. 'The Global Recession Reduced Canada's Mineral Production in 2009', *Information Bulletin*, Mar. At: <mmsd.mms.nrcan.gc.ca/stat-stat/prod-prod/PDF/ib2010_e.pdf>.

Northern Gas Pipelines. 2009. *Northern Gas Pipelines: Mackenzie Valley Pipeline Project*. At: <www.arcticgaspipeline.com/Delta%20Route.htm>.

Northwest Territories. 2009. *Statistical Quarterly* 31, 1. At: <www.stats.gov.nt.ca>.

Nunavut. 1999. *The Bathurst Mandate Pinasuaqtavut: What We've Set Out to Do*. Iqaluit: Legislative Assembly.

O'Neil, Peter. 2009. 'Jean's Seal Meal Draws Praise, Criticism', *StarPhoenix* (Saskatoon), 27 May, C12.

Page, Robert. 1986. *Northern Development: The Canadian Dilemma*. Toronto: McClelland & Stewart.

Rowley, Graham W. 1996. *Cold Comfort: My Love Affair with the Arctic*. Montréal and Kingston: McGill-Queen's University Press.

Shadian, Jessica. 2007. 'In Search of an Identity Canada Looks North', *American Review of Canadian Studies* 37, 3: 323–53.

Statistics Canada. 2004. 'Study: Diamonds Are Adding Lustre to the Canadian Economy', *The Daily*, 13 Jan. At: <www.statcan..ca/Daily/English/040113/d040113a.htm>.

———. 2008. 'Aboriginal Identity Population by Age Groups, Median Age and Sex, 2006 Counts, for Canada, Provinces and Territories'. At: <www12.statcan.gc.ca/english/census06/data/highlights/Aboriginal/pages/Page.cfm?Lang=E&Geo=PR&Code=01&Table=1&Data=Count&Sex=1&Age=1&StartRec=1&Sort=2&Display=Page>.

———. 2009a. 'Population and Demography: Population Estimates and Projections'. At: <cansim2.statcan.gc.ca/cgi-win/cnsmcgi.pgm?Lang=E&SP_Action=Result&SP_ID=3433&SP_TYP=5&SP_Sort=1&SP_Mode=2>.

———. 2009b. 'Main Demographic Indicators for Canada, Provinces and Territories, 1981 to 2007', *Report on the Demographic Situation in Canada: 2005 and 2006*. At: <www.statcan.gc.ca/pub/91-209-x/2004000/tabl0-eng.htm>.

Usher, Peter J., and George Wenzel. 1989. *A Strategy for Supporting the Domestic Economy of the Northwest Territories*. Report for the Legislative Assembly's Special Committee on the Northern Economy. Ottawa: P.J. Usher Consulting Services.

Widdowson, Frances, and Albert Howard. 2008. *Disrobing the Aboriginal Industry: The Deception behind Indigenous Cultural Preservation*. Montréal and Kingston: McGill-Queen's University Press.

Williams, Glyn. 2009. *Arctic Labyrinth*. Toronto: Viking Canada.

Williamson, Robert G. 1974. *Eskimo Underground: Socio-Cultural Change in the Canadian Central Arctic*. Occasional Papers II. Uppsala, Sweden: Almqvist & Wiksell.

Young, Oran. 2009. 'Whither the Arctic: Conflict or Cooperation in the Circumpolar North', *Polar Record* 45, 232: 73–82.

Chapter 11

Adams, Michael. 2007. *Unlikely Utopia: The Surprising Triumph of Canadian Pluralism*. Toronto: Viking Press.

Agriculture and Agri-Food Canada. 2009. *Overview of the Canadian Agriculture and Agri-Food System 2008*. At: <www4.agr.gc.ca/AAFC-AAC/display-afficher.do?id=1228246364385&lang=eng#a2>.

Baker, Liana. 2009. 'New Frontier in International Affairs', *CG Compass* (blog), 27 Aug. At: <www.canadiangeographic.ca/blog/posting.asp?ID=30>.

Conway, Sir Gordon. 2009. 'Geographical Crises of the Twenty-First Century', *Geographical Journal* 175, 3: 221–8.

Courchene, Thomas J., with Colin R. Telmer. 1998. *From Heartland to North American Region State: The Social, Fiscal and Federal Evolution of Ontario*. Monograph Series on Public Policy, Centre for Public Management. Toronto: University of Toronto Press.

De Blij, H.J., and Alexander B. Murphy. 2006. *Human Geography: Culture, Society, and Space*, 8th edn. Toronto: John Wiley.

Dobson, Wendy. 2009. *Gravity Shift: How Asia's New Economic Powerhouses Will Shape the Twenty-First Century*. Toronto: University of Toronto Press.

Florida, Richard. 2002. *The Rise of the Creative Class: And How It's Transforming Work, Leisure, Community and Everyday Life*. New York: Basic Books.

———. 2005. *The Flight of the Creative Class*. New York: HarperCollins.

———. 2008. *Who's Your City?: How the Creative Economy Is Making Where to Live the Most Important Decision of Your Life*. New York: Basic Books.

Frye, Northrop. 1971. *The Bush Garden: Essays on the Canadian Imagination*. Toronto: Anansi Press.

Garreau, Joel. 1981. *The Nine Nations of North America*. Boston: Houghton Mifflin.

Greenwood, John. 2007. 'Surging Prices Separate Wheat from the Chaff', *National Post*, 15 June, FP1.

———. 2009. 'Toronto Has Potential To Be Top Global Financial Centre, Report Says', *National Post*, 18 Nov. At: <network.nationalpost.com/np/blogs/toronto/archive/2009/11/18/toronto-has-potential-to-be-top-global-financial-centre-report-says.aspx>.

Hamilton, Graeme. 2007. 'Welcome! Leave Your Customs at the Door', *National Post*, 30 Jan., A1, A7.

Hare, F. Kenneth. 1968. 'Canada', in John Warkentin, ed., *Canada: A Geographical Interpretation*. Toronto: Methuen.

Ibbitson, John. 2010. 'House Reform Boosts Fastest-Growing Provinces', *Globe and Mail*, 2 Apr., A1, A6.

Innis, Harold. 1930. *The Fur Trade in Canada: An Introduction to Canadian Economic History*. New Haven: Yale University Press.

McCarthy, Shawn. 2009. 'Canada's Race for a High-Tech Strategy', *Globe and Mail*, 1 Aug., B1, B3.

Mohr, Patricia. 2009. *Scotiabank Commodity Price Index*, 23 Dec. At: <www.scotiacapital.com/English/bns_econ/bnscomod.pdf>.

Parkinson, David. 2009. 'Despite Some Price Woes, Commodities Bull Market Is Thriving', *Globe and Mail*, 17 July, B8.

Perreaux, Les. 2010. 'Asked to Remove Niqab, Quebec Woman Lodges Human-Rights Complaint', *Globe and Mail*, 3 Mar. At: <www.theglobeandmail.com/news/national/quebec/asked-to-remove-niqab-quebec-woman-lodges-human-rights-complaint/article1487526/>.

Saul, John Ralston. 1997. *Reflections of a Siamese Twin: Canada at the End of the Twentieth Century*. Toronto: Viking.

———. 2009. *The Collapse of Globalism and the Reinvention of the World*, 2nd edn. Toronto: Penguin Canada.

Scoffield, Heather. 2009. 'Recession Toll Harsh, but Easing', *Globe and Mail*, 1 June. At: <www.globeinvestor.com/servlet/story/RTGAM.20090601.wgdp0601/GIStory/>.

Statistics Canada. 2009. *Annual Demographic Estimates: Canada, Provinces and Territories 2009*. At: <www.statcan.gc.ca/pub/91-215-x/91-215-x2009000-eng.htm>.

Wallerstein, Immanuel. 1979. *The Capitalist World Economy*. Cambridge: Cambridge University Press.

———. 1998. 'Contemporary Capitalist Dilemmas, the Social Sciences, and the Geopolitics of the Twenty-first Century', *Canadian Journal of Sociology* 23, 2 and 3: 141–58.

Wei, Li. 2009. *Ethnoburb: The New Ethnic Community in Urban America*. Honolulu: University of Hawaii Press.

INDEX